AF307742

Acoustical Imaging

Volume 18

Acoustical Imaging

A Continuation Order Plan is available for this series. A continuation order will bring delivery of
each new volume immediately upon publication. Volumes are billed only upon actual shipment.
For further information please contact the publisher.

Acoustical Imaging

Volume 18

Edited by

Hua Lee and **Glen Wade**
University of California at Santa Barbara
Santa Barbara, California

SPRINGER SCIENCE+BUSINESS MEDIA, LLC

The Library of Congress cataloged the first volume of this series as follows:

International Symposium on Acoustical Holography.

   Acoustical holography; proceedings. v. 1- Springer Science+Business Media, LLC

     v. illus. (part col.), ports. 24 cm.

   Editors: 1967–   . A. F. Metherell and L. Larmore (1967 with H. M. A. el-Sum)
   Symposium for 1967–   held at the Douglas Advanced Research Laboratories, Huntington Beach, Calif.

   1. Acoustic holography — Congresses — Collected works. I. Metherell. Alexander A., ed.
II. Larmore, Lewis, ed. III. el-Sum, Hussein Mohammed Amin, ed. IV. Douglas Advanced
Research Laboratories, v. Title.
QC244.5.I.5   69-12533

ISBN 978-1-4613-6641-6   ISBN 978-1-4615-3692-5 (eBook)
DOI 10.1007/978-1-4615-3692-5

Proceedings of the 18th International Symposium on Acoustical Imaging,
held September 18-20, 1989, in Santa Barbara, California

© 1990 Springer Science+Business Media New York
Originally published by Plenum Press, New York in 1990
Softcover reprint of the hardcover 1st edition 1990

All rights reserved

No part of this book may be reproduced, stored in a retrieval system, or transmitted
in any form or by any means, electronic, mechanical, photocopying, microfilming,
recording, or otherwise, without written permission from the Publisher

PREFACE

How to produce images with sound has intrigued engineers and scientists for many years. Bats, whales and dolphins can easily get good mental images with acoustical energy, but humans have little natural ability for obtaining such images. The history of engineering and science, however, is an impressive demonstration that technological solutions can compensate, and then some, for deficiencies of nature in humans.

Thus with the proper technology, we too can "see" with sound. Many methods involving ultrasonic energy can be employed to enable us to do so. Few of these methods are at all reminiscent of the acoustic systems employed by animals. Pulse-echo, phase-amplitude and amplitude-mapping approaches constitute the conceptual bases for three fundamentally different types of acoustic imaging systems and can be used for categorizing the systems. However, by now systems exist that combine the approaches in such sophisticated ways as to make an unambiguous categorization of some of the more complicated systems difficult or impossible. Among the instruments so far produced are mechanically-scanning focused instruments, chirped pulse-echo instruments, and instruments involving holography, tomography, parametric excitation, phase conjugation, neural networks, random phase transduction, finite element methods, Doppler frequency shifting, pseudo inversion, Bragg diffraction and reflection, and a host of other principles.

The fifty-five chapters in this volume are selected from papers presented at the Eighteenth International Symposium on Acoustical Imaging which was held in Santa Barbara, California on September 18 - 20, 1989. These chapters were written by researchers from fifteen different countries, thus demonstrating the true international scope of the work going on. Three of these chapters, by Cain, Devaney, and Robinson, were invited papers in order to be sure to treat topics of unusual interest and scope adequately with a tutorial component. The other chapters fall into one or another of the categories of biomedical applications, imaging systems, advanced imaging techniques, nondestructive evaluation, acoustic microscopy, seismic imaging, parameter estimation and detection, multidimensional imaging, underwater imaging, and transducers.

The editors wish to thank a number of individuals and organizations for the important contributions they have made to the success of producing this volume. The symposium itself was sponsored by the National Science Foundation, under NSF Grant ECS 89-11412, in cooperation with the Acoustical Society of America, the IEEE Ultrasonics, Ferroelectrics and Frequency Control Society and the International Society for Optical Engineering. The Program Committee for the Symposium included besides the editors R. Waag, J. Powers, W. Chew, K. Wang, J. Greenleaf, J. Jones, P. Lewin, R. Weglein, B. Tittmann, P. Green, M. Fink, N. Chubachi, L. Kessler, E. Robinson, L. Ferrari, R. Algazi, P. Alais, and H. Ermert.

The editors would like to thank Dr. George Lea of the National Science Foundation for his enthusiastic support. We thank also the staff of the Conferences and Institutes of the University of Illinois for their considerable help in managing the operations of the symposium.

Hua Lee and Glen Wade

University of California at Santa Barbara

CONTENTS

SEISMIC IMAGING

PARAMETER ESTIMATION AND DETECTION

MULTI-DIMENSIONAL IMAGING

UNDERWATER ACOUSTIC IMAGING

TRANSDUCERS

DETECTION OF EARLY FATTY PLAQUE USING QUANTITATIVE ULTRASOUND METHODS

J.P. Jones*, P.A.N. Chandraratna**, T. Tak**, S. Kaiser*,
E. Yigiter*, and J. Gallet*

*Dept. of Radiological Sciences, UC Irvine, Irvine, CA 92717
**Cardiology, LA/USC Medical Center, Los Angeles, CA 90033

INTRODUCTION

Atherosclerosis, together with associated cardiovascular disease
is the number one health problem in the United States today. Although
currently available techniques, such as angiography and ultrasound imaging are useful in detecting advanced stages of the disease process (by
detecting luminal narrowing caused by atherosclerotic lesions), these
techniques have serious limitations in the detection of early atherosclerosis (since encroachment of the lumen is minimal in the initial
stages of the atherosclerotic process). Moreover, none of the current
diagnostic techniques are capable of characterizing the lesion as to
type. Such information, now only available through surgical intervention
is important to the planning of an appropriate clinical treatment course.
In addition, the invasive nature of angiography and its potential compli-
cations have inherent disadvantages when asymptomatic patients are to be
studied.

It is clear that there is a major need, as yet unfilled, for a
noninvasive technique capable of both characterizing atherosclerotic
plaque as well as detecting atherosclerosis at an early pre-plaque
stage. Experiments conducted over the past six years at the University
of California Irvine and the University of Southern California Medical
Center have been directed toward the development of just such a technique.
An extensive in-vitro study conducted by Dr. Joie Jones (UCI) and Dr.
P.A.N. Chandraratna (USC) on some 5000 samples of plaque has shown excellent correlation between the acoustical properties of plaque (such
as attenuation and impedance) and the chemical components of plaque (as
determined from histology). Early fatty plaque (the precurser to the
development of atherosclerosis) did not show significant changes in acoustical properties but was detected via specialized signal processing
techniques developed at UCI. These experiments strongly suggest that
in-vivo techniques could be developed that would permit the quantitative
differentiation of atherosclerotic plaque and the detection of early
fatty plaque. If such techniques were available, they would literally
revolutionize diagnostic medicine making current invasive procedures
unnecessary and lowering both the risks and the costs associated with
health care.

Materials and Methods

We have conducted extensive in-vitro acoustical measurements on

Acoustical Imaging, Vol. 18, Edited by H. Lee and G. Wade
Plenum Press, New York, 1991

atherosclerotic plaque taken from excised samples of human abdominal
aorta. In these studies, tissue samples were suspended in a water tank
and scanned using a rectilinear ultrasound scanner under computer control.
A data acquisition system, based on the TRW 8-bit, 20 MHz A/D converter
was used to digitize and record the RF ultrasound waveforms. A matrix of
A-lines was taken over a specified region of interest and RF waveforms
associated with each A-line recorded. A 13 mm non-focused 3.5 MHz transducer was used in all measurements. From this data, averaged values for
attenuation and impedance for each tissue sample could be computed (1,2).
Attenuation was measured using an insertion loss method in which the
echoes from a specular reflector were compared with the tissue samples
removed and between the transducer and the reflector. Impendance was
measured by first measuring the speed of sound in the sample and then
independently measuring the density. Here impedance is simply the product of density and sound speed. Based on some 5,000 different tissue
samples, significant differences in both impedance and attenuation were
found between normal aorta and ·fibrous, fatty, and calcified plaque. The
results are shown in Figures 1 and 2.

Samples of early fatty plaque (considered to be the precursor to
the development of atherosclerosis) however, did not show any measurable
differences in either attenuation or impedance from the normal aorta.
Early fatty plaque was also undistinguishable in the B-mode image. However, the application of a particular structure algorithm developed at
UCI and known as ECS (envelope correlation spectrum), clearly detected
the presence of early fatty plaque (3,4). The ECS is calculated by taking a cross correlation of the incident and the reflected waveform
(along a given A-line), envelope detecting the result, and, finally,
displaying the power spectrum of the result. The algorithm is particularly useful in extracting information about the structure or periodic
nature of reflectors (5). We believe this technique will be of value
in the detection of early aatherosclerosis as well as in the study of
the progression and regresson of the atherosclerotic plaque. Typical
in-vitro results are shown in Figure 3.

Recent studies by other workers (6,7) have demonstrated that alternate quantitative techniques (integrated backscattering) could also
differentiate between fibrosis and calcification in-vitro. However the
technique was unable to identify fatty lesions.

Future In-Vivo Studies

Based on the in-vitro studies reported above, we have undertaken
to replicate these measurements in-vivo. Going from in-vitro to in-vivo
requires (1) an upgraded data acquisition system, capable of interfac-
ing with a variety of commercial ultrasound units and (2) appropriate
software for the estimation of acoustical properties given the backscattered ultrasound waveform. We have recently implemented a new data acquisition system based on a "fast" IBM 286. We chose an ADC board with fast
buffer memory which would plug into the 286 computer and sample at 100M
samples/second. This speed was necessary since in the application of
interest, equipment tended to utilize transducers with center frequencies
between 5 and 10 MHz. The fast buffer memory permits us to sample at
100 MHz and use Direct Memory Access to transfer data from the buffer
to the computer as required. DSP chips were selected so the analysis
could be completed in near real time.

We hope to have this system for the collection of in-vivo data
operational in the near future. Our goal is to try to replicate our in-
vitro measurements in-vvo.

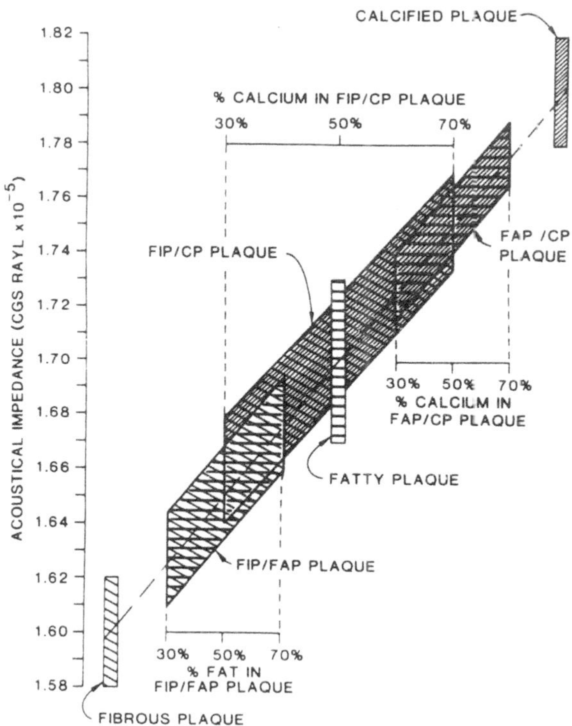

Figure 1. correlation of acoustical impedance with type of plaque. The boxes circunscribe all data points (some 5000). Mean values have a much tighter fit

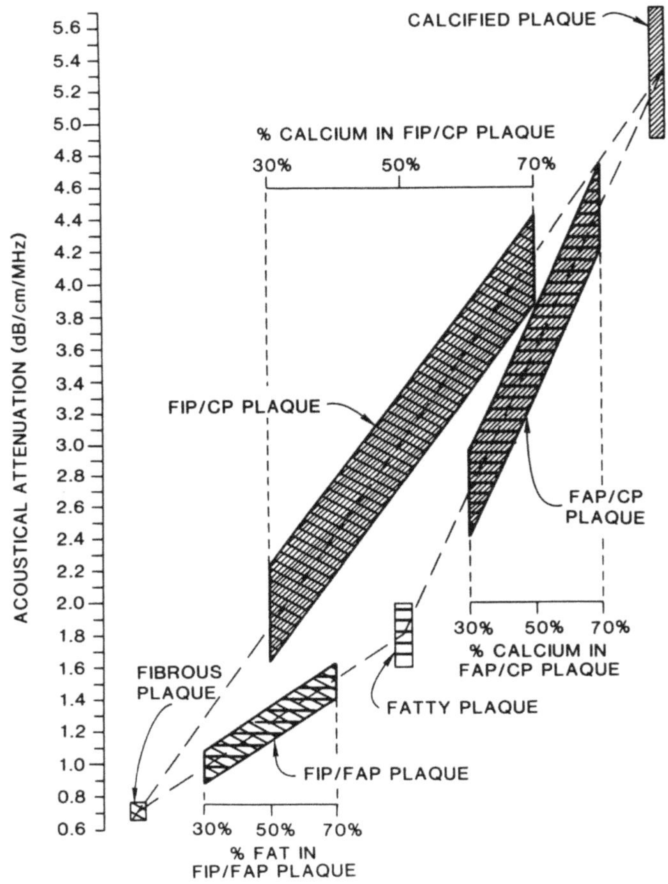

Figure 2. Correlation of
acoustical attenation with type of plaque

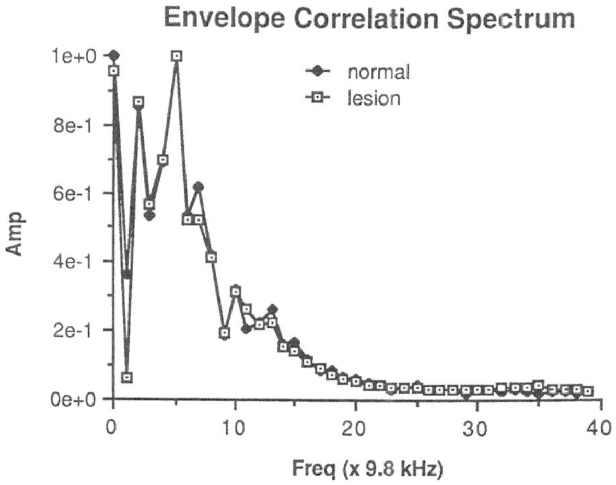

Figure 3. ECS for normal wall
and early fatty plaque

References

1. J.P. Jones, S. Kaiser, T. Tak and P.A.N. Chandraratna, "Quantitative
    Characterizatio of Atherosclerotic Lesions Using Ultrasound:, Ultra-
    sonic Imaging 9, 62(1987); also presented at the 12th International
    Symposium on Ultrasonic Imaging and Tissue Characterization, Arling-
    ton, VA(June 8-10, 1987).

2. P.A.N. Chandraratna, T. Tak, J.P. Jones, S. Kaiser and Shahbudin
    H. Rahimtooola, "Evaluation of Atherosclerotic Placees Using
    Quantitative Ultrasonic Methods" in Proceedings of the 60th Scientific
    Sessions, American Heart Association, Anaheim, California(November
    16-19, 1987).

3. P.A.N. Chandraratna, T. Tak, J.P. Jones, S. Kaiser and S.H. Rahim-
    toola, "Detection of Early Atherosclerosis by Quantitative Ultra-
    sonic Methods", presented at the 37th Annual Scientific Session,
    American College of cardiology, Atlanta, Georgia(March 27-31, 1988).

4. J.P. Jones, S. Kaiser, P.A.N. Chandraratna, C. Lohr, M. Thompson
    and T. Tak, "Evaluation of Atherosclerotic Plaque and Detection
    of Early Atherosclerosis Using Quantitative Ultrasound Methods",
    Ultrasonic Imaging, 10 69(1988); also, presented at the 13th Inter-
    national Symposium on Ultrasonic Imaging and Tissue Characterization,
    Arlington, VA(June 6-8, 1988).

5. J.P. Jones, R. Kovack, "A Computerized Data Analysis System for
    Ultrasonic Tissue Characterization", Acoustical Imaging. Vol. 9,
    (1980).

6. L. Landini, R. Sarnelli, E. Picano, M. Salvadori, "Evaluation of
    Frequency Dependence of Backscatter Coefficient in Normal and
    Atherosclerotic Aortic Walls", Ultrasound in Medicine and Biology,
    Vol. 12, No. 5, pp. 397-401(1986).

7. P. Pignoli, T. Longo, "Ultrasound Evaluation of Atherosclerosis,
    Methodological Problems and Technical Developments", Eur. Surg.
    Res. 18:238-253(1986).

PATTERN RECOGNITION ON HUMAN SKIN TISSUE

A. Pech[1], E.-G. Loch[1], W. v. Seelen[2]

[1]Gesellschaft zur Förderung der Forschung an der
Deutschen Klinik für Diagnostik, Wiesbaden, FRG
[2]Institut für Neuroinformatik, Ruhr-Universität,
Bochum, FRG

ABSTRACT

Using 20 MHz ultrasound the different layers of human skin
tissue can be separated. This process of pattern recognition is
done by extracting features within a twostep process including
an autoregressive signal model with a weighting factor for the
estimation of time dependant parameters and stationary first
order discrete markov chains. The statistic classification
process has to be trained by a preclassified set of training
data before classifying new data by a maximum likelihood
algorithm. The layers epidermis, corium, subcutaneous fat and
underlying structures can be separated. The results of the
classification process have been successfully tested in tests
of medicaments, diagnosis of osteoporosis and supervision of
therapy in psoriasis patients.

1.0 INTRODUCTION

Human skin is an organ of various medical interests. It
indicates processes inside the human body as well as it changes
its structure under external influences. Therefore the in-vivo
measurement of skin thickness is an important additional instrument to support the control of therapy in diseases e.g. psoriasis, diagnosis of skin tumors and osteoporosis as well as tests
of different medicaments.
The geometrical resolution of normal ultrasonic scanners is not

sufficient for the detection of human skin boundaries. To solve
this problem backscattered high frequency rf-signals with a
center frequency of 20 MHz are used to generate high resolution
pictures either in A-Mode or in B-Mode.

2.0 FEATURE EXTRACTION

The properties of the skin tissue at location s backscattering the ultrasonic signal may be described by different
features x_j comprehended in the feature vector $\underline{v}(x_j,s)$, where s
is evaluated using the ultrasonic velocity of the medium
inside the tissue and using time of flight. The elements x_j of
the feature vector $\underline{v}(x_j,s)$ are found by estimating the parameters of an autoregressive model and the transition probabilities of discrete markov chains.

2.1 AUTOREGRESSIVE MODEL

An autoregressive model describes a process $G(z)$ in the
discrete frequency domain

$$G(z) = \frac{b_0}{1 + a_1 z^{-1} + \ldots + a_m z^{-m}} = \frac{y(z)}{u(z)}$$

In this case the input signal is $u(z)$, $y(z)$ is the output of
the system. Making use of the properties of the z-transform and
combining the parameters a_1 and b_0 in a vector $\underline{\theta}$.

$$\underline{\theta}^T = [a_1 \; a_2 \; .. \; a_m \; b_0],$$

this definition may be brought into a difference equation of
the discrete time domain

$$y(k) = \underline{\alpha}^T(k) \cdot \underline{\theta} \; , \qquad\qquad k=nT_0 \; , \; n=1,2,\ldots \; ,$$

where k represents n times the sample interval T_0 and $\underline{\alpha}$ is
given by

$$\underline{\alpha}^T = [-y(k-1) \; -y(k-2) \; \ldots \; -y(k-m) \; u(k)].$$

The least squares estimate $\underline{\theta}(k)$ of these parameters is calculated by a recursive equation.

The different types of tissues and their characteristics lead
to a process with time dependant parameters. Thus the calcula-

tion of the autoregressive model has to consider this time
dependence. To do this, a weighting factor β is introduced,
where the influence of foregoing values decreases by an
exponential function. With $\underline{B}(k)$

$$\underline{B}(k) \;=\; \begin{bmatrix} \beta^N & 0 & . & . & 0 \\ 0 & \beta^{N-1} & . & . & . \\ . & & . & & . \\ . & & & \beta & 0 \\ 0 & . & . & 0 & 1 \end{bmatrix}$$

and $\underline{\Phi}(k)$

$$\underline{\Phi}(k) \;=\; \begin{bmatrix} \underline{\alpha}^T(1) \\ \underline{\alpha}^T(2) \\ . \\ . \\ \underline{\alpha}^T(k) \end{bmatrix}$$

$\underline{P}(k)$ is introduced by

$$\underline{P}(k) \;=\; [\underline{\Phi}^T(k)\underline{B}(k)\underline{\Phi}(k)]^{-1} \; .$$

The recursive estimation equations are

$$\hat{\underline{\theta}}(k+1) \;=\; \hat{\underline{\theta}}(k) \;+\; \underline{q}_B(k)\,[y(k+1)-\underline{\alpha}^T(k+1)\hat{\underline{\theta}}(k)]$$

$$\underline{q}_B(k) \;=\; \frac{1}{\underline{\alpha}^T(k+1)\underline{P}_B(k)\underline{\alpha}(k)\;+\;\beta}\;\underline{P}_B(k)\underline{\alpha}(k+1)$$

$$\underline{P}_B(k+1) \;=\; [\underline{I}\;-\;\underline{q}_B(k)\underline{\alpha}^T(k+1)]\;\underline{P}_B(k)\beta^{-1}$$

For the purposes of feature extraction on human skin tissue β
is fairly good at $\beta=0.95$, with $\beta=0.98$ the backward dependence
is too strong and leads to unusable results whereas $\beta=0.92$
produces estimates which are too noisy to be able to characterize a tissue type within a region.
A second order model worked out to be convenient for the
description of the behaviour of an ultrasonic pulse wave in
inhomogeneous tissue, that is the parameter vector is formed by
$\hat{\underline{\theta}} = [a_1\ a_2\ b_0]$. The elements of the parameter vector $\hat{\underline{\theta}}$ are
copied into the corresponding elements x_0, x_1 and x_2 of the
first part of the feature vector $\underline{v}(x_j,s)$.

2.2 MARKOV CHAINS

During the second step of feature extraction the process
is divided into subprocesses. Each one is covering a time

window which is half as long as the life time of the input
signal u(k). All subprocess are supposed to follow a first
order stationary discrete markov chain, that is the probability
of occurrence $p(w|k)$ of a value $y(k)=i$ depends only on the
value j of the chain at a time $k-1$

$$p[y(k)=i|y(k-1)=j,y(k-2)=l,\ldots y(0)=m] = p[y(k)=i|y(k-1)=j].$$

Since the process is described by a limited set of 256 values
corresponding to the 8 bit digitizer it may be described as a
markov chain with a limited number of states including reflecting boundaries. The markov chain is completely characterized
by its transition matrix $\underline{T}$ that consists of the transition
probabilities of all possible values. The probability of the
state i going to j within the next time interval is denoted by
p_{ij}.

$$\underline{T} = \begin{bmatrix} p_{oo} & .. & p_{no} \\ \cdot & \cdot & \cdot \\ \cdot & p_{ij} & \cdot \\ p_{on} & .. & p_{nn} \end{bmatrix}$$

The sum of all values in one row of the transition matrix is
calculated by $p_{i.}$,

$$p_{i.} = \sum_{j} p_{ij}$$

The probabilities p_{ij} have to be estimated out of the samples
formed by the subprocesses. The maximum likelihood estimate of
the p_{ij} is taken from the sampling distribution w_{ij}

$$p_{ij} = \frac{w_{ij}}{\sum_{j} w_{ij}}$$

This maximum likelihood estimate of the p_{ij} is taken for each
element of the sequentially filled transition matrix $\underline{T}$. There
are $256 \cdot 256$ transition probabilities, so we cannot put them
into a feature vector of a length that can be handled in a
usual computer without loss in numerical accuracy. In order to
reduce the data to a reasonable extent without losing the
desired information content a further processing step is
performed.

An abstraction is made by supposing the transition matrix
itself to be a two dimensional random process. The first order

expectation value $\underline{\mu}$ and the covariance matrix $\underline{M}$ of this random field are calculated as

$$\underline{\mu} = E \{ p_{ij} \}$$

$$\underline{M} = E \{ (p_{ij}-\underline{\mu})(p_{ij}-\underline{\mu})^T \}$$

Out of these expectation values μ_x, μ_y, σ_x^2 and σ_y^2 are transfered into the feature vector $\underline{v}$ forming the components x_3, x_4, x_5 and x_6 respectively.
The first components (x_0 to x_2) of the entire feature vector $\underline{v}(x_j,s)$ emphasize the deterministic part of the process whereas the components x_3 to x_6 reflect the statistical properties of the scattering process.

3.0 CLASSIFICATION PROCESS

The distribution of the feature vectors in feature space is estimated for a preclassified set of training data, estimating the probability distribution of the feature vectors corresponding to each tissue type to the classification system.

The feature vectors $\underline{v}$ are assumed to be normally distributed with a mean value $\underline{\mu}$ and covariance matrix $\underline{K}$

$$p(\underline{v}) = ((2\pi)^N \det\underline{K})^{-0.5} \exp(-0.5(\underline{v}-\underline{\mu})^T\underline{K}^{-1}(\underline{v}-\underline{\mu})) \quad .$$

with each class k characterized by its specific mean value $\underline{\mu}_k$ and its specific covariance matrix $\underline{K}_k$.

A decision theoretic approach is used to classify new input data. The mean risk R of the decision is evaluated as the sum over all classes k of all errors $e(\underline{v})$. The errors $e(\underline{v})$ are weighted by a cost function $\underline{C}(k,e(\underline{v}))$.

$$R = \sum_{\underline{v}} \sum_k \underline{C}(k,e(\underline{v})) \cdot p(k|\underline{v}) \cdot p(\underline{v})$$

Minimizing the mean risk R the decision is made.
For all k classes the probability vector $\underline{p}(\underline{v})$ is formed by $p(\underline{v})$ = $[p(1|\underline{v}) \ p(2|\underline{v}) \ \ldots \ p(k|\underline{v})]^T$. Introducing the conditional risk vector $\underline{R}(k)$ by $\underline{R}(k)$ = $[R(0) \ R(1) \ \ldots \ R(k)]^T$ it is given by

$$\underline{R}(k) = \underline{C}^T\underline{p}(\underline{v})$$

The optimal decision with respect to the costs is found by

searching the minimum of the risk vector $\underline{R}(k)$ for a given
feature vector $\underline{v}$. For an easier handling of this equation the
monotone transformation D is used

$$D \;=\; \sum_{k} C_k(k) \cdot p(k|\underline{v}) \;-\; \underline{R}(k)$$

D is called the decision variable and needs to be maximized in
order to find the optimal decision. Taking the natural logarithm of the decision variable leads to

$$D_k = -0.5 \ln \det \underline{K}_k - 0.5((\underline{v}-\underline{\mu}_k)^T \underline{K}_k^{-1}(\underline{v}-\underline{\mu}_k))$$

for each class k.
The decision is made by searching the maximum of D_k and thus
identifying the corresponding class k as the class of the
tissue type $\underline{v}$.
The decision is rejected if the feature vector $\underline{v}$ exceeds two
times the mahalanobis-distance Q_k of the class

$$Q_k = (\underline{v}-\underline{\mu}_k)^T \underline{K}_k^{-1}(\underline{v}-\underline{\mu}_k) \;\;.$$

The second criterion to supervise the consistency of the
classification result is a topological check. Beginning at the
start of the scan it is stated that each layer has to follow
the foregoing one without loss in continuity. A filter algorithm eliminates every single decision within a closed surrounding using the median accompanied by a variable counting the
actual class number. The regions for all classes are left
totally smoothed so that the thickness of the layers may be
computed simply by counting the pixels for each class and
computing time of flight from the sample interval T_0.

4.0 RESULTS

    The described process of feature extraction and classifi-
cation of signals backscattered from human skin tissue leads to
the following results:
The layers epidermis and corium can be separated from subcutaneous fat tissue which itself is distinguished from underlying structures. Out of this results skin thickness and
thickness of subcutaneous fat can be easily computed using the
time of flight to velocity relationship. Due to the topological
consistency check the algorithm is unsensitive against the

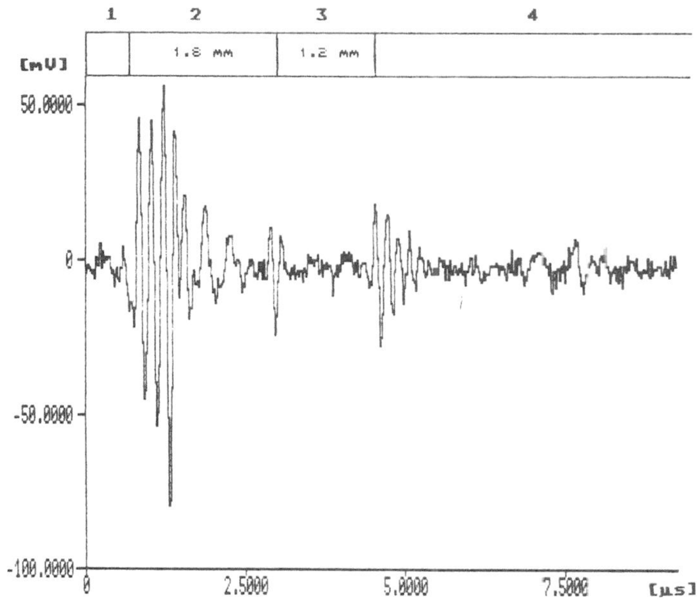

Fig. 1 Classified data of human skin tissue

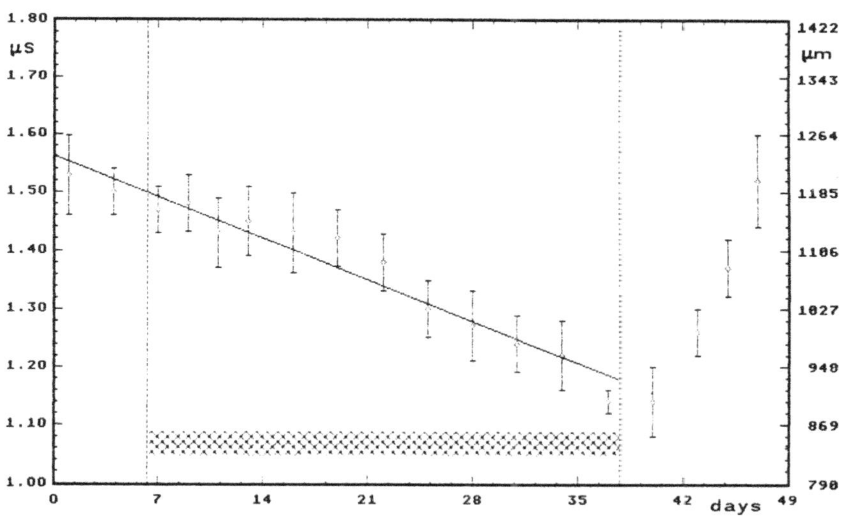

Fig. 2 Test of a corticosteroid

influence of hair bulges and blood vessels inside the subcutaneous fat layer. Fig. 1 shows the result of an in-vivo
measurement on human skin tissue of the forearm. It contains
the classified data showing the water path (1), the skin (2),
subcutaneous fat (3) and underlying structures (4). The skin
thickness is 1.8 mm whereas the subcutaneous fat has an
extension of 1.2 mm.

Picture two shows an application example of the measurement
technique in a medicament study. The dermal thinning of a
corticosteroid is tested over a period of 30 days. The marked
area indicates the period of application. Skin thickness
reduces during this period and regenerates rapidly after
stopping the treatment.

The measurement technique supports the therapy control in
psoriasis patients, who are exposed to ultraviolet light. The
effect of healing can be documented and the therapy can be
stopped when the thickness of the surrounding area is reached.

As a third example the examination of patients suffering from
osteoporosis is chosen to evaluate the quantitative relationship between bone density and human skin thickness. Though it
is well known that human skin tissue is thinning in osteoporosis, it has not been examined quantitatively up to now. Our
intention is to support and simplify early diagnosis of this
disease by an easy ultrasonic measurement technique in order to
avoid nowadays use of x-ray methods.

Introducing epidermis and corium as two different layers to the
system we made goals in separating them in the ultrasonic scan
although the classification rate of the one layer approach is
somewhat better than the results we get for the human skin as
two layers. This effect is important to detect whether the
effect of dermal thinning is effected by an atrophy of the
epidermis or a loss of fluid in corium, both of them are
possible effects during the application of corticosteroids.

REFERENCES

1. Isermann R., Digital Control Systems, Springer Verlag,
     Berlin (1981)

2. Immink W., Parameter Estimation in Markov Models and Dynamic
   Factor Analysis. Thesis. Utrecht (1986)
3. Langrock P., Jahn W., Einführung in die Theorie der
   Markovschen Ketten und ihre Anwendungen. B.G. Teubner
   Verlagsgesellschaft. Leipzig (1979)
4. Schürmann J., Polynomklassifikatoren für die Zeichen-
   erkennung, R. Oldenbourg Verlag. Munchen Wien (1977)

QUANTITATIVE IMAGING OF ACOUSTICAL AND

HISTOLOGICAL PROPERTIES OF EXCISED TISSUES

Jeffrey C. Bamber and Nigel L. Bush

Physics Department
Institute of Cancer Research:Royal
Marsden Hospital, Downs Road
Sutton, Surrey, SM2 5PT, U.K.

INTRODUCTION

Although ultrasound has become one of the most widely used and successful techniques for medical diagnosis, the precise physical basis for its use, particularly in cancer investigations, is still not fully understood. It is therefore unlikely that its full potential value has yet been achieved. To this end, and to obtain greater insight as to the potential value of quantitative approaches to tissue characterization, we have investigated the ultrasonic properties of normal and neoplastic human tissues[1,2,3].

During recent continuation of this work, effort has centered around the development of apparatus and techniques for the quantitative imaging of the ultrasound propagation and histological properties of excised tissue sections on both macroscopic and microscopic scales. These methods make it possible to study spatially inhomogeneous specimens, which are often the most interesting and clinically relevant. In this paper a specific case is used to illustrate the apparatus, methods and overall approach.

SPECIMEN PREPARATION

Each freshly excised specimen is degassed, embedded in gelatin and examined with a real-time B-scanner to select a section containing interesting pathology or normal tissue structure, at which the image is digitized. Video tape sequences are kept of parallel ultrasound sections on either side of this slice so as to assist matching and registration of an ultrasound image with the histological sections taken later. Fig. 1 shows a conventional ultrasound B-scan section through an excised human liver known to contain multiple metastases. A 7 mm thick slice, .centred on this scan plane, was cut and further divided so as to isolate the tissue in the 3 cm by 3 cm region of interest depicted by the white rectangle. This was then more thoroughly degassed and embedded in a disc of

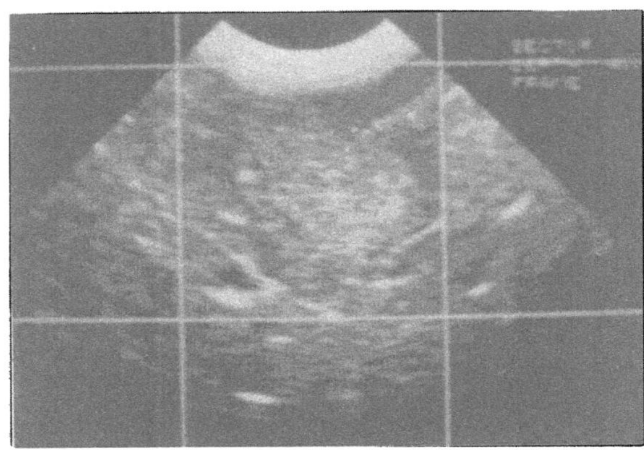

Fig. 1. Conventional B-scan section at 7.5 MHz through an excised human liver. A region of interest (ROI) has been specified around a metastatic deposit.

gelatin, also 7 mm thick, set between two latex rubber sheets which were stretched across a teflon annulus.

ACOUSTIC MEASUREMENTS

The teflon ring, containing the gelatin mounted specimen, is supported in a yoke which permits 2-dimensional angular alignment of the specimen with respect to the sound beam. The yoke itself hangs from a gantry which can be moved under computer control in the plane of the section, which is arranged to be perpendicular to the direction of sound propagation. This system of mounting has been developed over a number of years and has been found to provide a satisfactory way of achieving something close to a flat parallel-sided specimen.

The specimen is suspended at the focal plane of two weakly focused matched ultrasound transducers arranged co-axially and confocally. Either transducer may act as the transmitter or receiver of a short pulse of sound, which will pass through and be scattered by the specimen. Both the transmitted and the backscattered r.f. signals are digitised for 1024 equally spaced positions across the 3 cm by 3 cm specimen. Reference transmitted (specimen replaced by water) and backscattered (specimen replaced by a steel plate) signals are also digitised.

The 1024 transmitted pulse records are used to compute maps of sound speed and attenuation coefficient. Sound speed is computed from the pulse arrival time relative to that through water only. Various simple measures of pulse arrival time have been investigated for their immunity to the pulse distorting effects of frequency dependent attenuation and phase cancellation. Although phase velocity can be obtained from the Fourier transform of the received pulse, this is not worthwhile because soft tissue is close to being a non-dispersive medium and because of the pulse distortions which can occur. We have found from phantom studies that the mean time-of-arrival, $\bar{t}$, offers a good compromise between calculation time and immunity from artefact. This is computed from the pulse envelope, $E_V(t)$, as:

$$\bar{t} = \frac{\int_{t_1}^{t_2} t \cdot E_V \, dt}{\int_{t_1}^{t_2} E_V \, dt} \tag{1}$$

where t_1 and t_2 define time limits which include the whole of
the received voltage pulse. Having obtained the pulse arrival
time for 1024 positions (in a 32 by 32 array) on the specimen
the sound speed, c_t, is calculated according to:

$$\frac{1}{c_t(x,y)} = \frac{1}{c_W} - \frac{\bar{t}_W - \bar{t}_t(x,y)}{\delta z} \tag{2}$$

where c_W is the sound speed in water, $\bar{t}_W$ is the mean arrival
time for propagation through water only, $\bar{t}_t$ is the mean arrival
time with the tissue in the sound path and δz is the tissue
slice thickness. Fig. 3 shows the sound speed image for the 3
cm by 3 cm tissue section extracted from the excised liver.
Note that for visual presentation purposes only, this image and
the other acoustic maps have been enhanced by linearly
interpolating the 32 x 32 values up to 128 x 128 data points.

The attenuation coefficient, α, is obtained as a
continuous function of frequency, f, from the ratio of the
magnitude of the Fourier transform of the transmitted pulse to
that of the pulse received when there is no tissue in the sound
path. This is a commonly used method, expressed by:

$$\alpha(x,y,f) = \frac{20}{\delta z} [\log_{10} V_t(x,y,f) - \log_{10} V_W(f)] \tag{3}$$

Thus, at each position on the specimen, values of attenuation
coefficient are obtained for many different ultrasonic
frequencies in the range covered by the transducer bandwidth.
These data may be treated in a variety of ways. Here, for the
liver section under discussion, we have presented a frequency
averaged attenuation coefficient image (Fig. 4), obtained from:

$$\bar{\alpha}(x,y) = \frac{\int_{f_1}^{f_2} \alpha(x,y,f) \, df}{\int_{f_1}^{f_2} df} \tag{4}$$

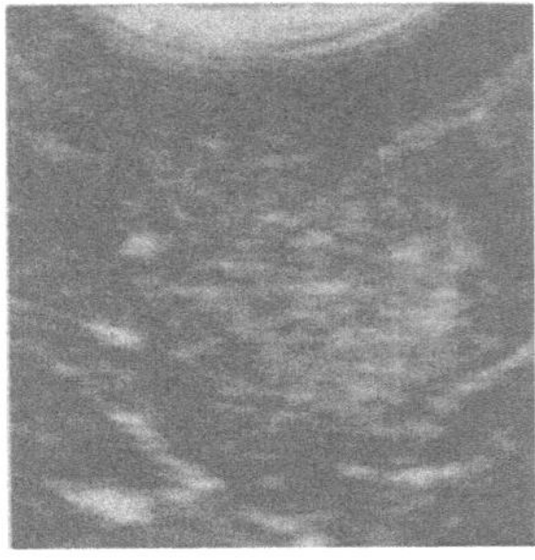

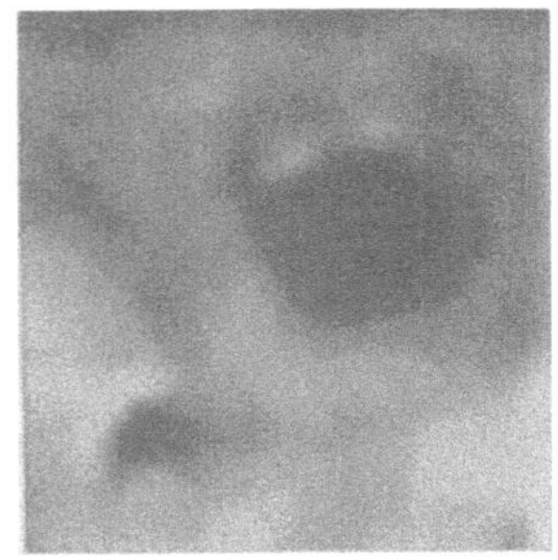

Fig.2. Magnified ROI from the
B-scan of Fig. 1.

Fig.3. Sound speed map.

and an image of the attenuation coefficient slope (Fig. 5),
which is the value of *m* obtained from a least squares fit to
the equation:

$$\alpha(x,y,f) = b(x,y)\, f^{m(x,y)} \tag{5}$$

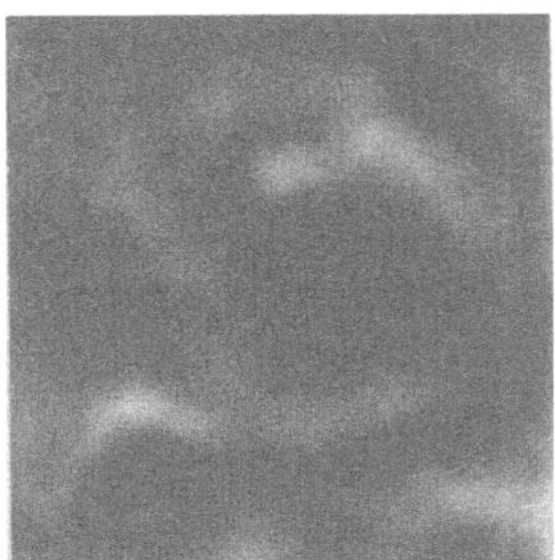

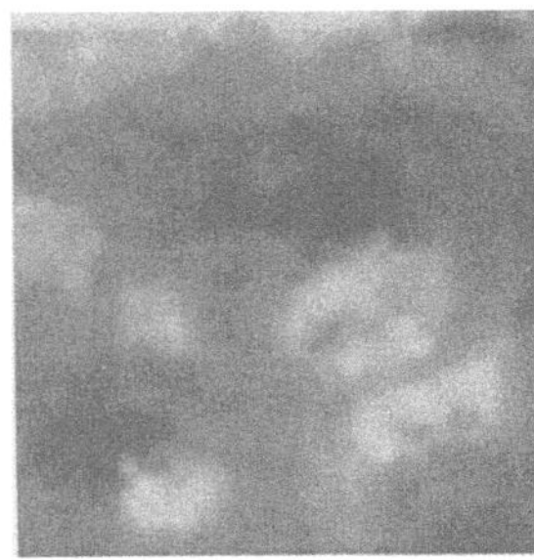

Fig.4. Attenuation coeff.
averaged over 3 to 7.5 MHz.

Fig.5. Frequency dependence
(*m*) of the atten. coeff.

Finally, the backscattering cross-section per unit volume
(or backscatter coefficient), μ_{bs}, is also measured as a
function of frequency and spatial position. This is obtained,
using a single transducer as transmitter and receiver, from the
ratio of the magnitude spectrum of a time segment of echoes
backscattered from within the specimen, $V_{bs}(f)$, to that of the
echo from a plane steel reflector, $V_r(f)$:

$$\mu_{bs}(x,y,f) = \left(\frac{V_{bs}(x,y,f)}{V_r(f)}\right)^2$$

$$\frac{4\alpha(x,y,f)\,R}{\Omega\, e^{-2\delta x\alpha(x,y,f)}\left(e^{\tau\alpha(x,y,f)\,c(x,y)} - e^{-\tau\alpha(x,y,f)\,c(x,y)}\right)} \tag{6}$$

where R is the intensity reflection coefficient for the steel reflector in water, Ω is the solid angle subtended by the receiving transducer at the centre of the specimen and τ is the time length of the echo segment. Some smoothing of the spectra is accomplished prior to this calculation, by multiplying the echo segment by a cosine tapered window function. This equation has been described previously[1], and is an approximate one which neglects the fact that energy is distributed non-uniformly across the width of the sound beam (according to the directivity function of the transducer). As demonstrated by D'Astous and Foster[4], however, this is a reasonable approximation for focused transducers.

Note from equation 6 that at each position, calculation of μ_{bs} requires the use of the sound speed image data to provide the value for c, and the attenuation image data to provide values for α at each frequency. In general, because the strength of the backscattered signal is much less than that of the transmitted signal, the backscatter coefficient can only be measured over part of the frequency range used for the attenuation image. Since the precise frequencies at which the backscattered spectrum is sampled are therefore generally not those for which attenuation data exist, attenuation coefficient values for use in equation 6 are obtained by interpolation.

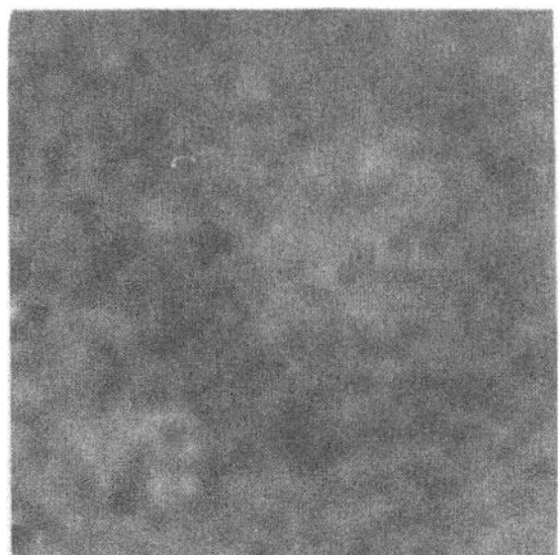

Fig.6. Backscatter coeff. reconstructed at 6 MHz.

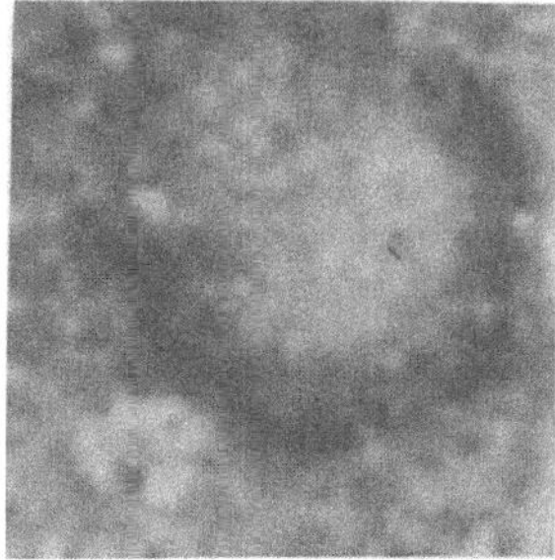

Fig.7. Backscatter coeff. averaged over 5 to 8 MHz.

The phenomenon of diffractive scattering results in pronounced fluctuations in the measured value of the backscattering coefficient from a given spatial position, as a function of both frequency[5] and orientation[1]. In B-scan imaging this effect is known as speckle and methods for reducing this form of random measurement error involve averaging over one or more of (a) spatial position, (b) orientation and (c) frequency. Limited averaging over orientation in the present system is possible by switching to use the other transducer as transmitter/receiver, providing backscatter data from an orientation which is at 180 degrees to that obtained using the first transducer. Fig. 6 shows the backscatter image of the liver section obtained at a single ultrasonic frequency. The choice as to whether to reduce the measurement noise by frequency averaging or by spatial averaging depends on whether spatial information is important or whether this can be sacrificed for information on the frequency dependence of scattering. Treating the backscatter data in a manner similar to that for the attenuation coefficient, one can produce images both of the backscatter coefficient integrated over the

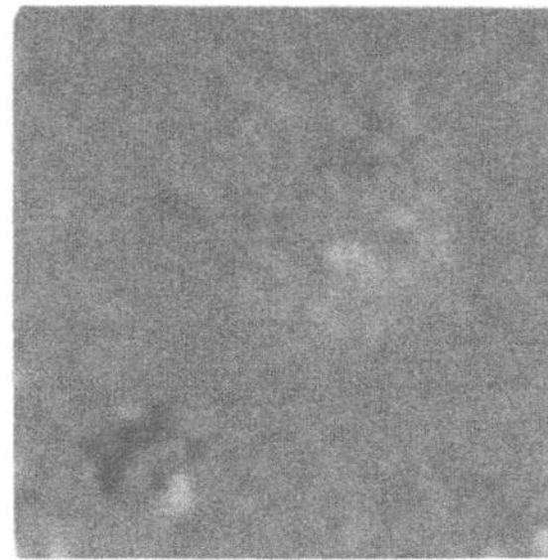

Fig.8. Frequency slope of the
backscatter coefficient.

measured frequency range (Fig. 7) and of the backscatter
coefficient slope (Fig. 8).

Data from an experiment such as described above may be
used in a variety of ways. As exemplified above, it is possible
to display any of the measured quantities (or derived
quantities such as the frequency dependence of a property) as
images and so observe their interdependence. In this regard it
is instructive to note the excellent agreement between the
integrated backscatter image of Fig. 7 and the original B-scan
of Fig. 1 (the relevant portion of which is reproduced at an
appropriate scale in Fig. 2), which exists even though the
backscatter signals for the two images were collected at
tissue/sound-beam orientations at 90 degrees to one another.
This is good evidence for the isotropy of the tissue scattering
structure in human liver and in this particular pathology. The
backscatter images also correspond well to the cut surface
appearance of the tissue section (not shown here) and
demonstrate that the metastatic tumour is imaged with good
contrast by its backscatter properties relative to normal
liver. Considerable tumour internal structure is indicated by
the backscatter data.

The speed of sound also images the tumour structure with
good contrast, demonstrating that efforts towards devising noninvasive methods for imaging speed of sound *in vivo* may one day
prove to be diagnostically important. On the other hand the
details in the attenuation images seem to be highly correlated
with the local gradient of the speed of sound image, suggesting
that phase cancellation artefact is primarily responsible.
Examination of the parts of the attenuation images which
correspond to more homogeneous velocity regions suggests that
in the absence of phase cancellation attenuation probably is
not a parameter which provides good contrast discrimination for
imaging this metastatic deposit. The frequency dependence of
attenuation looks perhaps a little more promising in this
regard, although in this example that are two many error points
(locations where the amount of attenuation relative to noise in
the measurement was two low to permit a reliable estimate of m)
to be completely sure. These observations are consistent with
those made previously[1].

It is clear from comparisons of the acoustic maps with the
appearance of the tumour on the cut specimen that the lesion

consists of two major regions which have substantially
different acoustic properties: a central nucleus (which
scatters a little more strongly than the surrounding normal
liver, has a relatively high frequency dependence of scattering
and a relatively low sound speed) and an outer annulus (which
scatters less strongly, has a moderate and homogeneous
frequency dependence of scattering and a moderately high sound
speed). This raises an important issue; that of being able to
study the correlation between the acoustic properties and the
spatial fluctuations of histological components of the tissues.
To facilitate such studies we have constructed a special system
for quantitative histological mapping of tissue sections, which
is described below.

HISTOLOGICAL MEASUREMENTS

After completing the acoustic measurements the tissue
slice is frozen and sectioned for optical microscopy. Sections
are taken from the central part of the specimen, the back
surface and the front surface, with the intention of eventually
summing these to obtain an approximation to the integral of the
histological properties through the whole 7 mm thick slice. All
sections are stained for components which are suspected as
being potentially important in determining the acoustic
characteristics. By scanning the section and using a process of
colour subtraction, a computerised microscope can build up a
quantitative picture of the spatial distribution of any
histological component for which a suitable colour stain
exists. Figs. 9 and 10 show maps (resolution = 128 by 128 true
data points) of the percentage fat and the percentage collagen
content in the specimen examined acoustically. At present these
maps correspond to a very thin (4 μm) section taken from the
centre of the 7 mm tissue slice.

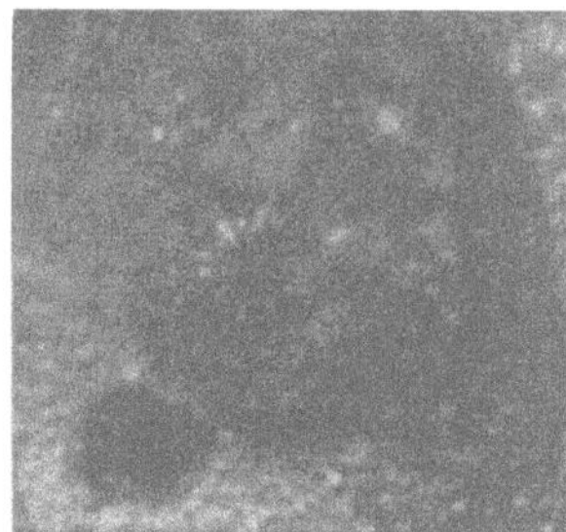

Fig.9. Map of percentage fat
content (by area).

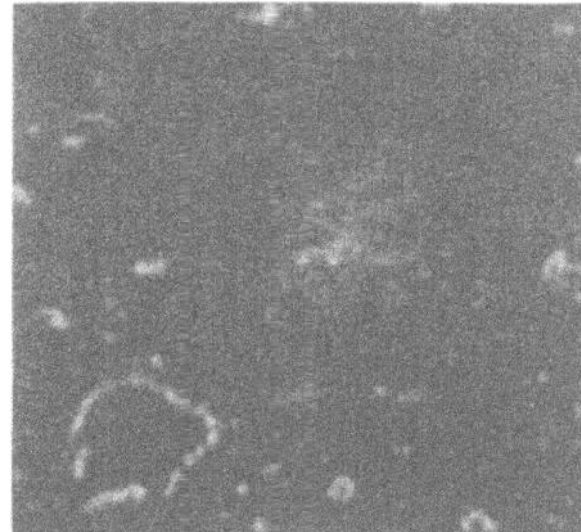

Fig.10. Map of percentage
collagen content (by area).

The mean fat content in the normal liver was 2.5% by area.
Although the fat content was much lower than this in parts of
the outer annulus of the tumour it was also a little lower in
the tumour nucleus, typically in the region of 1%. Both the
collagen and the fat maps appear to correlate well with the
integrated backscatter map and the B-scan but not with sound
speed, which may be influenced primarily by some other
component such as water content[2].

It is interesting to note that the collagen in regions of normal liver (mean collagen content = 0.5%) tends to line the microvasculature, whereas that within the tumour (mean = 1.6% but peaking in the centre of the tumour at 24%) has a fine mesh structure. It is possible that the fine distribution of many small scattering interfaces, which such a mesh structure would represent, causes the increase in the frequency slope of backscatter shown in Fig. 8, since this is an acoustic parameter already known to provide a measure of the size of scattering structures[6].

CONCLUSION

The results presented here are preliminary and part of a prospective long term study. In the future, histological maps which are the integral of the histological properties through the full tissue slice examined acoustically should enable us to obtain an understanding of the manner in which different histological components contribute to the various acoustic characteristics within single inhomogeneous specimens. Having classified tissue regions it should then be possible to go back to the original data for specific regions (which may be irregular or not continuous) and extract good averages of acoustic quantities such as the frequency dependence of scattering, which are likely to provide information about the microscopic distribution of histological components rather than their total content. It is intended that analysis of the stained sections under the computer driven microscope will then reflect this by mapping the distribution of parameters of tissue architecture which can be related to parameters of mathematical models of the ultrasound scattering structure derived from the acoustic measurements.

ACKNOWLEDGEMENTS

This work is supported by grants from the Cancer Research Campaign and the Medical Research Council.

REFERENCES

1.   J. C. Bamber and C. R. Hill, Acoustic properties of normal and cancerous human liver - I dependence on pathological condition, _Ultrasound Med. Biol._ 7:121-133 (1981)

2.   J. C. Bamber, C. R. Hill and J. A. King, Acoustic properties of normal and cancerous human liver - II dependence on tissue structure, _Ultrasound Med. Biol._ 7:135-144 (1981)

3.   J. C. Bamber, Ultrasonic propagation properties of the breast, _in_: "Ultrasonic Examination of the Breast," J. Jellins and T. Kobayashi, eds., John Wiley and ons, Chichester, (1984) pp.37-44.

4.   F. T. D'Astous and F. S. Foster, Frequency dependence of ultrasound attenuation and backscatter in breast tissue, _Ultrasound Med. Biol._ 12:795-808 (1986)

5.   M. O'Donnel, D. Bauwens, J. W. Mimbs and J. G. Miller,
     Broadband integrated backscatter: an approach to
     spatially localized tissue characterization in
     vivo, <u>in</u>: "Ultrasonics Symposium Proceedings," IEEE
     CH1482-9, (1979) pp.175-178.

6.   E. J. Feleppa, F. L. Lizzi, D. J. Coleman and M. M.
     Yaremko, Diagnostic spectrum analysis in
     ophthalmology:a physical perspective, <u>Ultrasound
     Med. Biol.</u> 12:623-631 (1986)

A FAST ALGORITHM FOR TRANSKULL BRAIN IMAGING

Zhengdi Qin[1], J. Ylitalo[1],
J. Koivukangas[2], J. Oksman[1]

[1]Department of Electrical Engineering
[2]Department of Neurosurgery

University of Oulu
90570 Oulu, FINLAND

INTRODUCTION

With the development of computer technologies, brain imaging itself has entered into a new era with progress in digital reconstruction, including X-ray computed tomography, nuclear magnetic resonance imaging and positron emission tomography. Each of them has practical drawbacks, for example, lack of portability and handling, high cost, and ionizing radiation. These factors restrict routine widespread use of these methods. Thus, other methods such as ultrasound imaging deserve serious consideration.

Ultrasound imaging has become an acceptable tool in many clinical fields. A number of studies have been done for visualizing human brain through intact skull using ultrasound for medical diagnostic purposes [1-6]. The results of past attempts suggested ultrasound of lower frequency for transkull brain imaging to avoid severe spatial and temporal pulse distortion at typical higher frequencies. In the recent studies, frequencies in the range of 1 MHz have been accepted using a phased array with transit time compensation for phase adjustment, providing better transkull imaging capability.

In the past few years, our laboratory has investigated the possibility of ultrasound imaging of adult brain through intact skull. Apparently there are two major difficulties for transkull brain imaging. One is the skullbone which exhibits an inhomogeneous layer with strong reflection and scattering for the ultrasound passing through. The results from our studies showed some analogy to those of earlier attempts. Using lower frequencies, range from 0.5 - 1 MHz, the image distortion and the decrease of S/N ratio due to the energy loss caused by the skullbone are acceptable. Nevertheless, if we attach the transducer to the surface of the skull and treat the skullbone as a special matching layer between the

transducer material and the brain tissue, the distortion and
the energy loss can be reduced to an even lower level.

For the 2D imaging of a slice of brain, a transducer array
is needed and another problem arises. If all the elements of
the transducer are attached to the surface of the skull, a
circular transducer array is needed for the spherical
structure of the skull. Few earlier attempts have studied the
application of a circular array for ultrasound transkull brain
imaging. Ultrasound computed tomography and compound B-
scanning can produce disc-like tomograms of the object, but
have never before been applied to the transkull brain imaging.

In our earlier papers [7-12], a circular ultrasound
imaging method (called Q imaging) and a frequency domain
compensation technique for an inhomogeneous layer were
reported. Combining these two techniques, a new method for
transkull brain imaging is reported in this paper. Using a
circular transducer array, the reconstruction process is based
on the linear array holography using the backward propagation
principle. A k-space compensation technique is introduced to
compensate for the higher velocity layer of skullbone and for
the circular geometry. The FFT algorithm can be directly used
in the reconstruction process and a real-time system is easily
achievable.

Q IMAGING

The detail of the Q imaging method was described in [12].
The basic idea of this method is that in a circular array 2D
imaging system, the circular area is cut by a radius and then
transformed to a rectangular area. The circular array is then
transformed to a linear array and the linear array imaging
reconstruction process is modified for the reconstruction of
circular 2D imaging. The reconstruction time for the circular
array imaging is the same as that for linear array imaging.

In a 1D linear array imaging system using the backward
propagation principle, the reconstruction process for the
image of an object line with distance z and parallel to the
hologram array is [13]:

$$\mathbf{u}_i(z, x_i) = \mathbf{F}^{-1}\{\mathbf{H}_l(z, f_x)\mathbf{F}\{\mathbf{u}_h(x_h)\}\} \tag{1}$$

where the waveform is supposed to be a plane wave
perpendicularly to the hologram plane. z is the distance
between the hologram plane and the object plane. $\mathbf{u}_h(x_h)$ is the
hologram data representing the wavefield on the hologram plane
and x_h is the coordinate of the hologram plane. $\mathbf{u}_i(z, x_i)$ is the
reconstructed image and x_i is the coordinate of the image
plane. $\mathbf{F}$ and $\mathbf{F}^{-1}$ denote the one-dimensional linear Fourier
transform and the inverse Fourier transform, respectively.
$\mathbf{H}_l(z, f_x)$ is the transfer function in which the hologram data is
transformed back a certain distance z to the image plane.

As we derived in [12], if the transducer is a circular arc
with radius R_t, the reconstruction process for the image of
the object on the circular arc with radius R_i is:

$$\mathbf{u}_i(R_i, \theta_i) = \mathbf{F}^{-1}\{\mathbf{H_c}(R_t, R_i, f_\theta)\,\mathbf{F}\{\mathbf{u}_h(R_t, \theta_h)\}\} \qquad (2)$$

where the waveform is supposed to be a cylindrical wave coming from the centre. $\mathbf{u}_h(R_t, \theta_h)$ is the hologram data representing the wavefield on the hologram arc and θ_h is the angular coordinate of the hologram arc. $\mathbf{u}_i(R_i, \theta_i)$ is the reconstructed image and θ_i is the angular coordinate of the image arc. $\mathbf{H_c}(R_t, R_i, f_\theta)$ is the transfer function in which the hologram data is transformed back from the transducer arc to the object arc.

$$\mathbf{H_c}(R_t, R_i, f_\theta) = \exp\{\eta d\phi\}\ \mathbf{H_l}(R_t - R_i, f_\theta) \qquad (3)$$

where η is the spatial expansion coefficient which is equal to R_t/R_i. $d\phi$ is the phase difference between the circular array system and the transformed linear array system. $\mathbf{H_l}(R_t - R_i, f_\theta)$ is the transfer function for the linear array imaging system.

Comparing (1) and (2), the reconstruction process and accordingly the time consumed for the circular array imaging are the same as that for linear array imaging. If the elements of the transducer are placed uniformly on the hologram arc. The FFT algorithm can be used for the Fourier transforms in the process. Expanding the circular arc to a full circle, a round circular array system is obtained. The analysis above is also valid to a complete circular array system to reconstruct a one-dimensional circumference image of the object.

For a two-dimensional image of a circular array using the pulse-echo model, another technique must be introduced to correct the near field curvature distortion. The curvature property due to the spherical wave-front is discussed in [7-12]. A so-called rearrangement operation technique is introduced to correct the curvature distortion in the frequency domain of the hologram.

VELOCITY LAYER

In our paper [11], a frequency domain compensation principle for an inhomogeneous plane layer in ultrasound holography was reported. The principle can also be applied to a spherical velocity layer with uniform thickness.

First, we suppose that the skull is a spherical layer with uniform thickness, and the ultrasound velocity of it is different from those of the material on both sides. The material of the transducer elements forms the outside layer on the skullbone if they are attached to the surface of the skull and the inside is the brain tissue which is to be imaged. The ultrasound velocity of the skull is in between those of both sides. So it can be treated as a special layer for the ultrasound energy to pass through. Of course, it is not a realistic matching layer because of the complexity and anisotrophy of the skull. But in our experiments we found that if the transducer elements are attached closely to the surface of the skull, the energy loss is smaller than that of in the case in which there is some material between the transducer and the skull. This also helps to reduce the multi-reflection error caused by the skull.

As we mentioned above, a 2D circular array system can be transformed to a linear array system. If there is a spherical layer in the circular array system, after the change of the geometry, the spherical layer is transformed to a plane layer in the transformed linear array system. Therefore all the theory we discussed in [11] can be applied here for the compensation of a spherical layer with uniform thickness in the circular array imaging system. The compensation can be done in the frequency domain of the hologram. The transfer function for the circular array imaging is modified by a simple multiplicative compensation factor B_s which can be merged into H_c to form a new transfer function for the new circular array imaging in which a spherical layer is present.

In a 2D imaging system, the rearrangement operation is also modified for correcting the distortions caused by the near field curvature property of the wavefield and the spherical layer of the skullbone. A look-up table is set up for the rearrangement of the spectrum of the hologram data.

To sum up, the whole reconstruction process of the circular array transkull brain imaging system is: first the one-dimensional linear Fourier transform (FFT can be used) is applied line by line along the depth direction to get a two dimensional spectrum of the hologram. In the frequency domain, the spectra are rearranged according to their depths using the look-up table and then multiplied by the compensation factor for the skullbone and the transfer function for the circular geometry. Finally, the inverse one-dimensional linear Fourier transform is used to get the image. The reconstruction time for a transkull image is practically the same as that for a linear array image. Thus a real-time system is easily achievable. On the other hand, if only a sector area is to be imaged, the hologram is not necessarily a complete circular data . In that case a circular arc transducer array can be used for a sector image. This is useful in many cases, for example, one can put an arc array on the forehead of the skull to get a sector brain image.

INSTRUMENTATION

The whole system is based on an IBM-AT computer. Fig. 1 shows the measurement system. The mechanical circular scanning of a single transducer was employed in our prototype system. The transducer was mounted on a holder which was attached to a step motor to rotate the transducer in specific increments. This rotation was controlled by the AT computer with equal steps programmed for a total rotation of 360° with 512 angles of view. The ultrasound frequency of 1 MHz was used in all of the measurements and the transmitting pulse was coherent with the pulse length of four wavelengths. Fig. 2 shows the drive waveform produced by a special coherent pulse generator.

The data acquisition of the hologram is similar to that of a traditional B-scanner. At each increment step of the transducer rotation, the coherent burst was sent and then the

Fig.1 Measurement system.

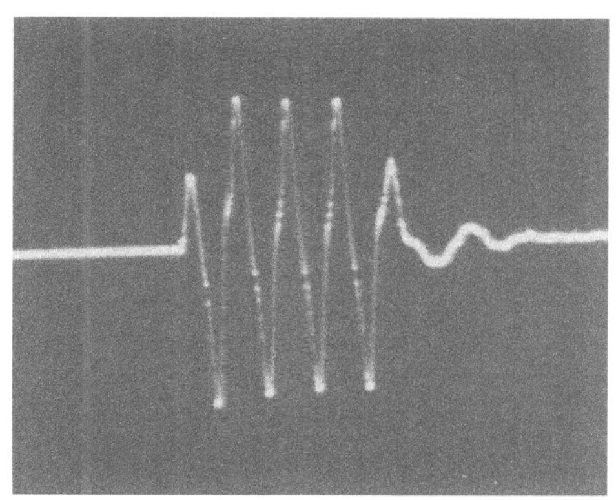

Fig.2 Transmitting pules
1 MHz, 4 λ, 150 V_{p-p}.

transducer was switched to the receiving mode to record the backscattered echoes as a function of time, representing various depth levels. Differing from a B-scanner, the data were recorded in a complex form and the transducer had a wide beam characteristic. 128 depth levels within the sampling window were sampled (the maximum depth was the radius of the transducer circumference). The data were digitized by an A/D converter in the computer (WAAG card: 8 BITS, 20 MHz). After the measurement at one position, the transducer moved to the next position along the circumference, driven by the step motor. When a complete set of data was acquired, the image was reconstructed by the AT computer. Using the machine level language, the reconstruction time for a 90° 128X128 pixel sector image is about 3 seconds without any special co-processor in the computer.

The objects were immersed in water and scanned by the transducer for 360° using 512 steps. In the transkull imaging

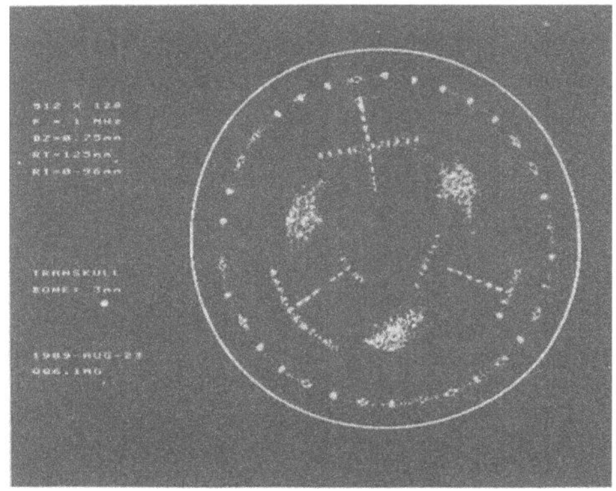

Fig.3 Image of
a test model.

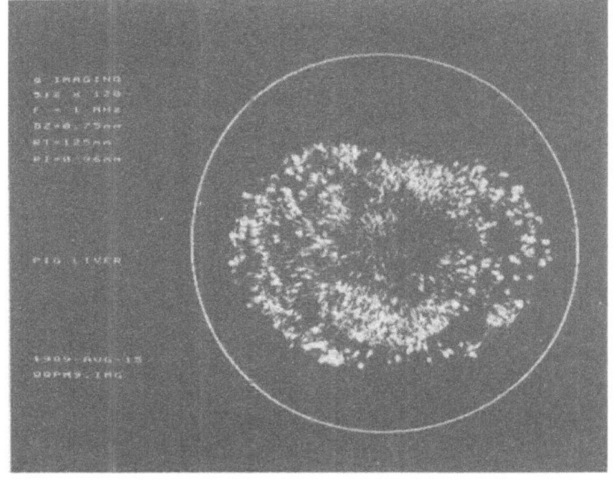

Fig.4 Image of a piece
of pig liver.

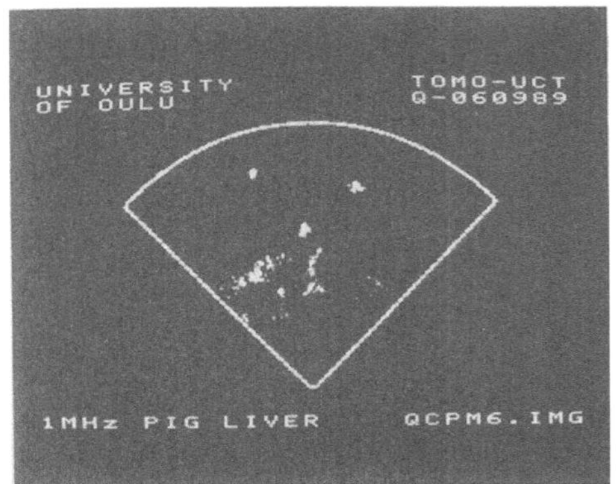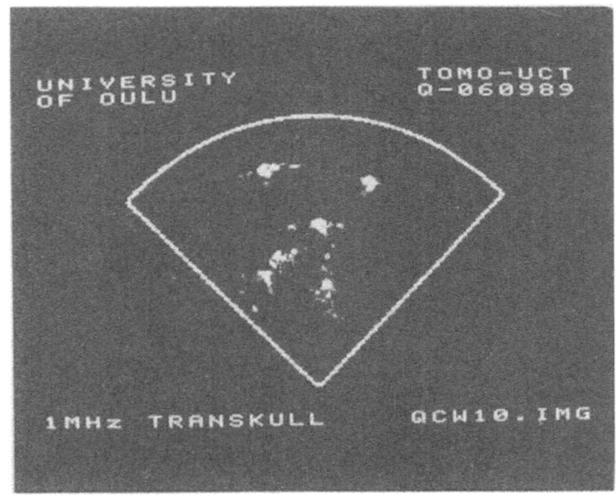

Fig.5 Images of a sorft tissue specimen with 3 pins in it.
left: imaging without skullbone,
right: transkull image.

experiments, a small piece of human skullbone (age 40, 3 mm thickness) was attached in front of the transducer to present a uniform thickness of a spherical skullbone.

Fig. 3 shows the transkull image of a test model [12]. The model has three different patterns composed of several points which were made from ϕ0.3 mm nylon monofilaments. Fig. 4 shows the transkull image of a soft tissue specimen (fresh pig liver). Fig. 5 shows the quarter images of a soft tissue specimen scanned by the transducer for 90° using 128 steps. There were three steel pins in the specimen. On the left is the image without the skullbone and on the right is the transkull image.

CONCLUSION

As a medical application of the ultrasound Q imaging principle, a circular array transkull brain imaging method is introduced in this paper. The reconstruction process is based on a linear array imaging process and the phase adjustment is performed in the k-space of the hologram. The major features of this method are a fast algorithm and a simple configuration. The result is a reflection tomogram. The preliminary experiments showed promising results, in which a piece of adult skullbone was attached in front of the transducer to present a spherical skullbone of uniform thickness. Of course, the spherical velocity layer of uniform thickness is only a rough approximation to an actual skullbone. For a practical system, a circular transducer array is needed and the compensation formula must be modified for the nonuniform thickness and the velocity variation of real skullbone.

ACKNOWLEDGMENT

This work was supported by the Academy of Finland. Support from the Technology Development Centre of Finland is acknowledged.

REFERENCES

1. K. T. Dussik, F. Dussik, L. Wyt, "Auf dem Wege zur Hyper-phonographic des Gehimes,"Medizinsche Wochenschrift, vol. 97, pp. 425-429,1947.
2. T. F. Hueter, R. H. Bolt, " An ultrasonic method for out-lining the cerebral ventricles,"J.Acoust. Soc. Am.,vol. 23, pp.160-167, 1951.
3. O. T. Ramm, S. W. Smith and J. A. Kisslo, "Ultrasound tomo graphy of the adult brain,"Ultrasound in Medicine, vol. 4, D. White and E. A. Lyons, Eds. New York: Plenum, 1976, pp. 261-267.
4. W. A. Erdmann, F. J. Fry, K. W. Johnston and N. T.Sanghvi, " Instrumentation for Ultrasonic Transkull Visualization," IEEE Trans. SU, vol. 29, no. 1, Jan. 1982, pp.5-11.
5. D.J.Phillips, S.W.Smith, O.T.von Ramm and F.L.Thurstone,"A phase compensation technique for B-mode echoencephalo-graphy", Ultrasound in Medicine, vol. 1, pp. 345/404,1975.
6. S.W.Smith, "Phased array ultrasound imacing through planar tissue layers," Ultrasound Med. Biol., Vol. 12, no. 3, pp. 229-243 Mar. 1986.
7. Z. D. Qin," Preliminary report on ultrasonic transkull brain imaging," Internal report, University of Oulu, Finland, May. 1987.
8. Z. D. Qin, "UHB imaging and the frequency domain compensa-tion principle,"Acta Acustica (in Chinese), accepted for publication. 1987
9. Z. D. Qin, " Frequency domain compensation in ultrasound holography," Doctoral dissertation, Zhejiang University, Hangzhou, China, June 1988.
10. Z. D. Qin, J. Ylitalo, E. Alassarela and J. Koivukangas, " Circular array ultrasound holography imaging ", Inter national Patent application, No.PCT / F189 / 00058, 1988.
11. Z. D. Qin, A. Tauriainen. J. Ylitalo, E. Alasaarela and W. X. Lu, "Frequency domain compensation for inhomogeneous layers in Ultrasound holography,"IEEE Trans. UFFC. Vol.36, no. 1, Jan. 1989, pp. 73-79.
12. Z. D. Qin, J. Ylitalo, J. Oksman and W. X. Lu, " Circular array ultrasound holography using the linear array approach," IEEE Trans. UFFC. Vol. 36, no. 5, Sep. 1989.
13. J. W. Goodman, Introduction to Fourier Optics. New York: Plenum Press, 1971.

DETECTION OF ROOT CARIES

Sidney Lees*, Thomas J. Nelligan+
and Robert Doherty+

*Forsyth Dental Center
140 Fenway
Boston MA 02115
+Panametrics, Inc.
221 Crescent Street
Waltham MA 02254

INTRODUCTION

The gums cover the roots of teeth in younger humans but recede in older people beginning with the fifth decade. Dental enamel covers the tooth to the gum margin but when the gums recede the roots are exposed, subjecting the exposed root surface to abrasion and decay. The root surface is softer than dental enamel and so structured that it is more susceptible to bacterial infection (Wefel et al. 1985). Root caries is more prevalent in the aging population whereas coronal caries is observed in younger people. Conventional methods for exploring dental enamel using a sharp probe may damage the softer root surface.

The chemical composition of the tooth root is significantly different from dental enamel which is 99% mineral and less than 1% organic. In the root as in the coronal dentin there is more organic matter and much less mineral. Typically the organic component is 12% by weight and 77% mineral. The rest is water. Since mineral is much denser than the organic substances, the density of enamel is about 2.9 g/cc while dentin is around 2.2 g/cc. (Lees et al. 1983) The carious attack begins with a loss of mineral whether dental enamel or dentin. The organic component can only be attacked after the mineral is removed. The loss of mineral changes the chemical composition of the tissue. In the early stages the organic component remains but the lost mineral is replaced by water, resulting in a decrease of the density of that region of the tissue.

The details of the early stages of the demineralization process in teeth is much more complicated than merely forming a pit by erosion from the surface (Margolis & Moreno 1985). The surface remains mineralized while the demineralization proceeds beneath. In the very beginning the surface is partially demineralized, but as the process continues the deeper part undergoes a mineral depletion while the surface layer actually increases its mineral content. A cross section of the lesion shows a dense surface layer about many micrometers thick and a demineralized region in the tissue with a depth depending on the duration of the attack.

It has been found that sonic velocity increases with mineralized tissue density, i.e. with increasing mineral content (Lees et al. 1983).

The specific acoustic impedance, Z, being the product of the density and
the sonic velocity, varies as the square of the tissue density. Typical
values for the sonic velocity and density of bovine incisor enamel are
6.03 km/sec and 2.9 g/cc yielding a Z value of 17.8. Typical corresponding
values for bovine incisor dentin are 3.4 km/sec, 2.1 g/cc and a Z of 7.14
respectively.

The reflection ratio, R, is the ratio of the pressure amplitude of
the reflected sonic wave to that of the incident wave normal to the
surface at the boundary of two dissimilar media as given by the well known
expression

$$R = \frac{Z_2 - Z_1}{Z_2 + Z_1} \qquad (1)$$

The variation of sonic velocity with density for the class of
mineralized tissues that includes dentin and enamel can be represented
with reasonable accuracy by the linear expression (Lees et al.1983)

$$c = 3.141p - 3.193 \qquad (2)$$

where p = density in g/cc and c = sonic velocity in km/sec. The
points plotted on the line in Fig 1 were taken from the cited reference.

The specific acoustic impedance of water at room temperature is
approximately 1.5. The reflection ratio from the surface of dentin-like
mineralized tissues in water as a function of wet density is shown in Fig
1. It can be seen that the reflected wave pressure amplitude is quite
sensitive to the mineral content of dentin-like tissue. The slope of R at
p = 2g/cc is 0.48.

This estimate shows that a small change in mineral content of the
tissue surface will markedly change the amplitude of the reflected
pressure wave. When the tissue surface is attacked, as already noted, the
local density decreases as the mineral is replaced by water. An area of
slightly demineralized dentin should be readily distinguished from the
surrounding normal tissue.

The slope of the R-curve in Fig 1 decreases as the density increases
The extension of the calculation to a density of 3.0 g/cc, about that of
dental enamel, yields a reflection ratio of 0.85 and a slope of only 0.12,
a fourth of that for dentin. The onset of root caries should be more
readily detected than coronal caries.

In addition to changes in the reflection ratio, the transit time of
a sonic pulse through the lesion should increase as the mineral is
depleted, according to Eq 2. It would be useful to determine the
subsurface mineral loss as well as a detailed picture of the subsurface
lesion but a single value of the transit time is inadequate. The increased
transit time can give an indication of the mineral loss which can be
useful to monitor the subsurface process The longer the transit time the
greater the mineral depletion. New techniques are required for more
detailed information.

MATERIALS AND METHODS

Bovine incisors were cut parallel to the tooth axis into slab-like 2
mm thick sections . Each section was covered with nail varnish except for
a well defined window where the lesion is produced. This is a standard
technique in this kind of

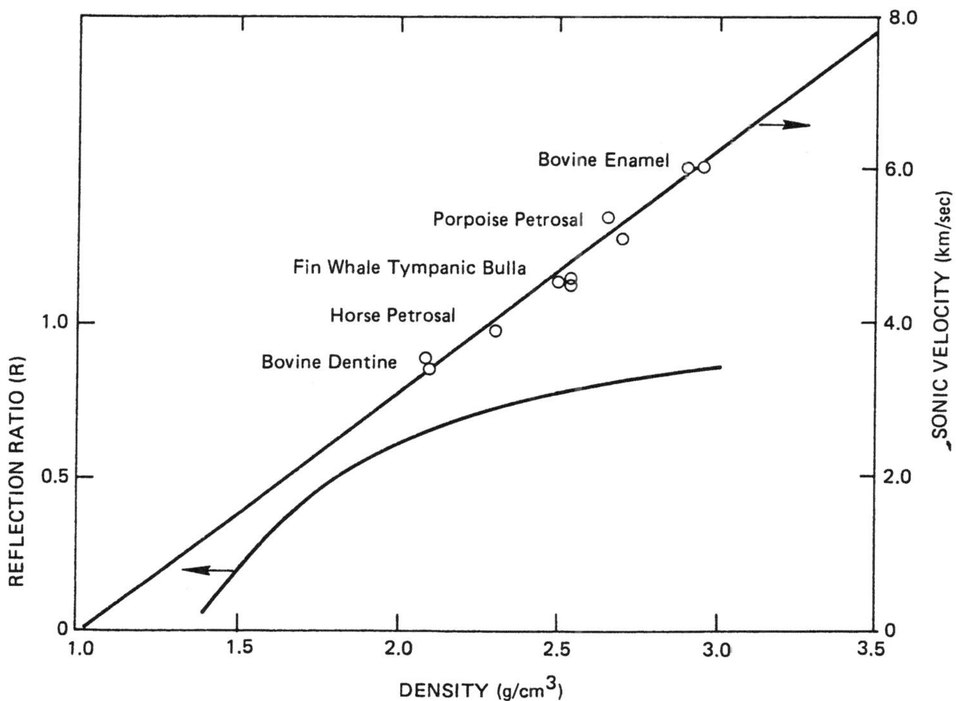

Fig 1. Sonic velocity of dentin-like hard tissues and the reflection ratio, R, in water, both as a function of wet tissue density.

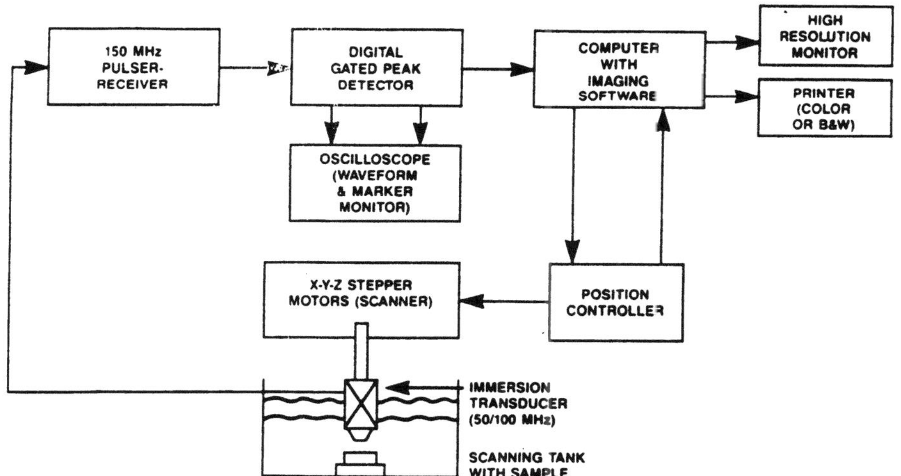

Fig 2. Block diagram of the HYSCAN System

research. The two optical images of Fig 4 are examples each showing
a white area on the tooth slab which was the window for that specimen. A
dentin lesion is induced by a specially formulated partially saturated
lactate buffer with a pH of 4.5. It is similar to the formulation
described by Margolis et al. (1985) for inducing carious-like lesions in
dental enamel. The compositions of the two kinds of buffers differ because
of the differences in mineral content between the two tissues. Each tooth
specimen was suspended from the screw cap in a polyethylene bottle. The
bottle was filled with enough solution to cover the top of the tooth
specimen and the demineralization allowed to proceed for a specific length
of time, in accordance with a time schedule: 1 hour, 24 hours, 96 hours, 1
week, 2 weeks, 3 weeks and 4 weeks. At the end of the specified time the
specimen was removed and washed in 0.15M saline. The nail polish was
removed with acetone and the tooth specimen again washed in saline, then
stored in saline (with sodium azide to inhibit bacterial and fungal
growth) at 4 C. The area of the lesion was barely visible in some of the
specimens when held at an angle but otherwise the lesion was not apparent.

The specimens were examined with the Panametrics ultrasonic HYSCAN
system using a focussed 50 MHz transducer in the pulse echo mode to
produce a C-scan image. The focal length in water was 1.24 cm, the
diameter of the piezoelectric element was 6.4 mm and the 6 db diameter of
the focal spot was 62 um.

A block diagram of the system is given in Fig 2. The transducer was
adjusted vertically by manual control to focus the ultrasonic beam on a
specific plane. The 400 lines per cm high resolution raster is generated
by a pair of programmable stepper motors. The well damped echo waveform
from a flat plate is shown in Fig 3. The digital gated peak detector in
Fig 2 converts the amplitude into a 4 bit (16 level) binary quantity which
is used to generate an image on a high resolution monitor, to drive a high
resolution printer or to be stored in memory as a digitized image for
later use.

The tooth slab was placed in the scanning tank and adjusted until
the entire surface produced echoes as observed on the oscilloscope
monitor. The mechanical scan was set into motion until the tooth image was
displayed on the monitor, the image transferred to memory and a print
produced. Each of the sixteen levels can be called up on demand,
illuminating the area of the image corresponding to that signal level. The
band of signal levels observed from normal dentin was recorded manually
and again those corresponding to the lesion. The two areas were found to
have different signal level bands.

The waveform oscilloscope monitor was used to look for subsurface
echoes as a means for estimating the depth of the lesions. It was also
used to measure the transit time through the entire thickness of the
specimen, front to back. This measurement was made at selected points in
the normal dentin and in the lesion. The instrument does not have the
capability to generate a map of the transit time.

RESULTS

Every tooth specimen yielded an ultrasonic image of a lesion, even
the one hour exposure specimen. Two examples of ultrasonic images of tooth
slabs are shown in Fig 5, to be compared with the optical images of Fig 4.
These represent the least and most exposed tooth specimens. The left image
shows the result after the specimen was in the buffer solution for one
hour, while the right image is an example of the effect after four weeks.

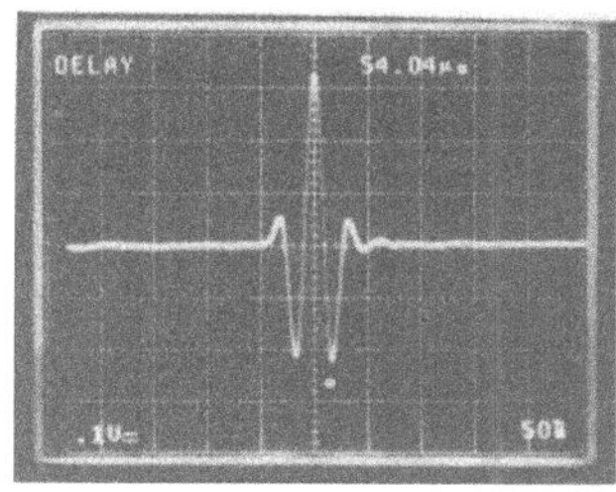

Fig 3. Echo waveform of the 50 MHz focussed transducer
       from a flat surface in water

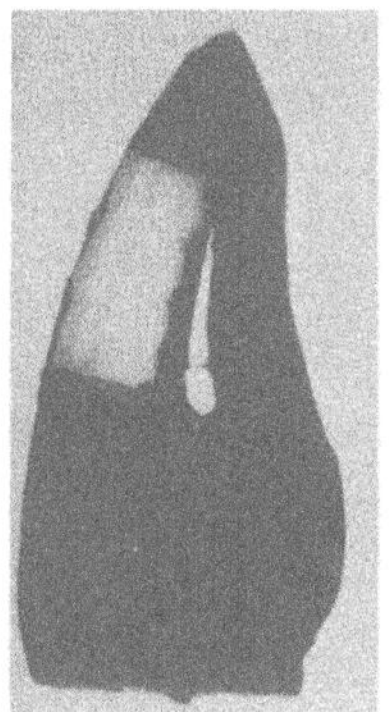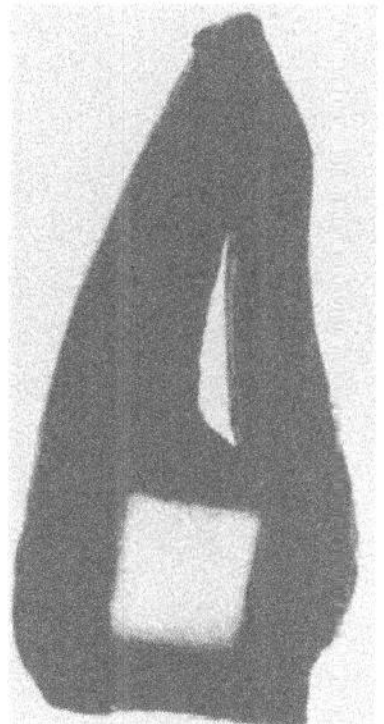

Fig 4. Examples of tooth slabs
     The tooth slab was covered with nail varnish except for
     the white area which provided a window for exposure to
     the buffered solution that induces a carious-like
     lesion in dentin.

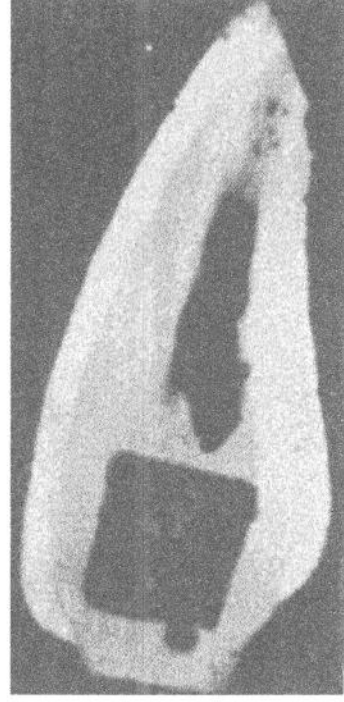

Fig 5. Examples of ultrasonic images of the tooth surface
     The lesion in each can be identified and matched with
     the optical image of Fig 4. The left specimen was
     exposed for one hour. The right one was exposed for
     four weeks.

First, it is to be noted how perfectly the ultrasonic image matches
the window, even for the shortest exposure. It shows that the surface is
rapidly changed by the buffer solution and that the reflected ultrasonic
pulse is responsive to small changes of the dentin surface as predicted
from Fig 1 in the Introduction.

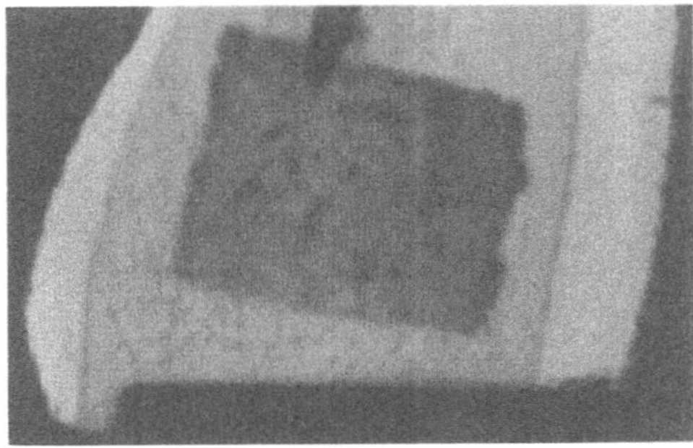

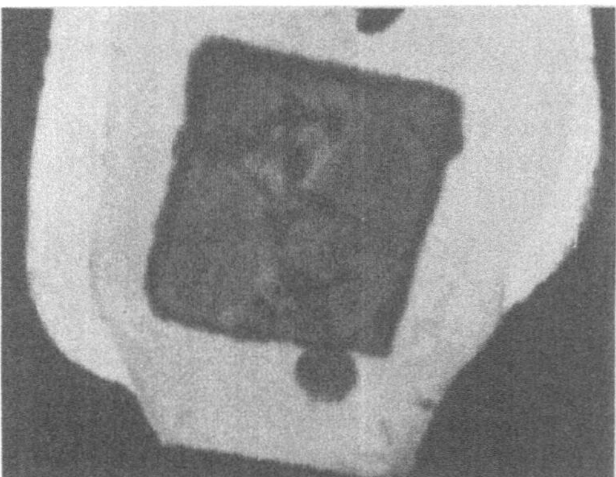

Fig 6.

Two detailed images of
the surface of a lesion.

The upper is an example
of one week exposure,
the right after four
weeks. It is obvious
that the attack is not
uniform. Some areas are
more susceptable to
attack. The darker the
area the greater is the
mineral loss.

Second, Fig 6 is an enlarged view of the lesion area. The darker the
image the greater is the mineral loss. The mottling shows that the mineral
loss is not uniform over the lesion, not surprising since the tissue is
inhomogeneous. Local variability of the mineral was also observed in the
normal dentin. As a consequence, the magnitude of the echo was found as a
band of values rather than a single number. One band was determined for
normal dentin and a second for the area of the legion. A similar
variability was experienced when measuring the transit time through the
specimen.

The reflection ratio of the area of the lesion can be compared to
that of normal dentin by taking the ratio of the signals from the two
regions. Since the magnitude of the signal spans a band of values for each
region, the ratio of the signals is also a band. Fig 7 shows the bands
from each sample as a log of the exposure time in hours. The band of
relative reflection ratios for the shorter exposures is lower than for the
samples exposed to the buffer solution for much longer times. The curve is
a guess as to how the effect varies with time, based on considerations
discussed below.

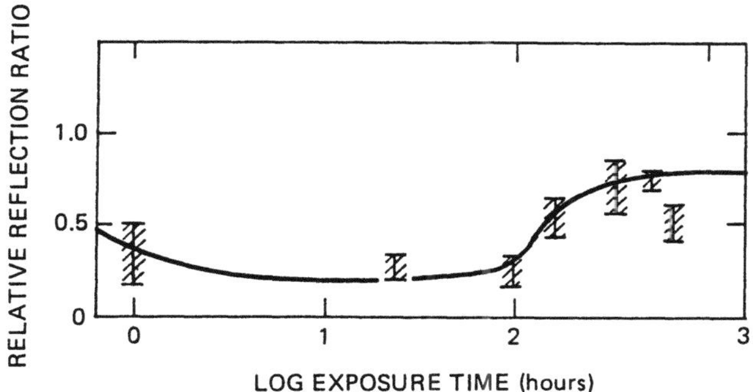

Fig 7. Relative reflection ratio of the lesion compared to
that from normal dentin

The ratio of the transit time through the region of the lesion to
that through normal dentin in the same specimen, as noted above, gives a
measure of the subsurface demineralization. The average value for this
ratio for each specimen is plotted in Fig 8 against the log of the
exposure time in hours. There is uncertainty due to the choice of specific
points for measuring the transit times and to variability in the
characteristics among the several specimens. The curve shows the relative
transit time, and presumably the depth of the demineralization, increases
continuously. Fig 7 suggests there is an initial decrease in the mineral
content of the surface and then an increase. Fig 8 shows a continuous
decrease in the subsurface mineral content.

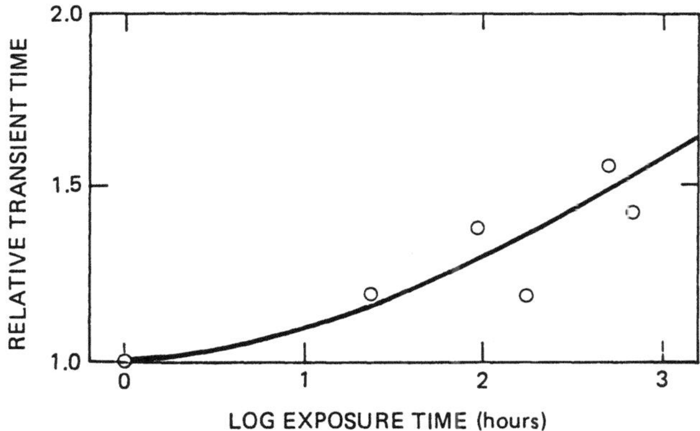

Fig 8. The relative transit time through the lesion with
respect to that through normal dentin

DISCUSSION

    The ultrasonic findings match the well known results found
previously by x-ray microradiography for enamel caries and subsequently
for root caries. Fig 9 is an example of the cross section of an
artificially induced enamel caries provided by Dr. H. Margolis which
illustrates the physical situation most clearly. The thin densely
mineralized surface layer can be readily observed and beneath it is the
more deeply demineralized lesion. It has been proposed that during the
transport of the calcium and phosphate ions through the tooth surface into
the free volume of the mouth outside the tooth, the mineral phase is
precipitated in the surface layer (Margolis & Moreno 1985).

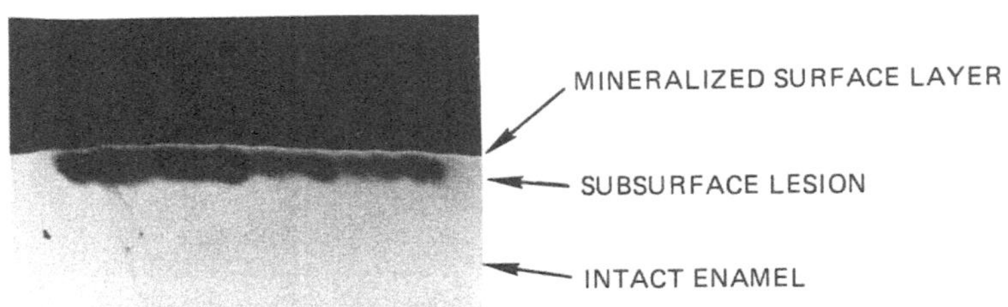

Fig 9. Microradiograph of a well developed enamel lesion.
This is a picture of the cross section of the lesion.
It shows a thin well mineralized layer and under this
layer is the darker demineralized subsurface lesion.
Dentin lesions develop the same pattern of
demineralization.

    The results given in Figs 7 and 8 are in agreement with this
hypothesis within the limitations of the data. The decrease in the
reflection ratio at the outset in Fig 7 indicates that the surface of the
lesion undergoes a loss of mineral using the information from Fig 1. The
subsequent increase in the reflection ratio would correspond to
redeposited mineral. The curve sketched in the figure indicates a flat
minimum in the reflection ratio over the interval from 10 to 100 hours, at
which time the reprecipitation becomes evident. More data are required,
particularly for the early stages of the lesion development, to clarify
the process.

    Fig 8 shows the subsurface lesion is deepening presumably both in
volume and in loss of mineral. There were no echoes from the bottom of
the lesion, indicating that the mineral loss varies in a continuous
pattern and that there is no sharp boundary. As a consequence there is not
enough information from which to make an estimate of the subsurface
mineral loss.

CONCLUSIONS

    It has been demonstrated that early detection of a lesion in dentin
is easily sensed by ultrasound. Since the root of a tooth has a very
similar composition and structure, the early detection of root caries
should be obtained with equal sensitivity. The effect of a partially
saturated lactate buffer, which can induce carious-like lesions, can be
unambiguously detected after the first hour of exposure and probably even
earlier. The changes in the reflection characteristics of the developing
lesion closely correspond to the effects predicted from a knowledge of the

composition of dentin and the theory of ultrasound at the boundary of dissimilar media.

There was no detectable echo from the bottom of the induced lesion. This indicates that there is no sharp boundary but rather a gradation of mineral content between the lesion and the sound tissue.

The continuous increase in the sonic transit time with exposure to the buffer shows that the lesion is increasing both in the loss of mineral and in the volume of the affected region.

ACKNOWLEDGEMENTS

Micheale Kearney-Glynn prepared the tooth specimens. Forsyth's contribution was supported in part by National Institute on Aging grant RO1-AGO2325.

REFERENCES

Lees,S., Ahearn,J.M. & Leonard,M.   1983
Parameters influencing the sonic velocity in compact calcified tissues of various species.
J. Acoust. Soc. Am. 74:28-33

Margolis,H.C. & Moreno,E.C.   1985
Kinetic and thermodynamic aspects of enamel demineralization
Caries Res. 19:22-35

Margolis,H.C., Murphy,B.J. & Moreno,E.C. 1985
Caries Research 19:36-45

Wefel,J.S., Clarkson,B.H. & Herman,J.R.   1985
Natural root caries: a histologic and microradiographic evaluation
J. Oral Pathology 14:615-623

REFLEX TRANSMISSION IMAGING: VISUALIZATION AND EVALUATION OF CALCULI FOR LITHOTRIPSY

Joel F. Jensen, Philip S. Green, Peter Schattner, Ajit S. Shah,
and Todd K. Whitehurst
SRI International, 333 Ravenswood Avenue, Menlo Park, CA 94025

Kenneth W. Marich
Diasonics, Inc. 1565 Barber Lane, Milpitas, CA 95035

INTRODUCTION

Reflex transmission imaging (RTI) is a method for producing orthographic images that depict focal-plane ultrasonic transmittance.[1-4] Unlike conventional transmission imaging methods, which require an opposing pair of transducers, RTI can be implemented using modified B-mode equipment and a bidirectional scan probe. Contemporaneously, an integrated reflection C-scan (IRCS) image can be generated, representing the focal plane reflectivity. The two images are in perfect registration, and reveal complementary information. In this study, we use these methods to image human gallstones and kidney stones in tissue phantoms, both before and after their disintegration by a piezoelectric lithotripter.

BACKGROUND

Referring to Figure 1, which depicts a dual-sector-scanned transducer, a *reflection* pixel value is produced by averaging the echo amplitudes from the focal region of the transducer after each pulse transmission. A complete image is produced by scanning the transducer in two dimensions. Generally, focal-zone averaging provides a better reflection image than does the conventional reflection C-scan technique, which uses only one or a few samples.

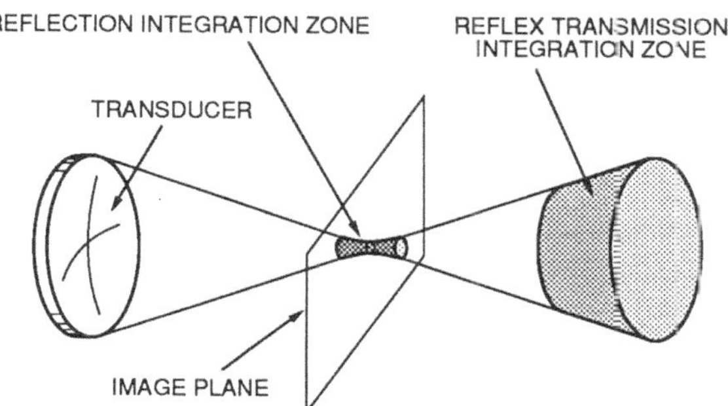

Fig. 1. Integrated reflection C-scans (IRCS) are formed by averaging the echo amplitudes from within the focal zone of the bidirectionally scanned transducer. The image plane may be normal to the central beam axis or oblique. Simultaneously, a reflex transmission image (RTI) is formed by averaging echo amplitudes from a region *beyond* the focal zone.

To produce a reflex *transmission* image, we integrate the echo amplitudes from a second range zone, which is situated beyond the focal plane. The amplitude of these echoes is strongly dependent upon the attenuation in the focal region, because the transmitted energy was substantially concentrated in that region. Moreover, because the transducer is focused at that depth on reception as well, only the energy passing back through the focus is received with high sensitivity. Thus, with each transmit pulse, two C-mode pixel values are obtained, one representing reflectivity and one transmissivity. A C-mode image pair is produced by scanning the transducer bidirectionally .

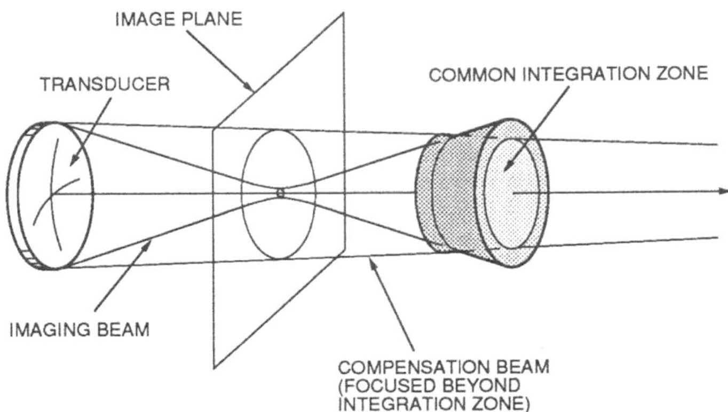

Fig. 2. To produce a more accurate image of focal-plane attenuation, we correct for variations in scatter cross-section within the RTI integration zone using a second set of pixel data, produced by refocusing the transducer to a point well beyond the integration zone. Each pixel value derived with this beam represents substantially the same echo-generating volume as does the corresponding image pixel. However, as the second beam is grossly out of focus in the image plane, it contains little representation of image-plane information; combining the two image fields removes background variations while leaving the focal-plane attenuation image substantially intact.

If the transducer f/number is low (say, 3 or less) the reflex-transmission pixel values depend primarily on the transmissivity of the focal region. However, the values also are influenced by attenuation variations in the whole focal cone and by variations of reflectivity in the integration zone, as previously described.[3] Reflectivity-variation error can be suppressed by extending the length of the integration zone—3 cm seems to provide good results for tissue imaging at several megahertz, and also provides ample suppression of multiplicative noise.[1] Moreover, if a variable-focus transducer is used, both sources of error can be further reduced by combining the reflex transmission image with a compensation image field, produced by first refocusing the transducer to a selected point well beyond the integration zone, then integrating the echo amplitudes from within the zone, as illustrated in Figure 2. This reference image is composed of echoes from essentially the same volume interrogated for RTI, but it is quite out of focus in the RTI focal plane. Much of the residual error is eliminated when these image fields are combined—say, by dividing the RT pixel values by the corresponding compensation values, usually, after local low-pass spatial filtering of the compensation image. Figure 3 demonstrates the effectiveness of compensation, even for an extremely inhomogeneous integration region.

Although this method of compensation has proved quite effective and is being incorporated into imaging systems currently under development, it was not used in the imaging studies reported below.

MATERIALS AND METHODS

A series of experiments were conducted in which gallstones and kidney stones were imaged in physical environments similar to those found in vivo. We placed human gallstones in simulated gallbladders (15-cm^3 saline-filled latex balloons) which were embedded in beef liver and covered by beef muscle, as diagramed in Figure 4(a). Similarly, human kidney stones were embedded in a lamb kidney, backed by beef liver, and covered by beef muscle, as in Figure 4(b). These tissue complexes were scanned using a computer-controlled laboratory scanner with a fixed-focus, 4-MHz, transducer (focal length: 173 cm, diameter: 63 cm). The image data were transferred to a microVAX computer for formatting and display.

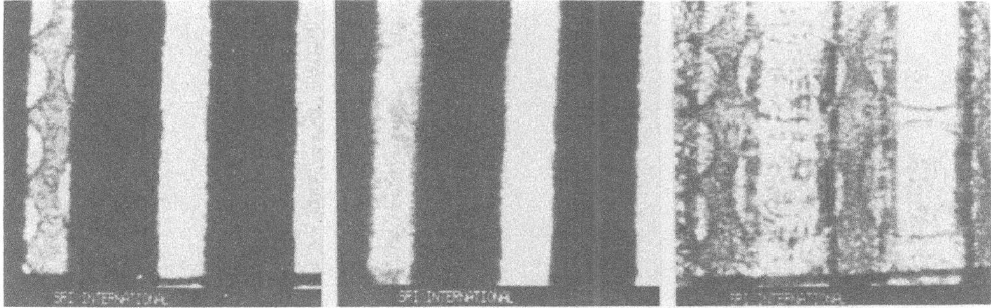

Fig. 3. Correction for inhomogeneous backscatter using a compensation image field produced by moving the transducer forward, rather than refocusing it as illustrated in Fig. 2. The object is a plastic drafting template; the backscatterer is a scatterer-filled gel with internal columns of very high scatterer concentrations. The uncompensated image is on the left. Note that the compensation field (center) shows the inhomogeneous backscatterer but not the template. The backscatterer variations are substantially reduced in the compensated image (right).

Images were first made with the stones whole. We had intended to then transfer the phantoms to the lithotripter—a Diasonics Therasonic™ system—for fragmentation. However, we discovered that, owing to interstitial gases formed by the ultrasonic heating of the unperfused tissues, we were unable to deliver enough energy to break the stones *in situ*. Instead, the stones were removed from the tissue, placed in water-filled containers, and then fragmented on the lithotripter. The stone debris was then placed in the tissue in approximately the same position the stones occupied originally. The tissue phantoms were then imaged again.

EXPERIMENTAL RESULTS

The pre- and post-lithotripsy RT and IRCS images of a pigmented gallstone are shown in Figures 5 and 6. One large fragment remained after lithotripsy. Images of a nonpigmented gallstone, produced under the same set of conditions, are shown in Figures 7 and 8. In this case, the stone was reduced to sand, which settled to the bottom of the bladder. The bladder had been replaced in the tissue matrix in a somewhat different orientation and, as a result, appears to be shaped differently. In the IRCS images, the integration range gate had been set so far back that the front edge of the stone was not imaged, leading to a donut-like appearance.

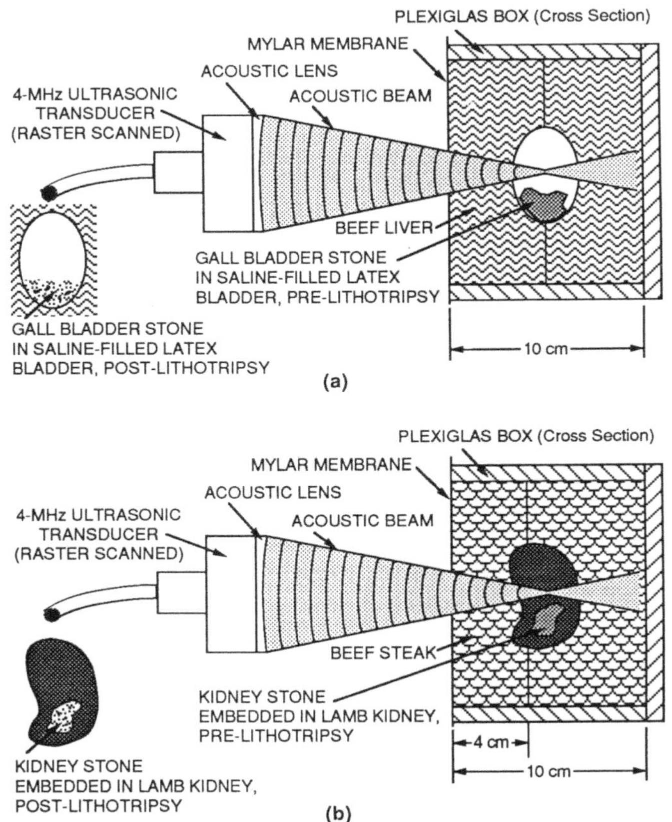

Fig. 4. Diagram of the tissue matrix containing (a) a human gallstone in a simulated gall baldder and (b) a human kidney stone implanted in a lamb kidney. The raster-scanned transducer was focused in a plane containing the stone.

The kidney-stone images are shown in Figures 9 and 10.   The stone is evident both by RTI and IRCS prior to lithotripsy, but only by RTI after lithotripsy. The RT image reveals a distribution of stone debris, whereas in the reflection image the stone particles and the soft tissue are indistinguishable, being of similar echogenicity.

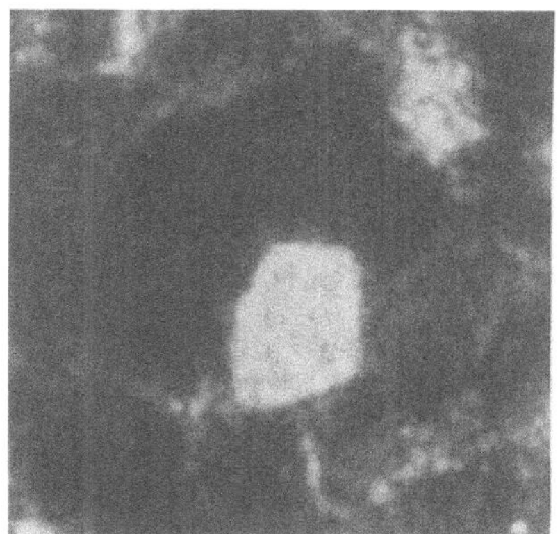

Fig. 5.   RT (left) and IRCS (right) images of a pigmented human gallstone prior to lithotripsy.

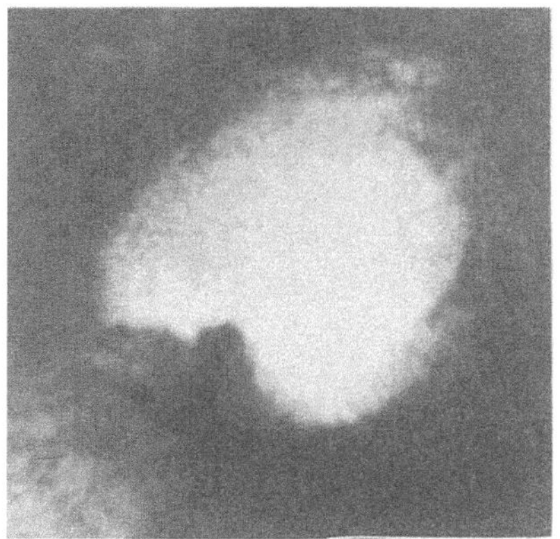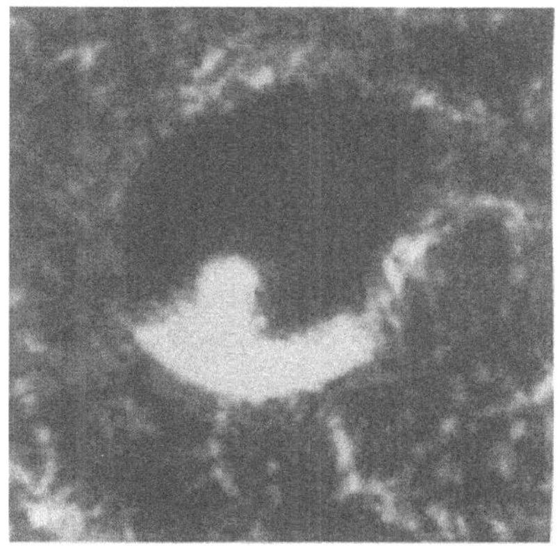

Fig. 6.   Post-lithotripsy RT (left) and IRCS (right) images of the pigmented human gallstone.

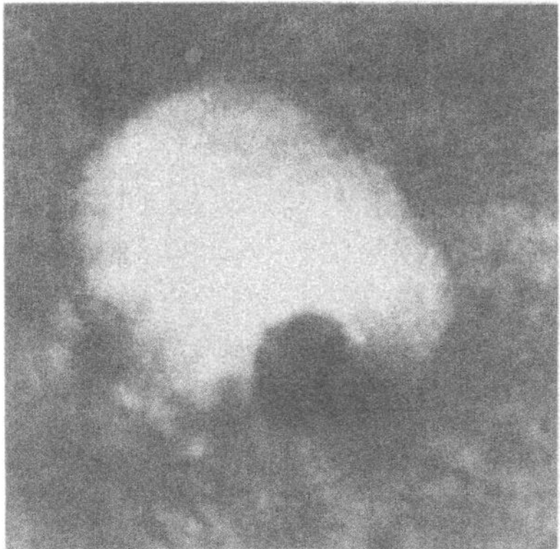

Fig. 7.	RT (left) and IRCS (right) images of an nonpigmented human gallstone prior to lithotripsy. The IRCS range gate did not quite encompass the front of the stone, which appears black.

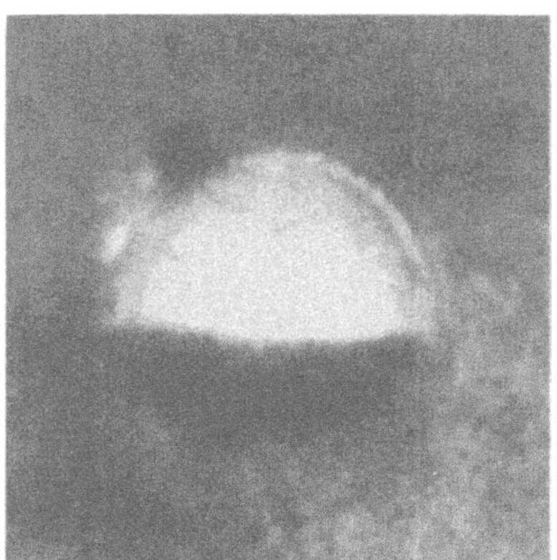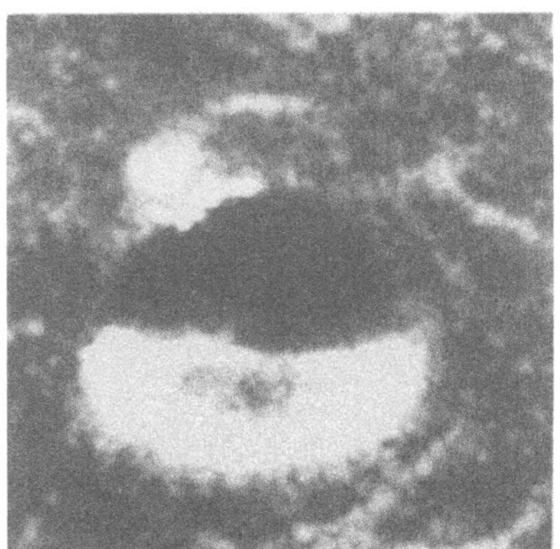

Fig. 8.	Post-lithotripsy RT (left) and IRCS (right) images of the nonpigmented human gallstone of Fig. 7. The IRCS range gate did not quite encompass the front of the stone debris.

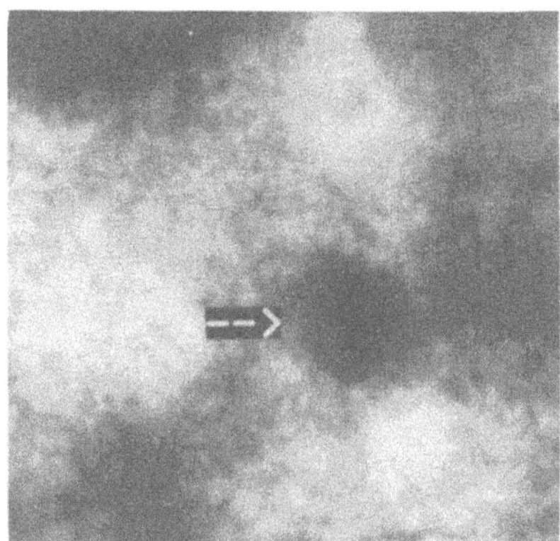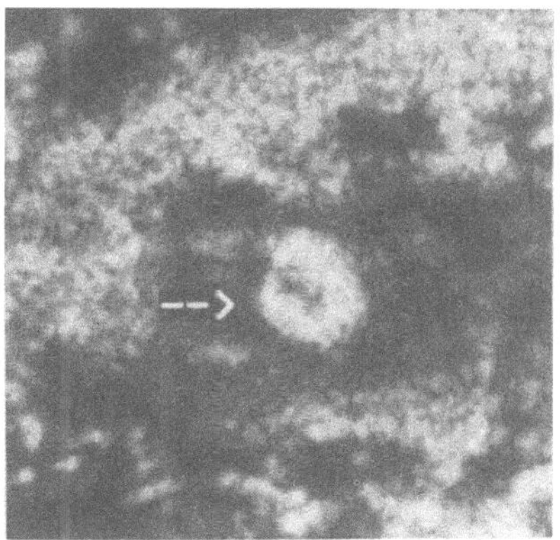

Fig. 9.   RT (left) and IRCS (right) images of a human kidney stone prior to lithotripsy.

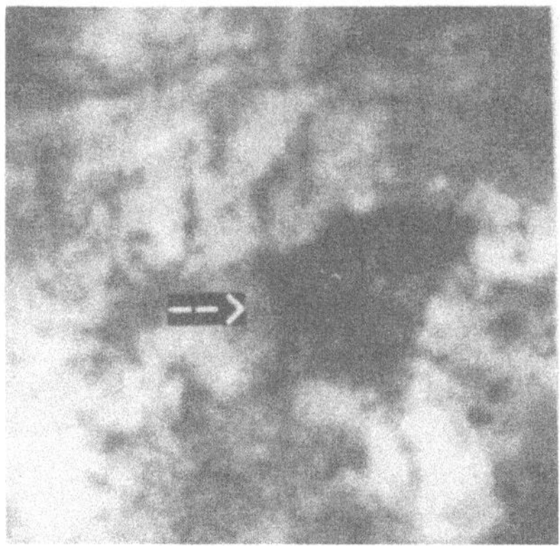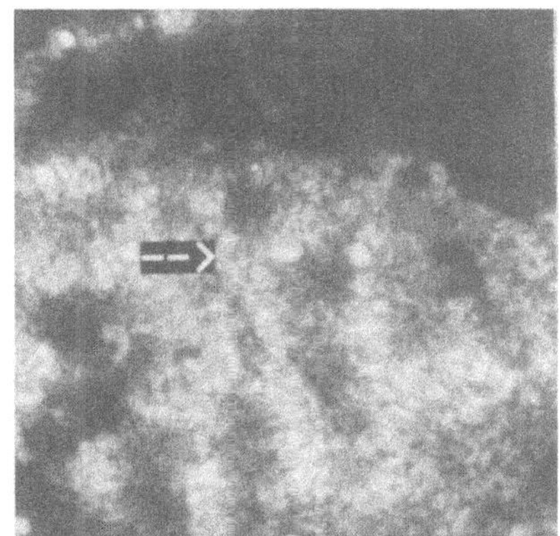

Fig. 10.   Post-lithotripsy RT (left) and IRCS (right) images of the human kidney stone of
Fig. 9. Only by RTI can the stone debris be differentiated from the surrounding echogenic
kidney tissue.

CONCLUSIONS

The potential applicability of reflex transmission and integrated reflection C-scan imaging to monitoring lithotriptic disintegration of renal and biliary calculi has been demonstrated in the laboratory. These two modalities provide complementary information. Used together, they should provide a high degree of assurance that the therapy has been satisfactorily completed.

REFERENCES

1. P.S. Green and M. Arditi, Ultrasonic reflex transmission imaging, *Ultrasonic Imaging*, 7:201-214 (1985).

2. P.S. Green, J.F. Jensen, and W. Ross, In-vivo reflex transmission imaging with a 10-MHz annular array scanner, *Proc. 32nd AIUM Convention* (1987) (Abstract).

3. P.S. Green, J.F. Jensen, and Z.C. Lin, Reflex transmission imaging, in *Acoustical Imaging,* 16: 1-13, L.W. Kessler, ed., Plenum Press, New York (1988).

4. P.S. Green, J.F. Jensen, Z.C. Lin, P. Schattner, and R. Edelman, Advances in reflex transmission imaging", *J. Ultrasound in Med.*, 7: 10 (Supplement):S199 (1988) (Abstract).

5. J.F. Jensen, P.S. Green, K.W. Marich, A. Stein, and J. Pell, Ultrasonic reflex transmission imaging enhances visualization of calculi for lithotripsy applications, *J. Ultrasound in Med.*, 7: 10 (Supplement):S256 (1988) (Abstract).

SELF FOCUSING WITH "TIME REVERSAL" ACOUSTIC MIRRORS

M.Fink, C.Prada, F.Wu

Groupe de Physique des Solides de l'E.N.S.
Université PARIS 7, Tour 23, 2 place Jussieu
75251 PARIS Cedex 05, FRANCE

INTRODUCTION

The focusing of an ultrasonic wave on a fixed or moving target, in an inhomogeneous medium is usually degraded. Such a problem occurs in different medical applications, such as hyperthermia or lithotripsy. Even when the target to be insonified is localized with high precision (absolute position determined by X-ray imaging for instance) , it is difficult to know which wavefront to transmit through the surrounding medium in order to obtain a good focusing. Ultrasonic imaging techniques also suffer from wavefront distortions through the abdominal layer. Dynamical focusing techniques suppose a constant sound velocity in the medium and neglect the wavefront distortions due to fat, muscles and collagen. The ultrasound velocity is usually taken equal to 1540 m/s, though it is 1410 m/s in fat and 1675 m/s in collagen. Previous publications [1][2] show that the detectability of low contrast tumors can be degraded by these distortions, and propose methods for distortion correction. The same kind of problems occurs in optics: wavefront distortions may be due to the aberrations of an optical system or to inhomogeneities of the refractive index in the transmitting medium (e.g. the atmosphere). Different methods allowing phase distortion compensation have been developed, among them the optical phase conjugation which has been studied since 1970 led to very impressive results. This technique applies to monochromatic fields and is performed with a phase conjugate mirror (PCM). An incident monochromatic wave distorted by an inhomogeneous medium impinges on the PCM, a new wave is generated by reversing the phase of the incident one on the mirror. The distortions of the wave are cancelled after the back propagation through the medium. Such phase conjugation can be physically realized in optics through non linear interactions that involve, for example, stimulated Brillouin scattering, three-wave mixing or the so-called degenerated four-wave mixing [3]. All these techniques require the incidence of several reference fields and only work with quasi-monochromatic waves as those produced by lasers. Methods of phase conjugation have also been developed in acoustics, but non linear effects in acoustics need high amplitude fields [4].

In ultrasound pulse-echo mode, broadband pulses are used and the above-mentioned methods of phase conjugation do not apply. Our approach to phase conjugation is quite different. We propose to replace the phase conjugation by a time reversal of the field, this method overcomes the restriction to monochromatic fields.

Acoustical Imaging, Vol. 18, Edited by H. Lee and G. Wade
Plenum Press, New York, 1991

In optics, the detected response depends on the time average intensity of the field. In acoustics, the transducer's linear response to the local field allows to develop a time reversal mirror (TRM). The time reversal mirror is made of an array (1D or 2D) of reversible transducers. The pulse echo signals due to a reflective target are measured with the elements of the array, each signal is stored on a shift register, reversed in time (Last In First Out technique) and then reemitted by the transducers. Such a parallel procedure allows to convert a divergent incident wave issued from a point like source into a convergent reflected wave, focusing on the source. Unlike the ordinary mirror which produces the virtual image of an acoustic object, the TRM produces the 'real' image. The method is efficient even if there are weak inhomogeneities between the target and the TRM. Then the principal interest of this time reversal process is that the distortions introduced by an aberrating medium located between the object and the mirror are corrected by the time reversal. Furthermore the process can be iterated and the iterative mode allows an automatic focusing on the most reflective target among a set of reflectors.

There are two principal limitations to the efficiency of the time reversal refocusing technique:

- The diffraction effects which act as a low pass filter on the spatial frequency spectrum of any object. The image of a point is a spot whose dimension depends on the wavelength.

- The sampling by a finite set of transducers and the limited aperture of the TRM. With such a spatial sampling by finite width elements one measures the average pressure over each element surface. These sampling effects are similar to those observed in usual focusing with an array of transducers. To get a accurate wavefront reversal, the optimum size of transducer elements is of the order of $\lambda/2$ where λ is the smallest wavelength of the pressure field. However such a fine sampling is not always necessary if the TRM is prefocused on the region of interest (cylindrical or spherical mirror).

In the first part, we review some theoretical results on time reversal and phase conjugate mirrors. We expose the theory of distortion corrections through a weakly aberrating medium (in the first Born approximation). In the second part, we describe experiments that support this theory and demonstrate the feature of our proposed time reversal method to automatically focus through an aberrating medium.

The experimental results have been obtained with a prefocused cylindrical mirror of 64 elements working at the central frequency of 3MHz. A complete parallel processing unit has been built allowing the time reversal process in real time on the 64 channels.

TRM THEORY

a. Time reversal and phase conjugation

An acoustical wave propagating in a lossless fluid medium of acoustic velocity $c(r)$, can be described by the pressure field $p(r,t)$, where r and t are the space and time coordinates, and $p(r,t)$ satisfies the wave equation :

$$\Delta p - \frac{1}{c^2(r)} \frac{\partial^2}{\partial t^2} p = 0. \tag{1}$$

If $p(r,t)$ is a solution of the equation, then $p(r,-t)$ is also a solution. In other words, the equation is invariant under time reversal. This property is due to the second order time derivative in Eq.(1). For a monochromatic field $P(r,t) = \Re e(A(r,\omega)exp(i\omega t))$ the time reversal $t \mapsto -t$ is equivalent to the phase conjugation $A(r,\omega) \mapsto A^*(r,\omega)$.

More generally a field $p(r,t)$ is caracterized by its temporal Fourier transform $A(r,\omega)$ and then $p(r,-t)$ is caracterized by $A^*(r,\omega)$.

In this description, the phase conjugation is performed in the whole space. In fact, the phase conjugation can only be physically done on a surface, with a phase conjugate mirror (PCM) or with a time reversal mirror (TRM) for broadband signals. The equivalence between time reversal and phase conjugation lead us to study theoretically the effect of the phase conjugate mirror at one given frequency. We shall see that under some restrictions, the PCM transformation is sufficient to make a phase conjugation of the whole field.

b. The phase conjugate mirror

A PCM produces a local phase conjugation followed by a reflection like that of an ordinary mirror. It is a convenient idealization that describes the effect of a true physical device, by means of which a field distribution $P(r)$ is replaced on the surface of the mirror by a new field distribution $P^*(r)$. We present some results discussed in detail in references [5][6], and adapt them to our case.

We introduce the following notations:

$$k = \frac{\omega}{c}, \; \vec{k} = (k_x, k_y, k_z), \; \vec{K} = (k_x, k_y), \; r = (x, y, z), \; \vec{X} = (x, y),$$

$P_c(r)$ is the field obtained after transformation of $P(r)$ on the PCM.

We suppose that the PCM covers the plane $\pi_m \{z = z_m\}$ and we consider waves propagating in the half space $\{z > z_m\}$. We introduce the plane wave decomposition of a wave $P(r)$, propagating towards the plane π_m. $P(r)$ can be determined from its 2-dimensional Fourier transform $A(\vec{K})$ in a plane orthogonal to the z axis. We have

$$P(\vec{K}, z) = \int_{R^2} A(\vec{K}) \exp(i(\vec{K}\vec{X} - k_z z)) d\vec{K}, \tag{2}$$

$$\text{where} \quad k_z = \begin{cases} \sqrt{k^2 - |\vec{K}|^2} & \text{if} \quad |\tilde{K}|^2 \leq k^2, \\ i\sqrt{|\vec{K}|^2 - k^2} & \text{otherwise.} \end{cases} \tag{3}$$

In this decomposition the plane wave is propagating when $|\vec{K}^2| \leq k^2$ and evanescent otherwise. To understand the effect of the PCM on the wave $P(r)$, we consider separately the cases of propagating and evanescent elementary plane wave.

Let $P(\vec{X}, z) = A \exp(i(\vec{K}\vec{X} - k_z z))$ be a plane propagating wave, with $k_z > 0$. The conjugation in the plane (π_m) leads to

$$P_c(\vec{X}, z_m) = A^* \exp(-i(\vec{K}\vec{X} - k_z z_m)), \tag{4}$$

and the reflection changes $\exp(-ik_z z)$ in $\exp(ik_z z)$. The resulting field is then

$$P_c(\vec{X}, z) = A^* \exp(-i(\vec{K}\vec{X} - k_z z)). \tag{5}$$

The PCM produces exactly the complex conjugate of $P(r)$.

Let $P(\vec{X}, z) = \exp(i\vec{K}\vec{X}) \exp(k_z z)$ be an evanescent plane wave. The conjugation in the plane π_m leads to

$$P_c(\vec{X}, z_m) = A^* \exp(-i\vec{K}\vec{X} + k_z z_m), \tag{6}$$

and the reflection changes $\exp(k_z z)$ in $\exp(-k_z z)$. The resulting field is then

$$P_c(\vec{X}, z) = A^* \exp(-i\vec{K}\vec{X} + 2k_z z_m - k_z z). \tag{7}$$

P_c is also an evanescent wave and is different from P^*.

Let $P(r)$ be the general wave given by Eq. (3). According to Eqs (5) and (7), we see now that the wave produced by the PCM can be written:

$$\begin{aligned}
P_c(\vec{X}, z) &= \int_D A^*(\vec{K}) \exp(-i(\vec{K}\vec{X} - k_z z))d\vec{K} \\
&+ \int_{R^2 - D} A^*(\vec{K}) \exp(-i(\vec{K}\vec{X} - 2k_z z_m + k_z z))d\vec{K}
\end{aligned} \tag{8}$$

$$\text{with } D = \{|\vec{K}|^2 \leq k^2\}$$

If $P(r)$ does not contain any evanescent component, we see that it is transformed by the PCM in

$$P_c = P^*. \tag{9}$$

The evanescent components are generally due to a source or scatterer smaller than the wavelength. They decrease exponentially and can be neglected at a few wavelengths from their origin.

c. Self-focusing with a PCM on a point like reflector in an homogenous medium

We consider a point like reflector located in $r_S = (x_S, y_S, z_S)$ with $z_S > z_m$ and due to an inhomogeneity of compressibility $\chi_o + \Delta\chi$ in a medium of compressibility χ_o.

The PCM focusing process can be described in three steps:
1. A first incident plane wave $P_{1i}(r)$ is transmitted by the mirror on the region of interest (transmit mode).
2. The wave $P_{1i}(r)$ is reflected by the target leading to $P_r(r)$.
The wave $P_r(r)$ is recorded in the plane of the mirror (receive mode).
3. It is then phase conjugated and reemitted by the mirror (transmit mode). The new incident wave $P_{2i}(r) = P_{rc}(r)$. In the following we show that $P_{2i}(r)$ focuses on the target.

For the incident wave $P_{1i}(r)$, the reflected wave is $P_r(r) = k^2 \dfrac{\Delta\chi}{\chi_o} P_{1i}(r_S)S(r - r_S)$

where $S(r - r_S) = \dfrac{exp(ik|r - r_S|)}{|r - r_S|}$ is the spherical wave issued from r_S.

The plane wave decomposition of $S(r) = \dfrac{exp(ik|r|)}{|r|}$ is (see [7])

$$S(r) = \int_{R^2} \frac{i}{2\pi k_z} exp(i(\vec{K}\vec{X} + k_z|z|))d\vec{K}, \tag{10}$$

where k_z is defined by Eq. (3).

$S(r)$ can be divided into 2 parts $S(r) = S^p(r) + S^e(r)$,

$$\text{where} \quad \begin{cases} S^p(r) = \displaystyle\int_D \frac{i}{2\pi k_z} exp(i(\vec{K}\vec{X} + k_z|z|))d\vec{K}, \\[2em] S^e(r) = \displaystyle\int_{R^2-D} \frac{i}{2\pi k_z} exp(i(\vec{K}\vec{X} + k_z|z|))d\vec{K}. \end{cases} \quad (11)$$

$S^p(r)$ corresponds to the propagating part and S^e to the evanescent part of S. $S^p(r)$ is different for $z > 0$ and $z < 0$ and can be written as (see Fig.1.a)

$$S^p(r) = H(z)S^p_+(r) + H(-z)S^p_-(r),$$

$$\text{where} \quad \begin{cases} H \quad \text{is the Heaviside function,} \\[1em] S^p_+(r) = \displaystyle\int_D \frac{i}{2\pi k_z} exp(i(\vec{K}\vec{X} + k_z z))d\vec{K}, \\[2em] S^p_-(r) = \displaystyle\int_D \frac{i}{2\pi k_z} exp(i(\vec{K}\vec{X} - k_z z))d\vec{K}. \end{cases} \quad (12)$$

From this decomposition we deduce the effect of the PCM on the spherical wave $S(r - r_S)$. We neglect the evanescent part of S. The wave incident on the mirror is $S^p_-(r-r_S)$. According to Eq.(9), the wave $P_{rc}(r)$ produced by the mirror is $S_c(r-r_S) = (S^p_-(r - r_S))^*$ in the whole half space $\{z > z_m\}$. (see Fig.1.b) Such a reflected field is close to the one of an ideal converging wave; however the lost of the evanescent waves smoothes the reflected wave and limits the efficiency of the refocusing process to a point spread function.

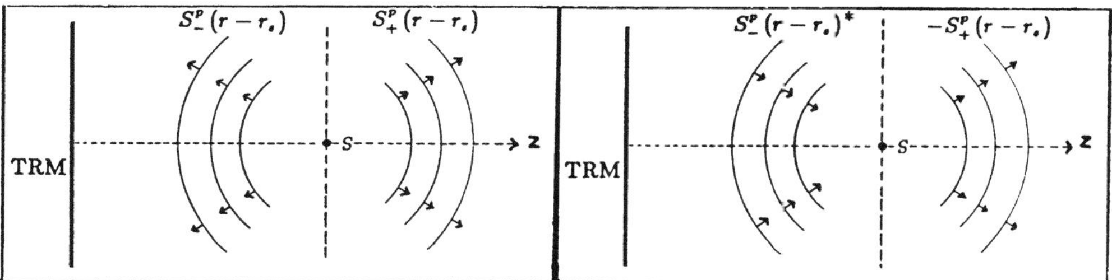

Fig. 1.a Spherical divergent wave. Fig. 1.b Wave reflected by the TRM.

d. Self-focusing on a reflector through an inhomogeneous medium with a PCM

In this paragraph we prove that the presence of a weakly aberrating medium between the reflector and the PCM does not deteriorate the quality of the focusing process under the following conditions:

- the evanescent waves are neglected,
- the medium is lossless,
- the aberrating medium is weak enough so that the first Born approximation is valid.

We suppose that S is a point like reflector located in r_S as described in the last paragraph and that between S and the PCM, the inhomogeneous medium is limited to the volume V bounded by the planes $\{z = z_1\}$ and $\{z = z_2\}$.

The inhomogeneous medium is caracterised by the compressibility $\chi(r) = \chi_o + \Delta\chi(r)$. A wave $P(r)$ propagating in V, satisfies the equation $\Delta P(r) + k^2 P(r) = s(r)P(r)$ in the volume V, where $s(r) = k^2 \dfrac{\Delta\chi(r)}{\chi_o}$ is the source density.

If $P_i(r)$ is an incident wave, the total wave $P'_i(r)$ satisfies the integral equation

$$P'_i(r) = P_i(r) - \frac{1}{4\pi} \int_V P'_i(r')s(r')\frac{exp(ik|r - r'|)}{|r - r'|} d^3 r' \tag{13}$$

In the case of the first Born approximation this equation reduces to

$$P'_i(r) = P_i(r) - \frac{1}{4\pi} \int_V P_i(r')s(r')\frac{exp(ik|r - r'|)}{|r - r'|} d^3 r' \tag{14}$$

Let us consider the PCM focusing process (fig.2).

1. A first incident plane wave $P_{1i}(r)$ is transmitted by the mirror on the region of interest (transmit mode). This wave is scattered by the aberrating medium leading to the wave $P_{1i_s}(r)$.

2. The total wave $P'_i(r) = P_{1i}(r) + P_{1i_s}(r)$ is reflected by the target leading to $P_r(r) = k^2 \dfrac{\Delta\chi}{\chi_o} P'_{1i}(r_s)S(r - r_s)$. This last wave is scattered by the aberrating medium leading to the wave $P_{r_s}(r)$. The total reflected wave $P'_r(r) = P_{r_s}(r) + P_r(r)$ is recorded in the plane of the mirror (receive mode).

3. It is then phase conjugated and reemitted (transmit mode). The new incident wave $P_{2i}(r) = P_{r_s c}(r) + P_{rc}(r)$ is scattered by the aberrating medium. Because of the first Born approximation, $P_{rc}(r)$ alone leads to the significant scattered wave $P_{rc_s}(r)$. The total wave is $P'_{2i}(r) = P_{rc}(r) + P_{r_s c}(r) + P_{rc_s}(r)$.

In the following we show that $P_{r_s c}(r) + P_{rc_s}(r) = 0$, which means that the wave scattered and then conjugated compensate the wave conjugated and then scattered.

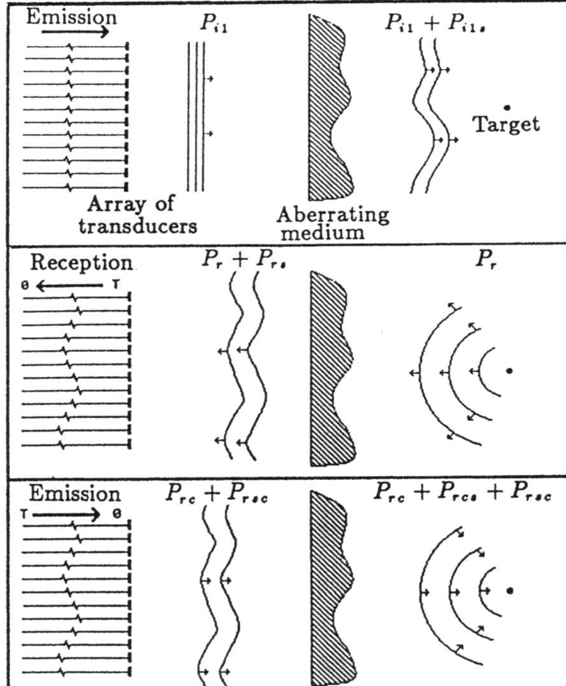

Fig. 2 Three steps of the TRM focusing process.

- Evaluation of $P_{rsc}(r)$:

With the first Born approximation $(P_{rs}(r) \ll P_r(r))$, the equation (14) leads to:

$$P_{rs}(r) = -\frac{1}{4\pi} \int_V P_r(r')s(r')\frac{exp(ik|r - r'|)}{|r - r'|}d^3r' \tag{15}$$

In front of the PCM we have $\dfrac{exp(ik|r - r'|)}{|r - r'|} = S_-^p(r - r')$ (see Eq.(12)) so that

$$P_{rs}(r) = -\frac{1}{4\pi} \int_V P_r(r')s(r')S_-^p(r - r')d^3r', \tag{16}$$

so according to the equation (9)

$$P_{rsc}(r) = -\frac{1}{4\pi} \int_V P_r^*(r')s(r')(S_-^p(r - r'))^*d^3r'. \tag{17}$$

- Evaluation of $P_{rcs}(r)$:

In the first Born approximation, the scattered wave due to $P_{rc}(r)$ is

$$P_{rcs}(r) = -\frac{1}{4\pi} \int_V P_{rc}(r')s(r')\frac{exp(ik|r - r'|)}{|r - r'|}d^3r \tag{18}$$

In this integral, the term due to $P_{rsc}(r)$ which is of the second order is neglected. According to the equation (9) $P_{rc}(r)$ is equal to $P_r^*(r)$ in the region of the scatterer. In the half space $\{z > z_1\}$, we have $\dfrac{exp(ik|r - r'|)}{|r - r'|} = S_+^p(r - r')$ (see Eq.(12)), so that

$$P_{rcs}(r) = -\frac{1}{4\pi} \int_V P_r^*(r')s(r')S_+^p(r - r')d^3r \tag{19}$$

As $(S_-^p(r-r'))^* = -S_+^p(r-r')$ (see Eq. (13)), Eqs. (17) and (19) show that $P_{rsc}(r)$ and $P_{rcs}(r)$ are opposite. The reconstructed wave is then $P_{2i}(r) = P_{r2}(r)$. The wave $P_{rc}(r)$ is exactly the wave that would be obtained without the aberrating medium (see c.). This is an outstanding property of the PCM that is confirmed by the experiments.

e. **Selective focusing by iterative mode in the case of an homogeneous medium containing several reflectors**

The time reversal process can be easily iterated. A first wave $P_{1i}(r)$ is transmit by the array of transducers. The reflected wave $P_{1r}(r)$ is recorded, time reversed and reemitted. The second incident wave $P_{2i}(r)$ generates a second reflected wave $P_{2r}(r)$ and so on. If the medium is homogeneous and contains a strong reflector among several weaker ones, the iteration produces a wave that focuses on the strong reflector.
We give a short demonstration for a monochromatic wave in the case of two reflectors (Fig.6.a). We suppose the first reflector (resp. the second) is located in r_1 (resp. r_2) and corresponds to an inhomogeneity of compressibility $\Delta\chi_1$ (resp. $\Delta\chi_2$). We neglect the multiple scattering between the two reflectors according to the first Born approximation. The first incident wave $P_{1i}(r)$ lead to the reflected wave

$$P_{1r}(r) = k^2 \frac{\Delta\chi_1}{\chi_o} P_{1i}(r_1)S(r - r_1) + k^2 \frac{\Delta\chi_2}{\chi_o} P_{1i}(r_2)S(r - r_2). \tag{20}$$

We put $a_1 = k^2 \dfrac{\Delta\chi_1}{\chi_o}$ and $a_2 = k^2 \dfrac{\Delta\chi_2}{\chi_o}$.

As shown in the paragraph c., the wave $S(r - r_1)$ is transformed by the PCM in $S_c(r - r_1) = S_-^p(r - r_1)^*$. Then the PCM transforms $P_{1r}(r)$ in $P_{2i}(r)$ with

$$P_{2i}(r) = a_1 P_{1i}(r_1) S_c(r - r_1) + a_2 P_{1i}(r_2) S_c(r - r_2). \tag{21}$$

$P_{2i}(r)$ is reflected leading to the wave

$$P_{2r}(r) = S_c(0)(a_1^2 P_{1i}(r_1) S(r - r_1) + a_2^2 P_{1i}(r_2) S(r - r_2)). \tag{22}$$

After N iterations the incident field is

$$P_{Ni}(r) = S_c(0)^{N-2}(a_1^{N-1} P_{1i}(r_1) S_c(r - r_1) + a_2^{N-1} P_{1i}(r_2) S_c(r - r_2)). \tag{23}$$

If $a_1 > a_2$, for N big enough, the second term of $P_{Ni}(r)$ is negligible, so that the wave focuses on r_1.

This property can be extended to the TRM as confirmed by the experiment. This is from our point of view the second important property of the TRM.

EXPERIMENTAL RESULTS

We have conducted different experiments in order to demonstrate the self focusing properties of TRM on different types of reflectors and through an aberrating medium. The acoustic mirror is a 40mm cylindrical array of 80mm radius of curvature made of 64 transmit and receive elements. The array pitch is 0.75mm and the central frequency of the elements is 3MHz. 64 programmable generators have been built, each generator can deliver a programmed electrical waveform, with a 175V peak to peak maximum voltage on a 50Ω transducer impedance. The sampling frequency is 25 MHz. The waveform duration can reach 4096 times 40ns. The dynamical range of each sample is 7bits. A 4K buffer memory allows to charge the time reversed waveform from A/D converters associated with each transducer element.

A first experimental verification was made to evaluate the programmable generator's efficiency in the time reversed mode. Figure 3.a shows the pulse echo signal $e(t)$ received by one transducer element from a plane steel mirror. The electrical excitation of the transducer was a numerical Dirac $\delta(t)$, $e(t)$ represents the complete acoustoelectric response in transmit-receive mode. Figure 3.b shows for the same

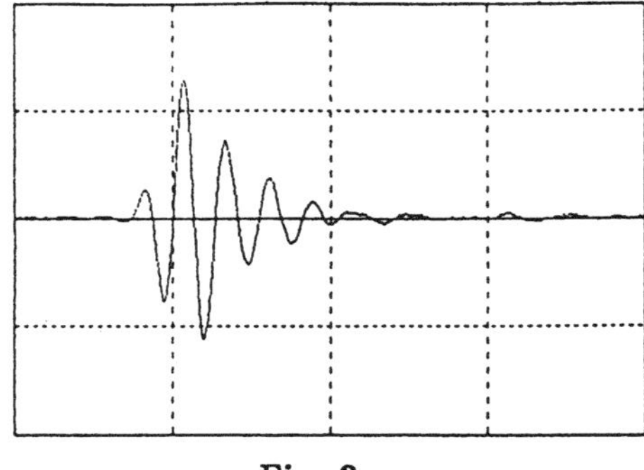

Fig. 3.a

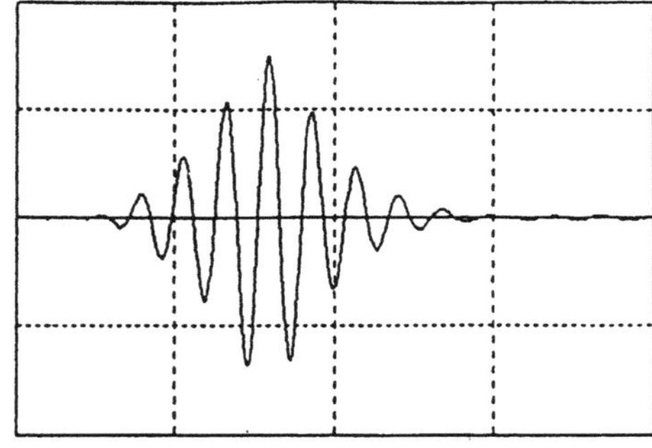

Fig. 3.b

transducer-mirror arrangement the new echo when the generator delivers as input the time reversed signal e(-t). Due to the linearity of the experiment, the output signal is expected to be proportional to e(t)*e(-t) which is a symmetrical signal. The symmetry of experimental echo confirms the efficiency of the time reversal generator.

The first experiment using the 64 elements of the TRM was done in a water tank with a Dapco hydrophone as a reflector point source. The TRM is prefocused at a depth of 80mm. The hydrophone was located at depth z=40mm from the mirror, and 10mm away from the array axis (Fig.4.a). A plane wave transmitted by the array insonifies the hydrophone. The reflected wave coming from the hydrophone target is recorded on the 64 elements, reversed in time and simultaneously reemitted. Figure 4.b shows the directivity pattern (maximum of the pressure field) of the reversed field, measured by scanning the z=40mm plane with the hydrophone. The pressure field reaches its maximum at the initial hydrophone location (source position). Note that this refocusing experiment is done far from the natural focus of the cylindrical array.

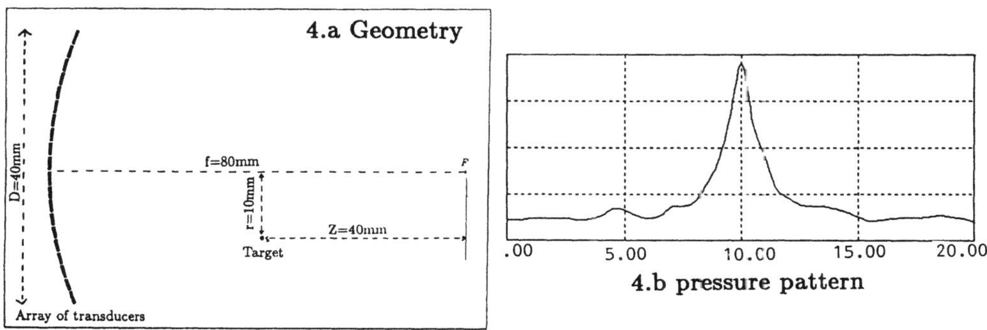

4.b pressure pattern

The second experiment was done in two steps, with a silicone aberrating layer. The layer is shaped like a prism with a weak parabolic curvature and has a variable thickness in the lateral direction with a maximum around 15mm and a uniform thickness in elevation (Fig.5.a). Silicone has a sound velocity of 1000m/s. The first step reveals the effect of the layer. Figure 5.b gives the classical focused pressure field pattern obtained in the focal plane of the cylindrical array with the silicone layer in front of the array. All the elements are excited together, the beam is distorted by the aberrator and focuses off axis.

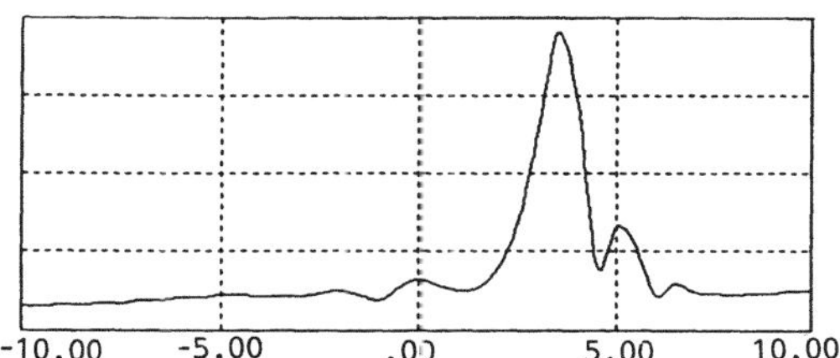

Fig. 5.b Classical focusing through the layer.

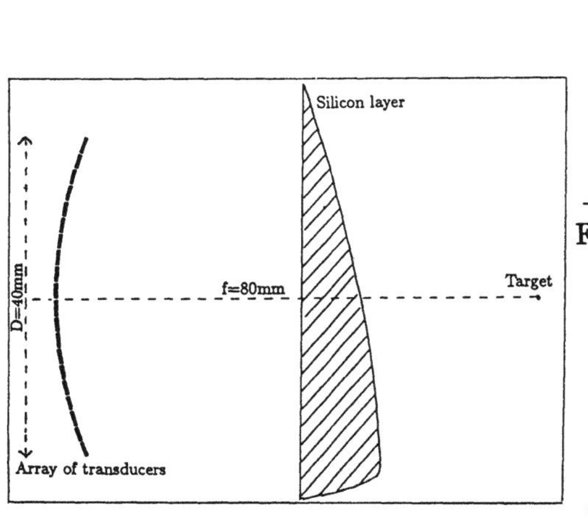

Fig. 5.a Geometry

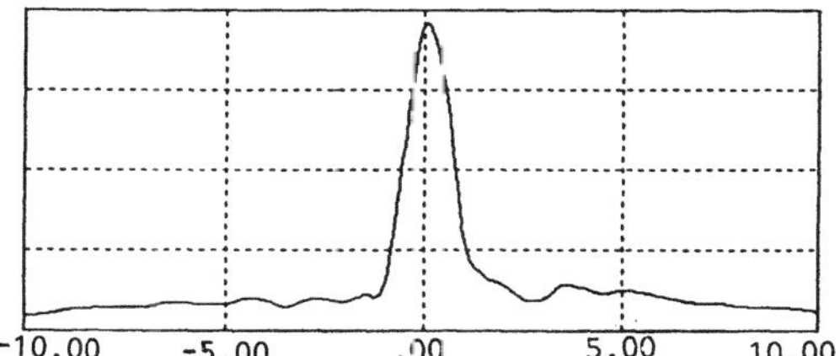

Fig. 5.c TRM focusing through the layer.

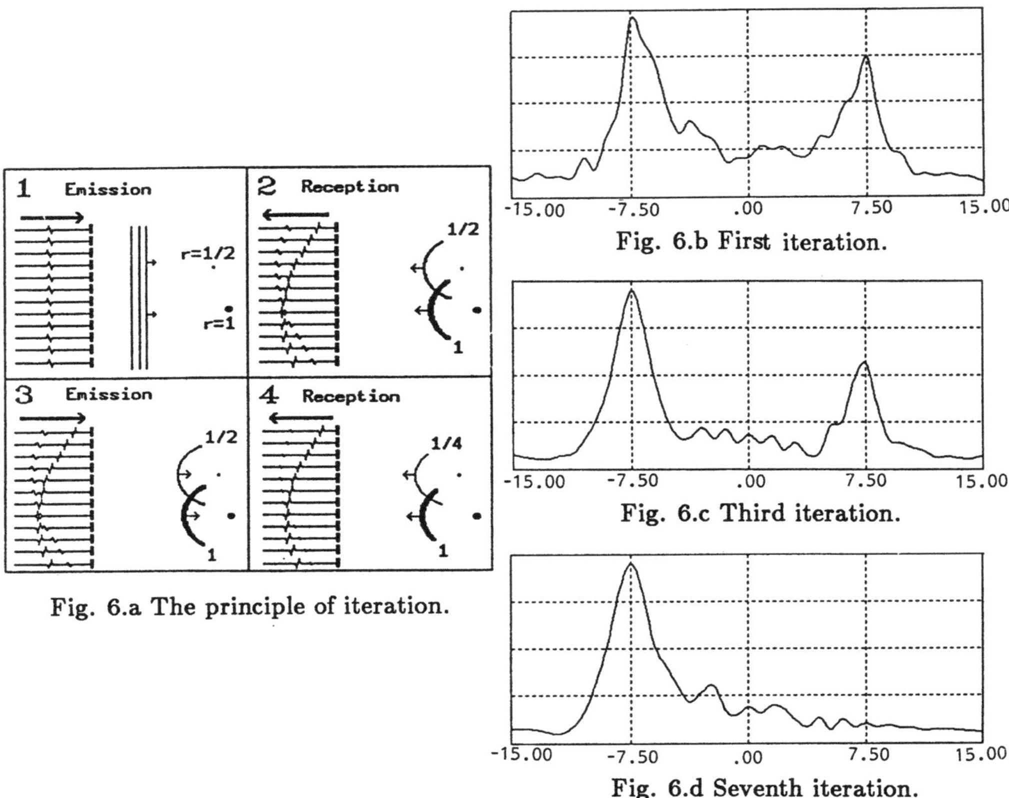

Fig. 6.b First iteration.

Fig. 6.c Third iteration.

Fig. 6.d Seventh iteration.

Fig. 6.a The principle of iteration.

The second step is illustrated by the figure 5.c. It is a time reversal experiment, where the hydrophone is located at the natural focus of the TRM. After a plane wave has been sent, the hydrophone acts as the source of a reflected wave. The reflected wave is recorded on the array, the corresponding signals are time reversed and reemitted through the silicon layer. The reversed wave is scanned by the hydrophone and the new directivity pattern is shown on the figure. We note that refocusing occurs exactly at the hydrophone initial position. This experiment supports the distortion correcting procedure.

The third experiment is done by iteration. Figure 6 shows an iterative experiment done with a target made of two different wires (Fig.6.a). After each iteration of the time reversal process, the hydrophone scans the plane of the reflectors. In the first step, a plane wave coming from the array illuminates the two reflectors. The reflected wave is stored, time reversed and reemitted. The first resulting pressure field pattern is observed on figure 6.b. We see the different reflectivity of the two targets. The process is then iterated : the new reflected wave is stored, time-reversed and reemitted, and so on... Figure 6.c shows the pressure pattern after the third iteration. We note that the field decreases at the location of the low reflector target. Figure 6.d shows the pressure pattern after the seventh iteration; the echo from the low reflector has disappeared.

CONCLUSION

To our knowledge, we have realized the first time reversal mirror working in pulse echo mode. We demonstrate two remarquable properties of the time reversal process:
-the ability to focus through an inhomogeneous medium on a reflective target whose position is unknown,
-the ability to iterate in order to make a selective focusing in the case of several targets.

REFERENCES

[1] O'Donnel and Flax. Phase aberration measurements in medical ultrasound human studies. Ultrason.Imag. 10 pp1-11 (1988).

[2] L.Nock, G.E.Trahey and S.W.Smith. Phase aberration correction in medical ultrasound using speckle brightness as a quality factor. JASA 85(5) p1819 (1989).

[3] D.M.Pepper. Non linear optical phase conjugation. Laser Handbook Vol 4, (North-Holland Physics Publishing, Amsterdam)

[4] F.V.Bunkin,Yu.a.Kravtsov and G.A.Lyakhov. Acoustic analogues of nonlinear-optics phenomena. Sov.Phys.Ups. 29(7) p607 (1986).

[5] G.S.Agarwal, A.T.Friberg and E.Wolf. Elimination of distortions by phase conjugation without losses or gains. Opt.Com. 43(6) p446. (1981)

[6] M.Nieto-Vesperinas and E.Wolf. Phase conjugation and symmetries with wave fields in free space containing evanescent components. JOSA 2(9) p1429 (1985)

[7] Bateman. Tables of integral transforms (Mc Graw-Hill 1954) Volume 2.

ACOUSTIC PHASE CONJUGATION USING NONLINEAR ELECTROACOUSTIC
INTERACTION AND ITS APPLICATION TO SCANNING ACOUSTIC IMAGING
SYSTEMS

M. Ohno

Olympus Optical Co., Ltd.
2-3 Kuboyama-cho, Hachioji, Tokyo 192, Japan

INTRODUCTION

Scanning acoustic imaging systems such as acoustic microscopes and acoustic flaw detectors are now widely used in industrial and medical researches. The advantage of acoustic imaging
is its ability of visualizing the inside of optically opaque
materials. These systems, however, are still accompanied with a
common drawback. That is, when we observe a sample which has a
non-flat surface, the surface image is always superposed on the
internal image. This is caused by the refraction of the acoustic
beam at the surface of the sample. It induces the deflection of
the beam and/or the distortion of the wavefront, and then the
shape of the wavefront of the detected beam becomes different
from the shape of the transducer. This fact restricts the reliable usage of acoustic imaging to flat-topped samples and the
removal of the surface roughness effect has been a target of
study in acoustic imaging [1].

In this paper, the author proposes a method to visualize
the inside of the sample without the influence of the surface
roughness. For this purpose, acoustic phase conjugation is used.

ACOUSTIC PHASE CONJUGATION

A phase conjugate wave [2,3] is defined as a wave whose
spatial amplitude function is complex conjugate of that of the
incident wave.

$$u_c(\mathbf{r},t) = \text{Re}\{U_i^*(\mathbf{r})\exp(i\omega t)\} \qquad (1)$$

Taking the complex conjugate of the spatial part of the function
is equivalent to changing the sign of time.

$$\text{Re}\{U_i^*(\mathbf{r})\exp(i\omega t)\} = \text{Re}\{U_i(\mathbf{r})\exp(-i\omega t)\} \qquad (2)$$

Therefore, the phase conjugate wave is a wave which has the same
shape of wavefront but propagates in the opposite direction to
the incident wave.

Several methods of generating acoustic phase conjugate waves
have been proposed by now [4-8]. In this study, a method using
nonlinear electroacoustic interaction (NEAI) in LiNbO$_3$ [9-13] is
adopted. This is a parametric three wave mixing process in a
material which has higher order piezoelectric and/or dielectric
tensors [14]. Suppose that an acoustic field

$$u_i(\mathbf{r},t) = (1/2)U_i(\mathbf{r})\exp\{i(\mathbf{k}_i\cdot\mathbf{r} - \omega t)\} + c.c. \tag{3}$$

and an electric field

$$e_p(\mathbf{r},t) = (1/2)E_p\exp\{i(\mathbf{k}_p\cdot\mathbf{r} - 2\omega t)\} + c.c. \tag{4}$$

are applied on the nonlinear material simultaneously. The parametric interaction between these two waves produces an acoustic
field

$$u_c(\mathbf{r},t) \propto (1/4)U_i^*(\mathbf{r})E_p$$
$$\times \exp[i\{(\mathbf{k}_p-\mathbf{k}_i)\cdot\mathbf{r} - (2\omega-\omega)t\}] + c.c. \tag{5}$$

Because of the very large difference between the sound velocity
and the light velocity, $\mathbf{k}_p$ can be neglected compared to $\mathbf{k}_i$. Then
we obtain

$$u_c(\mathbf{r},t) \propto (1/4)U_i^*(\mathbf{r})E_p\exp\{i(-\mathbf{k}_i\cdot\mathbf{r} - \omega t)\} + c.c \tag{6}$$

This implies that $u_c(\mathbf{r},t)$ is the phase conjugate wave of the inci
dent field $u_i(\mathbf{r},t)$.

Using the wavefront-reversing (or time-reversing) property,
the phase conjugate wave can be used to compensate the wavefront
distortion. When a wave suffers wavefront distortion after passing through an inhomogeneous medium, if the wave is returned by
a phase conjugator and propagates through the same medium again,
the phase shift is undone and the wavefront distortion is removed [15]. This process is the basis of the imaging method proposed in this paper.

IMAGING SYSTEM USING PHASE CONJUGATION

Figure 1 shows the schematic diagram of the scanning
imaging system using acoustic phase conjugation. The oscillator
1 generates a rf signal of angular frequency ω and this is
applied to the transducer as a tone-burst wave. The acoustic
wave radiated by the transducer propagates through the sample and
enters into the LiNbO$_3$ crystal. The pump electric field of
angular frequency 2ω from the oscillator 2 is applied on the
electrodes while the acoustic wave propagates in LiNbO$_3$. At this
time the phase conjugate wave of the incident wave is generated
via NEAI. According to its wavefront-reversing property, the
phase conjugate wave propagates back on the completely same path
as the incident wave has traced. When it arrives at the transducer, the wavefront distortion is cancelled and the shape of
the wavefront becomes the same as the shape of the transducer.
Therefore, the output signal of the transducer has no influence
of the wavefront distortion, which is brought about by the surface roughness of the sample. However, the attenuation of the

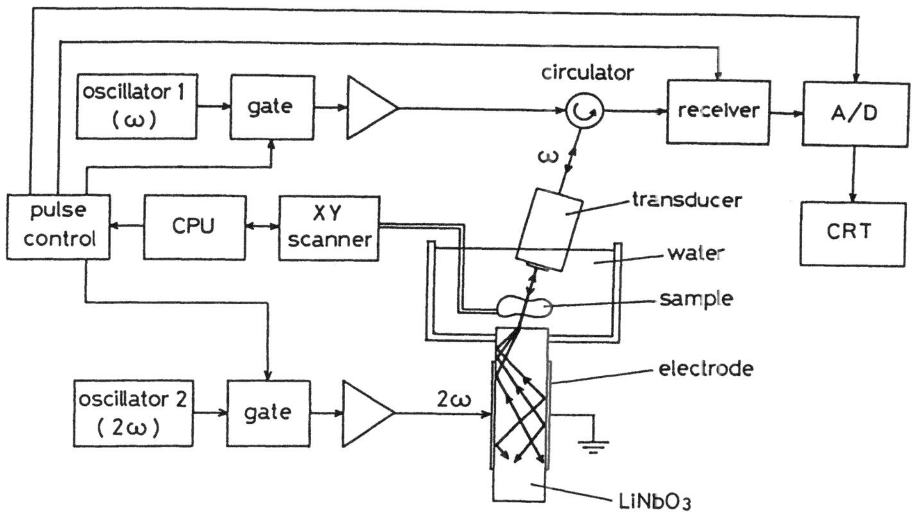

Fig.1   Scanning imaging system using acoustic phase
conjugation

acoustic wave cannot be recovered in any way. Therefore, if the
sample contains such sound attenuators as sound interceptors,
absorbers, or hard scatterers, the output signal will be modulated. By detecting the phase conjugate wave, we can get the
information about the integral of the sound attenuation over the
propagating path without the influence of the surface roughness
of the sample. The two-dimensionally scanned image will show the
attenuation property of the sample and the surface image will
not be superposed.

EXPERIMENTAL RESULTS

Using the system described above, several images for model
samples were obtained. Experimental conditions were as follows.
The size of the $LiNbO_3$ was 50(X)x10(Y)x10(Z) mm, acoustic fre-
quency=50 MHz, incident pulse duration=10 μs, pump electric fre-
quency=100 MHz, pump electric amplitude=1.9×10^4 Vm^{-1} (Y-direction), interaction length=34 mm, transducer diameter=3.0 mm, its
curvature radius= ∞ (non-converging type), and one-way path
length of the acoustic beam in water was 15 mm. A 30 μm-thick
SiO_2 layer was deposited on the entrance face of $LiNbO_3$ crystal
in order to reduce the surface reflection. The acoustic wave was
incident on $LiNbO_3$ at an angle of $15°$ to the X-axis in the XZ-
plane of the crystal. One reason of this oblique incidence was
to avoid the specularly reflected waves from the top surface of
the crystal, and the other reason was to get a large efficiency
of NEAI.
Three kinds of samples were prepared. Figures 2(a)-4(a)
show their profiles. The material of the samples was agar made
from NaCl solution. Its sound velocity was 1750 ms^{-1}. The sample

NO.1 has a two-dimensional periodic undulation pattern on its top
surface. The sample NO.2 also has the undulation, and contains a
sound interceptor made of vinyl. The sample NO.3 has a random
undulation on the top surface and contains an interceptor. For
these samples, conventional images and phase conjugate images
were obtained and compared. In taking the conventional images,
samples were put on a slide glass plate and the specularly reflected waves from the plate were detected. The same frequency
and transducer were used in both imagings.

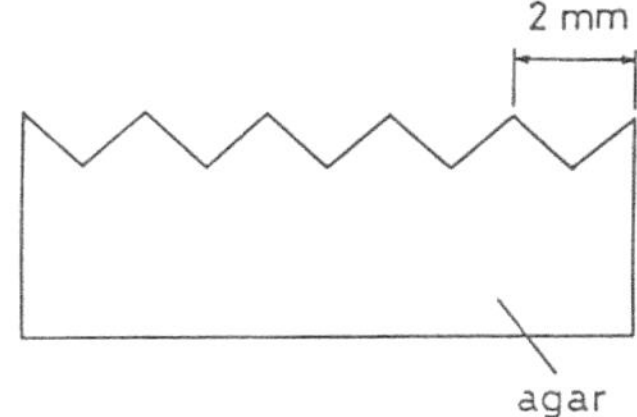

(a) profile of the sample

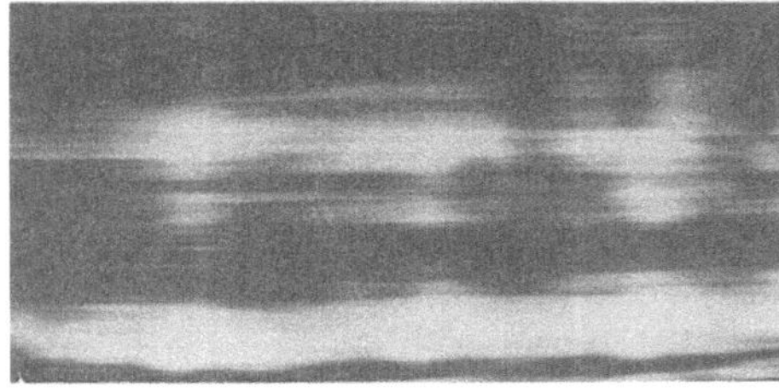

(b) conventional image   (c) phase conjugate image

Fig.2   Experimental results for sample NO.1

    Experimental results are shown in Figs.2(b),(c)-Figs.4(b),
(c). Figures 2(b) and (c) are the images of the sample NO.1. In
the conventional image (b), a mesh-like pattern is seen. This is
the influence of the surface undulation. On the other hand, the
phase conjugate image (c) shows almost no contrast. This fact
implies that the effect of the surface roughness was removed.
Figures 3(b) and (c) show the result for the sample NO.2. In
the conventional image, the mesh-like pattern is the surface undulation and the dark area in the center is the sound interceptor. They are mixed up and cannot be distinguished. However, in
the phase conjugate image, the surface roughness pattern was removed and only the sound interceptor is clearly observed. This
effect was also confirmed for the random undulation of the sample NO.3 (Figs.4(b) and (c)). The conventional image offers almost no internal information, while the phase conjugate image
reveals the existence of the sound interceptor.

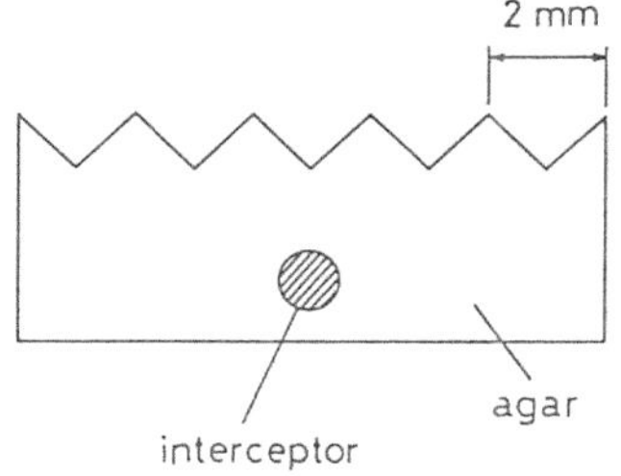

(a) profile of the sample

1mm

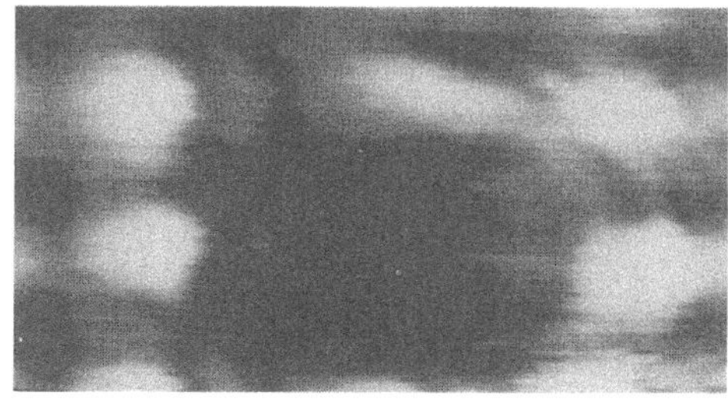

(b) conventional image

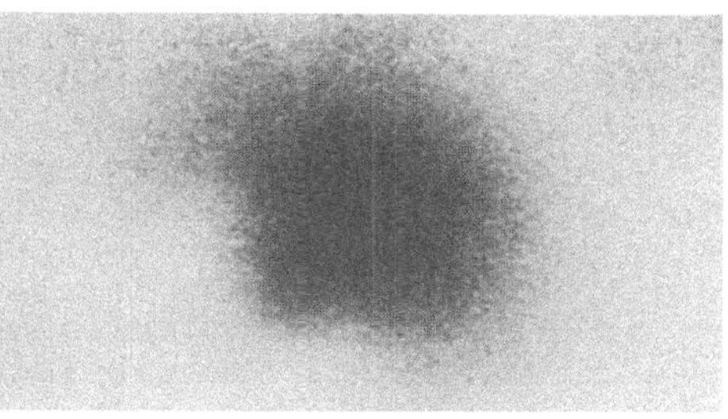

(c) phase conjugate image

Fig.3 Experimental results for sample NO.2

(a) profile of the sample

1mm

(b) conventional image

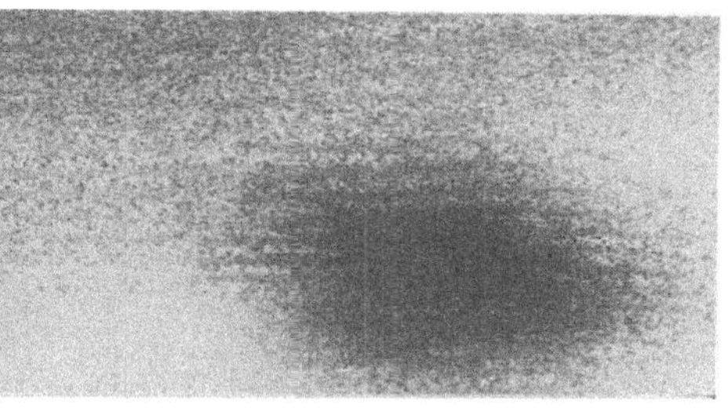

(c) phase conjugate image

Fig.4 Experimental results for sample NO.3

DISCUSSION

There are several other methods of acoustic phase conjugation than NEAI. Most of them are the holographic processes in
water, at water surface, or particle-suspended water. There are
two reasons why NEAI in $LiNbO_3$ was adopted in this scanning
imaging system. One is its operating frequency. The holographic
interactions in refs.[4-7] occur in water and thereby the
interaction length cannot be long enough for the frequency
range over 10 MHz because of the absorption. NEAI is free from
this restriction because it occurs in a solid crystal. The other
reason is the short response time of NEAI. In NEAI, the local
response is determined by the arising time of the nonlinear
polarization, and the total response is determined by the incident pulse duration. Therefore, the response time of the phase
conjugator in this system was of the order of μs, which is much
faster than that of holographic processes.
The wavefront-reversing property of the phase conjugator is
affected by the incident angle dependence of the phase conjugate
reflectivity. It must be independent of the incident angle in
order to reconstruct the ideal wavefront. However, in our system, it depends on the incident angle because of the incident
angle dependence of the transmissivity between water and $SiO_2/$
$LiNbO_3$ and the anisotropy of the nonlinear interaction coefficients. This drawback would be reduced by optimizing the cut angle of the crystal.
This system, at the present stage, can be used to observe
biological samples. The absorption property of roughly-shaped
samples would be investigated. In order to observe solid samples, considering that this system is a transmission type, it is
necessary to improve the S/N ratio of the system. The phase
conjugate reflectivity in this system was of the order of 10^{-3}
in amplitude. This can be made larger by magnifying the pump
electric field and/or using a higher frequency.

CONCLUSION

Acoustic phase conjugation was realized using nonlinear
electroacoustic interaction in $LiNbO_3$ and a scanning imaging
system using phase conjugate waves was constructed. This system
has visualized the inside of the samples without the influence
of the surface roughness.

ACKNOWLEDGEMENT

The author gratefully acknowledges the technical help of
K. Ishizaki and S. Iura in processing the $LiNbO_3$ crystal.

REFERENCES

[1] P. A. Reinholdtsen and B. T. Khuri-Yakub, Removing the
     effects of surface roughness from low-frequency acoustic
     images, in: "Review of Progress in Quantitative NDE",
     Plenum, New York (1988).
[2] R. A. Fisher, "Optical Phase Conjugation", Academic, New
     York (1983).

[3]   B. Ya. Zel'dovich, N. F. Pilipetsky, and V. V. Shkunov,
      "Principles of Phase Conjugation", Springer, Berlin (1985).
[4]   N. P. Andreeva, F. V. Bunkin, D, V, Vlasov, and K. Karshiev,
      Experimental observation of acoustic phase conjugation at a
      liquid surface, _Sov. Tech. Phys. Lett._, 8, 45 (1982).
[5]   E. A. Zabolotskaya, Phase conjugation of sound beams in
      connection with four-phonon interaction in a liquid
      containing gas bubbles, _Sov. Phys. Acoust._, 30, 462 (1984).
[6]   F. V. Bunkin, D. V. Vlasov, E. A. Zabolotskaya, and Yu. A.
      Kravtsov, Phase conjugation of sound beams in four-phonon
      interaction with temperature waves, _Sov. Phys. Acoust._, 28,
      440 (1982).
[7]   T. Sato, H Kataoka, T. Nakayama, and Y. Yamakoshi, Ultra-
      sonic phase conjugation using micro particle suspended cell
      and its application, _in_: "Acoustical Imaging vol. 17",
      Plenum, New York (1989), p.361.
[8]   N. S. Shiren, R. L. Melcher, and T. G. Kazyaka, Multiple=
      quantum phase conjugation in microwave acoustics, _IEEE J._
      Quantum Electron., QE-22, 1457 (1986).
[9] V. I. Reshetzky, Phase conjugate reflection and amplifica-
      tion of a bulk acoustic wave in piezoelectric crystals,
      J. Phys. C, 17, 5887 (1984).
[10] L. O. Svaasand, Interaction between elastic surface waves
      in piezoelectric materials, _Appl. Phys. Lett._, 15, 300
      (1969).
[11] R. B. Thompson and C. F. Quate, Nonlinear interaction of
      microwave electric fields and sound in $LiNbO_3$, _J. Appl._
      Phys., 42, 907 (1971).
[12] M. Ohno, Generation of acoustic phase conjugate waves using
      nonlinear electroacoustic interaction in $LiNbO_3$, _Appl._
      Phys. Lett., 54, 1979 (1989).
[13] A. P. Brysev, F. V. Bunkin, D. V. Vlasov, and Yu. E.
      Kazarov, _Pis'ma v Zhurnal Tekhnicheskoj Fiziki_, 8, 546
      (1982) (Russian).
[14] D. F. Nelson, Three-field electroacoustic parametric
      interaction in piezoelectric crystals, _J. Acoust. Soc. Am._,
      64, 891 (1978).
[15] M. Ohno, Wave front reversal in acoustic phase conjugation
      by nonlinear electroacoustic interaction in $LiNbO_3$, _Appl._
      Phys. Lett., 55, 832 (1989).

AN ECHOLOCATION AND IMAGING USING TRANSDUCERS OF DIRECTIONALLY DISTINGUISHABLE IMPULSE RESPONSE

Yasutaka Tamura and Takao Akatsuka

Department of Information Engineering, Yamagata University
Yonezawa 992, Japan

ABSTRACT

A new echolocation and imaging system is described which requires no mechanical nor electrical scan and uses small numbers of signal channels. This system acquires data for an image-forming with only a single transmitting and receiving process, and the image-forming can be performed entirely digitally. These properties are made possible by a new acoustical sensor with an impulse response from which we can distinguish the direction of a received or transmitted wavefront. We call the sensor an "Encoding Aperture"(EA). An implementation of the sensor and beam-forming characteristics are introduced. Finally, we demonstrate a point spread function of a 3-D imaging system using only two "Encoding Aperture Receivers" (EARs).

INTRODUCTION

Controlling and observation of a wavefront is needed for a beam-forming or imaging process using a coherent wave. Since a wavefront intrinsically has a parallel feature, we must handle a spatial and temporal signal. We therefore used an array of transducers and electrical or numerical scanning techniques when a high speed data acquisition was needed. However, it complicates the system extraordinarily.

Several techniques have been proposed which scan the beam electrically with a single signal channel. A frequency-sweep method for a RADAR has been proposed [1~3], and similar techniques for acoustical imaging systems have been reported [4]. An alternative is to use an spatial modulator for a wavefront. Sato has proposed a holographic system which uses a mechanically rotating phase mask [5], Hidaka has used a dispersion prism and frequency-sweep method [6], Kino and Shaw have proposed to provide a continuous phase shifted wavefront by the use of acoustic surface waves techniques [7,8]. It should be noticed that a wavefront is modulated on an aperture and combined to a single signal channel in these techniques. In the frequency-sweep approach, a certain frequency is related to a direction ([1 ~4, 7, 8]) or a point on an aperture ([6]). In Sato's approach, the wavefront on an aperture is encoded with a rotating phase mask, and reconstructed numerically.

However, the essence of the imaging with a coherent wave is to make wavefronts corresponding to distinct positions of targets distinguishable. Based on this idea, the authors have proposed a new sensor configuration for a coherent wave [9]. A single signal channel is connected to a sensor whose sensitivity is modulated on an aperture. If the spatial modulating function is so designed to have an impulse response from which we can distinguish the direction and delay of propagated wavefront, we can reconstruct an image using the detected waveforms. Such an aperture encodes a transmitted or received wavefront in order to distinguish the propagation delays and directions. We therefore call the aperture of the sensor an "Encoding Aperture". Tamura and Koyama have implemented an EA using a piezoelectric film with an electrode whose pattern is designed according to a spatial modulating function [10~12].

In this paper, we present the concept of the EA, and propose a new implementation with an array of wide-band transducers.

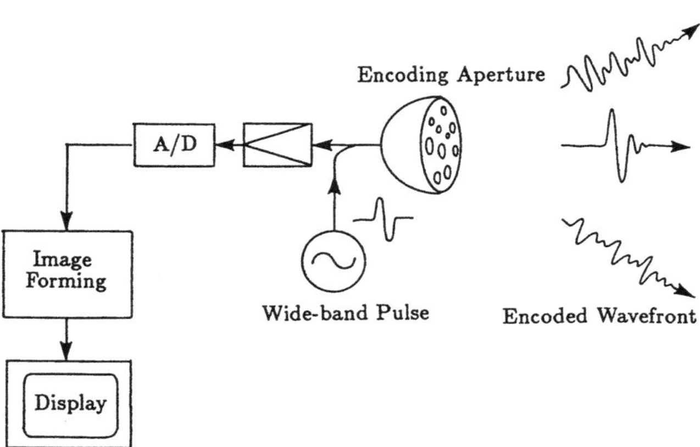

Fig.1. Conceptual figure of the encoding aperture.

PRINCIPLE

Encoding Aperture

Fig.1 shows the concept of the EA. Certain characteristics of the sensitivity is modulated spatially on an aperture of a transducer. For the characteristics, amplitude, phase, frequency response, or delay may be used. Due to the spatial modulation, the impulse response of this transducer changes with a propagation direction.

We can reconstruct an image when the waveforms of the impulse responses satisfy the following conditions.

i) They have non-zero energy at any direction within a observed region. Ideally, they have a constant value.

ii) They can be distinguished with an appropriate linear operation when their propagation directions are different. Ideally, they are orthogonalized, and can be distinguished with correlations.

When we drive the transducer with a wide-band signal, a wavefront of non-directional energy propagation is generated. The acoustical waveforms observed at different positions are different and mutually distinguishable. The same situation stands when the sensor is used as a receiver ; wide band wavefronts from different directions generate electrical waveforms from which we can distinguish the directions.

In short, the aperture which satisfies the condition encodes wavefront to a position distinguishable form. When we use this aperture for an echographic imaging system, we can acquire the information of targets in a whole observed region with a single transmitting and receiving process. Image formation is carried out by a decoding process.

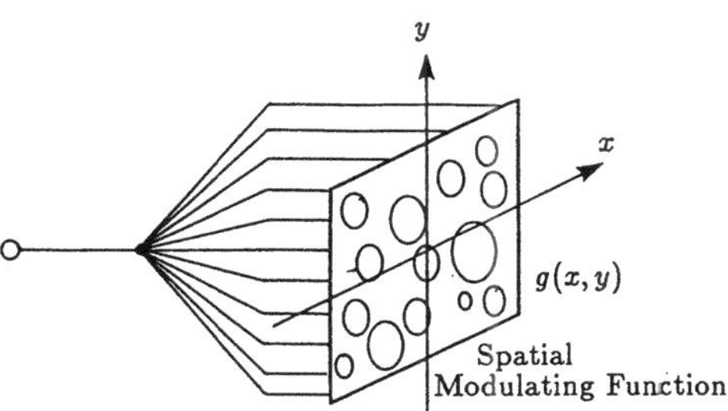

Fig. 2. Phase and amplitude modulation.

Spatial modulating function

Characteristics which are easy to implement are amplitude and phase. Mathematically, phase and amplitude modulation is represented as the wave field function is multiplied with a function $g(x, y)$ as in Fig.2.

The Fourier component $H(\theta, \phi; f)$ of the impulse response waveform of the sensor for the direction angle (θ, ϕ) and frequency f is related to the spatial Fourier component

$$G(u, v) = \int_{-\infty}^{\infty} g(x, y) e^{-j2\pi(ux + vy)} dx dy$$

(1)

of $g(x, y)$ as

$$H(\theta, \phi; t) \propto G(\frac{C \sin\theta\cos\phi}{f}, \frac{C \sin\theta\sin\phi}{f})$$

(2)

where, C is the velocity of sound. Note that the frequency response changes with the angle. From this property, it can easily be deduced that the spatial modulating function $g(x, y)$ has the following characteristics [9] :

i) $g(x, y)$ is a wide-band function.

ii) $g(x, y)$ is asymmetric.

iii) The phase of $G(u, v)$ rapidly changes with spatial frequency (u, v) .

Implementation using an array

We have proposed the use of a binary phase function which is generated by taking the polarity of a FM chirp signal [9] or random phase signal [10]. These modulating functions can be implemented as a piezo-electrical film and patterned electrode [10 ~ 12].

An alternative method to realize a wide band spatial function is to use a spatial train of impulses. This can be implemented using an array of small transducers. We adopted an array on an inclined plane [13] . The constant energy propagation and appropriate duration time of impulse response waveforms were confirmed by doing this. Fig.3 shows the geometry of the proposed array.

N wide-band and non-directional transducers are arranged on a circumference of an ellipse or a circle. The polarity of the transducers are coded with a binary maximal length sequence (M-sequence), and the transducers are connected into a single electrical signal channel. Finally, the normal axis of the array is inclined to the direction (θ_0, ϕ_0).

The proposed EA can be used as both a receiver and a transmitter, but we will consider the case of a receiver. The transducers of the array are used as receivers, and we assume that a transmitter is put at the center of the array. The transmitter emits a wide-band waveform denoted with $u(t)$, and the output of the array $r(t)$ is recorded. To simplify the discussion, we assume that the maximum difference of the arrival times of a wavefront at each of the transducers is larger than the reciprocal of the frequency band-width of the transmitted pulse. Thus, the impulse response of the EA becomes a train of pulses. The range of a point target is encoded into the delay of the train, and the direction is encoded into the intervals of each pulses.

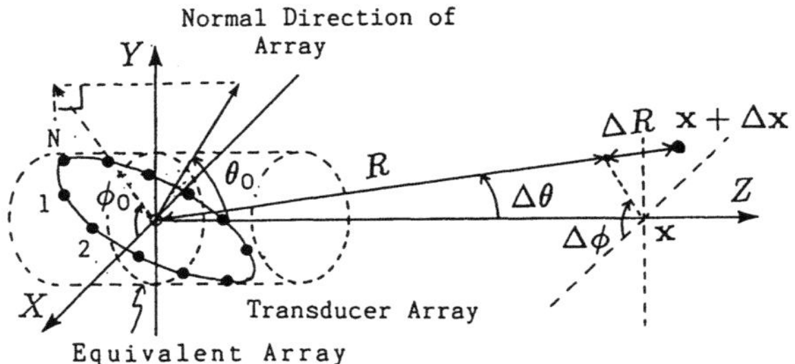

Fig. 3. Geometry of the array operates as an EA.

Image formation

The received waveform is composed of many echoes corresponding to targets in various ranges and directions. However, we can form an image with a correlation process, if the waveforms corresponding to different positions are uncorrelated.

Let $h(\mathbf{x}; t)$ is an echo waveform caused by a point target at position $\mathbf{x}$. Then, we get an image function $\hat{s}(\mathbf{x})$ by

$$\hat{s}(\mathbf{x}) = \int r(t) h(\mathbf{x}; t)^* dt \ .$$

(3)

For the $\mathbf{x}$ in the paraxial region of the array, the reference waveform $h(\mathbf{x}; t)$ is approximated as

$$h(\mathbf{x}; t) = \sum_{i=1}^{N} m_i \cdot u\left(t - \frac{|\mathbf{x} - \mathbf{x}_i| + |\mathbf{x}|}{C}\right) \ ,$$

(4)

where, $m_1 \sim m_N$ denote the M-sequence, $\mathbf{x}_i = (x_i, y_i, z_i)$ is a position of the i-th receiver.

The operator (3) is implemented as filters when the position $\mathbf{x}$ is in the paraxial region. So, we get the values of the image function on a certain direction using a corresponding matched filter [14].

ANALYSIS OF THE IMAGING SYSTEM

In this section we analyze the spatial characteristics of the imaging system by introducing the point spread function (P.S.F.) of the system.

The point spread function $g(\Delta \mathbf{x})$ is calculated with a correlation between two echoes corresponding to two distinct point targets at $\mathbf{x}$ and $\mathbf{x} + \Delta \mathbf{x}$ as

$$g(\Delta \mathbf{x}) = \int h(\mathbf{x};t)h(\mathbf{x} + \Delta \mathbf{x};t)^* dt \quad .$$

(5)

Now, we assume that $\mathbf{x}$ is on the z-axis, both $\mathbf{x}$ and $\mathbf{x} + \Delta \mathbf{x}$ are in the paraxial region of the array, and $|\Delta \mathbf{x}|$ is sufficiently small copaired with the range R. Then, the point spread function is approximated as

$$g(\Delta R, \Delta \theta, \Delta \phi) = \sum_{i=1}^{N} \varphi \left(\frac{2\Delta R - (x_i cos\Delta\phi + y_i sin\Delta\phi)\Delta\theta}{C} \right)$$
$$= \sum_{i=1}^{N} \varphi(\delta \tau_i) \quad ,$$

(6)

where, $\Delta R, \Delta \theta, \Delta \phi$ are the range and azimuthal differences between $\mathbf{x}$ and $\mathbf{x} + \Delta \mathbf{x}$,

$$\varphi(\tau) = \int_{-\infty}^{\infty} u(t)u(t - \tau)^* dt$$

(7)

is an autocorrelation function of the transmitted waveform $u(t)$, and

$$\delta \tau_i = \frac{2\Delta R - (x_i cos\Delta\phi + y_i sin\Delta\phi)\Delta\theta}{C}$$

(8)

represents a delay time between pulses caused by the two targets and received with the i-th receiver. Note that (6) equals to the P.S.F. of the 2-D array on the $z = 0$ plane which has receivers at $(x_1, y_1, 0) \sim (x_N, y_N, 0)$.

Fig. 4. shows the schematic of waveforms detected by the EA. In the figure, the transmitted pulse $u(t)$ is assumed as a short sinusoidal wave. Since we have assumed that the intervals of the arrival times of a wavefront from a point target are larger than the equivalent time-width of the transmitted pulse, we can divide the received echo into pulses caused by each receivers . This figure also shows waveforms detected by the equivalent 2-D array. In the case of the equivalent array, the P.S.F. is calculated by multiplication, integration and summation of received parallel signals. In the case of the EA, two serial signals are multiplied and integrated. Because the delay times corresponding to the i-th receiver have a same value $\delta \tau_i$, the P.S.F. of the two systems are same.

We should also note that the (6) is just an approximation for the neighborhood of the main peak of the P.S.F. at $\Delta \mathbf{x} = 0$. For large $|\Delta \mathbf{x}|$, undesired peaks may emerge. Hence, we

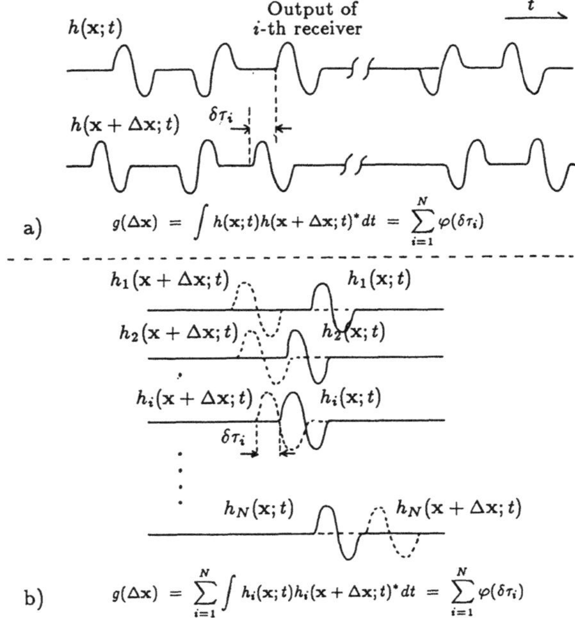

a)

$$g(\Delta x) \;=\; \int h(x;t) h(x + \Delta x;t)^* dt \;=\; \sum_{i=1}^{N} \varphi(\delta\tau_i)$$

b)

$$g(\Delta x) \;=\; \sum_{i=1}^{N} \int h_i(x;t) h_i(x + \Delta x;t)^* dt \;=\; \sum_{i=1}^{N} \varphi(\delta\tau_i)$$

Fig. 4. A schematic of the output waveforms of an EA and an equivalent array.
a) : EA,   b) : Equivalent array

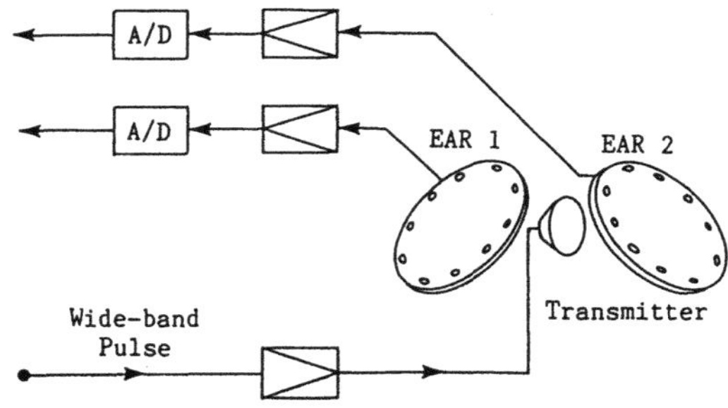

Fig. 5. Binaural sonar system using a pair of EARs.

should weight the output of the each element of the array with an appropriate code sequence such as M-sequence in order to reduce the artifacts.

As a consequence of the above discussion, we conclude that :

i) An array which has an extent in depth can operate as an EA, when the intervals of echo arrival times are larger than the equivalent time width of the transmitted pulses.

ii) The spatial resolution of the array is equal to that of the equivalent array which is the projection of the real array onto the plane normal to the direction of observation.

iii) The sensitivity of the elements of the array should be weighted with an appropriate sequence to reduce the artifacts.

The proposed EA is an array with a single signal channel which is approximately equivalent to a 2-D array with multiple signal channels. Hence, we can get an imaging system which is very easy to implement using the proposed EA.

SIMULATION

To demonstrate the ability of the proposed EA, we present the results of a computer simulation. The simulation assumes a 3-D imaging system with only two receivers ("Binaural Three-dimensional SONAR" : BaTS [13]). This system uses a pair of EARs, and a single wide band transmitter. In the simulation, we assume two circular arrays arranged symmetrically as shown in Fig.5.

The diameter of the array is 64λ (λ is a wavelength of the sound at the center frequency of the pulse). The arrays are inclined as $\theta_0 = \phi_0 = \pi/4$ rad. The distance between the centers of the arrays is $64\ \lambda$, and the numbers of the elements $N = 15$.

Fig.6 shows the distributions of the 3-D P.S.F. on three orthogonal cross sections. The P.S.F. has a main peak which changes sharply in the three cross-sectional planes. However some artifacts can be observed.

We are going to construct a trial system which operates in air. Fig. 7 shows the view of the binaural receiver of the system. Electret condenser microphones of 6 mm diameter are used as transducers. The array is designed for 50 kHz operation.

DISCUSSION

In the previous sections, we have pointed out that a coded 3-D array can operate as an EA. As an implementation, we present a system with a circular array and a M-sequence. However, another configuration of the array is possible. An asymmetric and 3-D structures, such as a spiral arrangement may operate as an EA. The code for the weighting of the transducers must also be reconsidered.

The proposed approach can be used for a 2-D imaging system. In this case, an inclined linear array which has coded transducers is one possible choice for the arrangement.

Artifact emerge in an image is the main problem of this system. The artifact is a result of the nonorthogonal correlations of encoded waveforms of the EA . One approach to reduce the artifact is to construct an imaging operator which uncorrelates the waveforms. The authors have reported this approach in [14]. We must, however, extend the result for a 3-D imaging system. The array configuration, code sequence, and operator must be designed with consideration of the criterion of the optimality of the whole system.

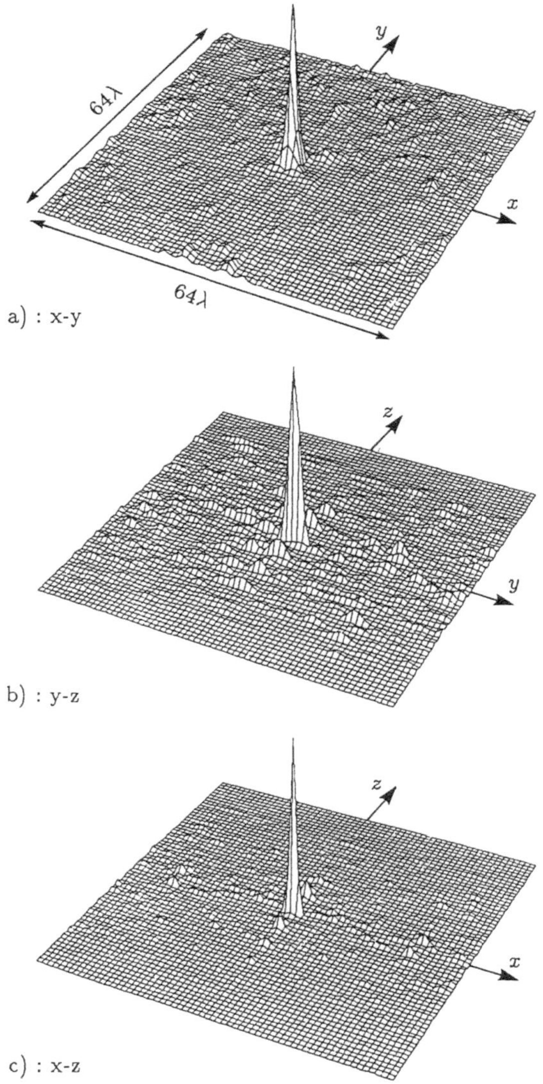

Fig. 6. Point spread function (P.S.F.) of the simulated system (Distance = 100λ)
a) : P.S.F. in a focal plane,   b),c) : P.S.F. on a cross-sectional planes.

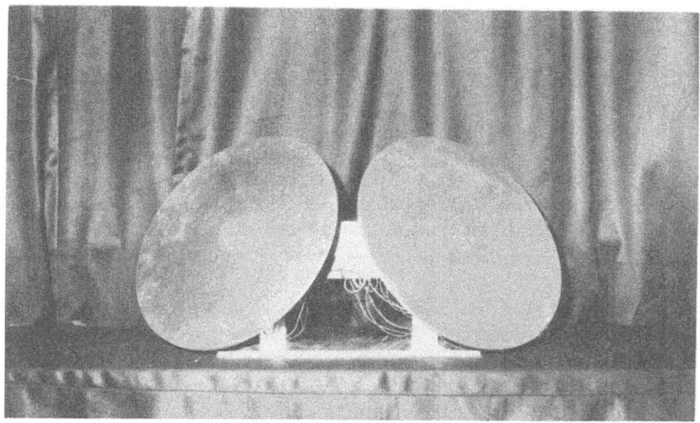

Fig. 7. View of the binaural receiver.

The proposed array has coded elements and operates as a signal generator. In other words, the array acts as a spatial and temporal filter. This feature is similar to the approach using a multilayer transducer [15,16]. Some interesting applications such as the compression and expansion of a pulse are possible with the array [16]. Furthermore, the similarity between the human ear and an EA is an interesting fact. The frequency response of the human ear changes depending on the range and direction of a sound source. This is considered as a cue for 3-D localization by the human ear [17] .

CONCLUSION

We have presented the concept of "Encoding Aperture" and have proposed an implementation with an array of wide band and nondirectional transducers. This array has a 3-D structure and each element is coded with an appropriate sequence. In this paper, we have presented an inclined circular array and a coding with a M-sequence.

We have introduced the P.S.F. of the imaging system and concluded that the EA is approximately equivalent to the projected 2-D array. The hardware of an imaging system can be simplified considerably with the EA. This has been demonstrated by the computer simulation of "BaTS".

To develop a practical imaging system using this concept, we must solve some problems. A design theory of an array and a code must be developed which generate near an orthogonal wavefront. An imaging operator which reduces the artifacts is needed. Finally, we must implement an experimental system and test the feasibility of the concept.

REFERENCE

1. R. W. Crosby and P. J. Kahrilas, Electronically scanned rader – An introduction, Sperry Engrg. Rev. (1965)
2. N. H. Ferhat, A new imaging principle, Proc. IEEE, 63-3, 377/380 (1976)
3. N. H. Ferhat, et al., Computer simulation of frequency swept hologram, Proc. IEEE, 64-9, 1453/1455 (1976)
4. K. Katakura, et al., Underwater acoustic imaging by frequency controlled beam steering, The Journal of the Acoust. Soc. of Japan, 31-12, 716/724 (1975)
5. T. Sato, W. Shusou, and J. Ishii, New ultrasonic imaging system using a moving random phase mask and a stationary point receiver, J. Acoust. Soc. Am., 62-1,102/107 (1977)

6. T. Hidaka, Image reconstruction by using a frequency-sweep method, J. Appl. Physics, 46-2, 786/794 (1975)

7. A. Rønnekleiv, et al., Grating acoustic scanners, Proc. Ultrasonics Symp., 91/93 (1975)

8. T. M. Waugh, et al., Acoustic imaging techniques for nondestructive testing, IEEE Trans., SU-23,313/316 (1976)

9. Y. Tamura, An echolocation using spatial modulation and wide-band pulse, Proc. of SICE, 26, 717/718 (1987)

10. K. Koyama, H. Ito, K. Tanaka and Y. Tamura, Acoustic imaging using polymer composite array transducer with directivity distribution, Jpn. J. Appl. Phys., 28, s28-1,251/253 (1988)

11. Y. Tamura, H. Ito, and K. Koyama, An acoustical imaging sensor with an encoding aperture – Transducer made of piezo-electrical polymer composite film –, Proc. of SICE, 28, 337/338 (1989)

12. K. Koyama, H. Itoh, K. Tanaka, and Y. Tamura, Polymer transducer system for imaging using pseudo inverse matrix detection, Acoustical Imaging, 18 (1989)

13. Y. Tamura and T. Akatsuka, An encoding aperture consists of transducer array, and its application for a binaural tree-dimensional SONAR, Proc. of SICE, 28, 335/336 (1989)

14. Y. Tamura and T. Akatsuka, An imaging operator for a high speed holographic sonar which uses an incompletely orthogonalized wavefront, Acoustical Imaging, 18 (1989)

15. O. M. Stuetzer, Piezoelectric pulse and code generators, IEEE Trans., SU-14-2, 75/88 (1967)

16. K. M. Sung, Piezoelectric multilayer transducers for ultrasonic pulse compression, Ultrasonics, 22-3, 61/68 (1984)

17. D. W. Batteau, The role of the pinna in human localization, Proc. Soc. London B168, 158/180 (1967)

AN ULTRASONIC ROBOT EYE USING NEURAL NETWORKS

Sumio Watanabe, Masahide Yoneyama

Ricoh Research and Development Center
16-1, Shin-ei-cho, Kohoku-ku, Yokohama
223 Japan

1.INTRODUCTION

Object recognition is an important aspect in the development of a robot eye system. The use of television cameras has been investigated, however no particularly effective method has been developed. We are developing a new method based on ultrasonic recognition and report some recent experimental results.

There are three advantages in using ultrasonic waves for object recognition. First, since sound travels more slowly than light, phase information is easily measured, resulting in the direct calculation of an object's three-dimensional structure. Second, the object's shape can be recognized despite differences in color, transparency or luminescence, making the recognition of glass or metal objects, for instance, easier. Third, objects can be recognized even in dark or smoky environments.

The practicability of ultrasonic recognition has been limited, however, by low image resolution. This is a result of a combination of factors, including long wavelength, a limited number of receivers and the use of receiver array with relatively small surface areas. To overcome these problems and increase the practicability of acoustic imaging, we have devised a system which combines existing techniques for acoustic holography with neural networks. Our experiments have shown that, using this system, two-dimensional objects can be identified and their improved images reconstructed[1,2]. Herein, we report the results from experiments conducted on the recognition of three-dimensional objects.

2.THEORY ON CONSTRUCTING ACOUSTICAL IMAGES

In this paper, the two axes for a receiver array are identified as (x,y), and the three axes for an object as (x',y',z'). The height of the receiver array is z. If a continuous plane wave $P_i(\vec{r}') = \exp(j\vec{K}_i \cdot \vec{r}')$ illuminates an object, and $\vec{k}_i = (k\sin\theta, 0, -k\cos\theta)$ is a wave number vector, it is assumed that the surface function of an object is

$$z' = \zeta(x', y') \quad (1)$$

and that the reflection coefficient is $\xi(x', y')$ (Fig.1). As a result, the sound pressure of scattered waves at the location $\vec{r} = (x, y, z)$ in a receiver is given by eq.(2) using the Kirchhoff approximation[3].

$$P(\vec{r}) = \frac{j\exp(jkr)F(\vec{r})}{4\pi r} \int dx' \int dy' \exp(j\vec{V}\cdot\vec{r}') \, \xi(x', y') \quad (2)$$

where

$$
\begin{cases}
\vec{V} = (V_x, V_y, V_z) \\[4pt]
V_x = -k\left(\dfrac{x}{r} - \sin\theta\right) \\[4pt]
V_y = -k\left(\dfrac{y}{r}\right) \\[4pt]
V_z = -k\left(\dfrac{z}{r} - \cos\theta\right) \\[4pt]
\vec{r}' = (x', y', z') \\[4pt]
F(\vec{r}) = \dfrac{(V_x^2 + V_y^2 + V_z^2)}{V_z}
\end{cases}
$$

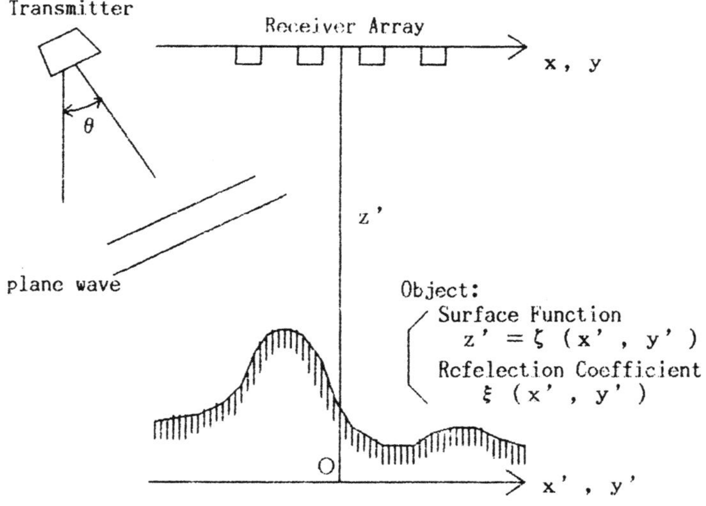

Figure 1. The Relative Positions of Object,
Receiver Array, and Transmitter

Eq.(2) can be rewritten as eq.(3),

$$P(\vec{r}) = \frac{j\exp(jkr)F(\vec{r})}{4\pi r} \int dx' \int dy'$$
$$\exp\{ jkx'\sin\theta - jk(1+\cos\theta)\zeta(x',y') \}\cdot\xi(x',y')$$
$$\exp(-j\frac{k}{r}(xx'+yy')) \qquad (3)$$

The inverse Fourier transform of the variables $(kx/r, ky/r)$ results in eq.(4).

$$\exp\{ -jk(1+\cos\theta)\zeta(x',y') \}\cdot\xi(x',y')$$
$$= \frac{(kz)^2}{\pi}\exp(-jkx'\sin\theta)\int dx \int dy \frac{P(\vec{r})}{jF(\vec{r})r^3\exp(jkr)}\exp(j\frac{k}{r}(xx'+yy')) \qquad (4)$$

The absolute value of the left hand side of eq.(4) represents the reflection coefficient and the phase term corresponds to the surface function.

(A) An Acoustical Imaging Method for 2-Dimensional Object

When only planar objects are considered, $\zeta(x',y')$ is equal to zero, and eq.(3) is reduced to $\xi(x',y')$.

$$\xi(x',y')$$
$$= \frac{(kz)^2}{\pi}\exp(-jkx'\sin\theta)\int dx \int dy \frac{P(\vec{r})}{jF(\vec{r})r^3\exp(jkr)}\exp(j\frac{k}{r}(xx'+yy')) \qquad (5)$$

If the material of an object is assumed to be homogeneous, the function $\xi(x',y')$ simply shows whether an object exists or not.

$$\xi(x',y') = \begin{cases} 1 & \text{if an object exists at } (x',y') \\ 0 & \text{otherwise} \end{cases} \qquad (6)$$

If the pressures $P(\vec{r})$ are measured and substituted into eq.(5) and then evaluated according to eq.(6), the shape of a planar object can be reconstructed. When the number of receivers is N×N, the image can be represented as an array of N×N pixels.

(B) Acoustical Imaging Method for 3-Dimensional Objects (1)

Let us consider two methods for calculating the 3-dimensional image of an object based on measured sound pressure $P(\vec{r})$. The first method, in which 3-dimensional inverse Fourier transform is used, is well known.

A 3-dimensional reflection coefficient $\eta(x',y',z')$ can be defined as

$$\eta(x',y',z') = \xi(x',y')\,\delta(z' - \zeta(x',y')) \qquad (7)$$

where $\delta(x)$ is Dirac's delta function. Using this coefficient, eq.(2) can be replaced by eq.(8).

$$P(\vec{r}) = \frac{j\exp(jkr)F(\vec{r})}{4\pi r}\int dx' \int dy' \int dz' \exp(j\vec{V}\cdot\vec{r}')\,\eta(x',y',z') \qquad (8)$$

$$= \frac{j\exp(jkr)F(\vec{r})}{4\pi r}\int dx' \int dy' \int dz'$$
$$\exp\{\,jkx'\sin\theta - jkz'\cos\theta\,\}\cdot\eta(x',y',z')$$
$$\exp\left(-j\frac{k}{r}(xx'+yy'+zz')\right) \qquad (9)$$

If the pressure $P(\vec{r})$ are measured using several different wave numbers and then substituted into eq.(9), the three-dimensional reflection coefficient $\eta(x',y',z')$ can be calculated by the inverse Fourier transform of the variables $(kx/r,\ ky/r, kz/r)$.

(C) Acoustical Imaging Method for 3-Dimensional Objects (2)

By the above method, 3-dimensional images can be obtained. Nevertheless, it takes long time to measure the sound pressures of several different wave numbers. So, in this paper, we propose a new method, by which 3-dimensional information can be obtained in shorter time, and use it in reconstructing images.

The surface function $\zeta(x',y')$ can be obtained by calculating the phase term of eq.(4). However, when ultrasonic waves are used, this phase term is periodic with cycle 2π, and only objects of very low heights can be measured. For example, if 40KHz sound waves are used and if $\theta=0.0$, the reconstructed region of $\zeta(x',y')$ is

$$0 \leq \zeta(x',y') \leq \frac{2\pi}{(1+\cos\theta)k} \doteqdot 4.25\text{mm} \qquad (10)$$

However, for taller objects, $P(\vec{r})$ can be measured for two different wave numbers, k and k+$\triangle$k and the phase term of k+$\triangle$k subtracted from that of k, thus extending the reconstructed region. If H(k) and H(k+$\triangle$k) each denote the left hand side of eq.(4) for their respective wave numbers, then

$$-k(1+\cos\theta)\,\zeta(x',y') = \arg(H(k)) + 2m\pi \qquad (11)$$

$$-(k+\Delta k)(1+\cos\theta)\,\zeta(x',y') = \arg(H(k+\Delta k)) + 2n\pi \qquad (12)$$

where m and n are integers and arg(x) is the argument angle of a complex variable x. Subtracting eq.(11) from eq.(12) results in eq.(13).

$$\zeta(x',y') = \frac{-1}{(1+\cos\theta)\Delta k}\{arg(H(k+\Delta k)) - arg(H(k)) + 2(n-m)\pi\} \qquad (13)$$

As Δk is small enough, height of the reconstructed region is extended,

$$0 \leqq \zeta(x',y') \leqq \frac{2\pi}{(1+\cos\theta)\Delta k} \qquad (14)$$

For instance, when 40KHz and 41.7KHz sound waves are used, the height of the reconstructed region becomes 100.0mm.

3. NEURAL NETWORKS

Using the above methods, 2- and 3-dimensional acoustical images are obtained. However, these images are generally very distorted, and it is difficult to recognize the object with this information. To overcome this problem, two neural networks are used; One to identify the objects, and the other to reconstruct improved images of objects. Each network is a 3-layered feed-forward neural network, taught by the error backpropagation algorithm[4].

(A) Neural Network used for Object Identification

Fig.2 shows the structure of the neural network used for object identification. When the acoustical image of an object is represented by N×N analogue values, the number of input layer units is also N×N. The number of output layer units is equal to the number of the categories of objects. The number of hidden layer units is variable. For recognition of 2-dimensional objects, the input patterns consist of the values of $\xi(x',y')$. For 3-dimensional objects, the values of $\zeta(x',y')$ are used for input patterns. The teaching pattern for an object in the first category is $(1,0,0,0,..,0)$, for an object in the second category $(0,1,0,0,0,..,0)$, and so forth. (Table 1).

Table 1. Teaching data values for object identification

N : the number of object categories

0,1 : teaching data values

Object category	Output unit number						
	1	2	3	4	5	...	N
1	1	0	0	0	0	...	0
2	0	1	0	0	0	...	0
3	0	0	1	0	0	...	0
N	0	0	0	0	0	...	1

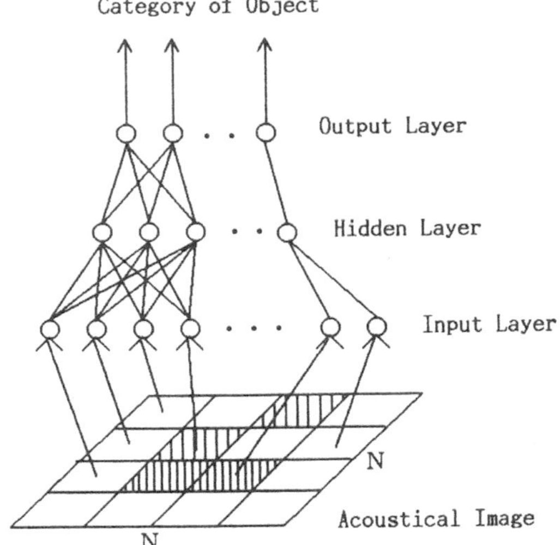

Figure 2. Neural Network for Object Identification
2-Dimensional Object Case
    Input Pattern: Reflection Coefficient
3-Dimensional Case
    Input Pattern: Height of Object

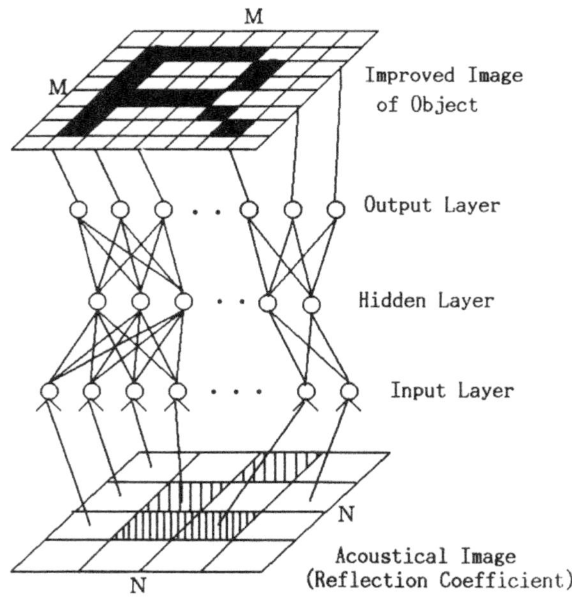

Figure 3. Neural Network for
2-Dimensional Image Reconstruction

Fig.3 and Fig.4 show the structures of the neural networks used to reconstruct improved images of 2-dimensional and 3-dimensional objects. When the acoustical image is given by N×N analogue numbers and the teaching patterns are represented by M×M, where M is an integer, then the number of input layer units is N×N. The number of output layer units is M×M in the 2-dimensional case, M×M×L in 3-dimensional case, where L is the number of pixels along the z-axis. In the 2-dimensional case, teaching patterns consist of improved images of the objects expressed by a value 0 or 1. In the 3-dimensional case, the teaching pattern is 1 if the surface crosses the corresponding pixel, and 0 if it doesn't.

4.SYSTEM STRUCTURE

In this paper, the results for 3-dimensional objects are reported. Experimental results for 2-dimensional objects are reported in reference[1,2].

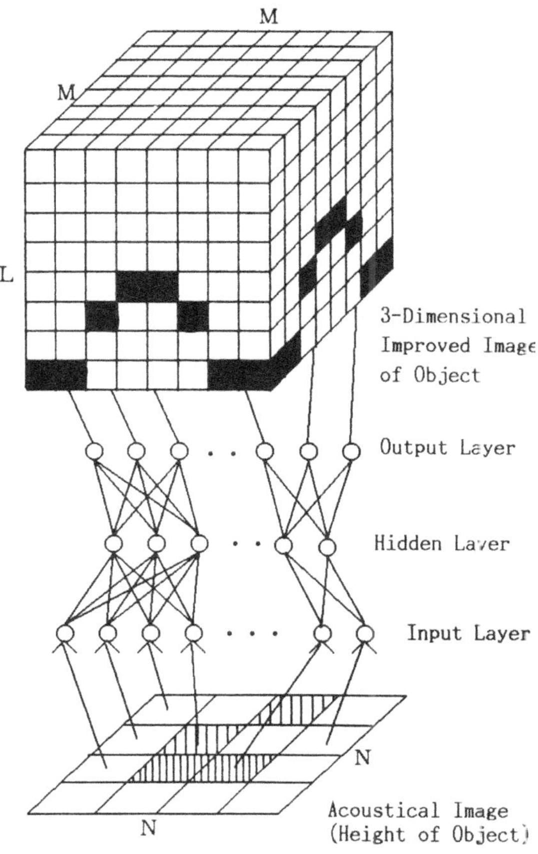

Figure 4. Neural Network for
3-Dimensional Image Reconstruction

We conducted two sets of experiments. In the first set,
the five simple objects were used as objects. The four of them
are shown in Fig.5. They were all made of aluminum. Each
object was placed at 9 different locations, resulting 5×9
samples. Each set of these 5×9 samples is referred to as "one
set". Two sets were collected; one set was used for teaching,
and the other was used for testing.

In the second set of experiments, the nine objects shown
were used as objects. The four of them are shown in Fig.6.
They were constructed of various materials, including
aluminum, wood and brass. It is unnecessary to include the
acoustic reflection coefficients for the differing materials
since only the phase term is used in recognition for 3-
dimensional objects. The size of each object was restricted
to 10cm along each of its three dimensions. By placing the
object at the center of the reconstructed area, and then
successively rotating it 15 degrees about its center, it was
possible to obtain 9×24 samples. Each set of these 9×24
samples is referred to as "one set", and three sets were
collected. Two sets were used for teaching, and the other was
used in testing.

Fig Figure 5. Photograph of
Simple Objects

Figure 6. Photograph of
Complex Objects

The block diagram of the system is shown in Fig.7 and a
photograph in Fig.8. Objects were placed on the surface z'=
0, and the receiver array was located on the plane z=24.5cm.
The number of receivers was 8×8, and the distance between
receivers was 2.0cm. Burst waves (sine wave, 10 cycles) were
sent from a transmitter onto the object at the angle θ=0.278
rad. If d is the distance between receivers, the calculation
of $(2\pi z)/(dk)$ indicates that the size of the reconstructed
area is 10.4cm×10.4cm. 40.0KHz and 41.7KHz ultrasonic waves
were used, resulting in a height of 10.0cm for the the
reconstructed region.

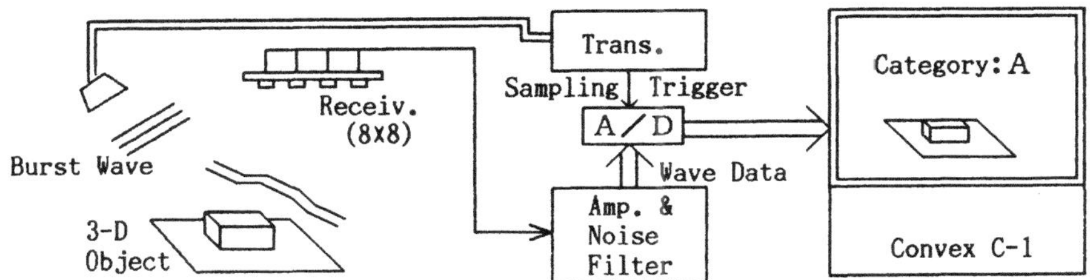

Figure 7. Block Diagram of System

Figure 8. Photograph of The System

Scattered waves were sampled every 1.0 microsecond and the wave data recorded. The real or imaginary part of $P(\vec{r})$ was calculated using the inner product of the measured wave and the referential cosine wave or the sine wave.

The Neural Networks were simulated in using software on a mini-computer, Convex, C-1.

5. EXPERIMENTAL RESULTS

(A) Acoustical Images

The 3-dimensional reconstructed images for the objects used in the first experiment are given in Fig.9. The figures (1), (2), (3), and (4) represent the values for $\zeta(x',y')$ for a cylinder, a cube, a pyramid, and a sphere. Since the number of receivers is 8×8, the images are also represented by an array of 8×8 pixels.

The reconstructed images for the objects used in the second experiments are shown in Fig.10. The figures (1), (2), (3), and (4) represent the values for $\zeta(x',y')$ for a spoon, a car, a bird, and a gas stopcock.

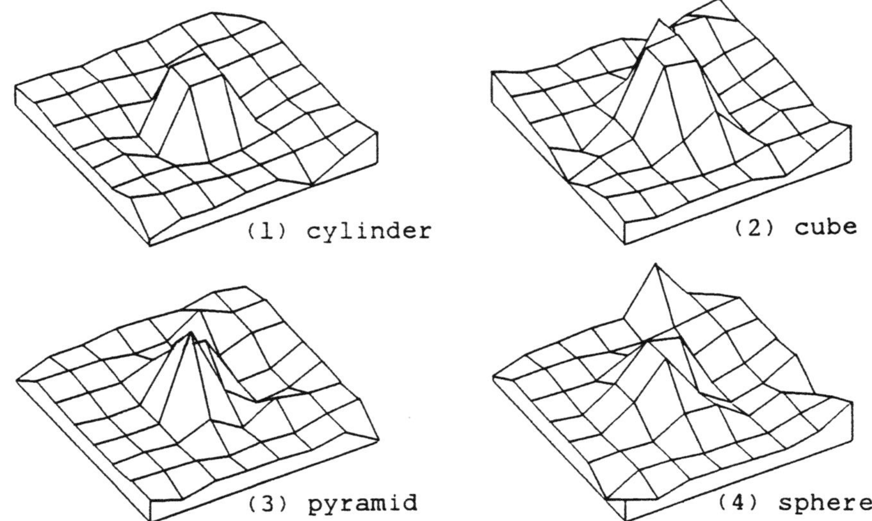

Figure 9. Acoustical Images

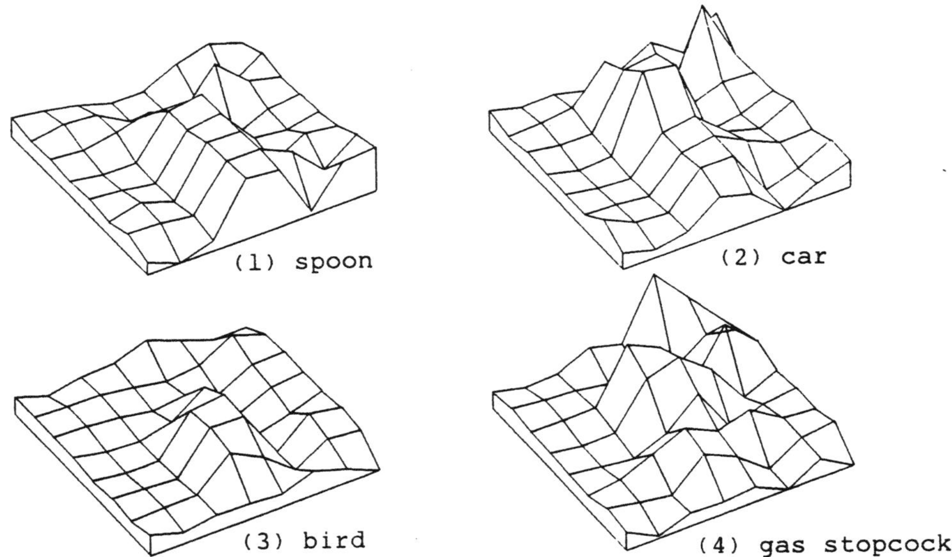

Figure 10. Acoustical Images

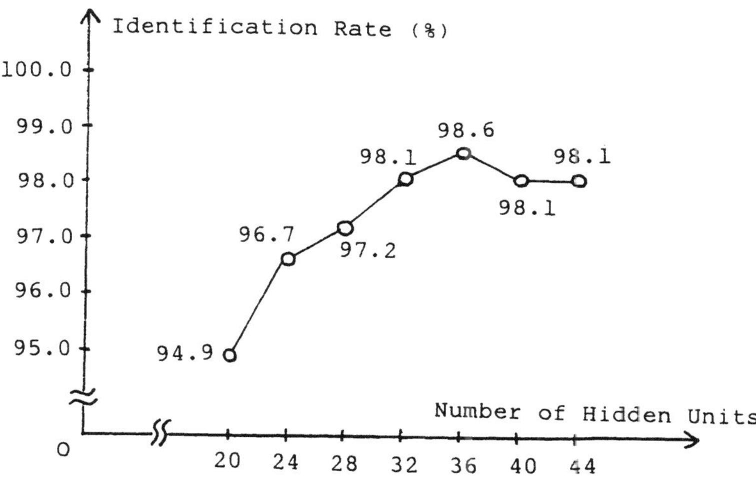

Figure 11. Identification Rate for Several
Numbers of Hidden Units

(B) Identification Experiments

In the first experiment, a neural network, with 8×8 input
layer units, 16 hidden layer units, and 5 output units, was
taught to identify the kinds of objects. In the teaching mode,
after about 800 learning cycles, the neural network learned to
output the correct answer for every pattern. In the
experimental mode, no incorrect pattern was output in a total
of 45 outputs.

In the second experiment, a neural network, with 8×8
input layer units, 32 hidden layer units, and 9 output layer
units, was taught to identify the kinds of objects. In the
teaching mode, after about 1500 learning cycles, the neural
network learned to output the correct answer for every
pattern. In the experimental mode, Only four incorrect
patterns were output in a total of 216 outputs. The results
using a different number of hidden layer units are shown in
Fig.11.

(C) Reconstruction of Images

In the first experiment, a neural network, with 8×8 input
layer units, 64 hidden layer units, and 24×24×24 output layer
units, was taught to reconstruct the 3-dimensional improved
images of objects. In the teaching mode, after about 1000
learning cycles, the network learned to output the correct
answer for every pattern. In Fig.12, (1),(2),(3), and (4) show
the experimental results for a cylinder, a cube, a pyramid,
and a sphere.
In the second experiment, a neural network, with 8×8
input layer units, 128 hidden layer units, and 24×24×24 output
layer units, was taught to reconstruct the 3-dimensional

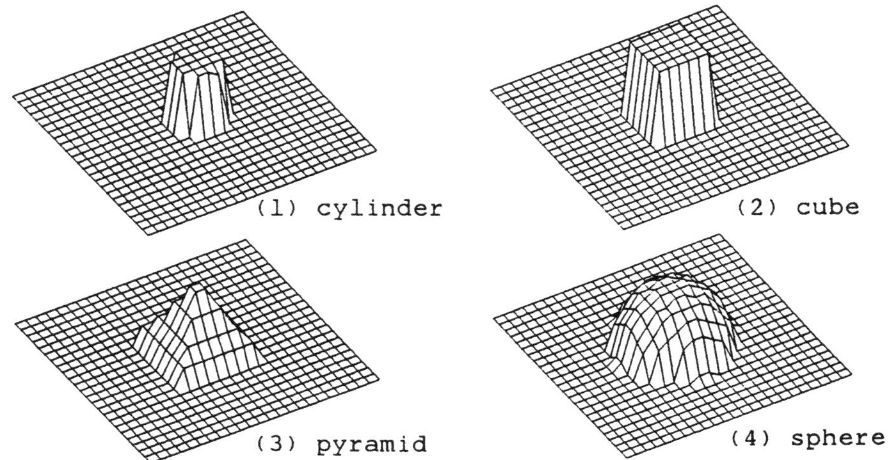

Figure 12. Reconstructed Images by Neural Network

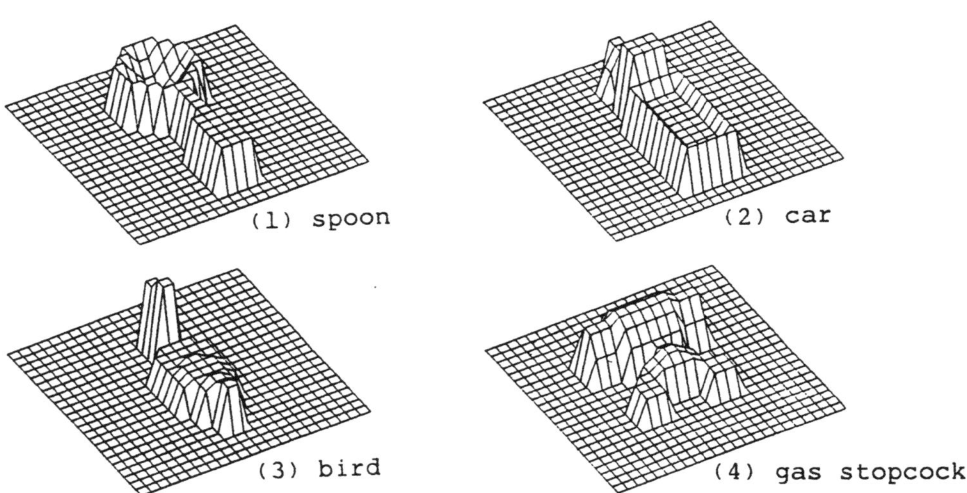

Figure 13. Reconstructed Images by Neural Network

improved images of objects. In the teaching mode, after about
1000 learning cycles, the network remained in a local minimum.
Even in the experimental mode, almost precise images were
obtained. In Fig.13 (1), (2), (3), and (4) show the
experimental results for a spoon, a car, a bird, a gas
stopcock.

6.CONCLUSION

     By combining acoustical imaging with a neural network,
an ultrasonic robot eye system has been devised for object
identification and image reconstruction. Experimental results
show that 3-dimensional objects can be identified and that
their images can be reconstructed. A topic for the future is
to improve this system to put it into a practical application.

7.REFERENCES

1. M.Yoneyama, S.Watanabe, H.Kitagawa, T.Okamoto, T.Morita,
   "Neural Network Recognizing 3-Dimensional Object through
   Ultrasonic Scattering Waves", IEEE Ultrasonics Symposium,
   (1988).
2. M.Yoneyama, S.Watanabe, "The Ultrasonic Robot Eye System
   Using Neural Network", 13th International Congress on
   Acoustics, (1989).
3. A.Ishimaru, "Wave Propagation and Scattering in Random
   Media Vol.2", Academic Press (1978).
4. D.E.Rumelhart, J.L.McCleland, and The PDP Research Group,
   "Parallel Distributed Processing Vol.1", The MIT Press
   (1986).

INVERSE SCATTERING AND DIFFRACTION TOMOGRAPHY

USING INTENSITY DATA

Anthony J. Devaney†

Department of Electrical and Computer Engineering
Northeastern University
Boston, Mass. 02115

INTRODUCTION

In linearized inverse scattering [1] and diffraction tomography [2] the complex index of refraction profile of a weakly inhomogeneous scattering object is reconstructed from measurements of the scattered field generated in a set of scattering experiments. In the so-called "classical" inverse scattering configuration the incident waves are monochromatic plane waves and the scattered field measurements are performed over the surface of a sphere, centered within the scatterer, and having a radius ρ that is much larger than the wavelength of the incident wave. This then amounts to having the complex valued scattering amplitude of the object as data and reconstruction algorithms exist [3-5] that allow the index of refraction profile to be estimated (reconstructed) from this data.

The requirement that the complex amplitude of the scattered wavefields be available as data to the reconstruction procedures of inverse scattering and diffraction tomography has limited the use of these reconstruction procedures in a number of practical applications. Even in applications such as ultrasound tomography where the complex amplitude of the transmitted wavefield is directly measurable it is not an easy matter to deduce the real phase in an unambiguous manner due to the well-known phase unwrapping problem [6]. In optical scattering experiments it is necessary to introduce an off-axis reference beam so as to measure the phase of the scattered wavefield (i.e., make a hologram of the scattered wavefield) and this complicates considerably the measurement system and ultimately limits the practicality of the method.

In a recent short paper [7] the current author addressed the inverse scattering problem defined above for the case where only the *intensity* (magnitude square) of the wavefields generated in the scattering experiments are available as data. It was argued on theoretical grounds that high quality approximate reconstructions could be obtained from such data as long as the radius ρ of the measurement sphere is much larger than the radius of the support volume of the scatterer. This theoretical argument was supported by a simple computer simulation of inverse scattering from a square well central potential.

---

†Also with A.J. Devaney Associates, 355 Boylston St., Boston, MA 02116

Acoustical Imaging, Vol. 18, Edited by H. Lee and G. Wade
Plenum Press, New York, 1991

In the study presented herein the results obtained in [7] are expanded upon and the computer simulation presented in [7] is discussed in some detail. The paper will limit its attention to the classical inverse scattering configuration. However, much of the theory and results can be readily extended to other measurement geometries such as the classical scan configuration of diffraction tomography [8].

INVERSE SCATTERING THEORY

We consider a scattering experiment where an incident plane wave $e^{iks_0 \cdot r}$ propagating along the direction of the unit vector s_0 illuminates a scatterer having a complex index of refraction profile $n(\mathbf{r}) = 1 + \delta n(\mathbf{r})$. Here, $k = \frac{2\pi}{\lambda}$ is the wavenumber of the incident wave in the background medium, (e.g., air in the case of optical scattering), with λ being the wavelength.. We will assume that the deviation δn of the index profile of the scattering object is concentrated in some finite volume about the origin $\mathbf{r} = 0$ and that its magnitude is small compared to the reference background value of unity so that the Born approximation [1] can be employed to compute the scattered wavefield. Using this approximation one finds that the field generated over the measurement surface can be expressed in the form [9]

$$\psi(\rho \mathbf{s}; \mathbf{s}_0) \sim e^{iks_0 \cdot \mathbf{s}\rho} + f(\mathbf{s}, \mathbf{s}_0)e^{ik\rho}/\rho \tag{1}$$

where $\mathbf{s}$ is a unit vector along the direction of observation and $f(\mathbf{s}, \mathbf{s}_0)$ is the *scattering amplitude* and is given by

$$f(\mathbf{s}, \mathbf{s}_0) = \frac{k^2}{4\pi} \tilde{\delta n}[k(\mathbf{s} - \mathbf{s}_0)] \tag{2}$$

where

$$\delta \tilde{n}(\mathbf{K}) = \int d^3 r \, \delta n(\mathbf{r}) e^{-i\mathbf{K} \cdot \mathbf{r}} \tag{3}$$

is the three-dimensional spatial Fourier transform of the index deviation.

When the field ψ can be directly measured the scattering amplitude $f(\mathbf{s}, \mathbf{s}_0)$ can be determined for some set of illuminating and measurement directions $\mathbf{s}_0$ and $\mathbf{s}$, respectively, and the inverse scattering problem reduces to estimating the index perturbation δn from its spatial Fourier transform $\tilde{\delta n}(\mathbf{K})$ specified over the set of points $\mathbf{K} = k(\mathbf{s} - \mathbf{s}_0)$ in Fourier space via Eq.(2). These points lie on the so-called *Ewald spheres* [1] well known from X-ray crystallography and a number of algorithms [1-5] are available for reconstructing δn from such data. One such algorithm is the *filtered backpropagation algorithm* [3] which can be expressed in the form [10]

$$\delta \hat{n}(\mathbf{r}) = \frac{-k^3}{4\pi^3} \int d\Omega_{s_0} \int d\Omega_s \, |\mathbf{s} - \mathbf{s}_0| f(\mathbf{s}, \mathbf{s}_0) e^{ik(\mathbf{s} - \mathbf{s}_0) \cdot \mathbf{r}} \tag{4}$$

where, $d\Omega_{s_0}$ and $d\Omega_s$ are differential solid angles associated with the unit vectors $\mathbf{s}_0$ and $\mathbf{s}$ and the integrals are over those portions of the unit spheres corresponding to the set of illumination and measurement directions for which the scattering amplitude is known.

The estimate $\delta \hat{n}$ of the index deviation generated by the filtered backpropagation algorithm is a low-pass filtered approximation of δn, having a three-fold Fourier transform $\tilde{\delta \hat{n}}(\mathbf{K})$ equal to $\delta \tilde{n}(\mathbf{K})$ over the region of $\mathbf{K}$ space covered by the set of

Ewald spheres and vanishing outside this region [10]. When a complete set of scattering experiments is performed so that s_0 and s completely cover the unit sphere then this region is the interior of a sphere, called the Ewald limiting sphere, that is centered at the origin and has a radius of $2k$. In this case then the estimate generated by the algorithm is given by

$$\delta \hat{n}(\mathbf{r}) = \frac{1}{(2\pi)^3} \int_{|\mathbf{K}| \leq 2k} d^3 K \, \delta \tilde{n}(\mathbf{K}) e^{i\mathbf{K} \cdot \mathbf{r}}. \tag{5}$$

The estimate given in Eq.(5) requires that both forward and backscatter measurements be performed for each experiment (for each value of s_0). If only forward scattering measurements are performed, corresponding to scattering directions s covering a half unit sphere, then the region of coverage in $\mathbf{K}$ space is the interior of a sphere of radius $\sqrt{2}k$ so that a reduced resolution image of the index perturbation is obtained.

The filtered backpropagation algorithm defined in Eq.(4) can be interpreted as being the coherent superposition of a set of *partial reconstructions* generated from data collected in each scattering experiment; i.e., for each direction of illumination s_0. This interpretation follows immediately upon rewriting Eq.(4) in the form

$$\delta \hat{n}(\mathbf{r}) = \frac{1}{4\pi} \int d\Omega_{s_0} \delta \hat{n}(\mathbf{r}, s_0) \tag{6}$$

where $\delta \hat{n}(\mathbf{r}, s_0)$ is the partial reconstruction generated from the scattering amplitude for direction of illumination s_0 and is given by

$$\delta \hat{n}(\mathbf{r}, s_0) = \frac{-k^3}{\pi^2} e^{-iks_0 \cdot \mathbf{r}} \int d\Omega_s \, |s - s_0| f(s, s_0) e^{iks \cdot \mathbf{r}}. \tag{7}$$

Although each partial reconstruction $\delta \hat{n}(\mathbf{r}, s_0)$ is, in general, a blurry, poor quality image of the index deviation, the coherent superposition of a number of these partial reconstructions yields a high quality reconstruction due to selective constructive and destructive interference between the partial reconstructions

INTENSITY DATA

We now consider the case where only the intensity of the wavefield ψ is known (measurable) over the measurement surface. This intensity is found from Eq.(1) to be given by

$$\mathcal{I}(s, s_0) = |\psi(\rho s; s_0)|^2$$
$$\sim 1 + \frac{1}{\rho^2} |f(s, s_0)|^2 + 2\Re\{e^{-iks_0 \cdot s\rho} f(s, s_0) \frac{e^{ik\rho}}{\rho}\} \tag{8}$$

where $\Re$ stands for the real part.

Now although, the intensity $\mathcal{I}$ is a real, non-negative quantity it contains information about the real phase of the scattering amplitude through the interference term $2\Re\{e^{-iks_0 \cdot s\rho} f(s, s_0) \frac{e^{ik\rho}}{\rho}\}$. Indeed, this intensity distribution over the measurement surface can be identified as being a *Gabor hologram* of the scattered wavefield over this surface [7], formed by the interference of the incident plane wave and the wave scattered by the index deviation δn. Unfortunately, there is no known method for uniquely retrieving the phase of the scattering amplitude from this intensity distribution so that the complete determination of the scattering amplitude and, hence, the solution of the inverse scattering problem given only intensity data remains an open issue.

In this paper our goal is less ambitious. We do not attempt to first deduce the scattering amplitude from the intensity distribution and *then* use this quantity in an inversion algorithm such as the filtered backpropagation algorithm. Rather, we bypass the phase retrieval step entirely and obtain an *approximate* solution of the inverse scattering problem directly in terms of the measured intensity distribution $\mathcal{I}$. The validity of this approximate solution will be found to depend on the measurement distance ρ being large compared with the radius of the support volume of the scatterer. We will not provide a detailed proof of the method but will simply present a plausability argument based on the filtered backpropagation algorithm discussed above.

The approximate method for reconstructing the index deviation from the intensity profile $\mathcal{I}$ is based on the observation that Eq.(8) can be formally solved for the scattering amplitude; i.e.,

$$f(\mathbf{s},\mathbf{s}_0) = D(\mathbf{s},\mathbf{s}_0) - \frac{1}{\rho}|f(\mathbf{s},\mathbf{s}_0)|^2 e^{ik(\mathbf{s}_0\cdot\mathbf{s}-1)\rho} - f^*(\mathbf{s},\mathbf{s}_0)e^{2ik(\mathbf{s}_0\cdot\mathbf{s}-1)\rho} \qquad (9)$$

where

$$D(\mathbf{s},\mathbf{s}_0) = \rho e^{ik(\mathbf{s}_0\cdot\mathbf{s}-1)\rho}\{\mathcal{I}(\mathbf{s},\mathbf{s}_0) - 1\}. \qquad (10)$$

It follows from Eq.(9) that if the scattering amplitude $f(\mathbf{s},\mathbf{s}_0)$ is approximated by the quantity $D(\mathbf{s},\mathbf{s}_0)$ in the filtered backpropagation algorithm (4) then the reconstruction generated by the algorithm will consist of three components: (i) the optimal reconstruction $\delta\hat{n}$ generated from f; (ii) an error component generated from $\frac{1}{\rho}|f(\mathbf{s},\mathbf{s}_0)|^2 e^{ik(\mathbf{s}_0\cdot\mathbf{s}-1)\rho}$; and (iii) an error component generated from $f^*(\mathbf{s},\mathbf{s}_0)e^{2ik(\mathbf{s}_0\cdot\mathbf{s}-1)\rho}$.

Before discussing the error terms it is important to note that for the proposed scheme to work it is necessary to form an interference pattern between the incident plane wave and the scattered wavefield. This then requires the scattering measurements to be performed in *forward* scattering directions where $\mathbf{s}_0\cdot\mathbf{s} \geq 0$. It then follows from the discussion presented under Eq.(5) that, at best, the proposed method will yield a low pass filtered estimate of the index deviation bandlimited to within a sphere of radius $\sqrt{2}k$ in $\mathbf{K}$ space.

Returning to Eq.(9) it is seen that the second error component is inversely proportional to ρ and, hence, vanishes in the limit $\rho \rightarrow \infty$. Let us then assume that ρ is sufficiently large that we can neglect this component and determine the conditions under which the third error component can be neglected. We consider then the error component generated by $f^*(\mathbf{s},\mathbf{s}_0)e^{2ik(\mathbf{s}_0\cdot\mathbf{s}-1)\rho}$. Denoting this term by $\mathcal{E}$ we conclude that

$$\mathcal{E}(\mathbf{r}) = \frac{1}{4\pi}\int d\Omega_{s_0}\mathcal{E}(\mathbf{r},\mathbf{s}_0) \qquad (11)$$

where

$$\mathcal{E}(\mathbf{r},\mathbf{s}_0) = \frac{-k^3}{\pi^2}e^{-ik\mathbf{s}_0\cdot(\mathbf{r}+2\rho\mathbf{s}_0)}\int d\Omega_s |\mathbf{s} - \mathbf{s}_0|f^*(\mathbf{s},\mathbf{s}_0)e^{ik\mathbf{s}\cdot(\mathbf{r}+2\rho\mathbf{s}_0)}$$

$$= \delta\hat{n}^*(-\mathbf{r} - 2\rho\mathbf{s}_0,\mathbf{s}_0) \qquad (12)$$

where $\delta\hat{n}^*(-\mathbf{r} - 2\rho\mathbf{s}_0, \mathbf{s}_0)$ is the complex conjugate of the partial reconstruction of δn generated by the scattering amplitude *but reflected about the origin and displaced along the $\mathbf{s}_0$ direction by an amount equal to 2ρ.*

Since each of the partial reconstructions $\delta\hat{n}(\mathbf{r}, \mathbf{s}_0)$ will be peaked in the immediate vicinity of the origin $\mathbf{r} = 0$, the partial reconstructions $\mathcal{E}(\mathbf{r}, \mathbf{s}_0)$ will be peaked about the point $-2\rho\mathbf{s}_0$. The total error component $\mathcal{E}(\mathbf{r})$ is a coherent superposition of these displaced partial reconstructions and can then be expected to be small in magnitude within the support volume of the scatterer so long as ρ is much larger than radius of this support volume. We conclude that the approximate reconstruction generated by naively employing $D(\mathbf{s}, \mathbf{s}_0)$ in the filtered backpropagation algorithm will be of high quality as long as ρ is much larger than the radius of the support volume of the scatterer.

A more quantitative determination of the error can be obtained by expressing the error component amplitude in the form

$$f^*(\mathbf{s}, \mathbf{s}_0)e^{2ik(\mathbf{s}_0 \cdot \mathbf{s} - 1)\rho} = \frac{k^2}{(4\pi)^2}\delta\tilde{n}^*[k(\mathbf{s} - \mathbf{s}_0)]e^{-i\frac{\rho}{k}|k(\mathbf{s} - \mathbf{s}_0)|^2} \tag{13}$$

where we have used the vector identity $|\mathbf{s} - \mathbf{s}_0|^2 = 2 - 2\mathbf{s} \cdot \mathbf{s}_0$ and have made use of Eq.(2). Substitution of Eq.(13) into the filtered backpropagation algorithm (4) will generate an inverse Fourier transform of the quantity

$$\tilde{\mathcal{E}}(\mathbf{K}) = \delta\tilde{n}^*(\mathbf{K})e^{-i\frac{\rho}{k}|\mathbf{K}|^2}$$

over the region of $\mathbf{K}$ space covered by the set of Ewald spheres generated by the incident and scattered wavevectors $\mathbf{s}_0$ and $\mathbf{s}$ (cf., Eq.(5)). As discussed above, this region is a sphere of radius $\sqrt{2}k$ so that the error $\mathcal{E}(\mathbf{r})$ will be given by

$$\mathcal{E}(\mathbf{r}) = \frac{1}{(2\pi)^3}\int_{|\mathbf{K}| \leq \sqrt{2}k} d^3K\, \delta\tilde{n}^*(\mathbf{K})e^{-i\frac{\rho}{k}|\mathbf{K}|^2}e^{i\mathbf{K}\cdot\mathbf{r}}. \tag{14}$$

We will employ Eq.(14) in the following section to quantitatively determine the contribution of the error $\mathcal{E}(\mathbf{r})$ in a computer simulation study. For the present we note that this term is the Fourier transform of the product of $\delta\tilde{n}^*(\mathbf{K})$ with the highly oscillatory factor $e^{-i\frac{\rho}{k}|\mathbf{K}|^2}$. This latter factor causes the integrand to be effectively bandlimited to the region $|\mathbf{K}| \leq \frac{2ka}{\rho}$ where a is the radius of the support volume of δn. It then follows from Parseval's theorem that the integral square of the error term will be approximated by

$$\epsilon = \int d^3r\, |\mathcal{E}(\mathbf{r})|^2 \approx \frac{4}{3\pi^2}\left(\frac{ka}{\rho}\right)^3 M^2 \tag{15}$$

where $M = \max|\delta\tilde{n}^*(\mathbf{K})|$. We then conclude from Eq.(15) that ϵ tends to zero in the limit where $\frac{a}{\rho} \to 0$.

COMPUTER SIMULATION

In this section we will illustrate the theory presented in the preceding section with a simple example. We consider the same example presented in [7]; i.e., a three-dimensional square well potential defined by

$$\delta n(\mathbf{r}) = \begin{cases} 1 & \text{if } |\mathbf{r}| \leq a \\ 0 & \text{elsewise.} \end{cases}$$

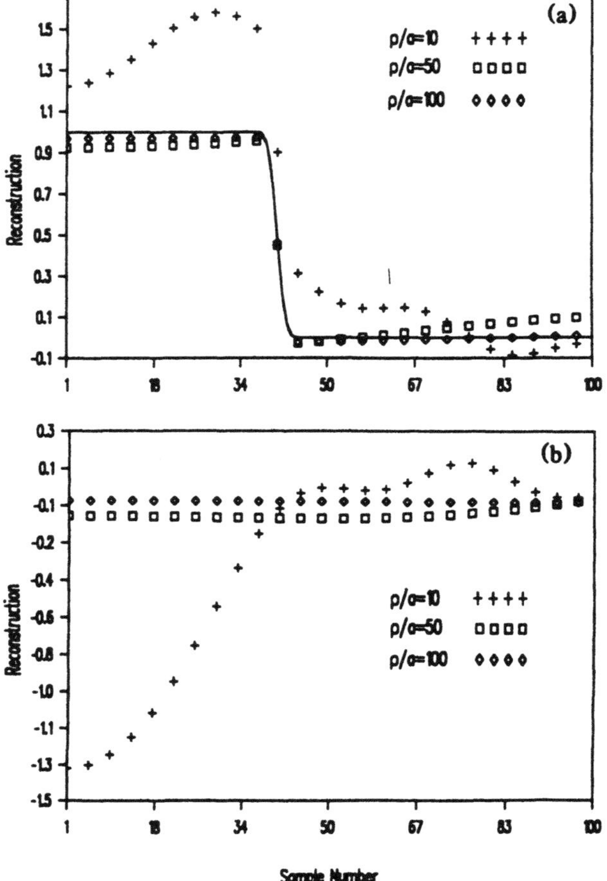

Real (a) and imaginary (b) parts of the approximate reconstruction of a spherically symmetric square well potential of radius $a = 10\lambda$ plotted as a function of the radius $r = |\mathbf{r}|$ using quarter wavelength spacing. The solid curve corresponds to the optimum reconstruction generated from the complex-valued scattering amplitude. (Taken from reference 7.)

The scattering amplitude is readily computed from Eq.(2) and we find that

$$f(\mathbf{s}, \mathbf{s}_0) = k^2 \{ \frac{a \cos Ka}{K^2} - \frac{\sin Ka}{k^3} \} \tag{16}$$

where $K = k|\mathbf{s} - \mathbf{s}_0|$.

The error term $\mathcal{E}(\mathbf{r})$ was computed using Eq.(14) for a potential having a radius $a = 10\ \lambda$ and for three different values of the ratio $\frac{\rho}{a}$. We show in Fig. 1 the real and imaginary parts of the sum of the ideal reconstruction generated from the scattering amplitude $f(\mathbf{s}, \mathbf{s}_0)$ and the error term $\mathcal{E}(\mathbf{r})$. It is seen that for values of the ratio $\frac{\rho}{a}$ greater than fifty the reconstruction is quite good. Similar results would be obtained for any linear combination of such square well potentials so long as the measurement distance ρ was large compared to the outer radius of the smallest sphere that completely surrounds all the potentials.

SUMMARY

We have in this paper reviewed the theory of inverse scattering using intensity data. It was shown that if the intensity of the total wavefield (incident plus scattered) is measured in the far field that high quality reconstructions can be obtained under conditions where the Born approximation holds. The success of the method is based on the fact that the measured intensity contains an interference term between the incident and scattered wavefields that is, in essence, a Gabor hologram of the scattered wavefield and thus contains the phase information that is required by inverse scattering algorithms. The error introduced by the other terms in the intensity distribution was investigated and shown to be negligible so long as the measurement distance is much larger than the support radius of the scatterer.

REFERENCES

1. E. Wolf, "Three-dimensional structure determination of semi-transparent objects from holographic data," *Opt. Commun.* **1**, p. 153, 1969.

2. R.K. Mueller, M. Kaveh, and G. Wade, "Reconstructive tomography and applications to ultrasonics," *Proc. IEEE*, Vol. 67, pp. 567-587, 1979.

3. A.J. Devaney, "A filtered backpropagation algorithm for diffraction tomography," *Ultrasonic Imaging*, Vol. 4, pp. 336-350, 1982.

4. S.X. Pan and A.C. Kak, "A computational study of reconstruction algorithms for diffraction tomography," *IEEE Trans. Acoustics, Speech and Signal Processing*, Vol. ASSP-31, pp. 1262-1275, 1982.

5. J.F. Greenleaf, "Computerized tomography with ultrasound," *Proc. IEEE*, Vol.71, pp.330-337, 1983.

6. J.M. Tribolet, "A new phase unwrapping algorithm," *IEEE Trans. Acoustics, Speech and Signal Processing*, Vol. ASSP-26, pp.170-177, 1977.

7. A.J. Devaney, "Structure determination form intensity measurements in scattering experiments," *Phys. Rev. Lett.*, Vol. 62, pp.2385-2388, 1989.

8. A.J. Devaney, "Diffraction tomography using intensity data," *IEEE Trans. Acoustics, Speech and Signal Processing* (to appear).

9. J.M. Cowley, *Diffraction Physics*. New York, North-Holland, 1984.

10. A.J. Devaney, "Inversion formula for inverse scattering within the Born approximation," *Opt. Letts.* **7**, p. 111, 1982.

Reconstruction of Two-Dimensional Refractive Index Distribution Using the Born Iterative and Distorted Born Iterative Method

Y. M. WANG AND W. C. CHEW
Electromagnetics Laboratory
Department of Electrical and Computer Engineering
University of Illinois
Urbana, IL 61801

Abstract. The Born iterative method and the distorted Born iterative method (DBIM) are used to solve two-dimensional acoustic inverse scattering problems. These methods were developed to solve the two-dimensional imaging problem when the Born and the Rytov approximations break down. Numerical simulations are performed using the both methods. Both of them give good reconstructed profiles when the first-order Born approximation fails. Meanwhile, the results show that each method has its advantages. The distorted Born iterative method shows faster convergence rate compared to the Born iterative method, while the Born iterative method is more robust to noise contamination compared to the distorted Born iterative method.

1. Introduction

Following the progress in the speed and capacity of the modern computer technology, computerized tomography first was introduced to generate the image of the human body using multiscan X-ray technique. In this case, straight ray propagation are assumed, which means that the refraction effect is small and can be neglected. For straight ray propagation, back projection tomography is an eligible technique for the image reconstruction. In the other cases, e.g. acoustic image, when the refraction effects are weak but not negligible, the diffraction tomography (DT) approach has been introduced and investigated within the framework of the Born and Rytov approximation [3,5,6,9]. Unfortunately, the conditions under which the Born and Rytov approximation can be applied are not frequently met in most practical problems [4,7,8]. To take the effects of strong diffraction into account, the inherent nonlinear integral equation of inverse scattering problems has to be solved beyond the Born approximation.

Two different approaches have been reported to solve the nonlinear inverse scattering problem. One of them is based on the inverse source method [13]. In this approach, the induced current source is first reconstructed. Then the object profile is recovered from the induced current source. The other approach is to solve for the object directly from the nonlinear integral equation of the inverse scattering problem by using the iterative method. Here, we will summarize two iterative methods we proposed recently to solve the two dimensional electromagnetic inverse scattering problem [1,2]. They are the Born iterative method and the distorted Born iterative method. Both of them fall into the later approach.

Acoustical Imaging, Vol. 18, Edited by H. Lee and G. Wade
Plenum Press, New York, 1991

2. Theory and Formulation

The geometry of the two-dimensional inverse problem is shown in Figure 1.
The cylindrical medium with an arbitrary cross section is inhomogeneous in the
xy plane but is homogeneous in z direction. The receivers are located around the
cylindrical object at finite discrete points. The object is illuminated by either a
plane wave or a field excited by a line source indicated as T in Figure 1. For
an acoustic wave, a scalar wave equation $(\nabla^2 + k^2)\phi(x, y) = 0$ will be used to
describe the field. The corresponding integral equation is

$$\phi(x, y) = \phi^i(x, y) + \iint_S G(\boldsymbol{\rho} - \boldsymbol{\rho}', n_b) k_0^2 \delta n^2 \phi(x', y') dx' dy', \tag{1}$$

where S is the scatterer cross section, and $G(\boldsymbol{\rho} - \boldsymbol{\rho}', n_b)$ is the solution of following
equation

$$\nabla_s^2 G(\boldsymbol{\rho} - \boldsymbol{\rho}', n_b) + k_b^2(x, y) G(\boldsymbol{\rho} - \boldsymbol{\rho}', n_b) = -\delta(\boldsymbol{\rho} - \boldsymbol{\rho}'). \tag{2}$$

Here,

$$\nabla_s^2 \equiv \frac{\partial^2}{\partial x^2} + \frac{\partial^2}{\partial y^2},$$

$$k_b^2 = k_0^2 n_b^2(x, y),$$

and n_b is the background refraction index. For homogeneous n_b, $G(\boldsymbol{\rho} - \boldsymbol{\rho}', n_b)$ in
a closed form is

$$G(\boldsymbol{\rho} - \boldsymbol{\rho}') = \frac{i}{4} H_0^{(1)}(k_b |\boldsymbol{\rho} - \boldsymbol{\rho}'|). \tag{3}$$

For an inhomogeneous background $n_b(x, y)$, $G(\boldsymbol{\rho} - \boldsymbol{\rho}', n_b)$ has to be solved numerically.

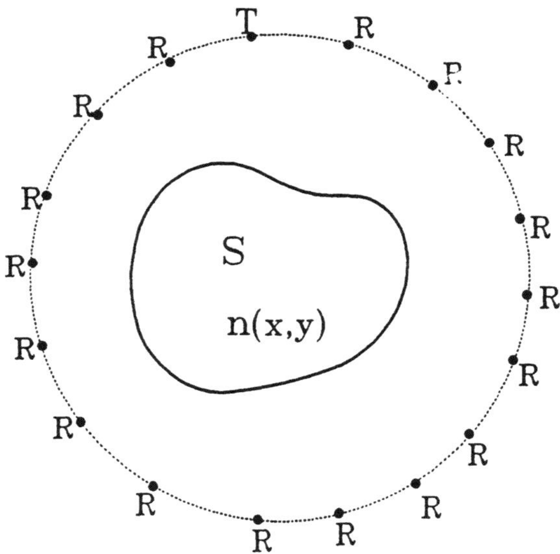

Figure 1. Geometrical configuration of the problem.

In solving the inverse scattering problem, the measurements are performed around the scatterer at R indicated in Figure 1. The integral equation (1) then becomes

$$\phi^s(x,y) = \iint_S G(\boldsymbol{\rho} - \boldsymbol{\rho}', n_b) k_0^2 \delta n^2 \phi(x',y') dx' dy', \tag{4}$$

and

$$\delta n^2 = n^2(x,y) - n_b^2(x,y) \tag{5}$$

is the refraction index profile to be reconstructed.

In weak scattering cases, where the scattered field is much smaller than the incident field, the integral equation (2) can be solved for $\delta n^2(x,y)$ under the Born or Rytov approximations. Unfortunately, the distortions of the reconstructed profile become intolerable under the first approximation when the criteria are not satisfied. In these cases, the strong diffraction effects have to be considered, which means that the inherent nonlinearity of the integral equation (2) has to be taken into account. The Born iterative method[1] and the distorted Born iterative method[2] were introduced to deal with the situations when the Born and Rytov approximations fail to generate the satisfied reconstructed profiles. The two methods are briefly discussed in the following two subsections.

2.1 Born iterative method

In the born iterative method, the Green's function, $G(\boldsymbol{\rho}-\boldsymbol{\rho}')$, remains unchanged in the iteration procedures. The outline of this approach can be summarized in the following steps.

(1) Solve the linearized inverse problem for the first order distribution function by using the Born approximation.

(2) Solve the scattering problem for the field in the object and at the observation points with the last reconstructed distribution function.

(3) Substitute the field in the object obtained in step (2) into the integrand in the integral equation and solve the inverse problem to recover the next order distribution function.

(4) Repeat step (2) and compare the field obtained by the reconstructed distribution function and the measured data, which in our case are the simulated fields for the exact distribution function at the observation points. If the difference is less than five percent of the scattered field, the iteration terminates. Otherwise, repeat the cycle until the solution is convergent.

As we see, in this method only the fields in the integrand are updated in each iteration step, while the Green's function remains unchanged throughout the iterations. We call this the Born iterative method because in each iteration, the kernel of the integral operator remains unchanged, only the field in the scatterer is updated.

2.2 Distorted Born iterative method

One immediate extension of the Born iterative method is to update the Green's function, the kernel of integration, as well as the field in the scatterer. We call this the distorted Born iterative method. The salient features of the new iterative procedure for solving the nonlinear integral Equation (4) is briefly sketched as follows:

(1) Solve the linearized inverse problem for the first-order object function by using the Born approximation where the homogeneous Green's function of Equation (3) with the unit refractive index is used.

(2) Solve the forward scattering problem for the field in the object and at the observation points. Next, calculate the point-source response in the object for every observation point with the last reconstructed object function. The second part of this step is to calculate the Green's function with the last reconstructed refractive index distribution as the background refraction index $n_b(x, y)$.

(3) Substitute the new Green's function and the field obtained in step (2) into the integrand, and subtract the scattered field at the receivers from the left-hand side of integral Equation (4). Then solve the above inverse problem for the corrections to the last reconstructed profile. Generate the new profile by adding the corrections to the previous profile.

(4) Repeat step (2) and compare the field scattered by the reconstructed distribution function and the measured data which, in our case, are the simulated fields for the exact distribution function at the receiver points. If the relative residual error (RRE) (see definition bellow) is less than a criterion which was given before or is larger than the RRE of the last reconstructed profile, the iteration terminates. Otherwise, repeat the cycle until the solution converges.

The definition of the relative residual error (RRE) in the jth iteration is

$$\text{RRE} = \frac{\sum_{i=1}^{M}(\phi^s(\boldsymbol{\rho}_i) - \phi^{s^{(j)}}(\boldsymbol{\rho}_i))}{\sum_{i=1}^{M} \phi^s(\boldsymbol{\rho}_i)} \tag{6}$$

where the summation is over the receiver points.

2.3 Direct scattering solution and the Green's function

To implement the above procedures on a computer, both the forward (Step (2) and Step (4)) and the inverse (Step (1) and Step (3)) problems need to be discretized. For consistency, we choose the same basis function $f_i(x, y)$ for both of them. For simplicity and comparison between two methods, the pulse basis function $f_i(x, y)$ has been used in discretizing both the direct and the inverse scattering problems. The point-matching method [16,17] is employed in solving the forward scattering problem and in calculating the Green's function with the refractive index at every iteration step.

The field and the refraction index of the object are represented as

$$\phi(x, y) = \sum_{i=1}^{N} \phi_i f_i(x, y), \tag{7}$$

$$\delta n^2(x, y) = \sum_{i=1}^{N} a_i f_i(x, y). \tag{8}$$

The equation of the forward scattering problem can be written as

$$\overline{\Gamma} \cdot \mathbf{a} = \mathbf{b}, \tag{9}$$

where $\mathbf{a}$ is an unknown column vector whose entries are the expansion coefficients of $\phi(x, y)$ expressed in the Equation (7), and $\mathbf{b}$ is the expansion coefficients of the incident field in the first term on the right-hand side of Equation (1). The elements of matrix $\overline{\Gamma}$ are

$$\Gamma_{ij} = \delta_{ij} - \iint_{S_i} G_0(\boldsymbol{\rho}_i - \boldsymbol{\rho}')k_0^2 \delta n^2(x',y')f_i(x',y')dx'dy'. \tag{10}$$

For the distorted born iterative method, the updated Green's function is obtained by solving the forward scattering problem. However, the linear system of equations for calculating the Green's function is the same as Equation (9) except that the entries in the column vector b is the incident field generated by a two-dimensional point source located at the corresponding receiver point.

At first glance, it seems that the distorted born iterative method needs much more computational effort than the born iterative method in each iteration step because of updating the Green's function. But it is not difficult to show that the matrix inversion involved in calculating the Green's function numerically in every iterative step is exactly that for solving the forward scattering problem at the same iterative step. Therefore, if the matrix has been inverted in solving the forward scattering problem, the same inversed matrix can be directly applied to calculating the Green's function so that the extra computational effort is proportional to MN^2 where M is the number of receivers and N is the number of basis functions used in the discretization of the problem. In most practical situations, the number of unknowns is much larger than that of receivers, i.e., $N \gg M$. Therefore, compared to the computational time spent on solving the forward problem which is proportional to N^3, the extra time spent on the calculation of the Green's function in each iterative step is not significant.

2.4 Inverse scattering solution

For a linearized version of the integral Equation (1), the linear system of equations for the inverse scattering problem is

$$\phi^{(l)^s}(x_j,y_j) = \sum_{i=1}^{N} a_i^{(l+1)} \iint_{S_i} G_l(\boldsymbol{\rho}_j - \boldsymbol{\rho}',n^l)k_0^2 f_i(x',y')\phi^{(l)}(x',y')dx'dy'$$

$$j = 1,...,M' \tag{11}$$

where S_i is the domain of the pulse function $f_i(x,y)$, and $\phi^{(l)}$ is the l-th forward scattering solution with the l-th refractive index distribution function. For $l = 0$, it is the incident field in the object, and Equation (8) has been used to express δn^2 in terms of the basis function $f_i(x,y)$. $G_l(\boldsymbol{\rho}_j - \boldsymbol{\rho}',n^l)$ is the Green's function with the refractive distribution n^l which has been obtained numerically.

Equation (11) for the inverse scattering problem can be written as the matrix equation

$$\mathbf{b} = \overline{\mathbf{K}} \cdot \mathbf{a} \tag{12}$$

where

$$\mathbf{b} = (\phi^s(\boldsymbol{\rho}_1), \phi^s(\boldsymbol{\rho}_2),...,\phi^s(\boldsymbol{\rho}_M))^T,$$
$$\mathbf{a} = (a_1,a_2,...,a_N)^T,$$

and $\overline{\mathbf{K}}$ is an $M \times N$ matrix whose elements are

$$K_{ji} = \iint_{S_i} k_0^2 G_l(\boldsymbol{\rho}_j - \boldsymbol{\rho}',n^l)\phi^{(l)}(x',y')f_i(x',y')dx'dy'$$

$$i = 1,...,N, j = 1,...,M$$

It is well known that Equation (12) of the inverse scattering problem is ill-posed [14,15,16,18]. In order to find an adequate solution of Equation (12), the regularization procedure [18,19] is employed to circumvent the instability of the problem. In the regularization procedure, instead of solving the matrix Equation (12) directly for a least-square solution, we solve an optimization problem which minimizes the cost function $C(\mathbf{a})$, defined as

$$C(\mathbf{a}) = \|\overline{\mathbf{K}} \cdot \mathbf{a} - \mathbf{b}\|^2 + \gamma\|\overline{\mathbf{H}} \cdot \mathbf{a}\|^2, \tag{13}$$

where γ is the regularization parameter, and $\mathbf{H}$ is the smoothing matrix. From Equation (13), one obtains the following matrix equation:

$$[\overline{\mathbf{K}}^\dagger \cdot \overline{\mathbf{K}} + \gamma\overline{\mathbf{H}}^\dagger \cdot \overline{\mathbf{H}}] \cdot \mathbf{a} = \overline{\mathbf{K}}^\dagger \cdot \mathbf{b}, \tag{14}$$

where $\overline{\mathbf{K}}^\dagger$ and $\overline{\mathbf{H}}^\dagger$ are the conjugate transpose of $\overline{\mathbf{K}}$ and $\overline{\mathbf{H}}$ respectively. In this paper, the zeroth-order regularization in which $\overline{\mathbf{H}}$ is the identity matrix of order N has been used to generate results given in the next section. Solution of Equation (14) is given by

$$\mathbf{a} = [\overline{\mathbf{K}}^\dagger \cdot \overline{\mathbf{K}} + \gamma\overline{\mathbf{H}}^\dagger \cdot \overline{\mathbf{H}}]^{-1} \cdot \overline{\mathbf{K}}^\dagger \cdot \mathbf{b}. \tag{15}$$

More attention is needed to choose an adequate regularization parameter γ in the solving above equation.

Instead of solving the matrix equation (14) for the image $\mathbf{a}$, the image $\mathbf{a}$ can also be obtained by directly using conjugate gradient method to optimize the cost function $C(\mathbf{a})$. With a linear smoothing term, conjugate gradient method theoretically gives the same result as Equation (15) in N iteration steps. Fortunately, for all the cases we had tried, the solution always converged to the solution accurately after twenty iterations using conjugate gradient method. In our case, N ranges from 200 to 400, this means a significant savings of computer time. Another potential advantage of using conjugate gradient method is that it could be used to solve for the image if a nonlinear smoothing term, e.g. maximum entropy method, is added to the cost function $C(\mathbf{a})$.

3. Numerical Results

Before presenting some numerical results, we define and review some terminologies which will be used in the following numerical analysis. First, we define the relative Mean Square Error (MSE) of the reconstructed refractive index profile as

$$\text{MSE} = \sqrt{\frac{\iint_S [(n^2)^{(i)}(\boldsymbol{\rho}) - n^2(\boldsymbol{\rho})]^2 dx\,dy}{\iint_S [n^2(\boldsymbol{\rho})]^2\,dx\,dy}} \tag{16}$$

where S is the scatterer's cross section, $(n^2)^{(i)}(\boldsymbol{\rho})$ is the reconstructed refractive index distribution in the i-th iteration, and $n^2(\boldsymbol{\rho})$ is the original refractive index distribution. In actual application, MSE is unknown since $n^2(\boldsymbol{\rho})$ is not known. Second, the Relative Residual Error of the j-th iterative reconstructed refractive index profile is defined in Equation (6).

3.1. Sin-like distribution with noise

Figure 2 shows the reconstruction of a sin-like refractive index distribution with

25 dB of the signal-to-noise ratio by using the distorted Born iterative method. The diameter of the object is 1λ. The peak value of the sin-like distribution is 1.8. Figure 2(a) is the reconstructed distribution of the first-order Born approximation. Figure 2(b) to Figure 2(e) are the iterative results from the second to fifth iterations. Figure 2(f) is the result after a filtering operation which we shall describe later. The algorithm terminates after five iterations because the relative residual error in the fifth iteration is larger than the RRE in the fourth iteration. The plot of the original refractive index distribution is given in Figure 3(a).

Figure 3 shows the reconstruction of the same problem given in the above by using Born iterative method. Figure 3(a) is the original distribution. Figure 3(b) is the reconstruction of the first-order Born approximation. Figure 3(c) to Figure 3(g) are the results from the second iteration to the sixth iteration. Figure 3(h) is the final result after 15 iterations. Here we have set the maximum number of steps as 15. Figure 3(i) is the filtered result.

When noise exists, unwanted artifacts are present in the reconstructed distribution. This image noise obscures the actual object features. For this reason, image filters are used to remove the noise so that the features can be identified [20]. The filter operates by passing a 5-cell window over the image, and replacing the center cell of the window with some function of all the cells in the window.

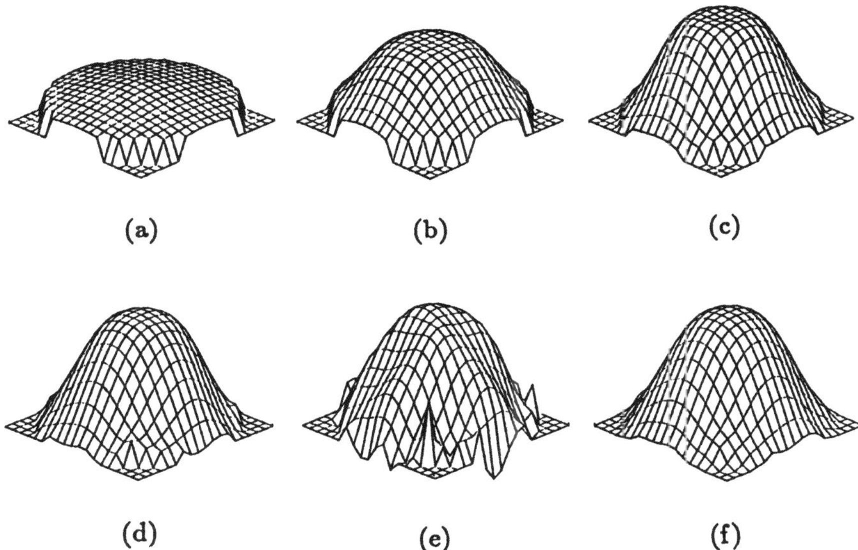

Figure 2. Reconstruction of a sin-like refractive index distribution with a 25db signal to noise ratio in the measurement field by using the distorted Born iterative method. The peak-value of the refractive index n^2 is 1.80. The diameter of the object is 1λ. (a) is the original distribution. (b) is the result of the first order approximation. (c) to (e) are the results from the second iteration to the fourth iteration. (f) is the final distribution after operating the filter function on (d).

3.2. Discussions

In the above, we give the examples using both the distorted Born iterative and the Born iterative methods to reconstruct the refractive index profiles. For the noisy cases, the Born iterative method is more robust than the distorted Born iterative method. The reason why the distorted Born iterative method is more susceptible to noise contamination is explained in the following paragraph.

The left-hand side of Equation (4) is unchanged in the iterative process for the Born iterative method which is the scattered field by the object with the free-space background. However, for the distorted Born iterative method the left-hand side of Equation (4) has to be subtracted from the scattered field of the last iterative reconstructed distribution with the background as the distribution before

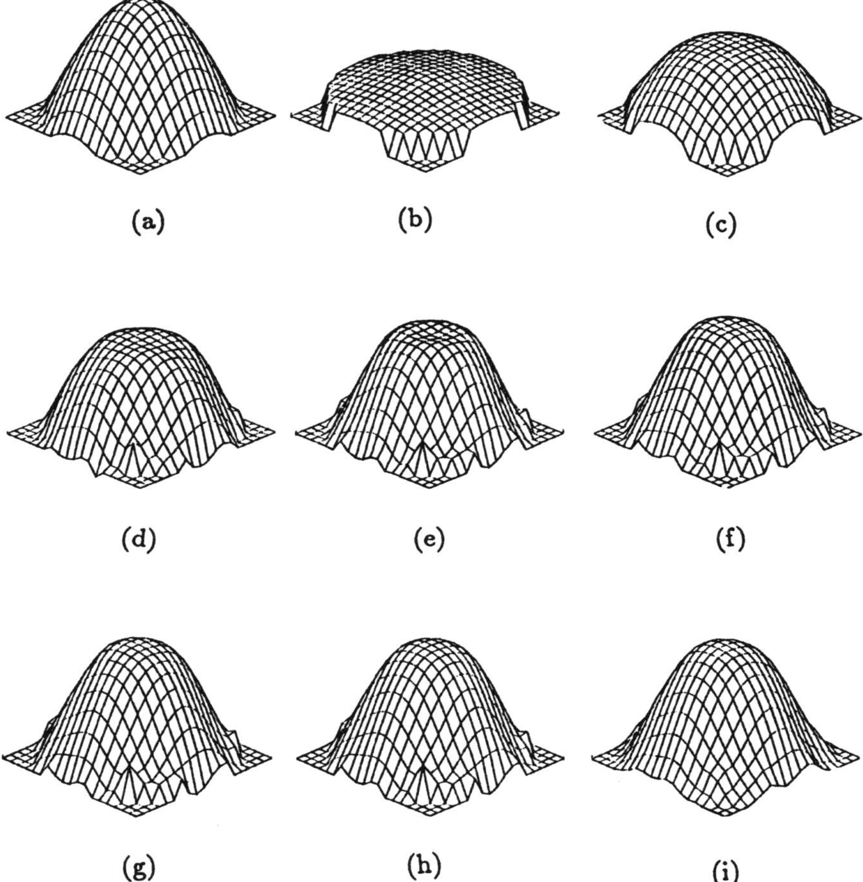

Figure 3. Reconstruction by using the Born iterative method for the same problem given in Figure 2. (a) is the original distribution. (b) is the result of the first order approximation. (c) to (g) are the results from the second iteration to the sixth iteration. (h) is the convergent distribution after the fifteen iterations. (i) the reconstructed distribution after operating the filter function on (h).

the last iteration. If the noise is added in the scattered field at the beginning (in our case, 25dB signal-to-noise ratio, which is equivalent to about 5.6% random noise, has been added in the examples given in Figure 10 and Figure 11), then after a few iterations, the noise will dominate the left-hand side of Equation (4). Consequently, the correction of the distribution after that step only contributes to the noise of the constructed distribution and no more information of the object could be derived.

4. Conclusion

Two iterative methods, the Born iterative and distorted Born iterative method, have been presented for solving the two-dimensional nonlinear inverse scattering problem. The comparison has been made between two methods. The results show that for the noiseless cases, the distorted Born iterative method is superior to the Born iterative method because of its faster convergent speed, while for the noisy cases, the Born iterative method is more robust than the distorted Born iterative method.

References

[1] Y. M. Wang and W. C. Chew, "An iterative solution of two-dimensional electromagnetic inverse scattering problem", *International Journal of Imaging Systems and Technology*, vol. 1, pp. 100–108, 1989.

[2] W. C. Chew and Y. M. Wang, "Reconstruction of two-dimensional permittivity using the distorted Born iterative method", Submitted to *IEEE Trans. Medical Imaging*.

[3] A. J. Devaney, "A computer simulation study of diffraction tomography," *IEEE Trans. Biomed. Eng.*, vol. BME-30, pp. 377–386, 1983.

[4] M. Azimi and A. C. Kak, "Distortion in diffraction tomography caused by multiple scattering," *IEEE Trans. Med. Imaging*, vol. MI-2, pp. 176–195, 1983.

[5] W. Tabbara, B. Duchêne, Ch. Pichot, D. Lesselier, L. Chommeloux and N. Joachimowicz, "Diffraction tomography: contribution to the analysis of applications in microwaves and ultrasonics," *Inverse Problem*, vol. 4, pp. 305–331, 1988.

[6] A. J. Devaney, "A filtered backpropagation algorithm for diffraction tomography," *Ultrasonic Imaging*, vol. 4, pp. 336– 360, 1982.

[7] J. B. Keller, "Accuracy and validity of the Born and Rytov approximations," *J. Opt. Soc. Am.*, vol. 59, pp. 1003–1004, 1969.

[8] M. Slaney, A. C. Kak, and L. E. Larsen, "Limitations of imaging with first-order diffraction tomography," *IEEE Trans. Microwave theory and Techniques*, vol. MTT-32, No. 8, pp. 860–874, 1984.

[9] D. K. Ghodgonkar, O. P. Gandhi, and M. J. Hagmann, "Estimation of complex permittivities of three-dimensional inhomogeneous biological bodies," *IEEE Trans. Microwave Theory Tech.*, vol. MTT-31, pp. 442–446, June 1983.

[10] M. M. Ney, A. M. Smith, S. S. Stuchly, "A solution of electromagnetic imaging using pseudoinverse transformation," *IEEE Trans. Medical Imaging*, vol. MI-3, No. 4, pp. 155–162, Dec. 1984.

[11] N. Bleistein and J. K. Cohen, "Nonuniqueness in the inverse source problem

in acoustics and electromagnetics," *J. Math. Phys.*, vol. 18, pp. 194–201, Feb., 1977.

[12] A. J. Devaney and G. C. Sherman, "Nonuniqueness in inverse source and scattering problems," *IEEE Trans. Ant. Propagation*, vol. 8, pp. 1034–1042, Sept., 1982.

[13] A. J. Devaney and E. Wolf, "Radiating and nonradiating classical current distributions and the fields they generate," *Phys. Rev. D*, vol. 8, pp. 1044–1047, Aug., 1973.

[14] S. J. Johnson and M. L. Tracy, "Inverse scattering solutions by a sinc basis, multiple source, moment method – part I: theory," *Ultrasonic Imaging*, vol. 5, pp. 361–375, 1983.

[15] S. J. Johnson and M. L. Tracy, "Inverse scattering solutions by a sinc basis, multiple source, moment method – part II: numerical evaluations," *Ultrasonic Imaging*, vol. 5, pp. 376–392, 1983.

[16] J. Richmond, "Scattering by a dielectric cylinder of arbitrary cross-sectional shape," *IEEE Trans. Antennas Propagation*, vol. AP-13, pp. 334–341, 1965.

[17] R. F. Harrington, *Field Computation by Moment Methods*. Malabar, Florida: Krieger Publishing, 1983.

[18] S. Twomey, *Introduction to the Mathematics of Inversion in Remote Sensing and Indirect Measurements*, New York: Elsevier Scientific, 1977.

[19] C. T. H. Baker, *The Numerical Treatment of Integral Equations*, Oxford: Clarenda 1977.

[20] M. Frank and C. A. Balanis, "Method for improving the stability of electromagnetic geophysical inversions," *IEEE Trans. Geoscience and Remote Sensing*, vol. 27, No. 3, pp. 339-343, May 1989.

DECISION-THEORETIC TREATMENT OF SUPERRESOLUTION BASED UPON

OVERSAMPLING AND FINITE SUPPORT*

J.M. Richardson and K.A. Marsh

Rockwell International Science Center

Thousand Oaks, CA 91360

ABSTRACT

It is well-known (at least in one dimension) that a function vanishing outside of a finite support domain has a Fourier transform that is analytic everywhere in frequency space. Consequently, if the transform is known exactly on a finite line segment in the complex frequency plane, it can by analytic continuation be determined everywhere and thus the original function can be recovered exactly. In this paper we consider realistic imaging problems in both one and two dimensions where the transform is imperfectly known from a set of noisy measurements at a discrete set of points in spatial frequency space. The true image in physical space is assumed to vanish identically outside of a specified support domain. The problem of estimating the image from the noisy measurements is approached within the well-established framework of linear Gaussian estimation theory.

INTRODUCTION

We have investigated the problem of the attainment of superresolution based upon _a priori_ information consisting of finite support of the "true" image in physical space. It is known that, in the absence of measurement noise, this _a priori_ information enables one to obtain a perfect reconstruction (i.e., infinite resolution) using analytic continuation in spatial frequency space. An additional requirement is that the observations represent an infinitely dense sampling in this space within a subdomain consistent with bandwidth limitations.

The problem of estimating the image from a non-dense set of noisy measurements is approached within the well-established framework of linear Gaussian estimation theory. We use a measurement model in which the possible images in the support domain are represented on a basis set of Fourier functions subject to cyclic boundary conditions corresponding to a reciprocal lattice of points in spatial frequency space called the

---

*This work was supported by the Independent Research and Development funds of Rockwell International.

representation grid. Measurements are assumed to be made at a given finite set of points (not necesssarily on the representation grid) in spatial frequency space. Measurement errors and the statistical ensemble of possible true images are assumed to be Gaussian with zero means and specified covariances. Using a least-mean-square error optimality criterion, the optimal estimate is given by a well-known closed-form expression, which can handle either overdetermined or underdetermined cases on a uniform basis. When the measurement set lies on the structure grid the solution reduces to a slightly modified pseudo-inverse with no superresolution. When the measurement set is more dense than the representation grid while still within a limited bandwidth domain superresolution is achieveable. Analytical expressions are derived for very high and very low signal-to-noise ratio (SNR) cases. Computational results using synthetic data showing the dependence upon SRN and oversampling will be presented.

ANALYTIC CONTINUATION

It is well known[1] that, if the bounded function $f(x)$ is zero everywhere outside of the interval $[-L/2, L/2]$ the Fourier transform of $f(x)$, i.e.,

$$F(k) = \int_{-\infty}^{\infty} dx\ f(x)\ \exp\ (-ikx) = \int_{-L/2}^{L/2} dx\ f(x)\ \exp\ (-ikx) \quad , \qquad (1)$$

is an analytic function of the spatial frequency k with no singularities except at ∞. Thus, the _precise_ knowledge of $F(k)$ on any line segment of nonvanishing length (e.g., an interval $[-b/2, b/2]$ on the real axis) can by analytic continuation determine the function $f(k)$ everywhere. Hence, the function $f(x)$ can be determined precisely in the interval $[-L/2, L/2]$, in fact everywhere, since by assumption $f(x) = 0$ everywhere outside of this interval. Certain segments of the imaging community believe that because of the finite bandwidth b, only a smoothed version of $f(x)$ can be determined, corresponding to a resolution length $2\pi/b$.

DECISION-THEORETIC FORMULATION

As stated above, we assume that the function $f(x)$ is zero everywhere outside of the interval $[-L/2, L/2]$ on the x-axis. Within this interval, the function can be represented by a linear combination of the Fourier functions $\exp\ (iKx)$ where K is a multiple of $2\pi/L$, i.e.,

$$f(x) = L^{-1} \sum_{K} F(K)\ \exp(iKx) \qquad (2)$$

where $F(K)$ is given by (1) with $k = K$ and where the sum on K includes all multiples of $2\pi/L$. Henceforth, the set of K values (i.e., $K = 2\pi m/L^{-1}$, $m = -\infty, \ldots, -1, 0, 1, \ldots, \infty$) is the explicit form of the representation grid. Now suppose we have a system that measured $F(k_n)$ with error at a preselected set of spatial frequencies $\{k_n\}$, where this set is invariant to reversal of sign of the k_n (i.e., if k_n is a member of the set, then $-k_n$ is also a member) and is not necessarily on the representation grid. We then consider the stochastic measurement model

$$g_n = F(k_n) + \nu_n, \quad n = 1, \ldots, N, \qquad (3)$$

where g_n is possible result of a noisy measurement, $F(k_n)$ is the possible result of a hypothetical noiseless measurement, and ν_n is the noise (i.e., measurement error). The function $f(x)$ is assumed to be a Gaussian random process with zero mean and with a covariance function $Ef(x)f(x')=\alpha\delta(x-x')$ or, equivalently, $EF(K)F(K')^*=\alpha L\delta_{KK'}$. The ν_n are assumed to be Gaussian random variables with zero mean and the covariance matrix $E\nu_n\nu_{n'}=\beta\delta_{nn'}$.

With the use of (1) and (2) we can write

$$\exp(ik_n x) = \sum_K \Gamma(k_n - K)\ \exp(iKx) \tag{4}$$

where

$$\Gamma(k) = L^{-1} \int_{-L/2}^{L/2} dx\ \exp(ikx) = \text{sinc}\ (Lk/2) \tag{5}$$

for an arbitrary value of k. An important property of $\Gamma(k)$ is that $\Gamma(K-K')=\delta_{KK'}$. The substitution of (4) into (3) gives

$$g_n = \sum_K \Gamma\ (k_n - K)\ F(K) + \nu_n \quad , \tag{6}$$

a more convenient form of the measurement model. Using either a least mean-square error or a best-score optimality criterion, we obtain the best estimate

$$\hat{F}(K) = \alpha L \sum_{nn'} \Gamma\ (k_n - K)\ C^{-1}_{g,nn'}\ g_{n'} \tag{7}$$

where $C^{-1}_{g,nn'}$ is the matrix inverse of

$$C_{g,nn'} = \sum_K \Gamma\ (k_n - K)\ \alpha L \Gamma\ (k_{n'} - K) + \beta\delta_{nn'} = \alpha L \Gamma\ (k_n - k_{n'}) + \beta\delta_{nn'} \tag{8}$$

It a set of measurements of the g_n are made with the actual measured values $\tilde{g}_n$, then the corresponding estimates of the $F(K)$ are given by (7) with the values $\tilde{g}_n$ substituted for the g_n.

The discussion so far has been limited to the one-dimensional (1D) case (i.e., x,k,K, etc. are all scalars). The two-dimensional (2D) case is easily obtained by replacing these quantities by 2D vectors (i.e., $x \to \underline{r} = \vec{e}_x x + \vec{e}_y y$, $k \to \underline{k} = \vec{e}_x k_x + \vec{e}_y k_y$, etc., where $\vec{e}_x$ and $\vec{e}_y$ are unit vectors in the x and y directions in a Cartesian coordinate system). An expression like kx is replaced by $\underline{k}\cdot\underline{r}$ and the differential dx in the integral (1) is replaced by the area element $d\underline{r}$ (= $dxdy$). The representation grid is now defined by the set of spatial frequencies given by $\underline{K}=2\pi\ (\vec{e}_x m_x L_x^{-1}+\vec{e}_y m_y L_y^{-1})$ where m_x and m_y take all integer values (positive, negative and zero) and where L_x and L_y are the length and width, respectively, of the rectangular support domain.

SPECIAL CASES

It is of special interest to note that if all of the measurements are made at k_n's that forms a subset of the representation grid (i.e., $k_n=K_n$) then the best estimates are $F(K)=0$, $K \neq K_n$, $n=1, \ldots, N$, and $\hat{F}(K_n)=\eta^2 g_n/(1+\eta^2)$, the simplest situation in one-dimensional imaging. The parameter η is the SNR given by $\eta^2=\alpha L/\beta$. The estimates for the K's corresponding to spatial frequencies $k_n=K_n$, where the measurements are made, are nontrivial and approach the g_n as the SNR η approaches ∞. The esti-

mates F(K) at other spatial frequencies are completely dominated by the
a priori statistics in the model and thus under the present assumptions
zero values are obtained.

When the k_n do not all lie on the representation grid, the situation
of particular interest here, we obtain the following results for very
small and very large SNRs, i.e.,

$$\hat{F}(K) = \eta^2 \sum_n \Gamma(k_n - K) g_n \quad , \quad \eta \ll 1, \tag{9}$$

$$\hat{F}(K) = \sum_{nn'} \Gamma(k_n - K) \Gamma(k_n - k_{n'})^{-1} g_{n'}, \eta \gg 1 \quad , \tag{10}$$

where $\Gamma(k_n - k_{n'})^{-1}$ is the matrix inverse of $\Gamma(k_n - k_{n'})$. The inverse
exists if the functions $\exp(ik_n x)$ are linearly independent in the interval $[-L/2, L/2]$.

To provide deeper insight into the nature of the two limiting cases
(9) and (10), it is desirable to consider noiseless test data, i.e.,
$g_n \to \tilde{g}_n = \int dx\, \tilde{f}(x) \exp(-ik_n x)$ where $\tilde{f}(x)$ is a given sample of the random
process $f(x)$. It is expedient to transform $F(K)$ to the original x-repre-
sentation. We obtain (9) and (10) in the forms

$$\hat{f}(x) = \eta^2 \int dx'\, L^{-1} \sum_n \exp(ik_n(x-x')) \tilde{f}(x') \quad , \tag{11}$$

$$\hat{f}(x) = \int dx'\, L^{-1} \sum_{nn'} \exp(ik_{n'} x) \Gamma(k_n - k_{n'})^{-1} \exp(-ik_n x') \tilde{f}(x') \quad . \tag{12}$$

The small SNR case given by (11) represents (aside from the factor η^2)
the process of finding the coefficient of <u>each individual</u> Fourier function $\exp(ik_n x)$ that provides the best approximation to $\tilde{f}(x)$ and then summing the results. The large SNR case given by (12) represents the process of finding the linear combination of the <u>total</u> set of Fourier functions $\exp(ik_n x)$ that provides the best approximation to $f(x)$.

A PARADOX

It is perhaps intuitive to suppose that a good approximation to a
function $f(x)$ with a large spatial-frequency bandwidth could not be obtained by taking a linear combination of Fourier functions $\exp(ik_n x)$ of
small spatial-frequency bandwidth. In the case where the k_n lie on the
representation grid (i.e., $k_n = K_n$) this intuition is correct. But when
this is not the case, the intuition is wrong. It can readily be shown
that the functions $\exp(ik_n x)$ with an infinite set of k_n values in the
interval $[-\varepsilon/2, \varepsilon/2]$ form a complete set of basic functions for the
representation of any function $f(x)$ in the interval $[-L/2, L/2]$ even with
ε chosen arbitrarily small. However, it is readily seen that the approximation procedure based upon a finite set of such functions becomes
increasingly noise vulnerable as ε approaches 0.

COMPUTATIONAL INVESTIGATIONS

We have carried out computational estimates of $f(\underline{r})$ based upon
noiseless synthetic data for the two-dimensional case with the objective
of understanding the dependence of the estimates upon SNR (assumed in the

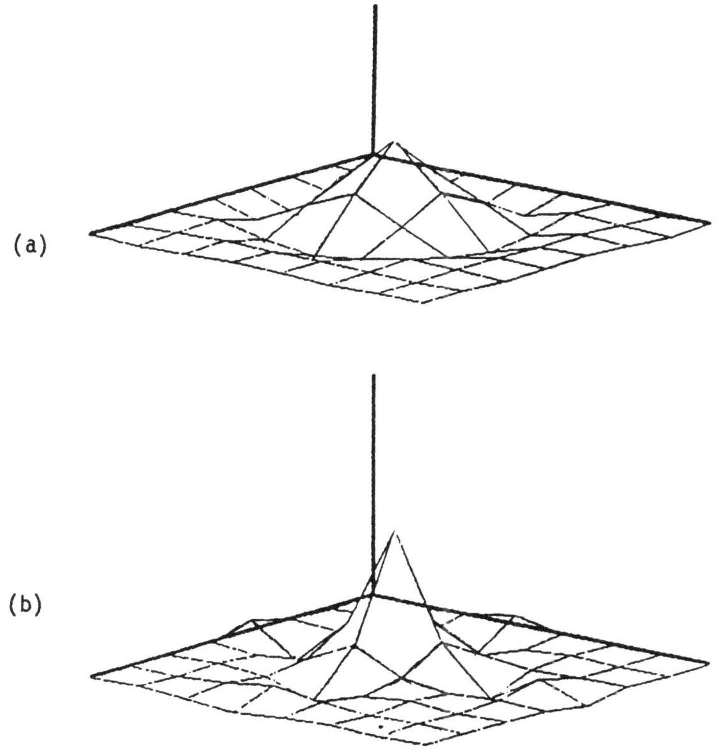

Fig. 1 (a) Reconstruction of a δ-function with ro oversampling.
(b) Reconstruction of a δ-function with an oversampling of
16 in each dimension.

algorithm) and the degree of oversampling. In Fig. 1a, we show a δ-function reconstruction with no oversampling and an SNR = 30. In Fig. 1b, we
show the same process with an oversampling of 16 in each dimension and an
SNR = 100. In the second case, the height of the peak has increased by
approximately 50%. Another case (not shown) with the same oversampling
but a significantly large SNR gives imperceptibly different results;
therefore, it is apparent that we have obtained essentially the maximum
performance with respect to SNR for the present oversampling factor.

CONCLUSIONS

It is clear that the ideal analytic continuation procedures for the
reconstruction of a function from bandwidth-limited data under the assumption of finite support is highly vulnerable to both noise and incom-
pleteness. This situation motivated a probabilistic approach based upon
decision theory. For the formulation of the problem presented here, it
was sufficient to use ordinary least-mean-square statistical estimation
theory leading to an optimal estimator that is linear. We found that
with synthetic data, corresponding to a finite bandwidth-limited set of
points in spatial-frequency space, the estimation error approaches a non-
zero limit as the SNR approaches infinity, but this asymptotic value is
approached in a sluggish manner. If the above set of points is on the

representation grid there is by definition no superresolution. One can
achieve perceptible superresolution when a sufficient number of points
are off the representation grid, a situation that is measured by the
oversampling factor in certain simple cases. In the limit where both the
SNR and oversampling factor approach infinity, one should obtain a result
equivalent to the analytic continuation procedure.

REFERENCES

1. R.P. Boas, "Entire Functions," Academic Press, (1954).

A GENERALIZED FRAMEWORK FOR INCOHERENT PULSE ECHO PROCESSING

AND IMAGING: THE RANDOM PHASE TRANSDUCER APPROACH

M. Fink*, R. Mallart*, P. Laugier**, S. Abouelkaram* **

*Groupe de Physique des Solides de l'E.N.S. Univ. Paris 7
2, place Jussieu 75251 PARIS Cedex 05, FRANCE
**Laboratoire de biophysique. U.A. CNRS 593. CHU COCHIN
24, rue du faubourg St-Jacques 75674 PARIS Cedex 14, FRANCE

INTRODUCTION

Spectral estimation and envelope detection are widely used tools in tissue characterization and ultrasound imaging. They are used in backscattered field and attenuation measurements. However, in multiscattering media such as biological tissue, interferences between echoes from randomly situated reflectors within the resolution cell lead to random errors in spectral estimation and to speckle noise in envelope detection. This speckle noise is linked to the spatially coherent behavior of piezoelectric transducers [1,2,3].

The problem of speckle reduction can be approached by the use of incoherent (phase insensitive) transducers [4,5]. Such transducers can be made in several ways. Power-sensitive receivers such as the acousto-electric transducer made of CdS exist. However, these transducers lack of sensibility. A similar approach consists in sampling the field with small-aperture piezoelectric elements laid out in a two-dimensional array. Square or envelope-detected echoes from individual elements are averaged. Averaging of the power spectrum may also be performed.

One limitation of these totally incoherent processing techniques is the loss of lateral resolution they yield. In order to reduce the speckle without loosing too much in resolution, we can use partially incoherent processing. This may be obtained by dividing the aperture into a number of relatively large coherent subelements whose individual directivity is rather good. Spatial compounding [6,7] is one way of achieving partial coherence. It requires the 1D or 2D scanning of a single transducer across the field.

In this paper we study the optimal choice of partially coherent processing in terms of the compromise between speckle reduction and directivity. We also present a new technique that allows the control of the degree of spatial coherence from total coherence to total incoherence. For this method, there is no need of dividing the aperture into subelements or of scanning the field.

We present applications of this technique to speckle reduction, detection of specular reflectors, attenuation measurements and imaging. We also show that one advantage of this technique is that it is not limited by diffraction effects. There is no need for diffraction correction in order to obtain unbiased attenuation estimates.

INCOHERENT PROCESSING OF PULSE ECHO SIGNALS

A simple approach to incoherent processing of pulse echo signals

When an ultrasonic beam illuminates a scattering medium, a pressure field is backscattered towards the receiving transducer. In the special (and simple) case of uniform plane wave illumination, each point on the receiving aperture of the transducer senses a pressure signal. This elementary pressure signal is random. In classical (coherent) pulse echo processing, the echo results from the coherent summation of the elementary pressure signals over the receiving aperture. Since the summation is coherent, constructive and destructive interference between random pressure signals appear and give rise to a noise that can be interpreted in terms of classical speckle theory [1,8,9]. In an analogy to the optical modality where the detectors are sensitive to the average light intensity over the response time of the detectors, and to get physical understanding of incoherent pulse echo processing, we now assume that a *totally incoherent* transducer is sensitive to the pressure intensity. Hence, the echo e_{inc} results from the summation of the intensity-detected elementary pressure signals.

$$e_{inc} = \int d\vec{X}\, O_r(\vec{X})A^2(\vec{X}, t) \tag{1}$$

where O_r is the receiving aperture function and $A(\vec{X}, t)$ is the envelope of the field.

SNR in totally incoherent processing

In classical (coherent) processing, the signal-to-noise ratio (SNR) is rather low: 1.91 or 1.0 respectively for the envelope and the intensity detected signals [1]. In the rest of this paper, we only consider the SNR of the intensity detected signals.

In this section we derive a general expression for the SNR obtainable in totally incoherent pulse echo processing. The SNR is defined as

$$SNR = \frac{<e>}{\left(<|e|^2> - <e>^2\right)^{\frac{1}{2}}}, \tag{2}$$

where $< >$ refers to the average over an ensemble of realization of a random variable and where e is the pulse echo signal. In a monochromatic approach, the backscattered pressure field can be written as

$$p(\vec{X}, t) = P(\vec{X}) \cos(2\pi f_0 t) \tag{3}$$

$P(\vec{X})$ is thus a complex random process whose statistics depend on those of the scattering medium. For such a monochromatic field, the envelope $A(\vec{X}, t)$ is the modulus of $P(X)$. Moreover, when the resolution cell contains a large number of scatterers, the pressure field is zero mean and has Gaussian statistics [1]. We have

$$<e_{inc}> = \int d\vec{X}\, O_r(\vec{X}) <|P(\vec{X})|^2> = <|P|^2> A_r = R_p(0)A_r \tag{4}$$

where A_r is the receiving aperture area, where O_r is the receiving aperture function and where Rp is the backscattered field autocorrelation function defined in the appendix. We also have

$$< |e_{inc}|^2 > = \int \int d\vec{X}_1 d\vec{X}_2 \ O_r(\vec{X}_1)O_r^*(\vec{X}_2) < |P(\vec{X}_1)|^2 |P(\vec{X}_2)|^2 > . \tag{5}$$

Now, since $P(\vec{X})$ is Gaussian, we have

$$< |P(\vec{X}_1)|^2 |P(\vec{X}_2)|^2 > = < P(\vec{X}_1)P^*(\vec{X}_1) >^2 + | < P(\vec{X}_1)P^*(\vec{X}_2) > |^2$$
$$= < |P(\vec{X})|^2 > + |R_P(\vec{X}_1, \vec{X}_2)|^2 \tag{6}$$

thus

$$< |e_{inc}|^2 > = < e_{inc} >^2 + \int \int d\vec{X}_1 d\vec{X}_2 \ O_r(\vec{X}_1)O_r^*(\vec{X}_2)|R_P(\vec{X}_2, \vec{X}_1)|^2$$
$$= < e_{inc} >^2 + \int d\vec{X} \ R_{O_r}(\vec{X})|R_P(\vec{X})|^2 \tag{7}$$

Therefore we have a simple expression for the SNR in incoherent processing of pulse echo signals:

$$SNR \ totally \atop incoherent = \frac{R_p(0)A_r}{\left(\int d\vec{X} \ R_{O_r}(\vec{X})|R_P(\vec{X})|^2\right)^{\frac{1}{2}}} . \tag{8}$$

The above formula for the SNR depends only on the receiving aperture function and on the autocorrelation function of the backscattered pressure field. It can be interpreted as follows. The SNR is low when the denominator is large. This happens when the backscattered field autocorrelation function is wide. Heuristically, this means that in order to increase the SNR, we have to receive a large number of uncorrelated informations. Over a surface of the order of the backscattered field autocorrelation, we can receive only one uncorrelated information. Therefore, to receive a lot of informations, we need a receiving surface whose area is many times larger than the field autocorrelation width.

The backscattered field autocorrelation

The previous section shows the importance of the autocorrelation function of the backscattered field. We now compute it in the special case of a continuous wave illumination. When the scattering medium is totally random, one can use the Cittert-Zernike theorem, a well-known theorem in the theory of coherent optics [10]. It says that the autocorrelation of the scattered field is the Fourier transform of the incident intensity distribution at the depth of interest (see appendix). In the case where the transmitting transducer is focused, and has an aperture function O_t, for $z = F$ (the focal length), the incident pressure field is equal to the Fourier transform of O_t taken at spatial frequency $\vec{x}/\lambda F$. Therefore, taking into account Fourier transform properties, the backscattered field autocorrelation is simply proportional to the aperture autocorrelation function:

$$R_P(\vec{x}) \propto R_{O_t}(\vec{x}). \tag{9}$$

It is worth noting that the size of the scattered field autocorrelation is of the same order than the transmitting aperture autocorrelation.

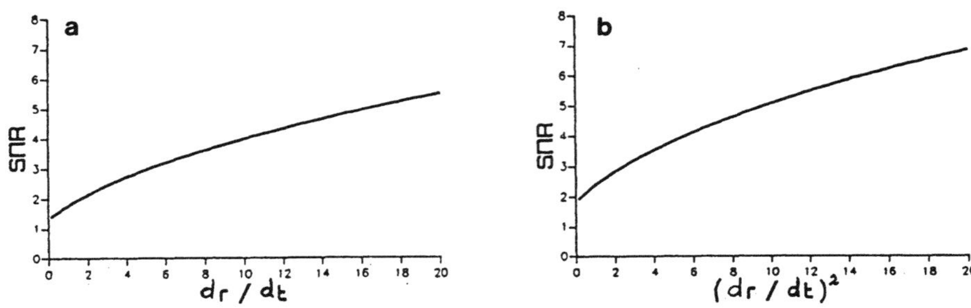

Figure 1.   SNR improvement with 1D (a) and 2D (b) apertures

Information grains

Underlying these findings is the concept of information grains. The size of an information grain is of the same order than the coherence length of the scattered field (i.e., the size of the autocorrelation of the backscatterd field). The number of received information grains is proportional to the square of the SNR improvement. This theory is independent of the method used to achieve incoherence, and sets the limits for the SNR improvement.

Let's consider for example the case of a focused transducer. If the transmitting and receiving apertures have the same size, we receive a very limited amount of uncorrelated data, and only a small SNR improvement is expected. In general, the amount of uncorrelated data depends on the relative sizes of the transmitting and receiving apertures.

In the case of 1D transducers (linear arrays), we have:

$$d_r > d_t \qquad SNR_{1D} = \sqrt{\frac{6}{Y(4-Y)}} \quad ; \quad Y = \frac{d_t}{d_r} \, , \tag{10}$$

where d_r and d_t are the aperture diameter in receive and in transmit mode respectively. When the same transducer is used in both receive and transmit modes, the aperture diameters are the same and the SNR _improvement_ is $\sqrt{2}$. For 2D transducers, we have

$$SNR_{2D} = \frac{X \ \pi^{3/2}}{\sqrt{2 \int rdr \ f^2(r) f(\frac{r}{X})}} \quad ; \quad X = \frac{d_r}{d_t} \, , \tag{11}$$

where the function f is defined by

$$f(r) = \frac{1}{2}\left[\cos^{-1}r - r\sqrt{1-r^2} \right] \qquad r \le 1 \\ = 0 \qquad \text{otherwise} \tag{12}$$

When the same transducer is used in both transmit and receive modes, the SNR _improvement_ is 1.94. Figure 1 shows the SNR _improvement_ for the 1D and 2D focused transducers.

This theory shows that the use of totally incoherent transducers is not a very efficient way of reducing speckle. Since only a few uncorrelated data can be received, it is better to divide the receiving transducer into coherent subelements whose sizes are well matched to the grain size. In that way, the maximal SNR improvement can still be achieved and the loss in lateral resolution is reduced. This is what we call partially incoherent processing. In the next section, we present a technique that allows a control of the degree of coherence from total incoherence to total coherence.

124

THE RANDOM PHASE TRANSDUCER APPROACH TO INCOHERENT PROCESSING

Principle

The random phase transducer approach to incoherent processing of pulse echo signals is illustrated in Figure 2. It consists in using a coherent transducer filling the entire receiving aperture. The degree of spatial coherence is controlled by moving a random phase screen (RPS) in front of the receiving transducer. Envelope detection or spectral analysis are performed on the pulse echo lines recorded for each RPS position. Averaging of these data is then performed over a set of RPS positions.

The RPS generates random phase shifts that depend on the position of the RPS as well as on the coordinates on the receiving aperture. These random phase jitters are superimposed on the natural phase relationships between different points of the backscattered pressure field. The final effect is to destroy the phase information contained in backscattered pressure field.

The random phase screen

The RPS is a phase shift generating object. It can be viewed as a transparency with transmittance function $t(X)$. Practically, it can be built as a plane object whose thickness varies with space and whose refraction index is different from the surrounding medium. The statistical properties of the RPS are characterized by the transmittance autocorrelation function, $R_t(X)$. In a model in which the screen thickness is a Gaussian random variable, the transmittance autocorrelation function can be defined by two parameters that determine the degree of spatial coherence. The first one relates to the magnitude of the phase shifts: σ_τ, the standard deviation of the phase shifts. It is clear that if the RPS introduces only small phase shifts (small σ_τ), a high degree of spatial coherence will remain even between distant points. The second parameter relates to the spatial coherence of the RPS itself: σ_L. Points on the RPS separated by more than σ_L will yield very different phase shifts provided that σ_τ is large enough.

Equivalence theorem

In the special case where the RPS transmittance autocorrelation function is proportional to a Dirac function, we show in [11] that the random phase transducer is equivalent to a totally incoherent transducer (i.e., sensitive to the pressure intensity). This is an important result because it shows that the RPS technique is an incoherent processing technique. Note, however, that it is more general because partial incoherence may also be obtained with this technique as will be shown later.

Equivalent coherent aperture function for partially coherent transducers

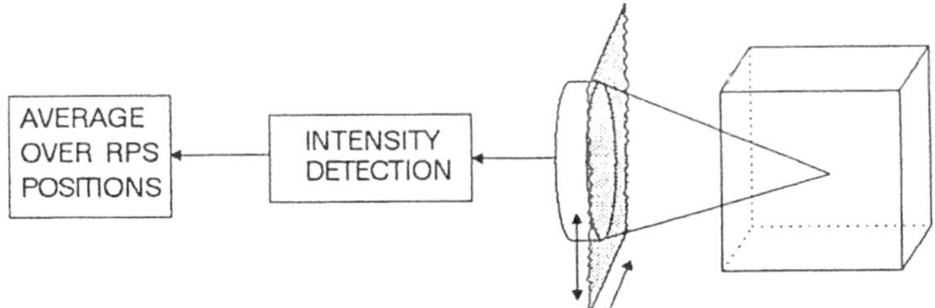

Figure 2.   The random phase screen approach: A random phase screen is moved across the beam of a coherent transducer. Averages over the screen positions are taken after envelope detection.

Without any loss in generality, the influence of the RPS on the directivity can be studied for the focused transducer. The derivations are extensively described in [11]. We show there that on the average the random phase transducer behaves like a focused transducer whose aperture function is an equivalent coherent aperture O_{eq} defined as

$$O_{eq}(\vec{X}) = R_t(\vec{X})R_0(\vec{X}). \tag{13}$$

An important property of the random phase transducer derives directly from the concept of equivalent aperture function. It relates to diffraction effects of transducers. It is well known [12] that the finite size of transducers induce diffraction effects that are responsible not only for their focussing properties but also for bias errors that are to be corrected when quantitative parameters are to be extracted from pulse echo measurements (e.g., attenuation measurements). The diffraction effects are strongly linked to the transducer size and disappear when the transducer is point-like. Now, for a random phase transducer, we have seen that the effective size is given by the equivalent aperture function and therefore, if the RPS autocorrelation function is quite narrow, we expect the diffraction effects to disappear, or at least to decrease (if the equivalent aperture function is not infinitely narrow). This will be shown on experimental results.

<u>Signal-to-noise ratio for partially coherent transducers</u>

<u>RPS in receive mode only</u>: Theoretical formulae for the SNR are derived in [11] when the RPS acts in receive mode only. These formulae depend only on the RPS autocorrelation function and on the geometry of the transducer. For a RPS defined by σ_τ and σ_L , the SNR improvement is

$$SNR = \frac{\int d\vec{X}\ R_t(\vec{X})R_{0r}(\vec{X})R_{0t}(\vec{X})}{\sqrt{\int\int\int\int \begin{array}{c} d\vec{X}_1 d\vec{X}_2 d\vec{X}_3 d\vec{X}_4\ O(\vec{X}_1)O^*(\vec{X}_2)O^*(\vec{X}_3)O(\vec{X}_4) \\ R_t(\vec{X}_2 - \vec{X}_1)R_t(\vec{X}_4 - \vec{X}_3)R_{0t}(\vec{X}_1 - \vec{X}_3)R_{0t}(\vec{X}_2 - \vec{X}_4) \end{array}}} . \tag{14}$$

Note that the SNR improvement depends only on the characteristics of the RPS and of the transmitting and receiving apertures. In general, we cannot find a simpler analytic expression for the SNR. However, it can be computed with the use of computers.

<u>RPS in transmit and receive mode</u>: The case where the RPS is present in both the transmit and receive modes is very difficult to analyse theoretically because it involves 8th order statistics of the RPS

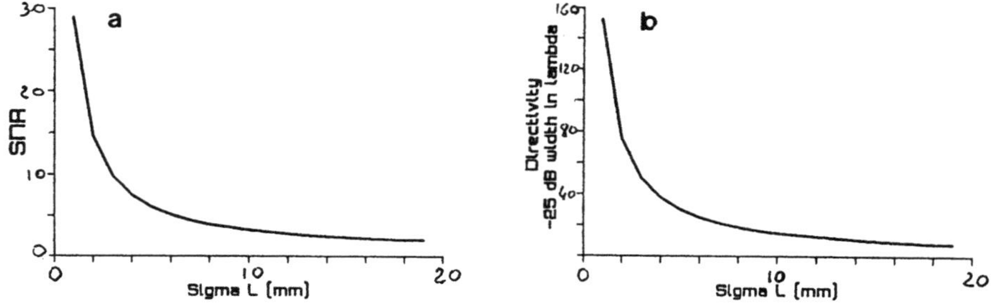

Figure 3. Compromise between directivity and SNR improvement: When the RPS is present in both transmit and receive modes, the SNR (a) and the directivity (b) are functions of σ_τ and σ_L: Here they are investigated for σ_τ of 0.4T. Note that very high SNR can be reached at the cost of a poor directivity.

transmittance. However, several results derived for the simpler configuration can be generalized.

We can consider that the main effect of the RPS in transmit mode is to modify the ultrasound beam. In average, the directivity cf the random phase transducer the one of a transducer with aperture furction O_{eq} defined earlier as

$$O_{eq} = R_t \cdot R_{0t} \tag{15}$$

To estimate the achievable SNR improvement, we just need to consider that the transmitting aperture is smaller than the true one. Its size is given by O_{eq}. Figure 3 shows for σ_τ of 0.4T the SNR we may achieve with different σ_L. This figure clearly shows the compromise between SNR and directivity.

EXPERIMENTAL RESULTS

In this section, we discuss some useful properties of random phase transducers for tissue characterization and echo imaging. These include (a) partial or total elimination of diffraction effects and (b) reduction of speckle noise.

Experiments are performed in pulse echo mode in a water tank. Parameters such as the location of the RPS in the beam (axial distance from the transducer), the shift of the RPS between acquisition of rf signals must be controlled with precision. The phantom target is a fine grain foam phantom. It has been verified that it follows Rayleigh statistics.

<u>Directivity</u>

Experimental procedures for directivity pattern estimation of random phase transducers in pulse echo mode have been discussed elsewhere [13]. Averaged directivity patterns are measured at the focal distance of a transducer (focus of 60 mm diameter of 13 mm and center frequency of 5 MHz) with two different RPS (RPS1 and RPS2). The parameters $\sigma_\tau(\mu s)$ and $\sigma_L(mm)$ were estimated for both screens: $\sigma_\tau,1 = 0.03$; $\sigma_L,1 = 0.50$; $\sigma_\tau,2 = 0.10$; $\sigma_L,2 = 1.25$ therefore, the RPS2 surface is rougher than that of the RPS1.

In these measurements, the RPS are located 40 mm from the transducer. Averaged directivity patterns are depicted in Figure 4, together with the directivity pattern of the focused transducer without RPS. The directivity

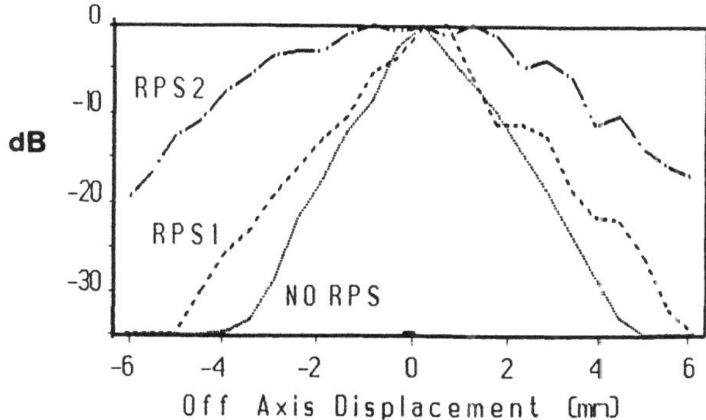

Figure 4. Directivity pattern: of the focused transducer at the focal distance without RPS, and averaged directivity patterns of the focused transducer at the focal distance with RPS1 and RPS2.

pattern is only slightly changed with RPS1, while a loss of directivity is clearly visible with RPS2. This is well explained by the concept of equivalent aperture function. For RPS1 the equivalent aperture function is larger than for RPS2, and thus the directivity loss is larger with RPS2.

SNR improvement

SNR improvements are estimated are estimated as follows: (a) acquisition of a 40 μs duration rf line for each position of the RPS; (b) envelope detection and attenuation correction of the envelopes; and (c) estimation of the SNR after averaging N envelopes corresponding to N shifts of the RPS. The RPS is shifted in steps of 2 mm; this step size was chosen because its gives the minimum correlation between echoes [13]. The averaged SNR is plotted versus the number of averaged lines. SNR improvement is calculated by dividing the measured SNR by 1.91 which is the expected SNR on a single line.

Figure 5 (a) represents the SNR improvement with the focused transducer for both RPSs versus the number of averaged A-lines. SNR improvement is better with RPS2 (improvement of 5) than with RPS1 (3 only). One immediately understands from Figure 4 that a better SNR improvement is reached at the price of loss of spatial resolution. These results can be related to the theory of information grains. For a focused transducer, the SNR improvement is limited to 1.94 when no loss of resolution in transmit mode is allowed. The SNR improvement can be better than that with a focused transducer only if the transmitted beam through the RPS is spread so widely that the grain size becomes smaller and that more grains are captured by the receiving aperture.

The effect of SNR improvement in the spectral domain is shown in Figure 5 (b-d), which compares the SNR improvement with spatial compounding and with the RPS technique. SNR improvement obtained with a RPS is equivalent to what is obtained by scanning the phantom. Therefore, the RPS could be used successfully in attenuation measurements; there is no need of scanning the transducer though the phantom if a RPS is moved across the transducer beam.

Diffraction effects

Time-varying diffraction effects alter the spectral content of the pulse echo signal and diffraction correction is required before attenuation estimation [12]. These diffraction effects can be measured following the method described in [12].

Figure 6 (a) shows the diffraction filter of both the coherent and incoherent transducer in a time-frequency representation. As might be expected, the maximum of energy due to the focusing effect of the coherent transducer disappears in the incoherent processing experiment. (The energy decreases with depth due to the $1/r$ beam spreading in the Fresnel approximation.) Further information on diffraction effects of incoherent transducers comes from the study of the centroid frequency versus depth. Figure 6 (b) shows the centroid frequencies corresponding to the calibration power spectra of Figure 6 (a). The classical variation of the centroid frequency with depth is totally removed when RPS2 is used and only partially removed when RPS1 is used. Note that the curves for the centroid variation with depth obtained with a RPS are below the one obtained without RPS. this is due to the frequency dependent attenuation of the RPS. Total removal of the diffraction effect means that the transducer is completely spatially incoherent, while partial removal of the diffraction effects means that some coherence is still present in the transmitted beam. The spatial coherence of the transmitted ultrasound beam can be chosen by the choice of the RPS.

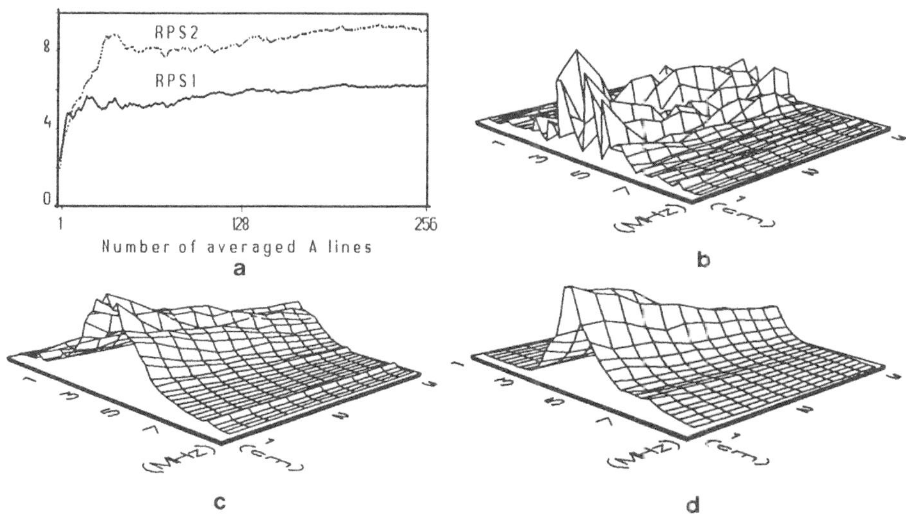

Figure 5.   SNR with the focused transducer: for both RPS1 and RPS2 versus the number of averages (a). (b) short time Fourier transform of a single A line, (c) averaged short time Fourier transforms of A lines obtained by moving RPS2, (d) averaged short time Fourier transforms of A lines obtained by scanning the phantom.

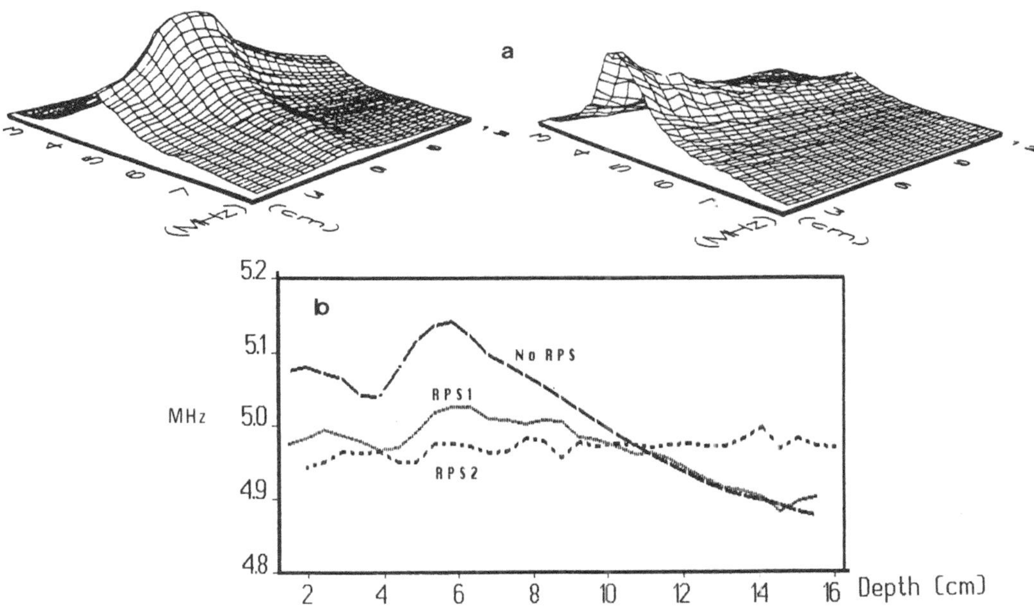

Figure 6.   Diffraction calibration: coherent focused transducer and incoherent transducer (RPS2) (a). Centroid frequencies corresponding to the calibration spectra for the focused transducer, for the same transducer with. RPS1 and with RPS2 (b).

Total removal of the diffraction effects is a very important result since in that case no more correction is required for attenuation estimation.

<u>Imaging</u>

All the previous results are summarized in Figure 7. It represents the pulse echo image of a phantom. A classical image obtained with a focused transducer is compared there to one obtained with the same transducer and a RPS. In the second image, each line results from the averaging of 32 envelope-detected A-lines corresponding to 32 RPS shifts. For this experiment, a specila probe was designed. The RPS is used in a rotating configuration that allows the probe to be compact. Speckle noise is reduced in the incoherent image by a factor of about 5. However this improvement is reached at the price of loss of lateral resolution.

CONCLUSIONS

A complete theory of incoherent processing of pulse echo signals has been developed. The information grain theory is well adapted to the prediction of the SNR improvement achieved with random phase transducers. The spatial coherence of a transducer can be controlled by the choice of the RPS. From this result, new possibilities are derived for the use of RPS in tissue characterization and in pulse echo imaging: (a) the removal of diffraction effects and (b) speckle noise reduction. Totally incoherent transducers completely remove diffraction effects. They might be used in applications where high SNR improvement are needed such as in the detection of very low contrast coherent scatterers or in attenuation estimation (in that case correction for diffraction effects is no longer required). The trade-off with spatial coherence is the degradation of the lateral resolution. However, another possibility of random transducers is to achieve partial incoherence. In the case of a partially incoherent transducer, we show that non-negligible SNR improvement may be obtained with a very limited loss of lateral resolution. Such transducers may be useful in pulse echo imaging.

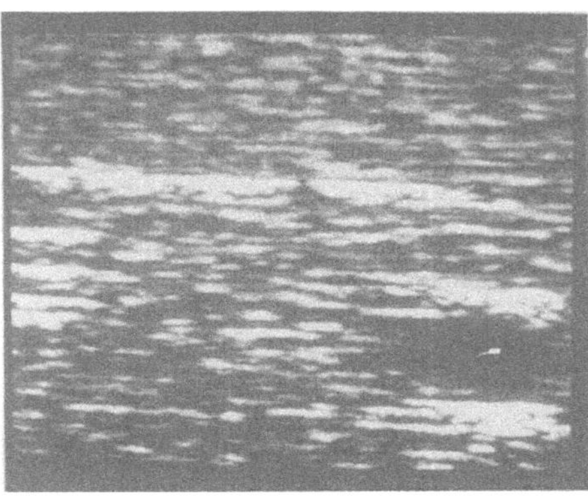

a

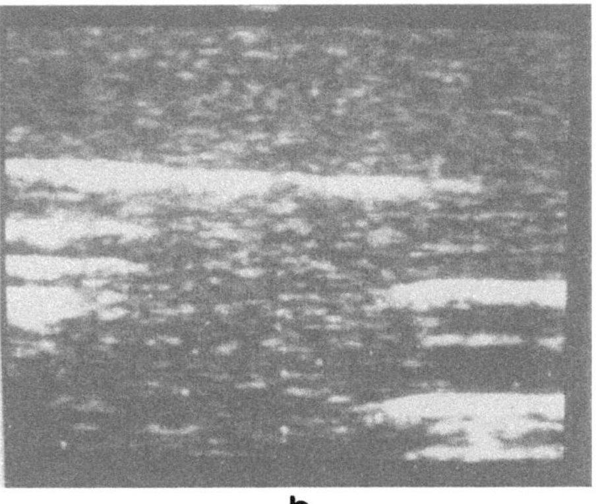

b

Figure 7. Echographic images of a RMI phantom: (a) Classical echographic image obtained with a focused transducer. (b) Echographic image obtained with an incoherent tranducer. Each line is the average of 32 envelope detected A lines obtained for 32 different positions of the RPS.

REFERENCES

[1]   Burckhart, C. B., Speckle in ultrasound B-mode scans, *IEEE Trans. Sonics Ultrasonics 25*, 1-6 (1978).

[2]   Abott, J. C. and Thurstone, F. L., Acoustic speckle: theory and experimental analysis, *Ultrasonic Imaging 1*, 303-324 (1979).

[3]   Wagner, R. F., Smith, S. W., Sandrik, J. M. and Lopez H., Statistics of speckle in ultrasound B-scans, *ISEE Trans. Sonics Utrasonics 30* , 156-163 (1983).

[4]   Heyman, J. S., Phase insensitive acoustoelectric transducers, *J. Acoust. Soc. Amer. 64,* 1370-1376 (1986).

[5]   Johnston, P. H. and Miller, J. G., Phase insensitive detection for measurement of backscattered ultrasound, *IESE Trans. Ultrasonics, Ferroelectrics, and Frequency Control 33,* 713-721 (1986).

[6]   Trahey, G. E., Smith, S. W. and von Ramm, O. T., Speckle pattern correlation with lateral aperture translation: experimental results and implication for spatial compounding, *IEEE Trans. Utrasonics, Ferroelectrics, and Frequency Control 33,* 257-264 (1986).

[7]   Shattuck, D. P. and von Ramm, O. T., Compound scanning with a phased array, *Ultrasonic Imaging 4,* 93-107 (1982).

[8]   Goodman, J. W., Some fundamental properties of speckle, *J. Opt. Soc. Amer. 66,* 1145-1150 (1976).

[9]   Goodman, J. W., Statistical Properties of Laser Speckle and Related Phenomena J. C. Dainty, Ed. 9, pp. 9-75 (Springer-Verlag, Heidelberg 1975).

[10]   Attia, E. H. and Steinberg, B. D., Self cohering large antenna array using the spatial correlation properties of radar clutter, *IEEE Trans. Antennas and Propagation* 37. 30-38 (1989).

[11]   Fink, M. and Mallart, R., The random phase transducer: a new technique for incoherent processing. Basic principles and theory, to be published in *IEEE Trans. Utrason. Ferroelect. freq. contr.* (march 1990).

[12]   Fink, M. and Cardoso, J. F., Diffraction effects in pulse echo measurement, *IEEE Trans. Sonic Ultrasonics 31,* 313-329 (1984).

[13]   Laugier, P., Fink, M. and Abouelkaram S., The random phase transducer: a new technique for incoherent processing. Experimental results´ to be published in *IEEE Trans. Utrason. Ferroelect. Freq. contr.* (march 1990).

APPENDIX

In the following, we introduce an approximation that greatly simplifies the theoretical approach to pulse echo processing. We assume that the pressure fields are *quasimonochromatic (continuous wave approximation)*. This approximation is partly justified because the behavior of pulse echo signals are strongly linked to their mean frequency. Furthermore, experimental results confirm the validity of this approximation.

This approximation is useful mainly because it eliminates the time dependence of the pressure fields. A monochromatic pressure field $p(\vec{X},t)$ can be totally described in terms of its amplitude $A(\vec{X})$, its phase $\Phi(\vec{X})$ and its frequency ω. The amplitude and phase can be combined to form the complex amplitude $P(\vec{X})$:

$$P(\vec{X}) = A(\vec{X})e^{j\Phi(\vec{X})}. \tag{16}$$

The pressure field is obtained from its complex amplitude and through the relation

$$p(\vec{X},t) = Re\left\{P(\vec{X})e^{-j\omega t}\right\}. \tag{17}$$

Therefore, once the frequency is set, the pressure field is completely described by its complex amplitude.

The scattering medium is modelled as a random distribution of scatterers located in the insonified volume. In pulse echo processing, the scattering volume is divided into elementary slices whose thickness is equal to the axial resolution cell. Under the monochromatic approximation, it is interesting to model the slice of scattering medium as a 2D random mirror. Note however that such a random mirror only describes a thin slice of the scattering medium. For each depth, a different random mirror has to be defined.

Let's call P_i the complex amplitude of the incident pressure field on the random mirror and P_r the complex amplitude of the pressure field reflected by the random mirror. P_i describes the illumination beam on the plane of the random mirror. The ratio between P_r and P_i defines the reflectance r of the mirror. At position $\vec{x}$ on the plane of the random mirror, we have

$$P_r(\vec{x}) = r(\vec{x})P_i(\vec{x}) \tag{18}$$

The reflectance $r(\vec{x})$ is a complex random function. In the realistic case where the scattering medium is uncorrelated (totally random) and contains a large number of scatterers per resolution cell, $r(\vec{x})$ is a Gaussian random function and its autocorrelation function is proportional to a Dirac function.

Note that coordinates on the plane of the random mirror are noted $\vec{x}$, while coordinates on the plane of the transducers are noted $\vec{X}$.

Since $r(\vec{x})$ is a random function, $P_r(\vec{x})$ is also a random function. Its autocorrelation function is

$$\begin{aligned}
R_{P_r}(\vec{x}_2, \vec{x}_1) &= \; <P_r(\vec{x}_1)P_r^*(\vec{x}_2)> \\
&= P_i(\vec{x}_2)P_i^*(\vec{x}_1)R_r(\vec{x}_2, \vec{x}_1).
\end{aligned} \tag{19}$$

Since the scattering medium is assumed to be uncorrelated, the above expression is equivalent to

$$R_{P_r}(\vec{x}_2, \vec{x}_1) = P_i(\vec{x}_2)P_i^*(\vec{x}_1)\delta(\vec{x}_2 - \vec{x}_1). \tag{20}$$

The diffracted wavefield originating from such a random mirror propagates as a complex random field whose phase and amplitude fluctuate in space. The scattered pressure field $P(\vec{X})$ observed on the plane of the receiver is obtained from the Huygens-Kirchoff formulation. In the Fresnel approximation, it can be reduced to the Fresnel transform of $P_r(\vec{x})$, which is defined by

$$P(\vec{X}) \propto \int d\vec{x}\ P_r(\vec{x}) \exp(\frac{j\pi}{\lambda z} \left| \vec{x} - \vec{X} \right|^2) = P_r(\vec{X}) \otimes \exp(\frac{j\pi}{\lambda z} \left| \vec{X} \right|^2) \tag{21}$$

where $\otimes$ is the convolution product and z the depth of the considered slice of scattering medium.

Our objective is to compute the autocorrelation function of the complex pressure field $P(X)$ on the plane of the receiver. That is, the autocorrelation of the output of a linear system whose impulse response is $\exp(\frac{j\pi}{\lambda z} \left| \vec{x} \right|^2)$ and whose input is $P_r(\vec{x})$. From linear system theory, we have

$$R_P(\vec{X}_2, \vec{X}_1) = R_{P_r}(\vec{X}_2, \vec{X}_1) \otimes \exp(-\frac{j\pi}{\lambda z} \left| \vec{X}_1 \right|^2) \otimes \exp(\frac{j\pi}{\lambda z} \left| \vec{X}_2 \right|^2)$$

$$= \left| P_i(\vec{X}_1) \right|^2 \exp(-\frac{j\pi}{\lambda z} \left| \vec{X} \right|^2) \otimes \exp(\frac{j\pi}{\lambda z} \left| \vec{X} \right|^2) \tag{22}$$

where $\vec{X}_2 = \vec{X}_2 - \vec{X}_1$. This can be rewritten, as the Fourier transform of $\left| P_i(X_1) \right|^2$ taken at the spatial frequency $X / \lambda z$:

$$R_P(\vec{X}_2, \vec{X}_1) = e^{\frac{j\pi}{\lambda z}(\left| \vec{X}_2 \right|^2 - \left| \vec{X}_1 \right|^2)} \mathrm{FT}\left(\left| P_i(\vec{X}_1) \right|^2 \right)(\frac{\vec{X}}{\lambda z}). \tag{23}$$

The above phase factor can be ignored in the study of backscattered field autocorrelation width. In the following, we ignore the phase factor and thus, we can consider that R_p is a function of $X = X_2 - X_1$.

$$R_P(\vec{X}_2, \vec{X}_1) \simeq \mathrm{FT}\left(\left| P_i(\vec{X}_1) \right|^2 \right)(\frac{\vec{X}}{\lambda z}). \tag{24}$$

The above derivation proves the Cittert-Zernike theorem.

SPATIAL PULSE RESPONSE COMPUTING FOR REFLECTION

- TIME DOMAIN APPROACH

Bogdan Piwakowski[*] and Bernard Delannoy

Laboratoire de Physique des Vibrations et d'Acoustique (CNRS,U.A.832
Valenciennes) Institut Industriel du Nord, BP48, 59651 Villeneuve d'Ascq
Cedex, France

INTRODUCTION

The time-domain spatial pulse response functions are largely applied for the analysis of transient fields in problems dealing with radiation (Harris,1981a), but the existing solutions are limited to the specific cases of boundary conditions when the radiating interface is assumed to be either rigid, free or soft. At the same time the problem of reflection is, in fact, generally not treated in the time-domain, except for the specific computations for the radiation coupling functions (Cassereau et Guyomar,1987).The problems of reflection of spherical waves from the planar interface is studied in ω-k domain by means of their decomposition into plane waves.

The paper presents an approach for the time-domain computing of the transient field, as radiated or reflected from plane interface, for arbitrary boundary conditions. The computational result assumes the form of the spatial-pulse-response, which is very useful for the analysis of the transient fields in the large-banded acoustical imaging systems.

MATHEMATICAL DEVELOPEMENT

Green's fonction for a half-space under arbitrary boundary condittions

In Fig.1.a. the plane surface **S** separatestwo non-absorbing media characterised by arbitrary acoustic impedances. Such a boundary can be described by means of plane wave reflection coefficient $b(\theta)$. We assume that $b(\theta)$ is defined for the longitudinal waves and is computed for the upper half-space containing point of observation **O**. Green's function G for the upper half-space can be found by means of the Sommerfeld method as a summation of the direct wave g_+ and its image-source wave g_- (see Guyomar;1985,1986) for the hypothetical source placed at distance z_1. The computation of the function G and its normal derivative can be thus presented as follows:

$$G = g_+ + g_- \tag{1}$$

$$\frac{\partial G}{\partial n} = \frac{\partial g_+}{\partial z_1} + \frac{\partial g_-}{\partial z_1} . \tag{2}$$

---

[*] Permanent affiliation: Institute of Telecommunication, Technical University cf Gdansk, Poland

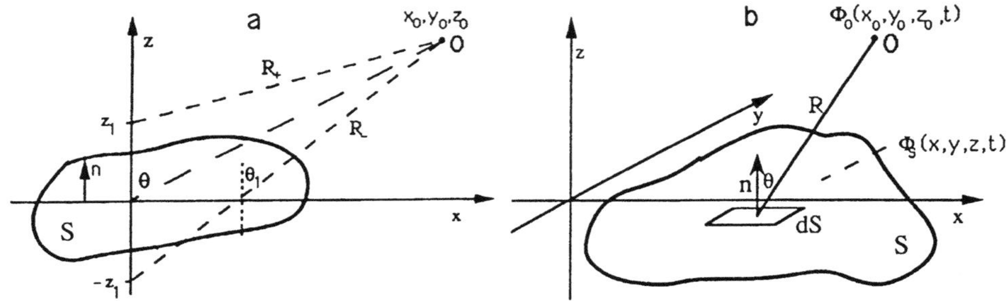

Fig.1.Geometry and notations for the determination of (a) half-space Green's fonction; (b) the pulse response for radiation..

Symbol g_+ denotes the free-space Green function: $g_+=\dfrac{\delta(t-R_+/c)}{4\pi R_+}$, function g_- is here assumed to be:

$$g_- = \frac{\delta(t-R_-/c)}{4\pi R_-}b(\theta_1),\qquad\qquad(3)$$

Thus we respectively have:

$$\frac{\partial g_+}{\partial z_1}= \frac{\partial g_+}{\partial R_+}\frac{\partial R_+}{\partial z_1},\text{ and }\quad \frac{\partial g_-}{\partial z_1}=\frac{\partial b}{\partial \theta_1}\frac{\partial \theta_1}{\partial z_1}\frac{\delta(t-R_-/c)}{4\pi R} + b(\theta_1)\frac{\partial}{\partial R_-}[\frac{\delta(t-R/c)}{4pR_-}]\frac{\partial R_-}{\partial z_1},$$

and by carrying out all the operations of deriviation, we obtain, for $z_1 = 0$,

$$G_= = (b(\theta) +1)\,\frac{\delta(t-R/c)}{4\pi R}\,,\qquad\qquad(4)$$

and

$$\frac{\partial G}{\partial n} = (1-b(\theta))\cos(\theta)\,[\frac{\delta'(t-R/c)}{4\pi Rc} + \frac{\delta(t-R/c)}{4\pi R^2}] + b'(\theta)\,\sin(\theta)\,\frac{\delta(t-R/c)}{4\pi R^2}.\qquad(5)$$

For the sake of this discussion, we will further consider Eq.(5) in two cases:
1. During the computational procedure coefficient $b(\theta)$ is nearly constant as a fonction of θ:

$$b(\theta) = \text{const.} = b\qquad\qquad(6)$$

We then have $b'(\theta) \approx 0$ and consequently the last term in Eq(5) can be disregarded.

2. The waves are assumed to be locally planar $(\frac{\partial g}{\partial R} \approx -\frac{1}{c}\frac{\partial g}{\partial t})$ which permits to disregard two terms factorised by $1/R^2$. In this case coefficient b can be considered as variable :

$$b(\theta)=\text{var},\qquad\qquad(7)$$

but its derivative should be reasonably limited.

Both conditions above mentioned justify the hypothesis (3) for which $b(\theta)$ was assumed to be the plane-wave coefficient . Such a coefficient obviously applies for the locally planar wave; when its value is constant, the same will hold true for all types of waves.

<u>Spatial pulse response for radiation from the boundary</u>

Employing the initial form of the Helmholtz-Kirchhoff equation

$$\Phi_o = \int_S [\Phi_s * \frac{\partial g}{\partial n} - g * \frac{\partial \Phi_s}{\partial n}] \, dS \tag{8}$$

where Φ_s denotes the velocity potential on the radiating surface (Fig.1.b) and taking :

$$g = G, \qquad \frac{\partial g}{\partial n} = \frac{\partial G}{\partial n}, \qquad -\frac{\partial \Phi_s}{\partial n} = v(x,y)\delta(t) \quad and \quad \Phi_s = cH(t)\, v(x,y)$$

we obtain the pulse response for radiation from a boundary expressed in terms of velocity potential:

$$h_\phi(t) = \frac{1}{4\pi} \int_S v(x,y) \left\{ \alpha_b \frac{\delta(t-R/c)}{R} + \beta_b \frac{cH(t-R/c)}{R^2} \right\} dS \tag{9}$$

We have here taken the commonly applied assumption that the radiating surface is locally plane (Lasota et al.1984); the symbol $v(x,y)$ denotes the normal component velocity distribution over S. Coefficient α_b is defined as

$$\alpha_b = 1 + b(\theta) + \cos(\theta)\,[1-b(\theta)] \tag{10}$$

and represents the generalised form of the boundary coefficient, previously defined by Delannoy et al.,1979) for specific cases of soft ,free and rigid boundary conditions. It is clearly seen that α_b takes the forms introduced by these authors for $b(\theta) = -1$,0 and 1 respectively:

$$\alpha_b = \begin{matrix} 2\cos(\theta) \\ 1+\cos(\theta) \\ 2 \, . \end{matrix}$$

Coefficient β_b is here defined in order to consider the boundary conditions for term "$1/R^2$" :

$$\beta_b = \cos(\theta)\,(1-b). \tag{11}$$

Integral (9) presents the generalised formula of the spatial pulse response for radiation for the arbitrary boundary conditions. Notice that taking $b=-1,0$ or 1 one finds the existing solutions for soft, free and rigid baffle, given by Guyomar and Powers(1985). Finally Eqs. (9),(10) and (11) give a complete time-domain description for transient radiation from the surface S for arbitrary boundary conditions.

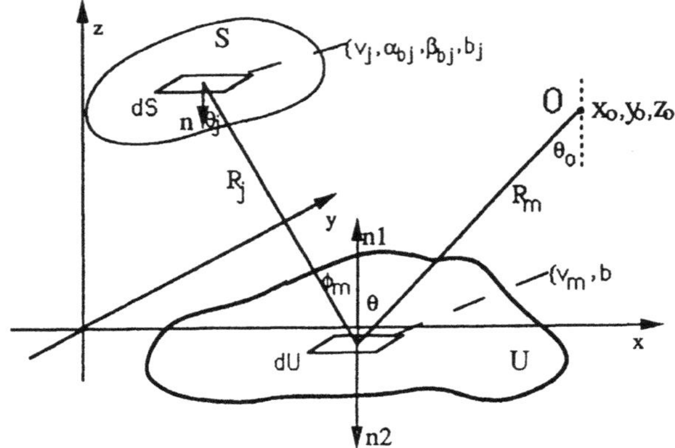

Fig.2. Geometry and notations for the determination of the pulse response for reflection.

Spatial pulse response for reflection

Let us now suppose that we are searching for the pulse response $h_\phi(t)$ of diffraction from an object which has been illuminated by means of the Dirac-type time-domain field. This problem is illustrated in Fig.2 which in reality presents an extention of Fig.1.b, incorporating the addition of the object surface U. Considering the diffraction as a secondary radiation from U, the pulse response $h_\phi(t)$ can be determined based on Eq(8) replacing Φ_s by Φ_m, where Φ_m is the potential of the field radiated at S and observed at U .(In the notation used index j refers to source S and index m to object U). Potential Φ_m can thus be determined based on Eq.(9) and Green's function and its normal derivative should be taken as given in Eqs(4) and (6). Finally combining them with Eq(8) and then carrying out the temporal convolutions (Piwakowski,1989) we obtain:

$$(12)$$

$$h_\phi(t) = \frac{v_j dS}{(4\pi)^2} \int_U \{ \frac{\alpha_{bj}}{R_j}[(b+1)\cos(\phi_m)M(t_i,R_m) - (1-b)D(t_i,R_m)] +$$

$$+ \frac{c\alpha_{bj}}{R_j^2}(b+1)\cos(\phi_m)[M(t_i,R_m)]^{I} +$$

$$+ \frac{c\beta_{bj}}{R_j^2}[(b+1)\cos(\phi_m)M(t_i,R_m) - (1-b)D(t_i,R_m)]^{I} +$$

$$+ \frac{2\beta_{bj}\,c^2}{R_j^3}(b+1)\cos(\phi_m)[M(t_i,R_m)]^{II} \} dU$$

in which :

- the symbols $[\ ]^{I}$ and $[\ ]^{II}$ represent the temporal integration simple and double, respectively,

- t_i expresses the delay for propagation on distance $R_m + R_j$ defined as: $t_i = \frac{R_j}{c} + \frac{R_m}{c}$;

- $h_\phi(t)$ is the pulse response of diffraction from object U illuminated by a source element dS by means of the Diract type velocity pulse; v_j denotes the normal component velocity exitation of dS ; the source and object boudary conditions are considered independently by means of coefficients b_j and b. This solution is applicable under condition (6).

- symbols M and D represent the action of the monopole-type and dipole-type primary sources and are defined as:

$$M(\tau,R) = \frac{\delta(t-R/c)}{R} \; ; \qquad D(\tau,R) = \cos(\theta)[\frac{\delta(t-R/c)}{R} + \frac{cH(t-R/c)}{R^2}] \; ; \qquad \tau = R/c.$$

The expression inside integral (12) represents the diffraction from the object surface element dU and includes four terms:

-the second and the fourth ones are involved by the sphericity of the illuminatig wave and disappear when the source is taken away (the wave becomes locally plane);

- the third and the fourth ones represent the response for the illumuination by term "$1/R^2$" emitted by the source (see Eq.(9)); they also disappear when the source is removed;

- finally only the first term is of importance for the illumination locally plane. If additionally the point of observation is at a sufficient distance to consider the diffracted wave to be also locally plane, this term takes a simple form:

$$h_\phi(t) = \frac{v_j dS}{(4\pi)^2} \int_U \frac{\alpha b_j}{R_j} \{[b(\theta)+1]\cos(\phi_m) - [1-b(\theta)]\cos(\theta)\} \frac{\delta(t-t_j)}{R_m} dU. \qquad (13)$$

It is shown (Piwakowski,1989) that when the exact solution (12) is replaced by its simplified version (13), the spectral precision of computations is better then 3dB for the frequences fullfiling the condition $kR>1.5$, where $R=\min[R_j,R_m]$ and $k=2\pi/\lambda$. Notice that solution (13) is applicable for $b(\theta)=\mathrm{var.}$(condition (7)).

The comparison of forms (9) ,(12) and (13) shows that for both, radiation and reflection, any modification of the boundary conditions produces the adequate modulation of the amplitudes of the monopole M and dipole-type D sources , which are factorised by terms (b+1) and (b-1). Notice the equilibrium which is maintained for M+D.

COMPUTATIONAL EXAMPLES

Application of integral (12)

Fig.3 shows the reflection of the Dirac-pulsed spherical wave from the plane infinite and rigid surface, for three classic boundary conditions of the point source. For these simplified cases integral (12) can be solved analytically by the method proposed by Lasota (1985). As an example, for case (a) if the source-emited pression has temporal form: $\dfrac{P_1\delta(t-R_j/c)}{R_j}$ we obtain :

$$h_\phi(t) = \frac{P_1}{\rho(R_s+R_o)}H(t-t_1) \quad \text{and} \quad h_p(t) = \frac{P_1}{R_s+R_o}\delta(t-t_1)$$

where h_p is the pression pulse response : $h_p = \rho\dfrac{dh_\phi}{dt}$ and $t_1 = \dfrac{R_s+R_o}{c}$ .These solutions seen in Fig.3.a. represent the reflected and delayed spherical wave which maintain its temporal form. As could be expected the reflection is perfectly large-banded the sphericity is maintained after reflection, the reflection is thus the "mirror type". For no rigid source the emited signal will always be accompaigned by its integral (step function response associated with the Dirac pression pulse).We point out that to obtain these solutions all terms of integral (12) should be considered for the computing.

Fig.4. shows the same responses as shown in Fig.3 except that in this case the reflecting surface U is limited, and assumes the shape of a piston (Fig.4.d). We observe that the "mirror" wave is followed by the edge waves and the amplitude of these waves depends on the source

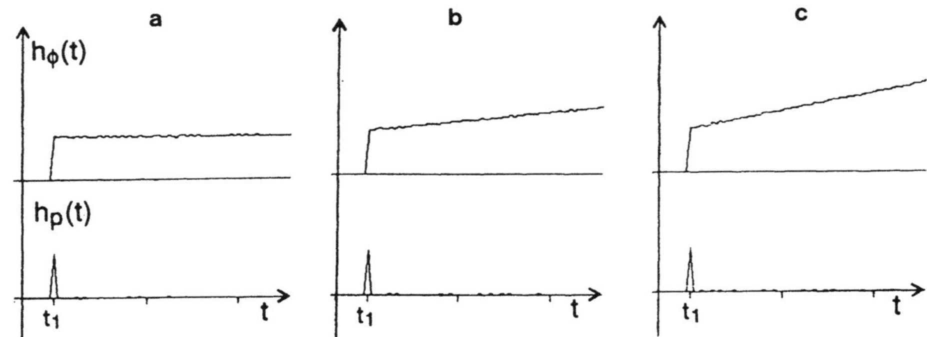

Fig.3. Spatial pulse responses for reflection from the plane rigid and infinite surface, for three classic source baffle conditions ; rigid (a), free-field (b), soft (c). Point source S and point of observation O are in the same location (as in Fig.4.d).

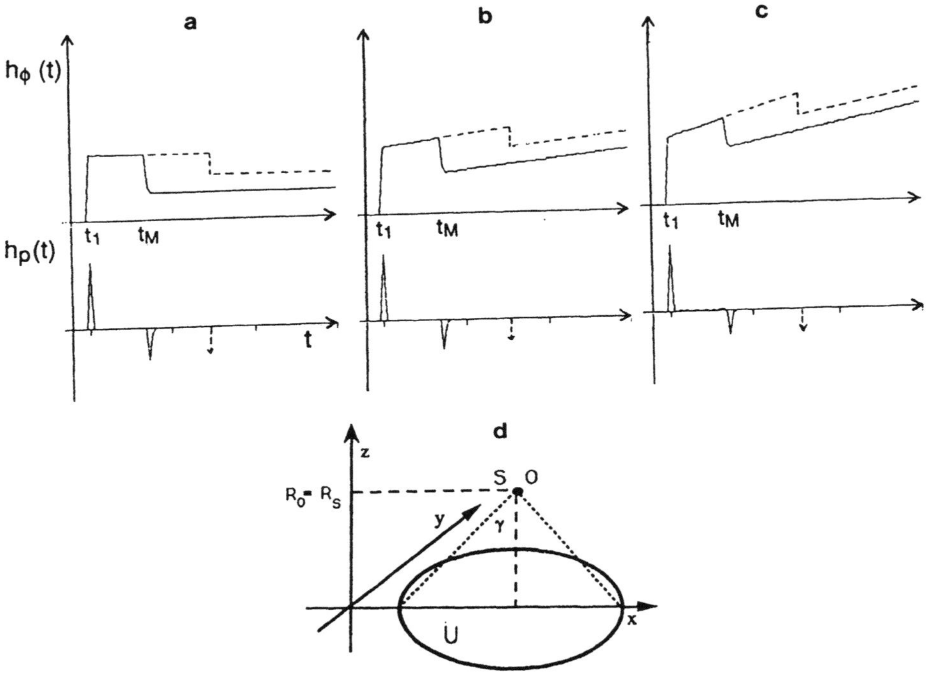

Fig.4.(a), (b), (c) The results for the same experiment as shown in Fig.3 except that the reflecting surface U is limited and assumes the form of a piston; (d) Point source S and point of observation O are located on the piston axis; angle $\gamma= 40°$.

baffle conditions. The pointed lines illustrate the responses behaviour when the piston diameter increases.

Fig.5. presents the influence of the boundary conditions on the reflecting surface. Notice that the reflected wave is directly factorised by the reflection coefficient without any modification of the temporal form of the reflected signal. This can be interpreted as a partial "absorption" by the reflector. For b=0 we thus have the free field conditions and there is no reflection.

Fig.6 presents the pulse response behaviour as a function of the distance of the observation point from the piston axis.We observe the "mirror" direct wave D followed by the edge waves E_1 and E_2. For $x_0=x_{03}$ the first edge wave E_1 begins to coinside with D ; for $x_0 > x_{03}$ the direct wave disappears. Between the edge waves we observe the diffraction response which can be compared with the "membrane" wave introduced by Harris(1981b) however the physical nature of this phenomena is different (amplitude distribution of M and D sources).

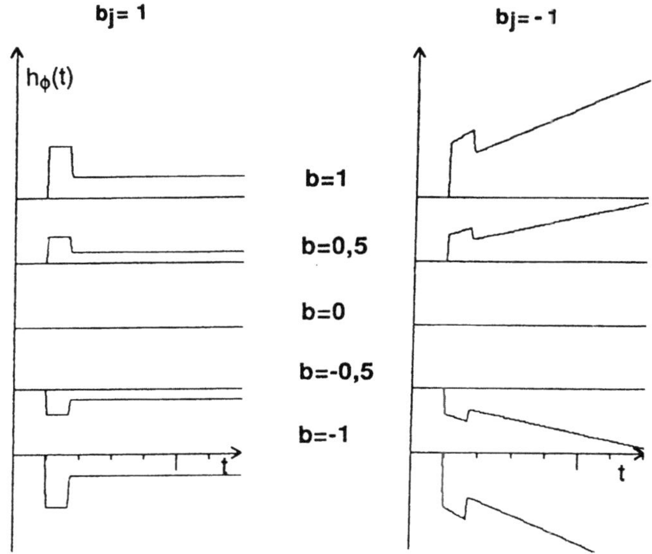

Fig.5. Pulse responses in terms of velocity potential for the same geometry as in Fig.4.d, presented as a function of the reflection coefficient b at the reflecting surface, for "rigid" (bj=1) and "soft" (bj=-1) point source S.

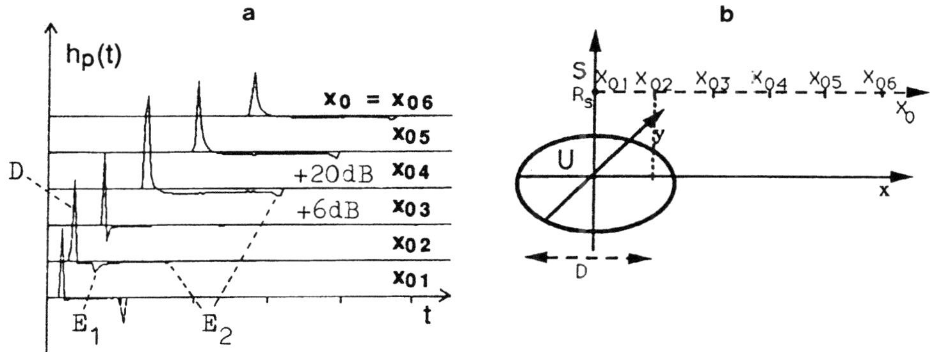

Fig.6 (a) Spatial pulse responses of diffraction from a perfectly rigid circular piston as a function of the distance of the observation point x_0 from the piston axis. Source position and reflector form as in Fig.4; (b) Geometry of the experiment.

<u>Application of integral (13)</u>

The computations based on Eqs.(9), (12) and (13) are applicable also for the non planar surfaces. In such a case the surface elements dU (or dS) should be approximated by the planar elementary surfaces which differ in location by amounts reasonably small, in comparison to the wavelength. Since, during the integration, each elementary surface is considered separately, the multiple scattering and shadowing at section of the diffracting (or radiating) surface should be disregarded. Fig.7 shows the responses of diffraction from the air-filled edge illuminated by the point source. The computations are based on Eq.(13) and we have assumed: $b(\theta)$=var. Notice the changes of polarisation of the responses due to the evolution of the $b(\theta)$ sign.

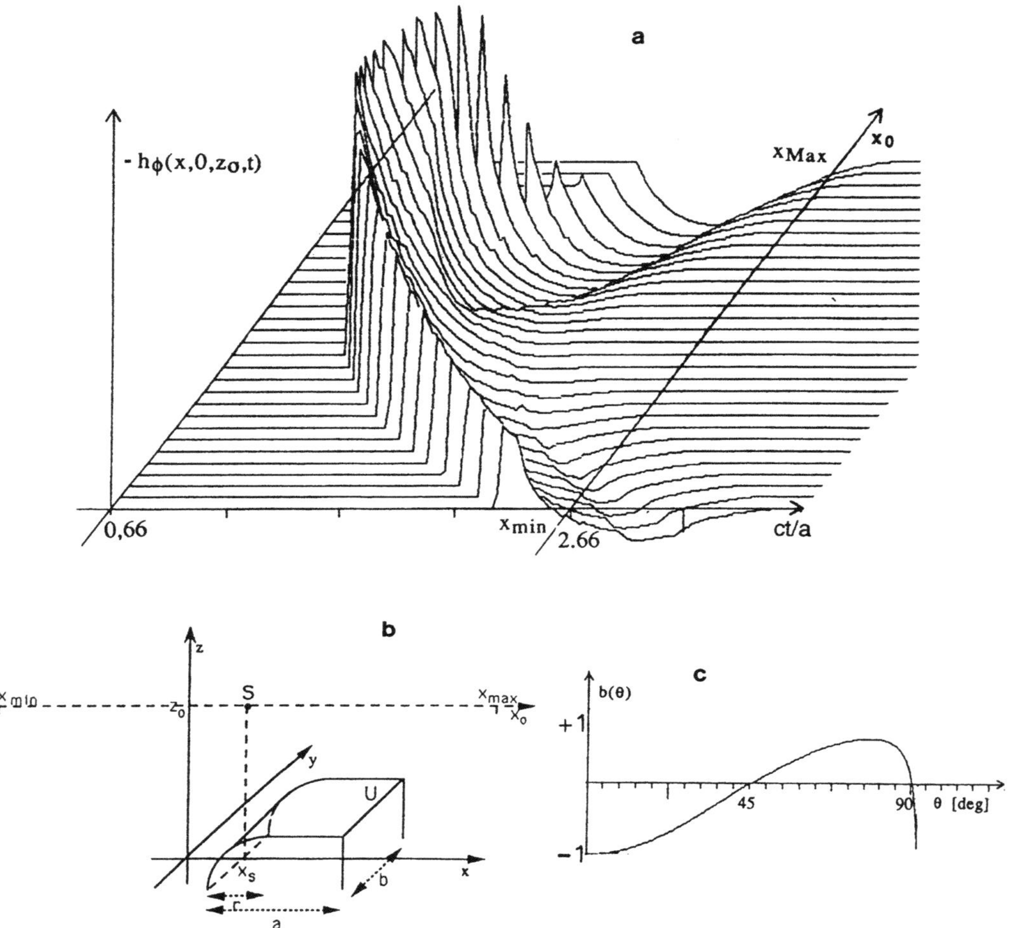

Fig.7. (a) Spatial pulse response of diffraction from air filled circular edge U in elastic non absorbing medium, for point source S placed in a rigid baffle at position $(x_s,0,z_o)$, versus ct/a and x/a; observation points are placed at $z=z_o$,$y=0$ and for $x_{min} < x < x_{max}$. (b) source-object-receivers geometry; $a/r=2$,$a/b=2$, $z_o/a=1$, $x_{min}=x_o-1,5a$, $x_{max}=x_o+1,5a$; (c) plane wave P-P reflection coefficient used for the computations   presented in section (a).

The absorption of the medium can be easily considered in Eq.(13) by replacing the free-space Green functions with its counterpart for the absorbing medium. Taking advantage of the hypothesis that the waves are locally planar, we have applied the approach commonly used in seismics (Aki and Richards,1980) for plane waves and for medium with linear attenuation. Green's function g_a is taken as:

$$g_a(t\text{-}R/c) = \frac{A(t\text{-}R/c)}{4\pi R}, \quad \text{where} \quad A(t\text{-}R/c) = F^{-1}\,[\,e^{\,-\pi fR/Qc + j(kR - \omega t)}\,]. \tag{14}$$

where k is assumed to be complex. Coefficient Q expresses the absorption and F^{-1} denotes the inverse Fourier transformation. The example of the computations using this model is presented in Fig.8

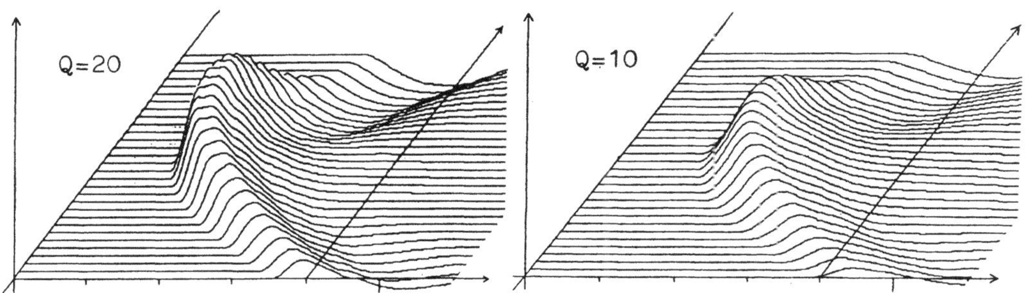

Fig.8. Results for the same experiment as the one shown in Fig.7 except that for the absorbing medium: Q=20 (α=1.32 dB/λ) and Q=10 (α=2.73 db/λ)

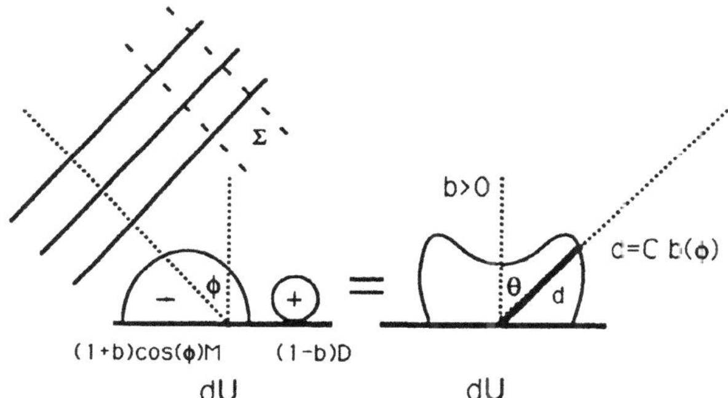

Fig.9. Time-domain interpretation of the diffraction of the locally plane wave from the elementary surface dU. The resulting directivity pattern is produced by the monopole and dipole sources

PHYSICAL MECHANISM OF THE REFLECTION FOR THE LOCALLY PLANAR WAVE

Let us suppose that the planar element dU from Fig.2 is illuminated by the locally planar wave with the $\phi_m=\phi$ and we are now searching for the wave reflected with the angle of geometrical reflection. We thus have $\theta=\phi$ and consequently: $\cos(\theta)=\cos(\phi)$, (Fig.9). Equation (13) becomes then:

$$h_\phi(t) = C\ dU\ \frac{b(\theta)\ \delta(t\text{-}t_j)}{R_m} \tag{15}$$

143

where C here represents a constant value. It is seen that the spherical wave radiated in the direction of the geometrical reflection is proportional to the reflection coefficient b(θ). This effect is produced by the action of monopoles and dipoles which "set itself" to produce a specific directivity pattern for diffraction from dU (Fig.9). Notice that this pattern is "controlled" by the angles ϕ_m, θ and the value of b(θ).This mechanism explains the classical case of plane wave reflection : when the plane wave illuminates an infinite planar surface U ,all the dU elements radiate the elementary sperical waves which form the plane wave-front and the reflected plane wave is factorised by b(θ). We point out that this model gives the complete description of the diffraction from element dU and should not be confused with the ray-methods.

CONCLUSIONS

An approach for the time-domain solution for diffraction from planar surface and for arbitrary boundary conditions is proposed. The modification of these conditions implies the modulation of the amplitudes of the monopole and dipole sources .

The solution is mathematically exact for the idealised case when the reflection coefficient b=const. When the radiated and reflected waves can be assumed to be locally planar, the solution simplifies and can be used in practical cases when b is not constant.

The computational result assumes the form of the spatial pulse response for a source radiation or diffraction from a given object. This form seems to be very useful for the analysis of the transient fields in acoustical imaging domain.The absorption can be easily taken into account.

If secondary diffractions are disregarded ,the solution is applicable for any irregular surface modelled as a summation of planar elements.

The obtained forms provide the time domain interpretation for the physical mechanism of the reflection of the plane wave : the amplitudes of monople and dipole-type secondary sources "set themself" in this way, resulting in a reflected wave proportional to b for the direction of geometrical reflection.

REFERENCES

Aki,K,.Richards,P,G.,(1980)"Quantitative seismology, Theory and Methods" W.H.Freedman Company
Cassereau,D.,Guyomar,D., (1987)"Determination of the pulse diffraction of an obstacleby ray modelling, *in* Acoustical Imaging vol.16,Plenum Press,NY 1988
Delannoy,B.,Lasota,H.,Bruneel,C.,Torquet,R. et Bridoux,E.,(1977)"The infinite planar baffles problem in acoustic radiation and its experimental verification" J.Appl.Phys.50,5189-5195).
Harris,G.R.,(1981a)."Review of transient theory for a baffled piston" J.Acoust.Soc.Am. 70 (1), 10-20
Harris,G.R.,1981b)."transient field of a baffled piston having an arbitrary vibration amplitude distribution" J.Acoust.Soc.Am. 70(1), 186-204.
Guyomar,D.,Powers,J;?(1985) "Boundary effects on the transient radiation fields from vibrating surfaces" J.Acoust.Soc.Am. 77(3), 907-915.
Guyomar,D., (1986) "Théorie et méthodes de la diffraction impulsionelle dans les milieux sans perte et dans les milieux attenuants" Thése d'état, Université de Paris 7.
Lasota,H.,Salamon,R., and Delannoy,B.,(1984) "Acoustic diffraction analysis by the impulse method: A line impulse response approach" J.Acoust.Soc.Am. 76(1), 280-290   .
Lasota,H., (1985) " Etude du champ acoustique des sources planes dans le domaine temporel" Thése d'état, Université de Valenciennes
Piwakowski.B.,(1989) In preparation, Thése d'habilitation à diriger de la recherche , Université de Valenciennes,

NDE OF DELAMINATIONS DURING PROCESSING OF CARBON-CARBON COMPOSITES

B. R. Tittmann* and J. R. Bulau+

Rockwell International Science Center
P.O. Box 1085
Thousand Oaks, CA 91358

ABSTRACT

The use of acoustic emissions (AE) sensors is reported for the detection of delaminations during the first carbonization of carbon-carbon panels. The onset of delaminations is characterized by high amplitude, long duration events which are distinct from the low amplitude, short duration events associated with microcracking which is a desirable evolution of the matrix structure during first carbonization. We also report on preliminary results on the observation of AE precursors to delaminations, so that the potential for warning and prevention of delaminations may be realized. The results open the door to the triangulation and 2-D imaging of the stress zone giving rise to the precursor.

INTRODUCTION

Delamination of composite materials either during manufacture or during subsequent use is a critical problem in advanced DoD systems applications. Materials used for these applications include carbon-carbon composites. In the manufacture of these composites, increased sophistication in the monitoring of the materials chemical state and its microstructure is needed to ensure performance, reliability, and low life cycle costs. New monitoring sensors are needed to operate in the high temperature environment associated with the manufacture of these materials. This paper addresses the use of an acoustic emission sensor to measure in-situ the chemical state and microstructure of carbon-carbon materials during manufacture and development with particular emphasis on the detection of delaminations.

BACKGROUND

The manufacturing process of carbon-carbon consists of many long-duration cycles of heating and cooling to effect the desired changes in material properties and composition. First carbonization refers to the first heat treatment after the initial cure cycle of the starting material. In this step the material is inside a furnace such as shown schematically in Figure 1 and the temperature is raised to 700° C following an empirically chosen temperature/time profile. The temperature of the component is monitored with thermocouples (TC), whereas the gas pressure/flow inside the furnace chamber is monitored with a gas flow meter. For resin-matrix composites this treatment entails major thermally induced physical and chemical changes with stresses related to both gas generation and the high shrinkage of the matrix [1-3]. After cool-down the material typically has a fine network of microfractures which are

---

* Dr. B.R. Tittmann is presently on a leave at Penn State University, University Park, PA.

\+ Dr. J. R. Bulau is now at Mobil Research Corp., Dallas, TX.

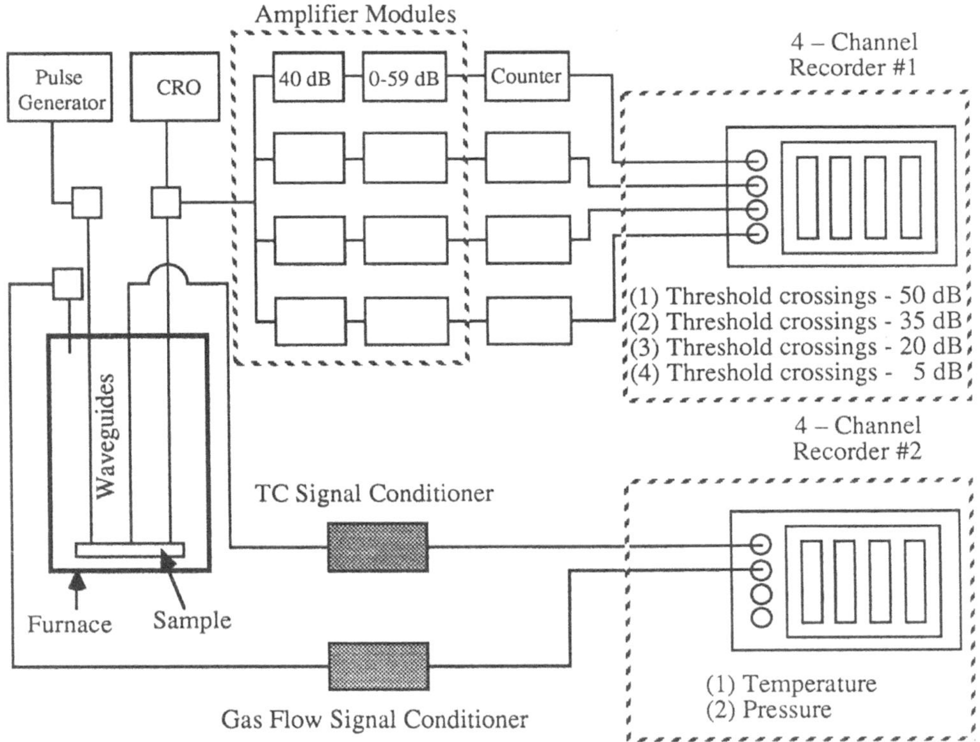

Fig. 1 Schematic of apparutus.

then refilled with matrix material in preparation for the next carbonization treatment. The use of acoustic emission sensors to monitor first carbonization is a technique naturally suited to keep track of the microcracking during the heat treatment [1-5]. To enable the use of conventional piezoelectric transducers as detector elements, we have found it useful to employ acoustic waveguides. These attach to the sample in the furnace on one end and lead the acoustic waves to the exterior of the furnace where the transducers are attached to the other end. From there the received acoustic emission signals are processed in specialized electronic circuitry to provide information for signal interpretation.

ACOUSTIC EMISSION APPARATUS

The apparatus assembled for use on this project is illustrated schematically in Figure 1. The function of most of the elements in this system is obvious. However, elaboration is required on a few points.

First, the acoustic waveguides are 1/8 inch diameter stainless steel rods. The rods are threaded at one end into the sample. This connection is bonded adhesively using a drop of K-641 resin, which is allowed to polymerize at approximately 160° C prior to each run.

Second, the AE transducers were built in-house. They are broad-band, designed to make use of many vibration modes spread over a broad frequency band. The spectrum for a typical transducer of this type has been discussed previously [1]. The transducers are clamped to a small platform, which is brazed to the top of the waveguide. The waveguide and the transducer are coupled acoustically using a high viscosity silicone oil, which thins somewhat, but does not evaporate, at elevated temperatures.

Third, in the instrumentation shown in Figure 1, the use of two waveguides and two transducers makes possible the application of a fixed amplitude pulse to one transducer which uses the other to measure the combined effects of coupling and transmission losses at various times during a run.

Temperature and several AE parameters described below were monitored continuously during all experimental runs. Net changes in sample dimensions following carbonization at different temperatures were documented by measuring the dimensions of the test specimen both before and after each run.

As shown in Figures 2 and 3, the system allows AE data to be acquired using a statistical approach. We record continuously four AE parameters; the total number of threshold crossings at each of four different amplitude thresholds. While RMS voltage is used commonly as a parameter in systems that monitor quasi-continuous emission activity, this was not useful in the present application because we always observed discrete events, though at a high event rate. We parametrized AE activity on a statistical basis which involved counting threshold crossings at each of several amplitudes, which are evenly spaced on a log scale. The threshold amplitudes selected were 31.6 microvolts, 178 microvolts, 1 millivolt, and 5.62 millivolts, all referenced to the transducer output. The most sensitive setting was approximately 7 dB above background noise. Figure 2 shows schematically a waveform in volts versus time for a typical transducer output. The figure identifies most of the important characteristics of the received waveforms, such as the maximum amplitude, the event duration, and the number of threshold crossings. The threshold amplitude is dictated by the adjustment of the detection circuitry. In Figure 3 we show schematically a high amplitude, long duration waveform typically associated with a delamination. In contrast, the low amplitude, short duration waveform is typically associated with microcracking.

In addition to the AE parameters several process parameters, always including temperature and time, and sometimes also including pressure and gas flow rate, were recorded and logged.

MECHANISM FOR ACOUSTIC EMISSION

AE is caused by a sudden release of localized stress. Sources of stress during carbonization are the transformation shrinkage of the matrix, differential thermal expansion and then entrapment of evolved gasses. The release of stress is typically associated with crack initiation and crack growth/extension [6].

In carbon-carbon pyrolysis microcracking is normal and desirable because the goal is to form an open, interconnected network of cracks to enable the escape of evolved gases. Desirable microcracks include fiber-matrix cracks and matrix-matrix cracks. Fiber-matrix

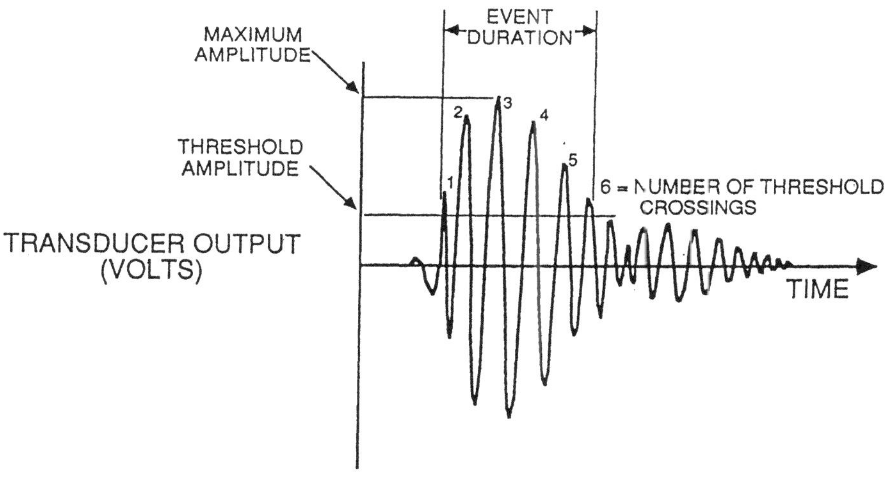

Fig. 2 AE terminology.

$$\text{Table 1. Source Duration.}$$

τ fiber breakage	=	$\dfrac{5 \times 10^{-6}\,m}{10^{3}\,ms^{-1}}$	=	5 nsec	>	200 MHz
τ delamination	=	$\dfrac{50 \times 10^{-3}\,m}{10^{3}\,ms^{-1}}$	=	50 μsec	>	20 kHz
τ microcracks	=	$\dfrac{50 \times 10^{-6}\,m}{10^{3}\,ms^{-1}}$	=	50 nsec	>	20 MHz

cracks arise as a result of the unbonding between the reinforcing fibers and the matrix. Fiber-matrix cracks developing between neighboring sites link up and produce matrix-matrix cracks which eventually extend through the matrix of a given ply typically in a direction perpendicular to the fibers. The AE waveforms associated with fiber-matrix cracks are typically short duration, high frequency, low amplitude events. This may be seen by following the crack growth and estimating the time and frequency as shown in Table 1. These values are order-of-magnitude estimates to demonstrate the difference between AE from microcracks and delaminations. Assuming a crack growth velocity of the order of the velocity of sound, i.e., 10^{-3} m/s, the total time for the crack to grow is given by its final length. As shown in Table 1, for microcracks these dimensions are the perimeter of the fiber or the thickness of the ply. In contrast, delaminations typically cover many centimeters, and the characteristic times are therefore comparatively long and the frequencies correspondingly low. Figure 3 shows schematically AE waveforms from a microcrack (Figure 3a) and a delamination (Figure 3b) respectively.

RESULTS AND INTERPRETATION

The results are presented in the form of a summary supported by Figures 4-6.

In Figure 4 the results of two runs are shown to contrast AE data between a normal "good"

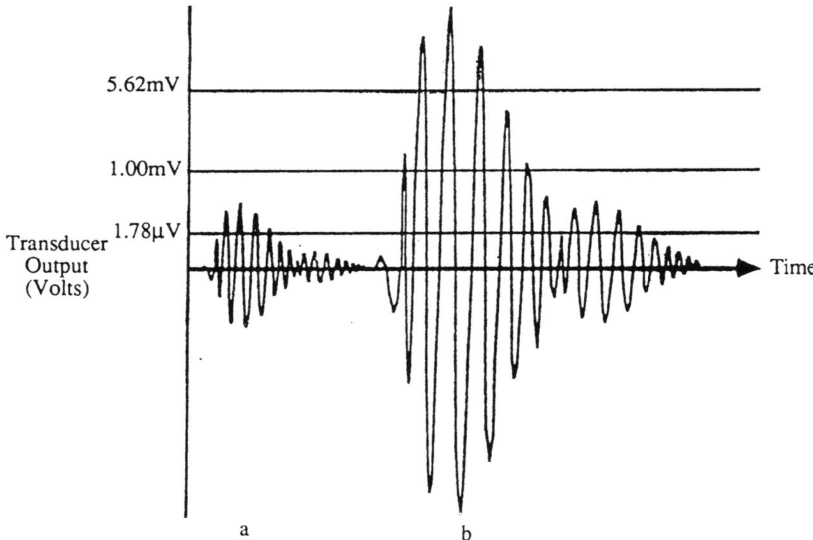

Fig. 3 Acoustic emission threshold counting concept and schematic drawings of waveforms corresponding to (a) microcracking and (b) delamination.

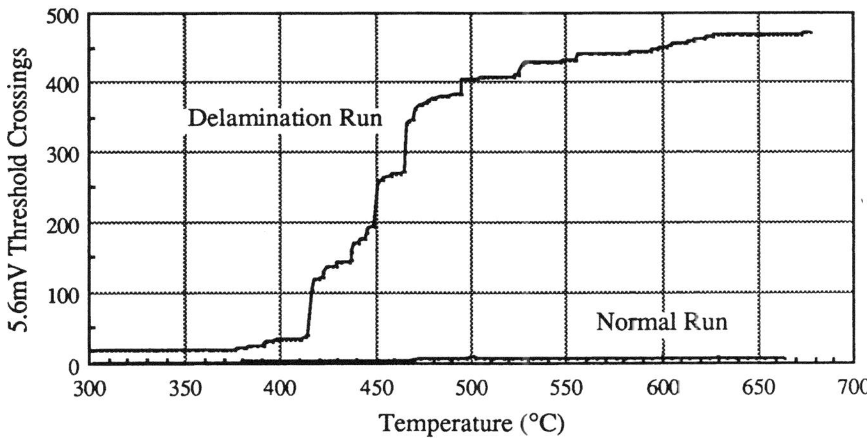

Fig. 4 Comparison of normal and delamination runs.

run and a "bad" delamination run. Presented are the cumulative crossings in the highest
threshold channel for a normal best treatment with a minimal heat-up rate of 10° C/hr. This
run gave very few cumulative crossings and therefore shows very few or no high amplitude
events. In contrast the abnormal or delamination run had a heat-up rate of about 350° C/hr
and produced massive delaminations which destroyed the sample. There were a significant
number of high amplitude events and the cumulative threshold crossings are shown as rising
sharply at about 420° C with events continuing to about 550° C. The abrupt increases at
420° C, 450° C and 470° C are thought to be the main events associated with the "unrip-
ping" of the plies. This sample was approximately 6 cm by 6 cm by 0.4 cm in dimensions.

Figure 5 presents gas analysis and AE data for another delamination run at a heat-up rate of
63.8° C/hr. Shown are the flow rate and the concentration of water vapor in the chamber. At
approximately 175° C, the sample, 30 cm by 30 cm by 0.3 cm, suddenly started to evolve
large amounts of water vapor. The onset of this gas flow coincided with sudden activity in
the highest two threshold channels leading to high amplitude counts shown here as bars indicating delamination events at 178° C, 187° C and 195° C. The first indications of impend-
ing damage occurred about 3 minutes prior to the delamination events; these indications
may be viewed as precursors or warnings for the coming delaminations.

In Figure 6 we present the AE data associated with a delamination run in which the precursors came substantially earlier in time and could have been used as an indicator to stop or alter the time/temperature profile of the carbonization run. Shown are the crossings averaged
over a 10 minute interval for the highest and the lowest threshold levels, 5.6 mV and 31.6
μV, respectively. Note that the background of low amplitude events representing the microcracking for the 31.6μV threshold dominates the few precursor events even though the latter
are of higher amplitude. This sample was about 6 cm by 6 cm by 0.4 cm in dimensions and
had a heat-up rate of 10° C/hr. For this size sample and this slow heat-up rate we do not normally expect delaminations and their actual occurrence suggest the presence of a flaw in the
sample most likely produced before the heat treatment in the lay-up or curing stages.

CONCLUSIONS

In this paper we have presented a short summary of work on the NDE of delaminations during the first carbonization of carbon-carbon composites. After a brief description of the process we discussed the acoustic emission instrumentation, the method of data ordering and
the physical mechanism for the acoustic emissions during processing. We present the results
of four runs, three of which produced delaminations and show how AE can distinguish between heat treatments with and without delaminations. We also report preliminary results of

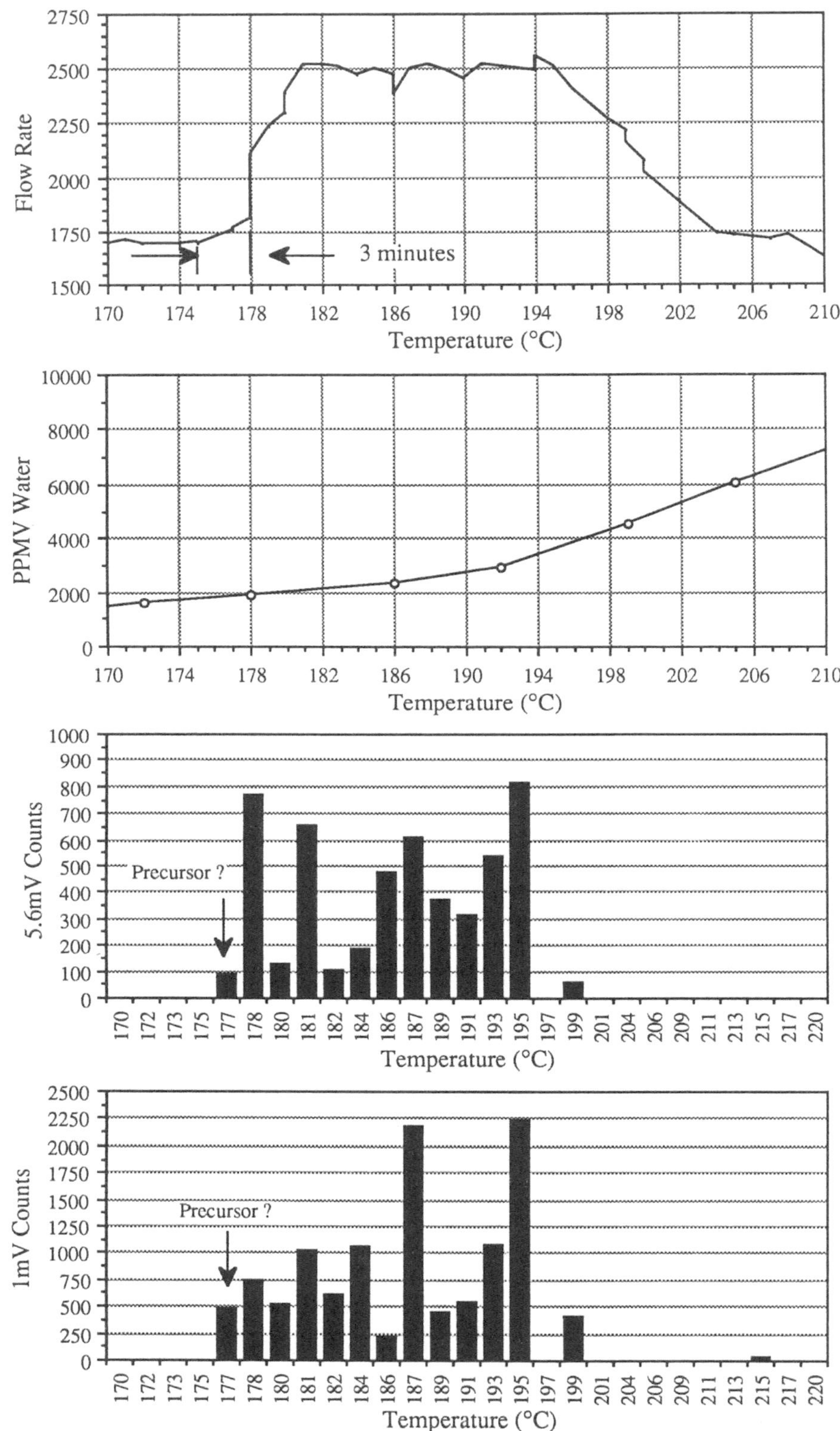

Fig. 5 Delamination run with precursor.

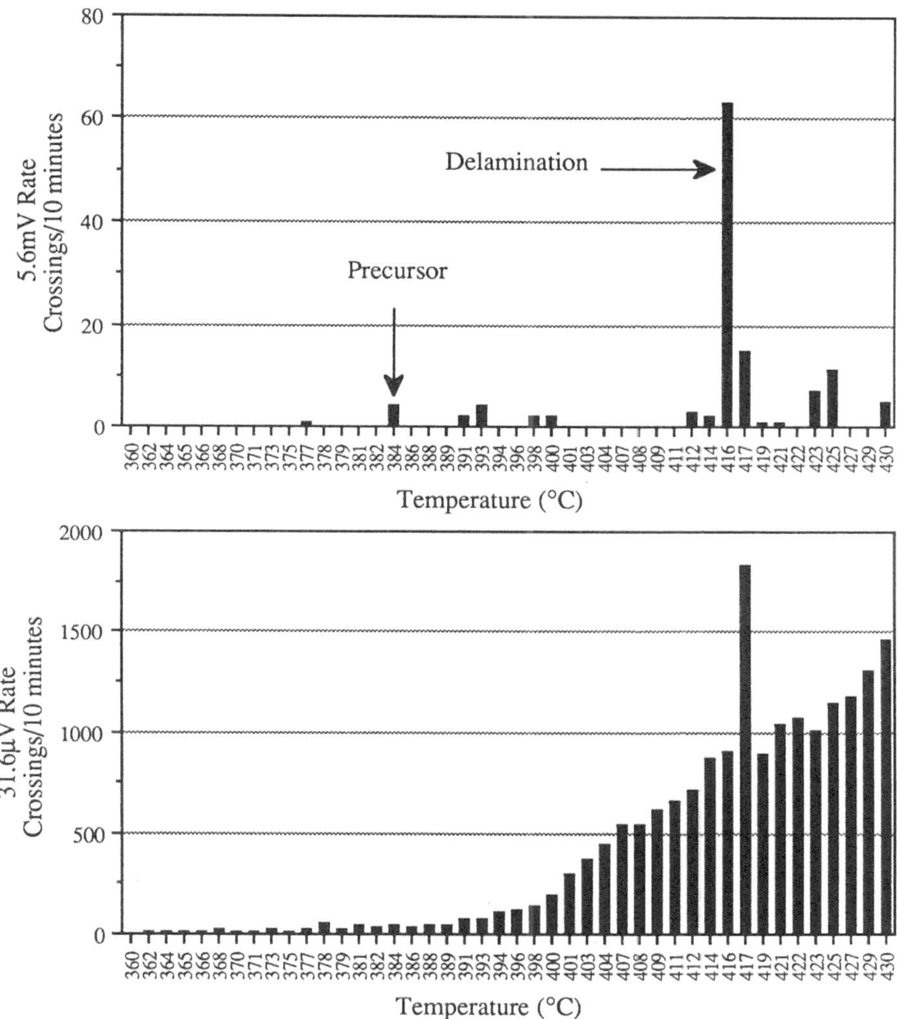

Fig. 6 Delamination run with precursor (10° C/hr. heat-up rate).

precursors to delaminations in the form of small AE events registering in the highest and next to the highest thresholds. These precursors, if recognized, may be used as warning signs to enable the operator or an expert system type temperature controller to alter the time/temperature profile of the heat treatment. The results set the stage for the use of commercially available triangulation and 2-D imaging software to locate and depict the affected stress zone giving rise to the precursor in order to locate critical areas of the component as it undergoes the heat treatment.

ACKNOWLEDGEMENT

The authors acknowledge the support of the Office of Naval Research under Contract N00014-87-C-0724.

REFERENCES

1. "Process Science for 2D Carbon-Carbon Exit Losses", prepared by Aerojet Strategic Propulsion Co., AFWAL Report No. TR 86-4028, July 1986.
2. K. T. Clemens and S. C. Brown, "Real-Time Acoustic Emission Monitoring/Control Technique for Carbonization of 2-D Phenolic-Carbon Composites", 8th JANNAF RNTS Mfg., Patrick AFB, October 1986.
3. W. J. Pardee, M. R. Mitchell, A. Gupta, F. Montgomery and J. Sheehan, "Effect of Carbonization Kinetics on In-Process Mechanical Properties", Proceedings of the ASM Conference, Indianapolis, IN, October 1989, in press.
4. J. R. Bulau, "AE Monitoring for Control of Carbon-Carbon Pyrolysis", Proceedings of IEEE 1988 Ultrasonics Symposium (B. R. McAvoy, Editor), IEEE Cat. No. 88 CH 2578-3, pp. 1057-1063 (1988).
5. B. R. Tittmann and J. R. Bulau, "Acoustic Emission and Ultrasonics Sensing for Carbon-Carbon Pyrolysis Monitoring", Proceedings of the ASM Conference, Indianapolis, IN, October 1989, in press.
6. B. R. Tittmann, "Acoustical Studies of Damage Mechanisms in Carbon-Carbon During First Carbonization", Proceedings of the IEEE 1989 Ultrasonics Symposium, (B.R. McAvoy, Editor), in press.

ULTRASONIC IMAGING AND FINITE ELEMENT ANALYSIS OF ADHESIVELY BONDED CYLINDERS

N.K. Batra, K.E. Simmonds, M.A. Tamm, and H.H. Chaskelis

Naval Research Laboratory
Washington, D.C. 20375-5000

INTRODUCTION

Adhesively bonded structures are increasingly being used for marine applications. One such application involves the use of adhesively bonded cylinders of polyethylene—rubber—steel. Steel forms the inner-most cylindrical lamina and polyethylene forms the outermost layer of this component. The adhesively sandwiched lamina, rubber, is subjected to axial shear loads—tangential to the curved surfaces. For this component to perform well under load it is necessary that the adhesive bonds at the polyethylene-rubber and steel-rubber interfaces be strong and free of any deleterious delaminations. Any surface areas which are devoid of adhesive, have trapped gas, are not chemically bonded by the adhesive or are simply in mechanical contact, form areas of delaminations detrimental to the performance of these components.

In this paper we discuss the effect of delaminations on the propagation of ultrasonic waves through such a component. We show how the variation in the intensity of transmitted ultrasound can be digitized, transformed into a two dimensional image and interpreted. The digitized data is wrapped to depict the three dimensional image of the transmitted amplitude. Image analyses, such as statistical histograms, are used to interpret the variation in the amplitude and the delaminations.

Since these adhesively bonded laminates are designed to bear shear stress, the effect of the disbonds under load must be predicted "a priori." We predict the strain energy distribution in the rubber at the adhesively bonded interfaces. Using finite element analysis and a mathematical model for such a multilayered structure, strain-energy density is computed and plotted for various simulated delaminations of known shape at the interfaces.

THEORY

We review briefly the effect of a delamination on the propagation of ultrasonic waves through cylindrical layers as shown in Fig. 1. Let us assume the incident wave is *normal* to surface of the multilayered structure, i.e. only compressional waves are propagated. It can be shown[1-3] that the reflection pressure amplitude from this entire set of layers is given by

$$\mathbf{R} = (Z_{in}^n - Z_{n+1})/(Z_{in}^n + Z_{n+1}) \tag{1}$$

where Z_{in}^n is the input impedance of the entire set of layers. The input impedance of the system is determined by the equation

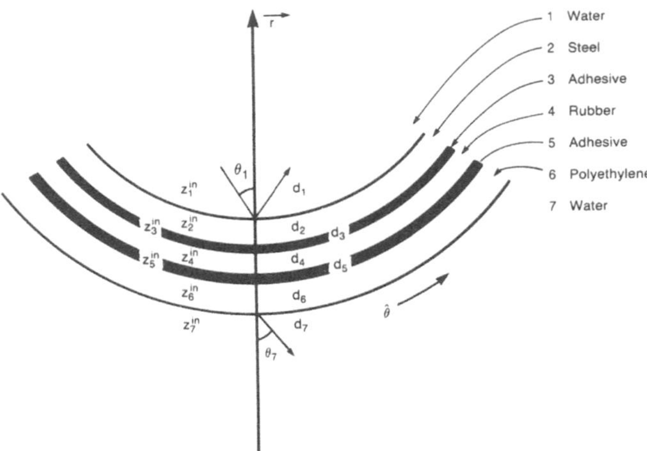

Fig. 1   Adhesively bonded cylinderical layers immersed
in acoustic couplant, water.

$$Z_{in}^n = [(Z_{in}^{n-1} - iZ_n \tan k_{nr}d_n)/(Z_n - iZ_{in}^{n-1} \tan k_{nr}d_n)] \, Z_n \tag{2}$$

where $Z_j = \rho_j c_j/\cos \theta_j$ and

$$k_j \sin \theta_j = k_{j+1} \sin \theta_{j+1} \quad j = 1, 2, \ldots 7. \tag{3}$$

ρ_j is the mass density of the jth layer, c_j is the velocity of propagation of the acoustic mode under consideration for the jth layer, θ_j is the angle of incidence at the jth and $j + 1^{\text{th}}$ layer interface and k_j is the complex wave vector of the layer, to include the effect of attenuation.

The transmission amplitude for these layers is given by

$$\mathbf{T}_{\text{system}} = \prod_{j=1}^{7} (Z_{in}^j + Z_j)/(Z_{in}^j + Z_{j+1})e^{i\phi j}. \tag{4}$$

where ϕ_j is the advance in phase in layer j. We assume that a debonded area or delamination consists of air or vacuum. Since $\rho_{\text{air}} \simeq 0$, $\rho_{\text{vacu}} = 0$, consequently $Z_3 \simeq \rho_{\text{air}} c_{\text{air}} = 0$, which implies $Z_{\text{input}}^3 \simeq 0$. Physically this implies that if there is a disbond in the layer 3 the incident energy from the previous layer is essentially reflected back, giving

$$T_{\text{system}} \simeq 0. \tag{5}$$

A similar argument applies to the delamination at layer 5. In practice, the transmission through a delamination is never completely zero. In the presence of a thin ($d_j \leq$ wavelength) delamination, some energy, though very small, does get transmitted due to the close proximity of the steel-rubber or rubber-polyethylene laminates. Consequently there is reduction in contrast between the images from ''good'' to ''bad'' regions.

Let us now consider the effect of delamination on the reflection coefficient. Since the thicknesses d_3 and d_5 of the adhesive layers are very small, it is to difficult resolve temporally the echoes reflected from the front and back surfaces of such layers. Consequently one depends, for the purpose of analysis, on the effect of loading of the incident acoustic field due to the presence of adhesive layer i.e. difference in amplitude due to presence or absence of adhesion. In pulse echo mode, the difference in amplitude of the echoes reflected from poly-adhesive interface with or without a good bond is comparable to the noise level. Therefore, a digital image of the amplitude variations of such an echo cannot clearly detect the areas of delamination (Fig. 2). However, a digital image formed by gating the echo from the rubber-steel interface can clearly detect the delamination between poly and rubber. Even though only

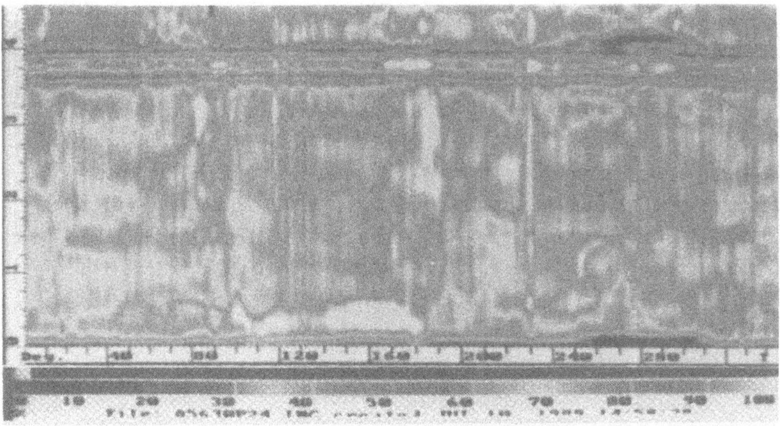

Fig. 2 — Digital image created from reflection amplitude
from rubber-polyethylene interface.

one transducer is used to perform pulse echo experiments, it can give information about transmission through interfaces also. For such a case the transmission through a delamination between poly-rubber would be almost zero and consequently the reflection from the rubber-steel interface in pulse echo mode would also be reduced to zero. The above argument can be extended to the detection of delaminations between the 2nd and 4th layers also (see Fig. 1).

For finite element analysis, we treat steel and polyethylene as rigid boundaries to the deforming rubber. Delamination faces are modeled by a lack of boundary constraints. The rubber itself is modeled as a hyperelastic material of Mooney-Rivlin form which allows large elastic strains for incompressible materials.[4] The elastic strain energy density at a point is given by

$$W = C_1(I_1 - 3) + C_2(I_2 - 3) \tag{6}$$

where W is the strain energy density , C_1 and C_2 are material constants and I_1 and I_2 are the first and second strain invariants. In this analysis, C_1 and C_2 are 0.24 MPa and 0.15 MPa respectively. The region of interest in the rubber is subdivided into a mesh of linearly interpolated, displacement-based 3-D continuum elements of a mixed formulation that includes pressure as an independent variable. A functional is formed in terms of the potential energies of the elastically deformed material and the boundary loads, with a Lagrange multiplier constraint imposed to enforce incompressibility. The variation of this functional is set equal to zero to obtain the equilibrium equations of the problem. This is given in an incremental form by

$$\int_{V_0} [C_1\delta\dot{I}_1 + C_2\delta\dot{I}_2 - p\delta\dot{J} - (\dot{J} - 1)\delta p] \, dV_0$$

$$- \int_{S_f} \vec{t}^* \cdot \delta\vec{v} \, dS = 0 \tag{7}$$

where p is the internal pressure, J is the ratio of deformed volume to original volume of material at a point, V_0 is the undeformed volume of the body, $\vec{v}$ is the displacement rate at a point, $\vec{t}^*$ are the applied tractions (loads) on the unconstrained part of the boundary, designated by S_f. The overscript dot refers to the rate of change of the relevant variable. Through displacement interpolation functions, all the kinematic variables can be expressed as functions of the element corner nodal displacements, which are the independent variables of the solution procedure[5,6] We assume that the displacement field satisfies both continuity throughout the body and the imposed displacement boundary conditions.

EXPERIMENTAL

Specimens used for this study are polyethylene-rubber-steel cylinders with outside diameter of 30.2 cm and height 12.5 cm. Thickness of steel, rubber and polyethylene laminates were 1.3 cm, 1.0 cm, and 1.4 cm respectively. The outer lamina of polyethylene has a circumferential groove of depth 0.5 cm and width 0.7 cm. For the purpose of this study, three slot-like delaminations were created. The size and location of these are as follows: (1) $\theta = 126°$, depth 5.0 cm and angular width 5.6° between rubber-steel interface, (2) $\theta = 54°$, depth 4.9 cm and width 3.5°, also between rubber-steel interface and (3) $\theta = 86°$, depth 7.2 cm and angular width 3.5° at rubber-poly interface. The purpose of these artificial disbonds is to simulate the delaminations that can occur and establish the measurement criteria for a component.

Transmission data was acquired by using a PC-based interactive ultrasonic imaging system consisting of a pulser (maximum output ~500 volt spike), broadband receiver, gates, display screen, etc. The display screen has a resolution of 450 × 512 pixels and 256 grey scale levels. The system is capable of updating the display buffer for visual observation during the data acquisition. The scanner can be programmed to control the rotational motion of the specimen and position of the scanning transducers. The cylinder can be scanned at 16 rpm and it takes about 10 minutes to acquire the digitial data for each specimen. The scanning resolution is $\Delta\theta = 0.70°$ and $\Delta z = 0.8$ cm. The system can create two-dimensional images from digital acoustic data, and also perform statistical analysis of the data. Various broadband transducers of center frequencies ranging from 2.25 to 10.0 MHz were used to excite an acoustic pulse. Some experiments were performed using a pulse-echo technique, i.e. using only one transducer placed approximately ~5.0 cm away from the outer poly surface and gating the signal from the back of the laminate next to the interface under evaluation—which essentially can evaluate the disbonds at a particular interface of a multilayered structure. Data is also taken using two matched transducers. The transmitting and receiving transducers are fixed relative to each other along the radius vector of the multilayered cylinder centered on a turntable. For transmission experiments, the outer transducer, used as a receiver, is placed as close to the outer lamina as is possible. The peak detected received signal for each (z, θ) position of ultrasonic beam is digitized at a sampling rate of 25 MHz using an 8-bit digitizer and stored on a hard disk. This data can be displayed as a 2-D unwrapped acoustical image of the cylinder.

Three dimensional images (to scale) of the acoustical properties of the structure were created using a 4D/70GT Silicon Graphics System. An algorithm was written to enable to rotate 3-D ribbon-like projections of the transmission amplitude. This facilitates the viewing of the multilayered cylinder for analysis of location and extent of the disbonds.

RESULTS AND DICUSSION

Figure 2 shows a 2-D unwrapped digital image of the variation in the amplitude for 2.25 MHz ultrasonic pulse reflected from the polyethylene-rubber interface. As discussed in the previous sections, the change in amplitude due to a simulated delamination located at $\theta = 86°$ and 5.6 cm $\leq z \leq 12.7$ cm is hardly distinguishable. This is due to the fact that the change in the reflection coefficient due to the absence of bond (i.e. delamination) is very small. The horizontal lines near the top are due to scattering losses in energy from the edges of the outer surface circumferential groove in the polyethylene lamina. However, if one images (Fig. 3) the variation in the amplitude of an echo reflected from the next inner interface i.e. rubber-steel, the notch at 86° is visible clearly. The other two notches—one located at $\theta = 54°$, 7.6 cm $\leq z \leq$ 12.7 cm and other at $\theta = 126°$ and $0 \leq z \leq 5.1$ cm are also visible in the ultrasonic image. In order for the reflection to take place from this interface (rubber-steel), the energy must be transmitted from the previous interface. Any delamination at the previous interface reduces drastically the energy transmitted to this interface and thus provides a good contrast (i.e. change in energy) in the image of the "so called" reflected echo. Essentially it is the transmission through an interface which is useful for detection of delaminations as predicted by the

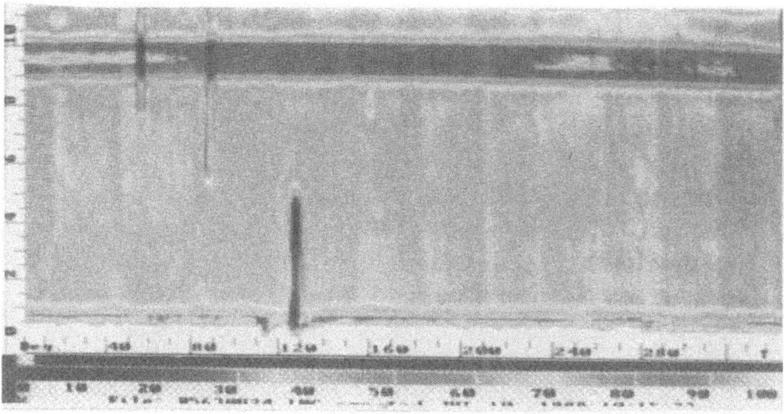

Fig. 3 — Digital image formed by reflection of 2.25 MHz pulse from steel-water interface. Note that notch-like disbonds invisible in Fig. 2 are clearly distinguishable from the background.

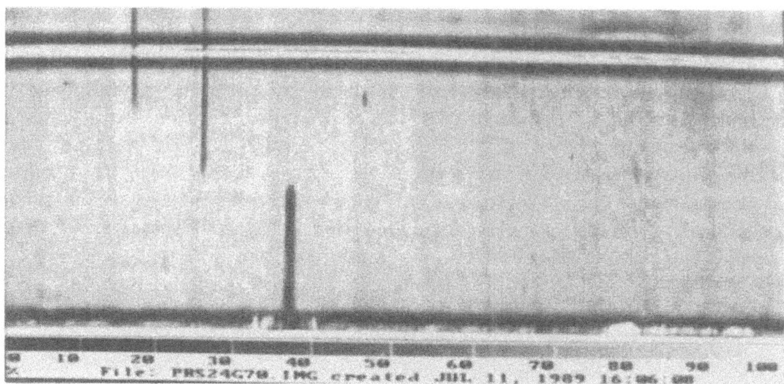

Fig. 4 — Digital image created by a 10 MHz
transmitted ultrasonic pulse.

mathematical model of propagation of ultrasound through a multilayered structure. The ultrasonic atttenuation due to scattering is frequency dependent. Consequently, at a higher frequency (for example 10 MHz) the contrast and sharpness of the edges increases (Fig. 4).

A computer routine was written to plot the digital transmitted amplitude as a function of (z, θ) in cylindrical coordinates (Fig. 5). Figure 5b represents the variations in the amplitude as seen by the receiving transducer for the cylinder shown in Fig. 5a. The algorithm is capable of rotating the image around any specified axis. This allows the viewer to focus the attention on areas of disbonds. A movie has also been produced showing this motion of the ribbon-like 3-D image.

The severity of the disbond is determined by analyzing the 2-D unwrapped image. One can split the image into a set of disjointed regions, each with constant acoustic transmission properties, separated by well defined boundaries.[7] One can then define a class component, i.e. value of transmission amplitude within a region and a position component (z, θ). These images

Fig. 5a — 3-Dimensional view of an adhesively bonded specimen.

Fig. 5b — 3-D ribbon-like projection image of transmitted amplitude through the multilayers of the cylinder.

represent the variations in acoustic amplitude and ignore the irrelevant variations, such as relative position of transducer, noise fluctuations, fine texture, etc. The statistics of the total population of pixels is a mixture of the variation of acoustic properties of the component. One can estimate the relevant parameters, such as mean and variance, for the total digital signal element array and then select optimal thresholds for separating the pixels belonging to different populations based on the acoustic property criterion, i.e. transmission coefficient tends to drop in the regions of delaminations. Such a method of clustering by color or spectral band histogram helps in the determination of disbond areas of the component—even though the histogram presents no information whatsoever on the location of the regions. Our computer-based instrumentation allows us to generate such image statistics. Figures 6(a,b) show the histogram of 2-D images of two specimens. From these one can compute the percentage of the total area for which transmission is poor, i.e. possibly disbonded areas. Figure 7 shows a plot of such fractional area against the number of specimens. It is interesting to note that the curve is a typical S-shape curve, similar to the distribution function of a standard normal distribution.

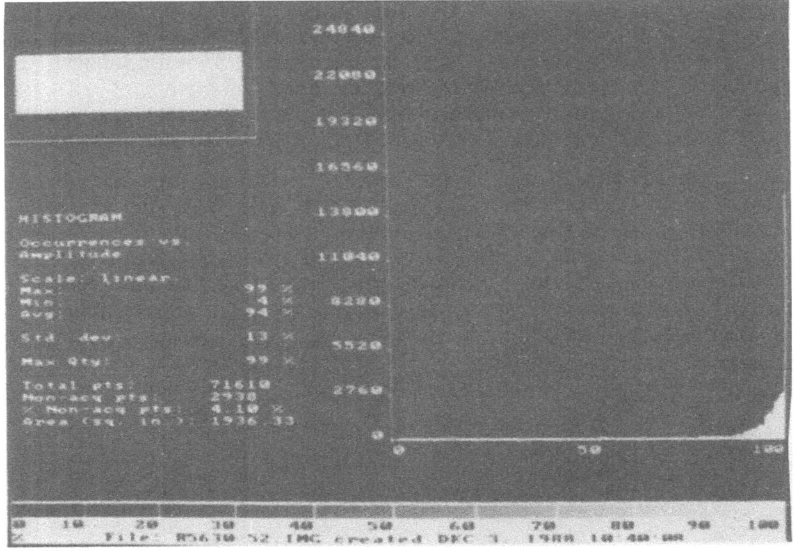

Fig. 6a — Histogram of digital transmission data
for a cylinder with disbonds.

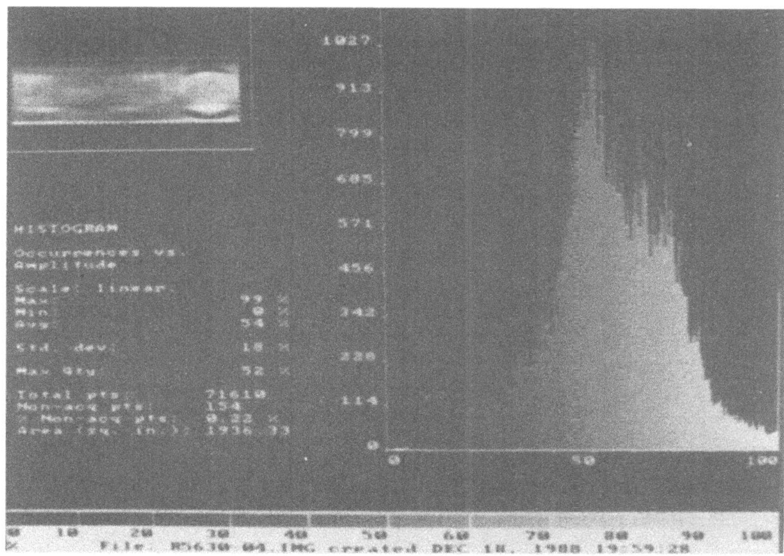

Fig. 6b — Histogram of digital transmission amplitude data
for a cylinder without disbonds.

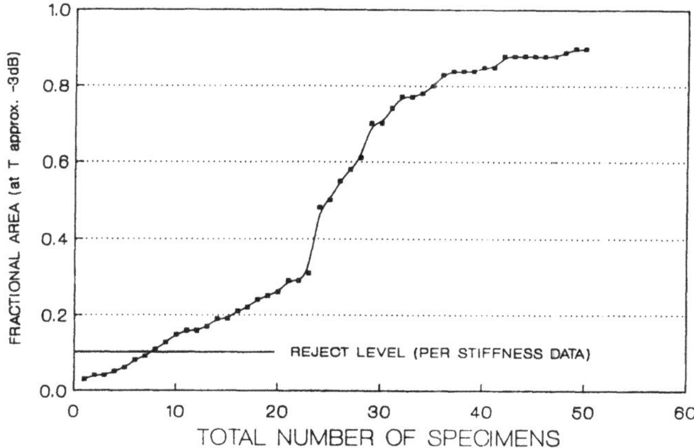

Fig. 7 — Fractional area (A) of the specimen with good ultrasonic tran-
simittance (area with transmission amplitude greater than -3 dB of the
maximum transmitted amplitude through a cylinder without disbonds) vs
the specimen population. Note that measured fractional quantity "A"
looks like a distribution function of a standard normal distribution.

Attention is focused on modeling the strain-energy density in the rubber due to the simulated disbonds at the interfaces. Figure 8 shows strain-energy density distribution around the
notch at $\theta = 86°$ at the rubber-polyethylene interface under a static loading. The analysis was
performed using the ABAQUS finite element program, with the model and graphics generated
by PATRAN software. The loading is a combination of an axial displacement of the inner surface of the rubber to a magnitude of 0.55 cm with a pressure loading of the lower edge to a
magnitude of 3.9 MPa. Notice that there is an increase in the energy around the boundaries of

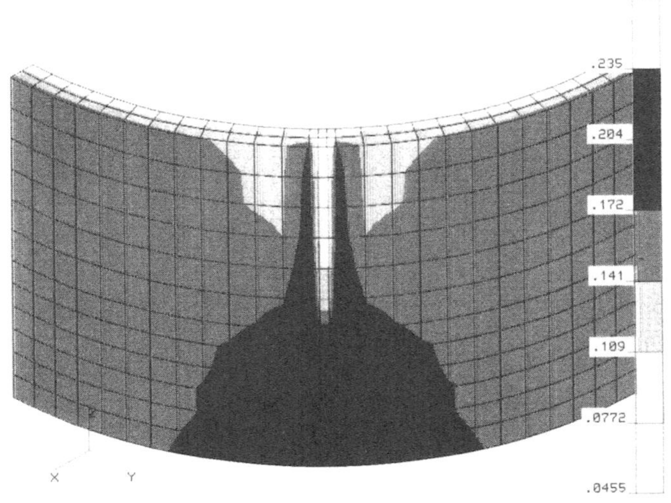

Fig. 8 — Finite element analysis of strain-energy density distribution
around a simulated and ultrasonically imaged disbond.

the notch, particularly around the tip. For simplicity, we modeled only a portion of the material surrounding the flaw. However, due to the overall symmetry of the cylinder and loading, the analytical results away from the flaw can be extended to the rest of the unmodeled region.

In this study, all the flaws are physically simulated on one specimen. However, for mechanical analysis, we assume that these flaws are isolated from each other. This approach is taken in order to determine the effect of each disbond individually on the mechanical performance of the component. In general, the effect of flaw interaction must also be considered. Each disbond produces a different pattern of stress intensification due to different loading conditions. The severity of the disbonds can be ranked by the maximum value of the strain energy produced. Under the loading conditions considered, the flaw shown in Fig. 8, results in the most severe straining of the rubber for the three delaminations simulated in the specimen. A determination of acceptability of a given disbond based on the corresponding calculated maximum value of strain energy requires experimentally produced values of critical strain energy for catastrophic failure of the component. Even without experimental correlation, the values of the strain energy give a measure of damage potential within the component. This forms the basis of an accept/reject criteria of the component.

SUMMARY

In this paper we have demonstrated how the transmission of ultrasound through adhesively bonded cylindrical layers can be used to create a 3-D projection image of the acoustical properties of the interfaces. The rotation of the 3-D ribbon-like image enables us to identify and view the disbonds at the interfaces from different viewing angles. From finite element analysis, we can interpret the severity of the delaminations for the component performance. Work is continuing to create solid 3-dimensional images of the structure depicting defects and debonds at correct (r, θ, z) coordinates instead of their projections in a plane normal to the radius vector.

ACKNOWLEDGMENTS

The authors would like to thank Dr. John Michopoulos of Geo-Center, Washington D.C. for his valuable assistance in writing the computer code for creating 3-D images.

REFERENCES

1. N.K. Batra and H.H. Chaskelis, Determination of Delaminations in Adhesively Bonded Layers Under Compression, in: Review of Progress in Quantitative NDE, Vol. 7B, D.O. Thompson and D.E. Chimenti, ed., Plenum Press, New York (1988).
2. D.L. Folds and C.D. Loggins, Transmission and reflection of ultrasonic waves in layered media, J. Acoust. Soc. Am. 62: 1102-1109 (1977).
3. L.M. Brekhovskikh, "Waves in Layered Media," Academic Press, New York (1980).
4. R.S. Rivlin, Large Elastic Deformations, in: "Rheology, Vol. 1," Frederick R. Eirich, ed., Academic Press, New York (1956).
5. "ABAQUS Theory Manual, Version 4.6," Hibbitt, Karlsson, and Sorensen, Inc., Providence, R.I. (1987).
6. J.T. Oden, "Finite Elements of Nonlinear Continua," McGraw Hill, New York (1972).
7. Roland Wilson and Michael Spann, "Image Segmentation and Uncertainty," Wiley, New York (1988).

APPLICATIONS OF HIGH RESOLUTION DECONVOLUTION TECHNIQUES TO ULTRASONIC NDE

C.H. Chen

Department of Electrical and Computer Engineering
Southeastern Massachusetts University
N. Dartmouth, MA 02747

S.K. Sin
Information Research Lab., Inc.
N. Dartmouth, MA 02747

ABSTRACT

The technique of ultrasonic pulse echo measurement is a very cost-effective method of NDE (Non-Destructive Evaluation). These measurements are, however, usually masked by the characteristics of the measuring instruments and the propagation paths taken by the ultrasonic pulses. With a proper modeling of the pulse echo, these effects can be reduced by deconvolution. In this paper, several high resolution deconvolution algorithms that have been popularly used in other areas such as seismic explorations are revised and adapted to ultrasonic NDE applications. Simulation results are presented to support the feasibility of using these algorithms to extract impulse responses from the ultrasonic pulse echoes for use in defect classifications. The performances and computational complexities of these algorithms are analyzed and compared. In addition, application of these deconvolution techniques to B-scan enhancement is also discussed.

INTRODUCTION

Ultrasonic pulse echo measurement is a very cost-effective method for the Non-Destructive Evaluation (NDE) of materials. This method (simply referred to as Ultrasonic NDE hereafter), which involves the coupling of short pulses at ultrasonic frequencies into the specimen and recording the echoes reflected from any acoustic discontinuity or acoustic mismatch, has been used extensively for the detection of hidden flaws embedded in materials [1,2]. In order to gain deeper insights into the characteristics of the internal flaws, more and more sophisticated signal processing techniques are now being incorporated in the analyses of these ultrasonic returns [3]. A very good example is the application of power spectral analysis; the received signals are first lowpass filtered and digitized at rates fast enough to avoid aliasing, and then their power spectra are computed by algorithms such as FFT. These spectra have been shown empirically as well as analytically to be closely related to the geometry and orientation of the ultrasonic reflectors [4,5].

The effectiveness of these processing techniques depends, of course, largely on the quality of the received signal and therefore, the quality of the interrogating ultrasonic pulse. The shape of this pulse is affected by electronic transmitter and receiver circuitries, the transducers and the propagation paths between the transducers and the reflectors in addition to the material tested. One way to alleviate the external effects is by using the technique of deconvolution. By means of deconvolution, a large portion of the irrelevant informations contained in a flaw echo, as far as the characteristics of the flaw is concerned, can be removed. In addition to being a signal enhancement step, the result of deconvolving a flaw signal, which we shall regard as the "impulse response" of the flaw, can also be used directly in pattern recognition for classification of the flaw into one of certain defect types.

Despite the many-fold benefits in the application of deconvolution to ultrasonic NDE, there are, however, no dedicated algorithms developed for this purpose. On the other hand, the technique of deconvolution has been applied successfully in other areas such as speech and seismic signal processings. In fact, many elegant algorithms have been developed especially for seismic explorations during the last few decades. Seismic and ultrasonic signal processings are very similar in many aspects. It seems logical to conceive that the spectrum of deconvolution techniques developed for the former can still be applied to the latter.

In this paper, we investigate and evaluate several popular seismic deconvolution algorithms in search of a well-behaved and yet efficient method for applications in ultrasonic NDE. Emphasis has been placed on certain modifications of these algorithms tailored for our applications and some practical considerations on their implementations. The relative performances and computational complexities of these algorithms are then compared in terms of simulation results on some ultrasonic signals. Based on the idea of seismic signal processing, examples of the application of deconvolution on B-scan enhancement are also presented.

THE APPLICATION OF DECONVOLUTION IN ULTRASONIC NDE

A minimum system for making ultrasonic measurements consists of a pulse generator (or pulser), a pulse receiver, a transmitting transducer, a receiving transducer and an oscilloscope. Most commercially available ultrasonic transducers can be used both as transmitter and receiver of ultrasonic waves. The pulse generator (or pulser) sends out an electrical pulse which is converted into a mechanical pulse in the transmitting transducer. The range of frequency of this pulse extends from about 0.2 to 100 MHz. The pulse thus generated is then coupled into the test specimen. An echo reflected from any acoustic mismatch in the material is converted into an electrical pulse which is received by the pulse receiver. The received echoes are amplified and displayed on the oscilloscope.

If the velocity of ultrasonic waves in the tested material has previously been determined, then the distance between the defect and the test surface can be deduced from the time of flight of the acoustic pulse. The defect location can therefore be derived by scanning through the surface of the test specimen with the ultrasonic transducer and noting the position at which the defect echo amplitude is the largest. Furthermore, if the defect is relatively greater than the width of the ultrasonic beam used for the scanning, the procedure will even yield an approximate geometric outline of the defect.

Theoretical studies which try to explain mathematically the relation between the pulse echo and the geometry of the reflector turns out to be exceedingly complicated even for a simple but non-trivial case [5]. As an alternative to theoretical approaches, Chen [3,6] has suggested the application of digital signal processing (DSP) and pattern recognition techniques to solve this flaw characterizing problem. In his approach, a large set of sample pulse echoes from different classes of defect types are first measured. Features are then extracted from each of these pulse echoes to form a feature-vector or pattern. These labeled patterns (i.e. patterns of known class) are collected to form a data-base for the classification of unknown patterns. The process of building up such data-base is commonly known as "training". This new "training". This new approach is both general and versatile. Although the selected features are usually physical parameters associated with the pulse-echo, this method is far less demanding in the understanding of the detail mechanism of the interactions between the ultrasonic waves and the reflectors than by theoretical analyses. However, by a suitable modeling of the ultrasonic signals, there is a potential improvement in the pattern recognition approach to NDE by using the technique of deconvolution.

In the pulse echo technique, measurements are obtained when the interrogating pulse is reflected from some acoustic mismatch along its propagation path. The shape of a pulse echo therefore depends on the geometry of the acoustic reflector from which it originates. Generally speaking, given a particular reflector, there is a specific input/output relationship between the input pulse and the echo. Therefore, if we model the reflector as a linear time-invariant (LTI) system with impulse response denoted by $h_d(t)$, this input/output relation can be represented by the following equation,

$$y(t) = x(t) * h_d(t) \tag{1}$$

where x(t) and y(t) represent the input pulse and the pulse echo respectively. With this model, defect of a particular geometry would be completely characterized by its impulse response.

In practice, a complete deconvolution operation which we discuss in the following sections starts by measuring the reference signal. The flawless specimen mentioned can be regarded as a kind of calibration block for this purpose. Results will be the best if the depth of the defect to be investigated is the same as the thickness of the calibration block.

The basic idea here is that different class of defects would be characterized by a different impulse response. By means of deconvolution, those irrelevant informations originally contained in the pulse-echoes are reduced to a minimal provided a good reference is available. From this standpoint, pattern recognition based on features extracted from the more informative impulse responses instead of the unprocessed pulse echoes provides a gateway to improve the performance of defect classification.

The usage of impulse response in ultrasonic NDE can be summarized [7] as follows: (1) Determination of Born diameter by using the Born Inversion Algorithm [8], which is related to defect size. The algorithm makes use of the impulse response, (2) Determination of the size of volumetric material defects by using a satellite-pulse observation technique [9] which makes use of the impulse response, (3) A signal enhancement step for spectral analysis in ultrasonic spectroscopy [4]. (4) Feature extraction from impulse response for defect classification, and (5) Enhancement of B-scan ultrasonic images.

DECONVOLUTION ALGORITHMS AND PRACTICAL CONSIDERATIONS

Four types of algorithms which have been used in seismic exploration [10] are considered here; namely, Wiener Filtering, Spectral Extrapolation, Curve-Fitting Methods and Minimum Variance Deconvolutions (MVD). Before going into the details, let us first summarize the notations that will appear repetitively in this section. Unless stated otherwise, these notations will be used consistently throughout the rest of this paper.

x(k):   the seismic wavelet in reflection seismology or the reference signal in ultrasonic NDE.

h(k):   the impulse response (also known as reflection series in seismology) to be uncovered.

$\hat{h}$ (k):   an estimate of h(k).

y(k):   the observed signal (or simply called the observations). In seismology, this is a single seismic trace; in NDE, this is an ultrasonic pulse echo.

$\hat{y}$ (k):   an estimate of y(k) as computed from $\hat{h}$ (k) by the following equation,

$$\hat{y}(k) = x(k) * \hat{h}(k)$$

n(k):   a noise sequence.

In addition, $X(w)$, $H(w)$, $\hat{H}(w)$, $Y(w)$ and $\hat{Y}(w)$ are used to denote the Fourier transforms of x(k), h(k), $\hat{h}$ (k), y(k) and $\hat{y}$(k) respectively.

WIENER FILTERING

The convolutional model for discrete-time signals and in the presence of noise can be expressed as,

$$y(k) = x(k) * h(k) + n(k) \tag{2}$$

Our goal is to find a least mean square error estimate of h(k) via the frequency domain and the solution is given by (e.g. see [11]),

$$\hat{H}(w) = \frac{Y(w) . X^*(w)}{|X(w)|^2 + S_n(w)/S_h(w)} \tag{3}$$

where $S_n(w)$ and $S_h(w)$ are the power spectral densities of n(k) and h(k) respectively. We may modify this solution by introducing a constant p in eq. (3) such that

$$\hat{H}(w) = \frac{Y(w) . X^*(w)}{|X(w)|^2 + p . S_n(w)/S_h(w)} \tag{4}$$

Walden [12] has shown that, if

$$p = \left(1 + \lambda_2\right)/\lambda_1 \quad \text{where} \quad \lambda_1 > 0 \text{ and } \lambda_2 \geq 0,$$

then eq. (4) is the solution to the minimization of

$$\lambda_1 . E\left\{(h(k) - \hat{h}(k))^2\right\} + \lambda_2 . E\left\{\tilde{n}^2(k)\right\} \tag{5}$$

where

$$\tilde{n}(k) = n(k) * F^{-1}\left\{\frac{X^*(w)}{|X(w)|^2 + p . S_n(w)/S_h(w)}\right\}$$

is the filtered noise.

Therefore, by varying the parameter p, we are varying the percentage of efforts put into minimizing the estimation error and the filtered noise. In practice, to avoid the estimation

166

of the power spectral densities in the above equations, we may simply replace $S_n(w)/S_h(w)$ by a constant and eq. (4) becomes,

$$\hat{H}(w) = \frac{Y(w) \cdot X^*(w)}{|X(w)|^2 + q} \tag{6}$$

where q is a positive real constant to be decided. This constant is sometimes known as the "noise-desensitizing factor". We call the Wiener Filter formulation in eq. (6) the Wiener filter A.

The ultrasonic signals of our interests are combinations of damped sinusoids. Therefore we can model the signals x(k) and y(k) as AR (auto-regressive) processes [13] with Fourier transforms given by,

$$X_{AR}(w) = \frac{1}{1 + \sum_{k=1}^{p} a_k \exp(-jkw)} \tag{7}$$

$$X_{AR}(w) = \frac{1}{1 + \sum_{k=1}^{o} b_k \exp(-jkw)} \tag{8}$$

where P and Q are the auto-regressive orders while a_k and b_k the auto-regressive coefficients of x(k) and y(k) respectively. We may replace X(w) and Y(w) in eq. (6) by $X_{AR}(w)$ and $Y_{AR}(w)$ to obtain,

$$\hat{H}(w) = \frac{Y_{AR}(w) \cdot X^*_{AR}(w)}{|X_{AR}(w)|^2 + q} \tag{9}$$

our task is then to extrapolate $\hat{H}(n)$ from $N_1 - 1$ to 0 and from $N_2 + 1$ to N/2. To do this, let

$$\hat{H}(n) = -\sum_{k=1}^{p} d_k^f \cdot \hat{H}(n-k) \qquad , N_2 < n \le N/2 \tag{10a}$$

$$\hat{H}(n) = -\sum_{k=1}^{p} a_k^b \cdot \hat{H}(n+k) \qquad , 0 \le n < N_1 \tag{10b}$$

where $\qquad d_k^f$ and a_k^b

are the P-th order forward and backward prediction coefficients of $\hat{H}(n)$ respectively. These coefficients can be computed by using Burg's Technique [14].

<u>**Curve-Fitting Methods**</u>

Let

$$\vec{h} = (h(0), h(1), h(2), \ldots\ldots)^T$$

$$\vec{y} = (y(0), y(1), y(2), \ldots\ldots)^T$$

$$X = \begin{pmatrix} x(0) & 0 & \cdot & \cdot & \cdot & 0 & 0 \\ x(1) & x(0) & \cdot & \cdot & \cdot & 0 & 0 \\ x(2) & x(1) & \cdot & \cdot & \cdot & 0 & 0 \\ \cdot & x(2) & \cdot & \cdot & \cdot & \cdot & \cdot \\ \cdot & \cdot & \cdot & \cdot & \cdot & x(0) & \cdot \\ \cdot & \cdot & \cdot & \cdot & \cdot & x(1) & x(0) \\ \cdot & \cdot & \cdot & \cdot & \cdot & x(2) & x(1) \\ \cdot & \cdot & \cdot & \cdot & \cdot & \cdot & x(2) \\ \cdot & \cdot & \cdot & \cdot & \cdot & \cdot & \cdot \\ \cdot & \cdot & \cdot & \cdot & \cdot & \cdot & \cdot \end{pmatrix}$$

The convolution model can then be expressed in matrix form as,

$$\vec{y} = X \cdot \vec{h} \tag{11}$$

where h and y are vectors and x is a non-square matrix. Our goal is to find an estimate of the vector h such that the error e, which is defined as,

$$e = ||\vec{y} - X \cdot \hat{\vec{h}}|| \quad , \tag{12}$$

is minimized.

This is equivalent to a curve-fitting problem. In particular, we are interested in minimizing the L1 norm and L2 norm in eq. (12).

<u>**1. L2 Deconvolution**</u>

In this method, the L2 norm in eq. (12) is minimized. The solution to this problem (see. e.g. [15] [16] is given by,

$$\left(X^T \cdot X \right) \cdot \hat{\vec{h}} = X^T \cdot \vec{y} \tag{13a}$$

or

$$\hat{\vec{h}} = \left(X^T \cdot X \right)^{-1} \cdot X^T \cdot \vec{y} \tag{13b}$$

The matrix $(X^T X)$ is Toeplitz and therefore fast algorithms such as the Levinson Recursion [17] can be used to obtain $\hat{h}$ in eq. (13) without explicitly finding the inverse matrix of $(X^T X)$. Furthermore, instead of using y(k) in the above equations, we use $y(k-k_0)$. The solution of $\hat{h}$ is quite sensitive to the choice of the parameter k_0. From our experiences with ultrasonic pulse echoes, k_0 is chosen such that the two signals x(k) and y(k) are aligned.

In this method, the L1 norm in eq. (12) is to be minimized. This is a more time-consuming optimization process [18]. The estimate thus obtained is, however, more robust in the sense that the error vector (y -X$\hat{h}$) will contain more zeros than that obtained by L2 deconvolution [19]. This property is very desirable in some applications such as seismic exploration. Two approaches to this problem will be discussed: 1. By Linear' Programming, 2. By iteratively refining an L2 solution.

The application of Linear Programming [18] to solve for the system defined by eqs. (11) and (12) in the L1 norm has been studied extensively by Barrodale, et al [20] [21]. It is, however, very computational expensive to find a global optimal solution for the vector h. In many cases, this forces us to be contended with a sub-optimal solution. The algorithm of "One-at-a-time Spike Extraction" suggested by Barrodale et al [22] is such an algorithm which can be summarized as follows.

Initially, assume there is only one non-zero element in h(k). The optimal location of this single spike is then found by brute force (i.e. by computing the error for each possible position using linear programming an choosing the one with minimum error). This position is then fixed and carried to the next iteration. At the j-th iteration, assume there are j+1 spikes in h(k), the positions of j of which are carried from the j-th iteration. Our task during this iteration is to look for the optimal (j+1)-th spike location, again, by brute force. The iteration continues until the required number of spikes have been extracted. This algorithm is locally optimal in the sense that both the new spike location and all the amplitudes of $\hat{h}$(k) at the (j+1)-th iteration are chosen to yield the maximum possible drop in e, the estimation error, in moving from the j-th to the (j+1)-th iteration.

Minimum Variance Deconvolution (MVD)

In this method, a state-space approach is used which enables the application of Kalman Filtering in deconvolution, to deal with time-varying or non-stationary processes. The readers are referred to Refs. 7 and 23 for mathematical details of the method.

SIMULATION RESULTS AND DISCUSSIONS

The algorithms described in the last section, except the L1 Spike-Extraction algorithm, have been implemented on an IBM PC compatible running under DOS. The L1 algorithm is implemented on a VAX 785 machine running under the VAX/VMS.

The ultrasonic signals used for these simulations are shown in Fig. 1; namely, T15A0, T15A1, T15A2 and T15A3. These are part of a larger data set obtained from the Army's Material Technology Laboratory (Watertown, MA). T15A0 is the reference signal, while T15A1, T15A2 and T15A3 are respectively the pulse echoes from a flat-cut, an angular-cut and a circular-hole (see Ref. 6 for an illustration of these defect geometries). The ultrasonic frequencies used in these measurements are centered at 15 MHz. Each trace contains 512 data points digitized at a rate of 100 megasamples/second. The impulse responses estimated by the algorithms Wiener filter A, spectral extrapolation and L2 are shown in Figs. 2, 3, 4 respectively. These three are considered most effective for ultrasonic NDE. Computer results on other algorithms are available in Ref. 7.

Some Discussions

For seismic applications, we are mostly interested in the location and amplitudes of the spikes extracted from a seismic trace by deconvolution. In ultrasonic NDE, however, we are more interested in the characteristics of the spikes themselves, since they contain informations about the corresponding reflectors. We believe that the hidden impulse responses, if exist, should be in the form of smooth transients of short durations (comparable

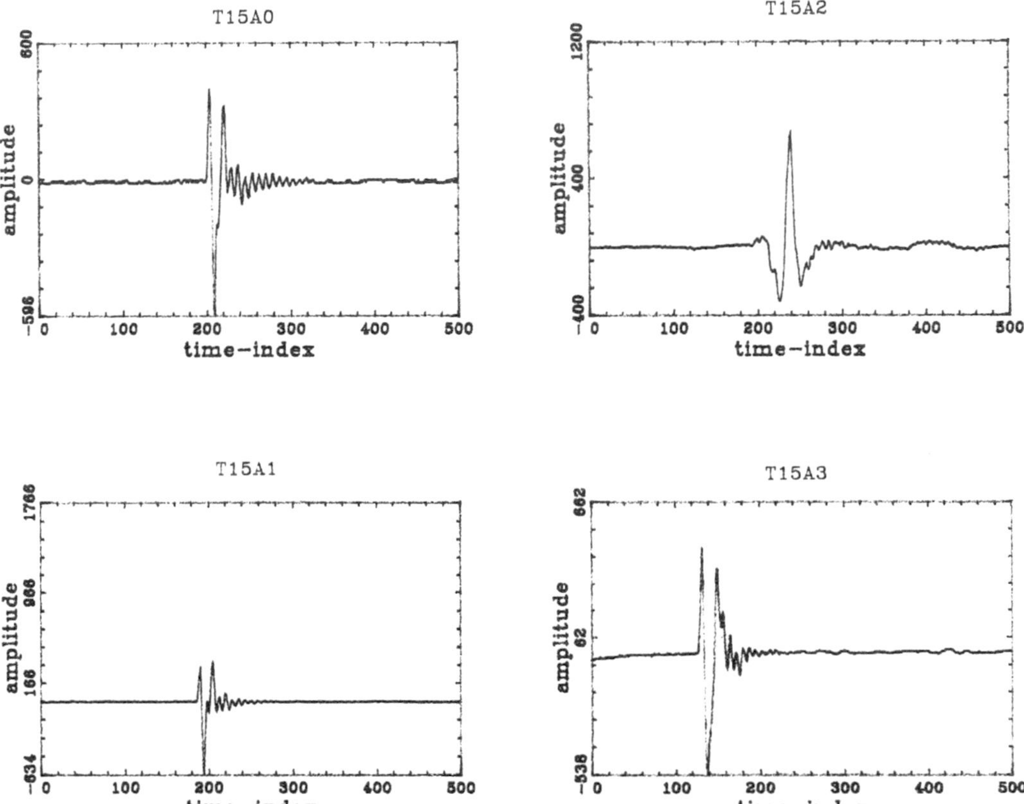

Fig. 1. Typical ultrasonic signals considered.

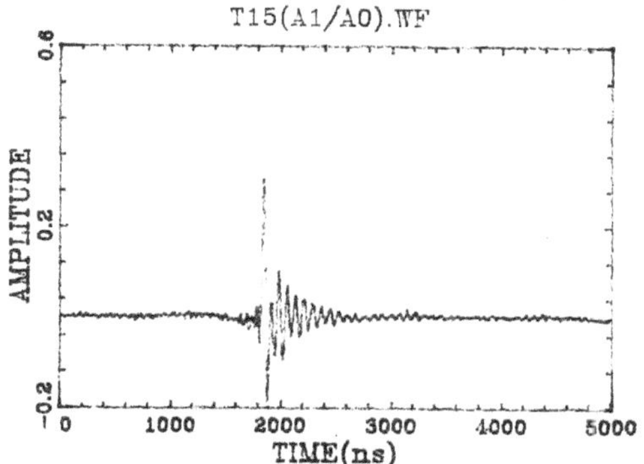

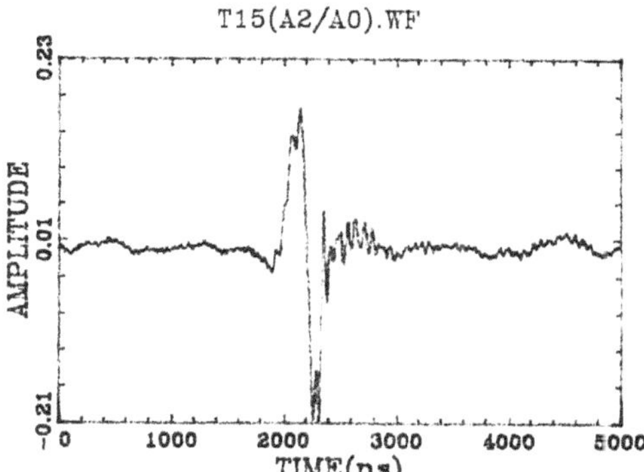

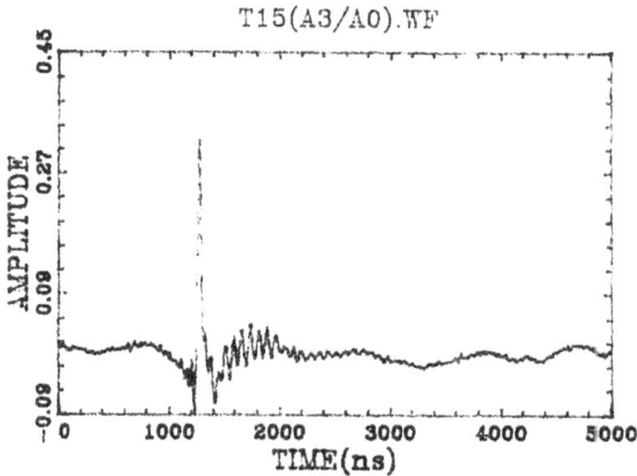

Fig. 2.   Deconvolution by Wiener filter A.

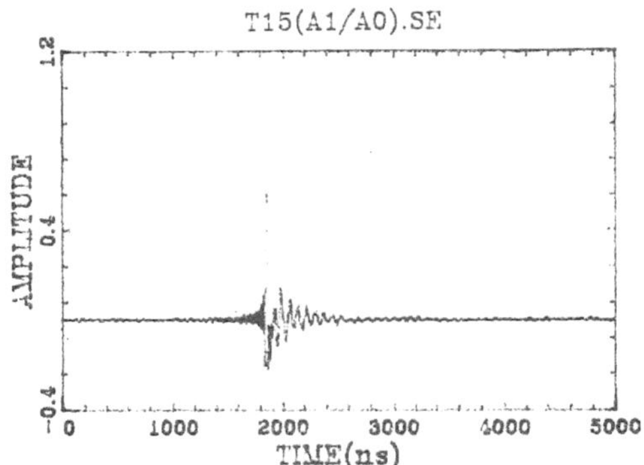

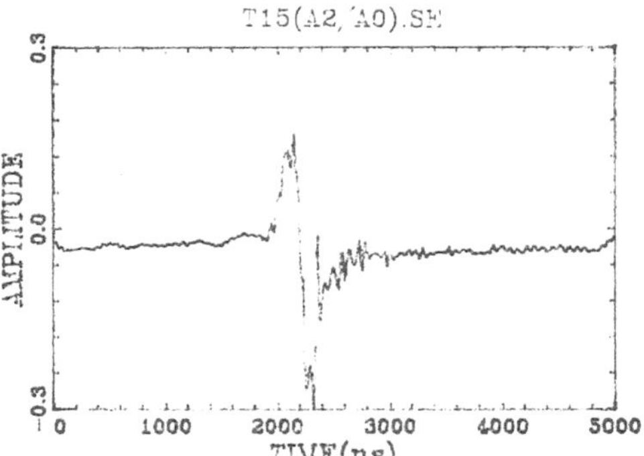

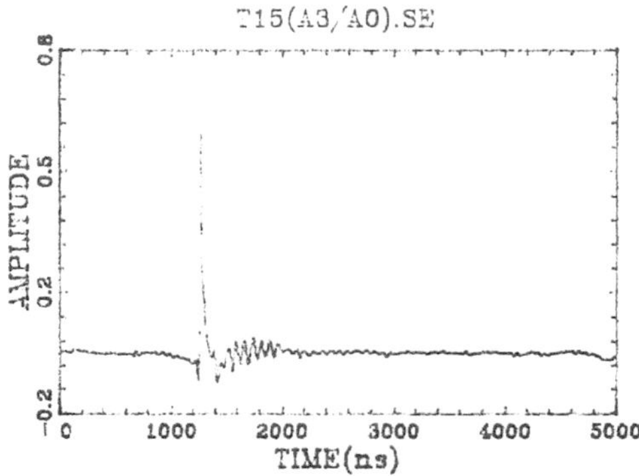

Fig. 3. Deconvolution by Spectral Extrapolation.

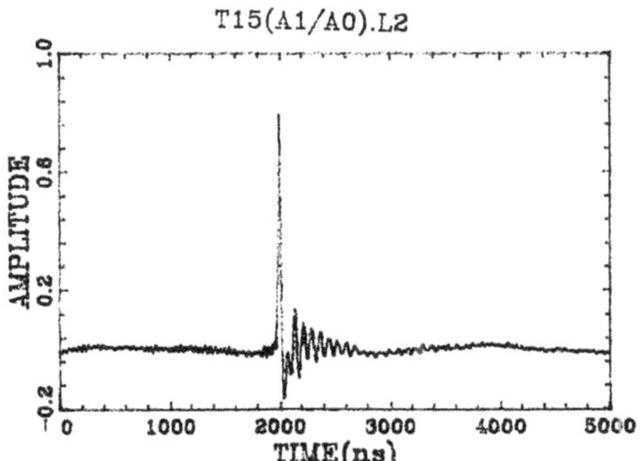

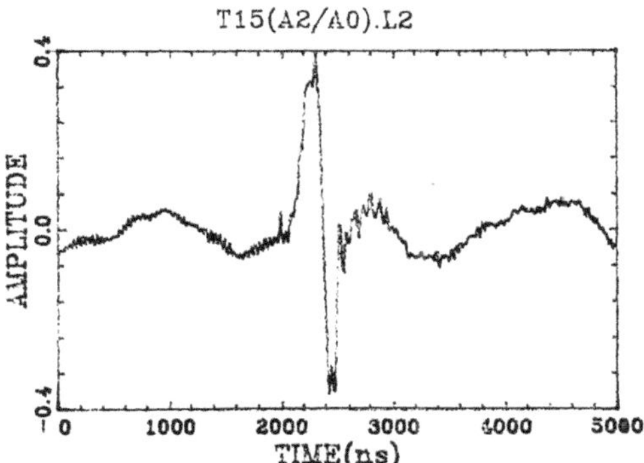

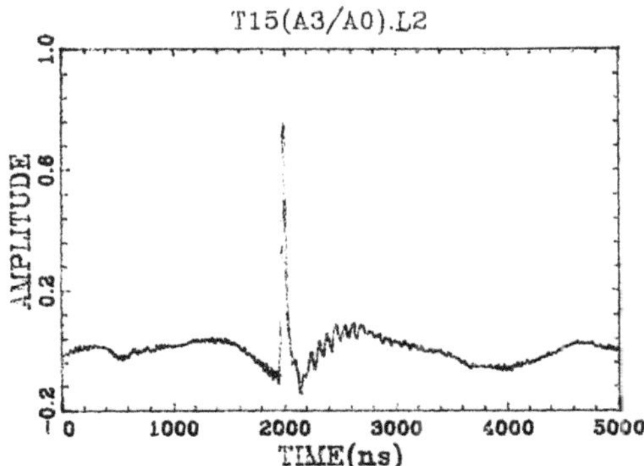

Fig. 4. L2 deconvolution.

to a single spike in a reflection series in seismology). In fact, the motivation for the formulation of the Wiener Filter B, in which X(w) and Y(w) in eq. (6) are replaced by $X_{AR}(W)$ and $Y_{AR}(W)$ (defined in eq. (7) and eq. (8)), is partly based on this argument. The Spectral Extrapolation is also superior to the conventional Wiener filtering by comparing Fig. 2 with Fig. 3. The two signals x(k) and y(k) have been carefully aligned before running the L2 and L1 deconvolution algorithms. The results obtained by the L2 method usually need some smoothing.

On Computational Complexities

The purpose of this section is to give some idea about the relative efficiencies of the algorithms discussed in this paper. Computational complexities are expressed in terms of the number of real multiplications required by each method. In addition, the following simplifications have been made:

1)   The record lengths of x(k) and y(k) are both equal to N.

2)   The number of real multiplications required by an N-point FFT is the same as that by an Inverse FFT of the same order, which is approximately equal to $(3N/2) \text{Log}_2(N)$.

3)   A real division has the same complexity as a real multiplication.

4)   One complex multiplication is equivalent to four real multiplications.

5)   The complexity of Burg's Technique for computing the P-th order AR coefficients from an N-point sequence is dominated by the number of computations required to compute the reflection coefficient a_k (k = 1, 2, 3 ..., P) at each step of the Levinson recursion algorithm [17], which is approximately given by NP. If the sequence is complex, this number becomes 4NP.

For Wiener Filter A, 2 N-point FFT's are required to compute X(w) and Y(w). To compute $\hat{H}(w)$ from X(w) and Y(w) requires N complex multiplications and 4N real multiplications which add up to BN real multiplications. To compute $\hat{h}(k)$ from $\hat{H}(w)$ requires another N-point FFT, which makes the total number of real multiplications required by this method equals to $(9N/2)\text{Log}_2(N) + 8N$.

If the Wiener Filter B is used, we require to run Burg's algorithm twice in order to compute the AR coefficients for both x(k) and y(k). $X_{AR}(w)$ and $Y_{AR}(w)$ are obtained from these coefficients through 2 N-point FFT's. $\hat{H}(w)$ is then computed in a similar way as in Wiener Filter A. If the AR order used for x(k) and y(k) are both equal to P, the total number of real multiplications required is given by $2NP + (9N/2)\text{Log}_2(N) + 8N$.

For the Spectral Extrapolation technique, we first run the Wiener Filter A algorithm to obtain $\hat{H}(w)$ in the "reliable frequency region". Let the number of elements outside this region be d. Extra computations are then required to compute: (a) the linear prediction coefficients, say of order P, from this complex sequence using Burg's Technique (which needs about $4(N/2 - d)P$ real multiplications), (b) the rest of the frequency components of $\hat{H}(w)$ by extrapolation (which needs 4dP multiplications). Total number of real multiplications required is then about $2NP + (9N/2)\text{Log}_2(N) + 8N$.

L2 deconvolution requires the Toeplitz matrix (X^TX) and the vector (X^Ty) which may be computed by FFT [24]. The total number of multiplications required is about $(6N)\text{Log}_2(N) + 8N + 2.5N^2$.

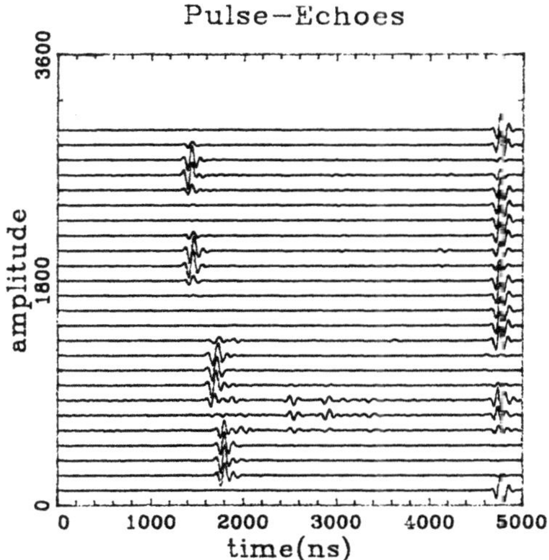

Fig. 5.   An unprocessed B-scan image.

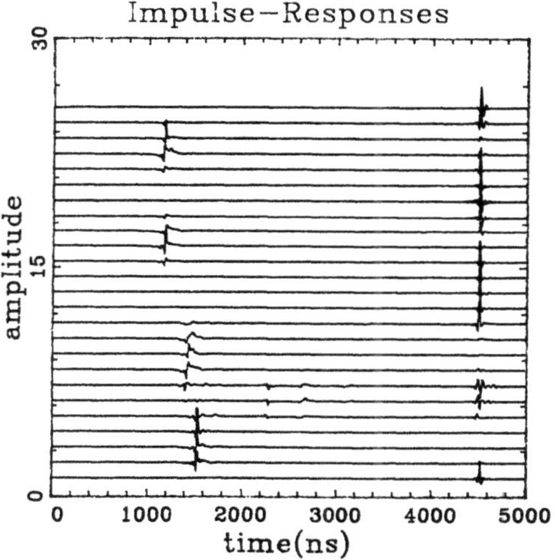

Fig. 6.   The enhanced image of Fig. 5.

<u>**Time-Domain versus Spectrum-Based Techniques**</u>

It is interesting to note that the algorithms we have discussed so far can be divided into two classes. The first class is what we called time-domain techniques, i.e., computations are carried out directly in the time domain without, or only implicitly, using Fourier transforms. L1, L2 and MVD can be put into this class. The second class of algorithms are totally computed via the frequency domain and/or by transformation back and forth between the time and the frequency domains. Wiener Filtering (A and B) and Spectral Extrapolation belong to this class.

Recall the oscillatory effects found in the results of Wiener filtering A. The anomalous structures are due to the band-limited nature of the input signals. The "lost band-widths" are partially recovered by a suitable modeling of the input signals, as that done in Wiener Filter B, or by extrapolation algorithm, as that done in the Spectral Extrapolation method. On the other hand, there are no band-width problems in the time-domain techniques such as L1 and L2 deconvolutions. One drawback of these 2 algorithms are that larger variances are usually associated with the estimates, and if the input sequences are not properly shifted relative to one another, these methods collapse.

In considering the computation requirements of these algorithms, spectrum-based techniques are generally more efficient than time-domain techniques. Since the ultimate goal of ultrasonic NDE is to provide real-time diagnostics in many situations, efficiency is a very important issue in the selection of a computation algorithm. Therefore, if a good way for band-width correction is available, spectrum-based algorithms, such as the spectral extrapolation algorithm, are usually preferred.

APPLICATION OF DECONVOLUTION FOR B-SCAN ENHANCEMENT

The method of ultrasonic B-scan is very similar to the idea used in seismic signal processing. Therefore, a B-scan image can also be enhanced by using the technique of deconvolution. Fig. (5) shows an unprocessed B-scan of an aluminum with some internal defects. Fig. (6) shows the enhanced result obtained by performing deconvolution on each trace.

ACKNOWLEDGEMENT

This work was supported by Contract DAAL04-88-C-0003 to Information Research Laboratory, Inc. The authors gratefully acknowledge the advices, suggestions and encouragement of Dr. Otto R. Gericke of US Army Materials Technology Laboratory in performing this work.

REFERENCES

1.   R. Halmshaw, "Nondestructive Testing", Edward Arnold (Publishers) Ltd., 1989.

2.   J. Szilard (Ed.), "Ultrasonic Testing", John Wiley & Sons, 1982.

3.   C.H. Chen, "A signal processing study of ultrasonic nondestructive evaluation of materials", Technical Report MTL TR 87-11, Material Technology Laboratory, Watertown, Feb., 1987.

4.   O.R. Gericke, "Determination of the geometry of hidden defect by ultrasonic pulse analysis testing", Journal of the Acoustical Society of America, Vol. 35, No. 3, Mar., 1963.

5.   L. Adler, K.V. Cook and W.A. Simpson, "Ultrasonic frequency analysis", in "Research Techniques in Nondestructive Testing Vol. 3" by R.S. Sharpe (Ed.), Academic Press, London, NY, 1977.

6.   C.H. Chen, "High resolution spectral analysis NDE techniques for flaw characterization, prediction and discrimination", in "Signal Processing and Pattern Recognition in Nondestructive Evaluation of Materials" by C. H. Chen (Ed.), NATO ASI Series, Springer-Verlag, 1988.

7.   C.H. Chen and S.K. Sin, "High resolution deconvolution techniques and their applications in ultrasonic NDE", unpublished report of Information Research Lab., Inc. April 1989. A major part of the report will appear in Journal of Imaging Systems and Technology, vol. 1, no. 2, 1989.

8.   L.J. Bond, J.H. Rose, S.J. Wormley and S.P. Neal, "Advances in Born Inversion", in "Signal Processing and Pattern Recognition in Nondestructive Evaluation of Materials", C.H. Chen, ed., Springer-Verlag, Berlin Heidelberg, New York, 1988.

9.   G.T. Gruber, G.J. Hendrix and T.A. Mueller, "Exploratory development for a high reliability flaw characterization module", Technical Report AFWAL-TR-4172, Material Laboratory, Air Force Wright Aeronautical Laboratories, Ohio, March 1985.

10. M.T. Silva and E.A. Robinson, "Deconvolution of Geophysical Time Series in the Exploration for Oil and Natural Gas", Elsevier Scientific Publishing Co., 1979.

11. J.B. Thomas, "An Introduction to Statistical Communication Theory", Wiley, 1967.

12. A.T. Walden, "Robust deconvolution by modified Wiener filtering", Geophysics, Vol. 53 No. 2, Feb., 1988.

13. S.L. Marple, Jr., "Digital Spectral Analysis with Applications", Prentice-Hall, Inc., 1987.

14. J.P. Burg, "Maximum Entropy Spectral Analysis", Proceedings of the 37th Meeting of the Society of Exploration Geophysics, 1967.

15. C.L. Lawson and R.J. Hanson, "Solving Least Squares Problems", Prentice-Hall, Inc., 1974.

16. G.H. Golub and C.F. Van Loan, "Matrix Computation", Baltimore: Johns Hopkins University Press, 1983.

17. N. Levinson, "The Wiener R.M.S. (root mean square) error criterion in filter design and prediction", Journal on Mathematical Physics, Vol. 25 pp. 261-278, Jan., 1947.

18. M.J. Best and K. Ritter, "Linear Programming", Prentice-Hall, Inc., 1985.

19. R. Yarlagada, J.B. Bednar and T.L. Watt, "Fast algorithm for L_p deconvolution", IEEE Trans. on Acoustic, Speech and Signal Processing, Vol. ASSP-33 No. 1, Feb., 1985.

20. I. Barrodale and F.D.K. Roberts, "An improved algorithm for discrete L1 linear approximation", SIAM Journal on Numerical Analysis, Vol. 10, No. 5, Oct., 1973.

21. I. Barrodale and F.D.K. Roberts, "Solution of an over determined system of equations in the L1 norm", Communications of the Association of Computing Machinery, Vol. 17 No. 6, June 1974.

22. I. Barrodale, et al, "Comparison of the L1 and L2 norms applied to one-at-a-time spike extraction from seismic traces", Geophysics, Nov., 1984.

23. J.M. Mendel and J. Kormylo, "New fast optimal white noise estimators for deconvolution", IEEE Trans. on Geoscience Electronics, Vol. GE-15, No. 1, Jan. 1977.

24. R. Yarlagada and B.N.S. Babu, "A note on the application of FFT to the solution of a system of Teoplitz normal equations", IEEE Trans. on Circuits and Systems, vol. CAS-27, no. 2, Feb. 1980.

CHARACTERIZATION OF LAYERED STRUCTURES BY A LIQUID WEDGE TRANSDUCER AND A CORNER REFLECTOR

A. Atalar, H. Köymen[†], and O. Yemişçiler[†]

Electrical and Electronics Engineering Department
Bilkent University, Ankara, Turkey
† Middle East Technical University, Ankara, Turkey

ABSTRACT

A layered solid structure immersed in a liquid supports leaky interface waves in the form of Rayleigh waves or layer waves such as generalized Lamb waves. The modes are unique to the elastic and geometrical properties of the layered structure and they can be used to characterize the structure. We introduce a special measurement geometry consisting of a liquid wedge transducer and a corner reflector that can detect the presence of the leaky modes in the layered structure. The received signal amplitude as a function of the incidence angle depends highly on the elastic parameters and thickness of the layers. The same problem is treated also theoretically. The experimental results are compared with theoretical predictions and they are found to be in good agreement.

INTRODUCTION

Solid structures having layers made up of different materials are technologically very important and their characterization is not very easy. A layered solid structure immersed in a liquid supports acoustic leaky interface waves in addition to the acoustic bulk waves. The leaky waves can be in the form of Rayleigh waves or layer waves such as generalized Lamb waves. These modes can be excited by plane waves in the liquid incident at a critical angle. The excitation and detection of leaky modes are responsible for the principal contrast mechanism in the scanning acoustic microscope [1]. We introduce a special geometry to measure the excitation efficiency of the leaky modes supported in the layered structure for the purpose of characterizing it. By recording the received signal amplitude as a function of the incidence angle, a curve is obtained, which depends highly on the elastic parameters and thickness of the layers. This curve has peaks at angular positions corresponding to critical angles where the coupling to leaky modes is maximum. Hence, the amplitude of the peaks will be proportional to the excitation and detection efficiency of the leaky modes. The excitation problem is investigated also theoretically by computing the reflection coefficient of acoustic plane waves at the liquid-layered-solid interface. This reflection coefficient exhibits phase transitions corresponding to the leaky modes. Using an angular spectrum approach it is possible to calculate the behavior of the proposed measurement geometry. The agreement between the experimental results and theoretical predictions is discussed.

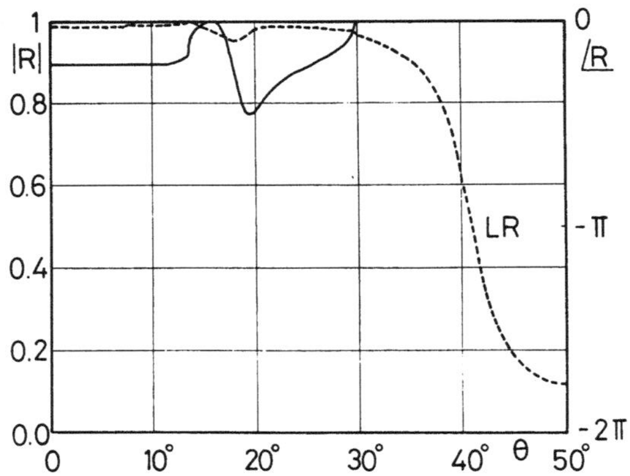

Figure 1. Plane wave reflection coefficient amplitude (solid) and phase (dashed) for a 0.05 mm copper layer on an aluminum substrate at 5.4 MHz ($k_{ts}d$=0.557).

EXCITATION OF LAYER MODES

It is well known that the Rayleigh waves can be excited on the planar surface of a solid material immersed in a liquid by bulk waves incident on the surface from the liquid side, provided that the angle of incidence is at the Rayleigh critical angle [2]. This forms the basis of the wedge transducers [3,4]. The excitation of a Rayleigh wave on the surface exhibits itself as a phase transition at the critical angle in the plane wave reflection coefficient as a function of incidence angle [5].

Waves excited in a layered solid structure will be either bulk waves or leaky interface waves. The leaky waves can be in the form of leaky Rayleigh waves [6] or layer waves such as *generalized Lamb waves* [7]. The reflection coefficient at the liquid-layered-solid interface has phase transitions corresponding to these modes. The angular positions of the phase transitions can be used to predict the incidence angle of the bulk wave in the liquid for efficiently exciting the respective layer modes.

An analytic expression for the reflection coefficient at the liquid-layered-solid interface for a single layer is given in the literature [8,9]. We evaluated this reflection coefficient for some layered structures. Fig. 1 shows the reflection coefficient amplitude and phase between water and a thin copper layer on an aluminum substrate for $k_{ts}d$=0.557, where k_{ts} is the *shear* wavenumber of the *substrate* and d is the layer thickness. The elastic constants of materials involved were taken as follows: for water, $c_{11} = .2277 \cdot 10^{10}$ Nt/m^2, $\rho = 1000$ kg/m^3; for aluminum, $c_{11} = .111 \cdot 10^{12}$, $c_{44} = .25 \cdot 10^{11}$, $\rho = 2695$; and for copper, $c_{11} = .22335 \cdot 10^{12}$, $c_{44} = .4599 \cdot 10^{11}$, $\rho = 8500$ As seen in the figure, for a thin layer of copper there exists a single transition in the phase which corresponds to a leaky Rayleigh-like (LR) wave. For this particular case, this mode is faster than the Rayleigh wave on the free copper surface. As the layer thickness is increased (see Fig. 2), a second ($M2$) and a third mode ($M3$) appear as indicated by the phase transitions in the reflection coefficient where the amplitude is unity. These are generalized Lamb wave modes that are leaky due to the presence of the fluid. More generalized Lamb wave modes ($M4$, $M5$) emerge when the layer thickness is further increased as depicted in Fig. 3. Notice that, in all cases the amplitude of the reflection coefficient is unity for $\theta > 29.7°$ and remains close to unity for $\theta < 29.7°$, indicating that there is not much power transfer into the solid from the liquid side due to impedance mismatch. Moreover, from one case to the other there is not much difference between the amplitude curves. On the other hand, the phase of the reflection coefficient varies considerably as the layer thickness is changed, making it the more interesting component of the reflection coefficient.

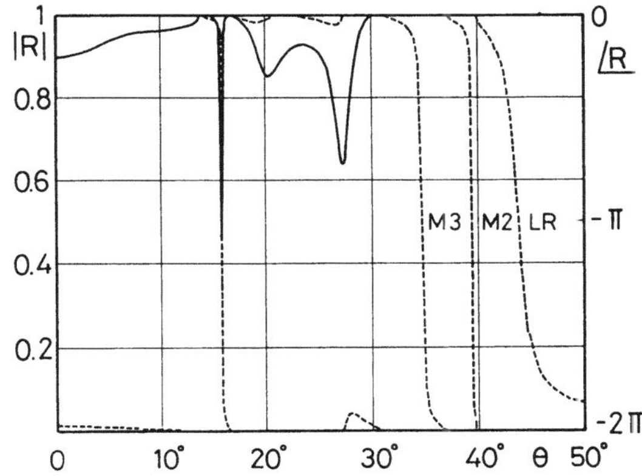

Figure 2. Plane wave reflection coefficient amplitude (solid) and phase (dashed) for a 0.9 mm copper layer on an aluminum substrate at 5.4 MHz ($k_{ts}d$=10.0).

The existence of various modes and their excitation angles are compactly presented in Fig. 4 as a function of the $k_{ts}d$. This figure is actually a different way of displaying the dispersion relation [10]. It depicts only the leaky waves, it does not show bulk waves which may also be excited [11]. LR mode has a wave speed close to that of Rayleigh wave in the substrate material when the layer is very thin. As the layer thickness is increased the phase velocity decreases rapidly, and it reaches a minimum with a value lower than that of the Rayleigh wave in the layer material. As the layer thickness is further increased, the wave gets faster and it approaches to the speed of the Rayleigh wave in the layer material. $M2$ mode exists only for layer thicknesses above a certain critical value. At the cutoff point the wave speed is equal to that of the shear wave in the substrate material, and for a thicker layer it approaches to the shear wave speed of the layer material. $M3$ and higher modes have a very similar behavior to $M2$ mode except that they exist for even thicker layer widths.

It was shown that only a single mode (LR) can exist when the layer material has a higher transverse speed than that of the substrate [12]. Hence, an aluminum layer on a copper substrate would not support generalized Lamb wave modes. Here, we will consider structures which support those higher order modes.

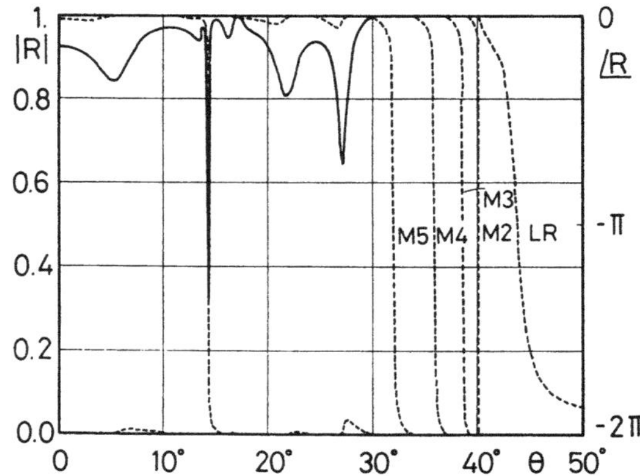

Figure 3. Plane wave reflection coefficient amplitude (solid) and phase (dashed) for a 1.5 mm copper layer on an aluminum substrate at 5.4 MHz ($k_{ts}d{=}16.7$).

MEASUREMENT TECHNIQUE

It is well known that if a bounded beam of bulk wave is incident at a critical angle corresponding to a leaky mode, the reflected wave will have a nonspecular component which originates from a spatially displaced location on the reflecting surface. If the incidence is far from a critical angle, such a nonspecular component will not exist. This fact can be utilized to determine the presence of a leaky layer mode as excited by a wedge transducer. We can consider a bulk wave transducer with varying angle to generate the incident wave in the liquid. Another wedge transducer at a distance can act like the receiver if the inclination of the wedge is at the same angle (see Fig. 5(a)). The separation of the transducers is such that the specularly reflected beam will not be intercepted by the receiver, but only those waves which couple to a mode traveling in the layers will cause an output in the receiver. If the detected signal amplitude, V, is recorded as a function of incidence angle, θ, a characteristic curve, $V(\theta)$ [13], will be obtained. This curve will have peaks at angular positions corresponding to critical angles where the coupling between the transducers is maximum. Hence, the amplitude of the peaks will be proportional to the excitation and detection efficiency of the leaky modes. The efficiency depends on the size of the transducers, the wavelength, the distance between the transducers, and most importantly the behavior of the reflection coefficient at the critical angle. Since the angular dependence of the reflection coefficient is a unique property of the solid structure, $V(\theta)$ can be used for characterization purposes.

The measurement technique described above has the inherent alignment difficulty of the transmitter and receiver wedge angles. This problem may be overcome if one makes use of a *corner reflector* [14]. A corner reflector has the property that it reflects the incoming rays at the same angle as they are incident. So, it performs the same operation in the angular spectrum domain as a mirror performs in the space domain. As shown in Fig. 5(b), the receiver transducer is replaced by a corner reflector positioned in such a way to avoid interception of specularly reflected rays.

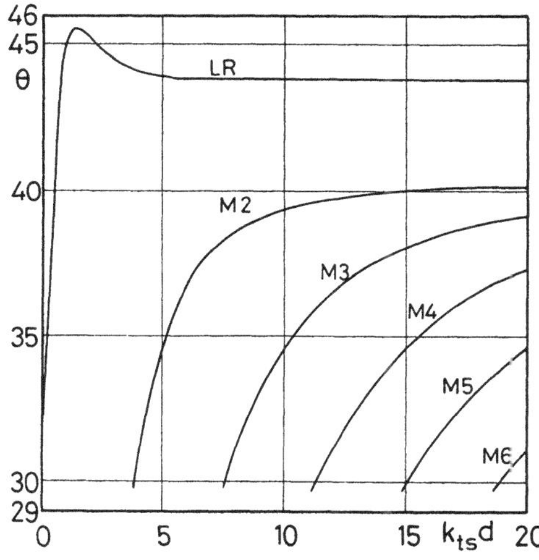

Figure 4. Excitation angle of various modes from the liquid side as a function of copper layer thickness-wavenumber product on an aluminum substrate

In this case the transmitting transducer acts also as the receiver and the system must be used in a pulse-echo arrangement. By adjusting the angle of inclination of this single transducer it is possible to excite the modes selectively and hence measure their excitation efficiency. Referring to Fig. 5(b) the transducer inclination (θ) is adjustable about the point O with a radius R. The point O is placed a distance D below the surface which is set equal to the distance of the corner of the corner reflector (point B) from the surface. The corner reflector, whose minimum edge length is t, is positioned a lateral distance X away from the point O. For $X = 0$ the specularly reflected rays will be directly incident on the corner reflector. To avoid the interception of these rays, X must be chosen to be greater than $t + W/(2\cos\theta)$, where W is the width of wedge transducer. This arrangement also makes sure that the nominal distance that the waves must travel in the solid (path $A'B'$) is always equal to X regardless of the inclination angle θ. Notice that, with corner reflector setup, the path in the solid material is traversed twice, as opposed to one way traversal in the two wedge transducer configuration.

The main advantage of this technique is the fact that the phase of the reflection coefficient is detected without having to make a phase measurement. Since the phase of the reflection coefficient is the more interesting component compared to the magnitude, the proposed technique is inherently more sensitive than a pure magnitude based measurement.

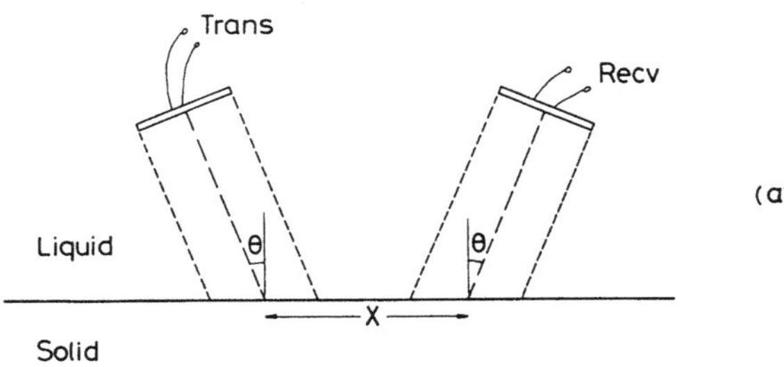

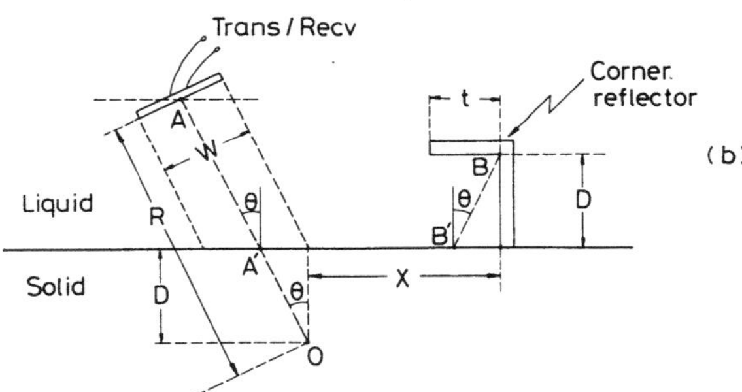

Figure 5. Geometry for (a) $V(\theta)$ measurement, and (b) $V(\theta)$ measurement with corner reflector

The amplitude of the received signal can be found by an application of the angular spectrum technique. For simplicity we assume a two-dimensional geometry. The angular spectrum of waves produced by a transducer of width W with a wavelength λ referenced at the midpoint of the transducer with respect to x axis is given by

$$\frac{\sin[\pi W/\lambda\ (k_x/k_0)]}{\pi W/\lambda\ (k_x/k_0)}$$

If this transducer is inclined at an angle θ with respect to the x axis, the angular spectrum should be rotated by the angle θ. This rotation can be achieved by replacing k_x/k_0 with $\sin(\sin^{-1}(k_x/k_0) - \theta)$ and the angular spectrum at point A with respect to x axis can be obtained as

$$\Psi_A^+(k_x) = \frac{\sin[(\pi W/\lambda)\ \sin(\sin^{-1}(k_x/k_0) - \theta)]}{(\pi W/\lambda)\ \sin(\sin^{-1}(k_x/k_0) - \theta)} \tag{1}$$

Propagation of angular spectrum is taken care of by an exponential multiplier factor. For the given geometry the propagation between points A and B is equivalent to multiplying by

$$\exp[j k_x(X + R\sin\theta) + k_z R\cos\theta]\mathcal{R}(k_x/k_0) \tag{2}$$

Here, the reflection process at the surface of the layered structure is included through a multiplication by the reflection coefficient, $\mathcal{R}(k_x/k_0)$. Hence the angular spectrum at point B, Ψ_B^+, can be written as

$$\Psi_B^+(k_x) = \frac{\sin[\pi W/\lambda\ \sin(\sin^{-1}(k_x/k_0) - \theta)]}{\pi W/\lambda\ \sin(\sin^{-1}(k_x/k_0) - \theta)}\ \exp[j k_x(X + R\sin\theta) + k_z R\cos\theta]\ \mathcal{R}(k_x/k_0) \tag{3}$$

The field at this point is truncated by the finite aperture of the corner reflector. To take this truncation operation into account, we transform the angular spectrum, Ψ_B, to space domain and and multiply the field by a pupil function of width $2t$, where t is the length of shortest arm of the corner reflector as seen in the figure. Hence the reflected angular spectrum is given by

$$\Psi_B^-(k_x) = F\{rect(x/2t)F^{-1}\{\Psi_B^+(k_x)\}\} \tag{4}$$

where

$$rect(x) = \begin{cases} 1 & \text{if } |x| < 1/2 \\ 0 & \text{otherwise} \end{cases}$$

The propagation of this angular spectrum back to point A is similarly achieved by multiplying it by the factor given in Eq. 2. Hence, the received angular spectrum at point A is

$$\Psi_A^-(k_x) = \Psi_B^-(k_x)\exp[j k_x(X + R\sin\theta) + k_z R\cos\theta]\ \mathcal{R}(k_x/k_0) \tag{5}$$

The received signal, V, by the transducer as a function of θ can be expressed in terms of the incident, $\Psi_A^+(k_x)$, and reflected, $\Psi_A^-(k_x)$, angular spectra as follows:

$$V(\theta) = \int_{-\infty}^{+\infty} \Psi_A^+(k_x)\Psi_A^-(k_x)\ dk_x \tag{6}$$

To find $V(\theta)$, the quantities inside the integral must be obtained from Eqs. 1, 3, 4 and 5. Hence, the evaluation of a single point in $V(\theta)$ curve involves two Fourier transformations and an integration operation. Note that, the reflection coeffiecient $\mathcal{R}(k_x/k_0)$ needs to be calculated only once per $V(\theta)$ curve because of the way we formulated the problem.

Eq. 6 is numerically evaluated for some specific cases. Figs. 6 and 7 show the calculated $V(\theta)$ curves for 0.9 mm and 1.5 mm thick copper layers on aluminum substrate at 5.4 MHz. It is clearly seen that the peaks of the $V(\theta)$ curves correspond to the phase transition angles of the reflection coefficient given in Figs. 2 and 3. On the same figures the measurement results are also depicted. We have chosen the various geometrical parameters as follows: $W = 10$ mm, $R = 150$ mm, $X = 17$ mm,

$t = 10.6$ mm. The agreement between the theoretical and experimental results is reasonably good. But note that the relative amplitude of the peak corresponding to the Rayleigh-like wave is higher in the experiments than those in theory. The discrepancy may be due to ignorance of the attenuation or the three dimensional nature of the problem in the theory. If the attenuation of Rayleigh-like wave is less than the Lamb wave modes, the discrepancy can be explained.

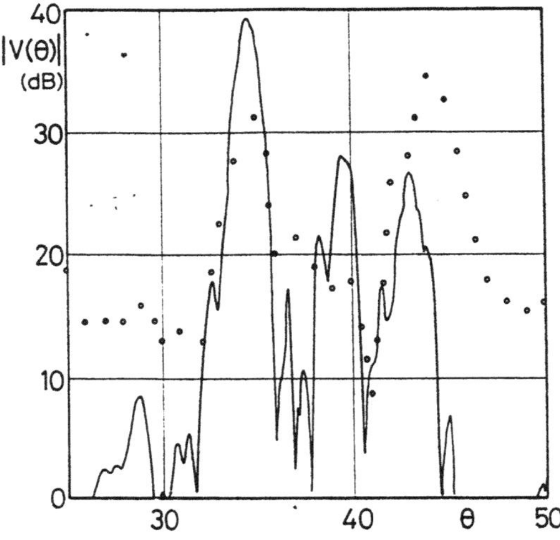

Figure 6. Calculated (solid) and measured (dots) $V(\theta)$ curves at 5.4 MHz for Cu layer on Al substrate for a layer thickness of 0.9 mm

DISCUSSION and CONCLUSIONS

We have introduced a method to measure the excitation sensitivity of layer modes in a layered structure by a wedge transducer and a corner reflector. This method results in a curve called $V(\theta)$ which has peaks corresponding to the phase transitions of the reflection coefficient at the liquid-layered-solid interface. The shape of the curve is highly dependent on the elastic parameters and thicknesses of the layers. The fact that the path in the layered material is traversed twice, enhances the dependence of $V(\theta)$ on the reflection coefficient and it is probably the most important reason for the high dependence. The presence of narrow peaks at the critical angles makes the differentiation between different structures easy when they have only slightly differing critical angles. The angular position of critical angles are especially sensitive for higher order Lamb wave modes with respect to mechanical properties. This can

be easily understood from the slope of dispersion curves given in Fig. 4: The curves
have the highest slope for angles close the mode cut-off angle. Therefore, with any
given thickness the highest ordered mode will give the highest sensitivity to parameter
variations. Luckily, this mode is also the one with highest excitation efficiency, since
the corresponding $V(\theta)$ peak is the highest.

The sensitivity of $V(\theta)$ curves to elastic and geometric parameters can be exploited
for NDE and NDT purposes. A delamination, a change in the elastic parameters or in
the thickness of the layers, or other flaws in the layer structures can be detected, since
the obtained $V(\theta)$ curve will be substantially different than the original. Obviously,
the dimensions of the system and the frequency can be scaled to accomodate thicker
or thinner layer thicknesses.

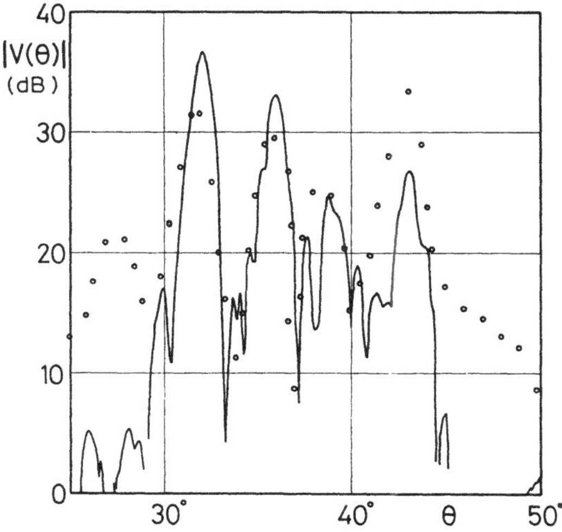

Figure 7. Calculated (solid) and measured (dots) $V(\theta)$ curves at 5.4 MHz for Cu layer
on Al substrate for a layer thickness of 1.5 mm

For good differentiation the peaks of the $V(\theta)$ curves need to be narrow. This requires that the angular spectrum of the wedge transducer must be sufficiently narrow,
that is, the width of the transducer must be sufficiently wide. If this is not satisfied,
the peaks will be wide and they will interfere with each other, reducing the parameter
sensitivity of the technique.

Numerical simulations have shown that the parameter X must be sufficiently large
to get $V(\theta)$ curves with clear critical angle peaks. However, making X large reduces
the signal level and the dynamic range. So, the parameter X must be properly chosen
not to cause dynamic range reduction.

We have treated the structures which has only a single layer. But, the measurement technique is more general, and it can be easily extended to materials with multi layers. But, the usefulness of this technique may be limited to a class of layered structures. For example, the layer must be a softer material than the substrate to be able to support the Lamb wave modes. In particular, the shear wave velocity of the layer must be less than that of the substrate. If this condition is not satisfied, only Rayleigh-like waves are excited. In such a case, $V(\theta)$ curves will have a single peak corresponding to the Rayleigh like wave. The shape of such $V(\theta)$ curves are less sensitive to the variations in the parameters of the layered structure.

References

[1] W. Parmon, and H.L. Bertoni, "Ray interpretation of the material signature in the acoustic microscope", *Elect. Lett.*, vol. 15, p. 684, 1979.

[2] H.L. Bertoni and T. Tamir, "Unified theory of Rayleigh angle phenomena for acoustic beams at liquid—solid interfaces", *Appl. Phys.*, vol.2, pp.157–172, 1973.

[3] H.L. Bertoni and T. Tamir, "Characteristics of wedge transducers for acoustic surface waves", *IEEE Trans Sonics Ultrason.*, vol. 22, pp. 415–420, 1975.

[4] J. Fraser, P.T. Khuri-Yakub and G.S. Kino, "The design of efficient broadband wedge transducers," *Appl. Phys. Lett.*, vol. 32, pp. 698–700, 1978.

[5] L.M. Brekovskikh, *Waves in Layered Media*, (Academic Press, New York, 1960).

[6] D.E. Chimenti, A.H. Nayfeh and D.L. Butler, "Leaky Rayleigh waves on a layered halfspace," *J. Appl. Phys.*, vol. 53, pp. 170–176, 1982.

[7] D.E. Chimenti and A.H. Nayfeh, "Leaky Lamb waves in fibrous composite laminates," *J. Appl. Phys.*, vol. 58, pp. 4531–4538, 1985.

[8] D.L. Folds and C.D. Loggins, "Transmission and reflection of ultrasonic waves in layered media," *J. Acoust. Soc. Am.*, vol. 62, pp. 1102–1109, 1977.

[9] D.B. Bogy, S.M. Gracewski, "Reflection coefficient for plane waves in a fluid incident on a layered elastic half-space," *J. Appl. Mech.*, vol. 50, p. 405, 1983.

[10] G.W. Farnell and E.L. Adler, "Elastic Wave propagation in thin layers," in *Physical Acoustics*, edited by W.P. Mason and R.N. Thurston, (Academic, New York, 1972) vol. IX, pp. 35–127.

[11] E.L. Adler, and I-H. Sun, "Observation of leaky Rayleigh waves on a layered half-space" *IEEE Trans. Son. Ultrason.*, vol. 18, pp. 181–184, 1971.

[12] A.H. Nayfeh, and D.E. Chimenti, "Reflection of finite acoustic beams from loaded and stiffened half-spaces," *J. Acoust. Soc. Am.*, vol. 75, pp. 1360–1368, 1984.

[13] A.Atalar, H. Köymen, O. Yemişçiler, "Measurement of sensitivity of different wave modes to subsurface defects" *Proc. of IEEE 1988 Ultrason. Symp.*, p. 771.

[14] F.R. Rollins, "Ultrasonic examination of liquid-solid boundaries using a right-angle reflector technique," *J. Acoust. Soc. Am.*, vol. 44, pp. 431–434, 1968.

NON-DESTRUCTIVE EVALUATION OF ENGINEERING CERAMICS BY HIGH-FREQUENCY

ACOUSTIC TECHNIQUES

S. Pangraz, H. Simon*, R. Herzer, and W. Arnold
Fraunhofer-Institute for Non-Destructive Testing
Bldg. 37, University,
D-6600 Saarbrücken 11, FRG
*present address: AKO-Werke, D7988 Wangen/Allgäu, FRG

INTRODUCTION

In order to detect defects by ultrasound efficiently, it is necessary to make the wavelength comparable to the defect size. Due to fracture mechanics considerations, it is necessary to detect defects smaller than 100 μm in engineering ceramics by NDT-techniques [1]. This requires high-frequency ultrasonics (UT) due to the large sound velocity in ceramics and, in order to obtain sufficient spatial resolution, pulses shorter than 100 ns are needed. A large bandwidth for both the electronics and the transducer are then necessary. The transducer employed may be excited by either an exponentially decaying step-pulse (broadband excitation) or by a rf-carrier pulse.

PHYSICAL BACKGROUND

Depending on the ratio of the defect size a to the employed ultrasonic wavelength λ, one can distinguish three regimes [2]: Rayleigh scattering ($a/\lambda \ll 1$), resonance scattering ($a/\lambda \sim 1$) and geometric scattering ($a/\lambda \gg 1$). If we employ 50 MHz ultrasonic waves in ceramics, we are in the Rayleigh-regime for the order of defect size mentioned above. Neglecting diffraction, the intensity I_{sc} backscattered from a given spherical defect can be estimated to be:

$$I_{sc} = I_o (f(\pi))^2 \pi a^2 \exp(-\alpha r) \, d\Omega \qquad (1)$$

where $d\Omega = \Delta F/4\pi r^2$. Here, I_o is the ultrasonic intensity irradiated on the defect, $f(\pi)$ is the backscattered amplitude [3], α is the attenuation coefficient in the material, ΔF is the area of the transducer, r is the depth of the defect, and $d\Omega$ is the solid angle subtended by the probe as seen from the defect. Because $f(\pi) \propto a^2$, $I_{sc}/I_o \propto a^6$! This means that very efficient transducers have to be used in order to get a high signal/noise ratio (SNR) for a given excitation voltage of the transducer. In the case of other defect shapes also expressions similar to Eq. (1) hold good [2].

TRANSDUCERS

We have examined various ultrasonic transducers and found that transducers made out of Polyvinylidene fluoride (PVDF) and the copolymer Polyvinylidene fluoride Trifluorethylene (P(VDF-TrFE)) are well suited for C-scan imaging [4]. They have sufficiently large bandwidth, and by carefully matching the tranducers electrically, one can obtain insertion losses as small as 11 dB in immersion techniques. Additionally, they can be manufactured as focussing probes with well matched acoustic impedance to water [5].

ELECTRONIC EQUIPMENT

Besides using broadband excitation of the transducer by a step-like pulse with a decay-time of approx. 10 ns (USH 100 made by Krautkrämer-Branson Köln-Hürth, FRG) we also employ sine-burst excitation of the transducer (Fig 1). From various laboratory experiments a prototype was designed with the following characteristics: A contineous wave-signal is generated by voltage-controlled oscillators (VCO) and is then amplitude modulated at low power levels. This allows one to keep the leakage power at minimum by using two Pin-diode switches in series for modulating. After this, the pulses are amplified to approximately 20 W peak power. The system built by us has an on/off ratio of at least 90 dB in the frequency range from 10-200 MHz. The minimum pulse-width contains one rf-cycle and the pulses are switched at their zero crossings. The maximum number of cycles is 256. Particular emphasis is given to obtaining maximum signal to noise ratio (SNR) of both emitting and receiving parts of the electronics allowing maximum sensitivity for defect detection. All parameters of the system can be controlled by a microcomputer or PC [6].

EXPERIMENTAL RESULTS

A-Scans

In order to test the performance of our system, A-scans were obtained from defects with well-defined size in densely sintered and small grained SiC and Si_3N_4 ceramics. Table I shows our results on the measured SNR. In this case we employed a focussing transducer with a focal length Z = 25 mm. The diameter D_L of the piezoelectric element was 3 mm. The influence of the attenuation in the water path and in the ceramic was of minor importance on the measured SNR.

It is striking to note that the difference in SNR as measured for various defects of same size is very small, eventhough the differences in acoustic impedance of the defects to the host material are quite large. This reflects to the fact that in the Rayleigh regime, the amplitude of the backscattered signal depends, in a rather complicated way, on the elastic parameters of the host material and the defect [3]. Due to

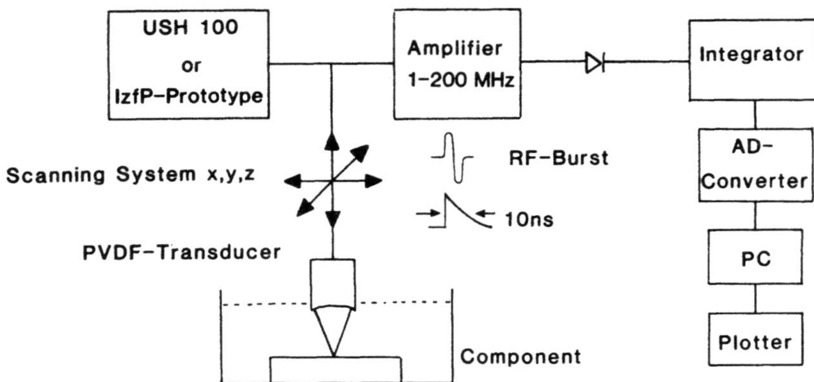

Figure 1. Block diagram of the system used for high-frequency ultrasonic C-scan imaging. The immersion technique is used for coupling.

Table 1

Type of defect	Diam. (μm)	SNR (dB)	Depth of defect (mm)
Pores	500	47	2.9
	100	29	2.8
	50	14	2.7
Si-Inclusion	100	26	2.9
	50	14	2.6
Fe-Inclusion	100	27	2.6
	50	13	2.5
WC-Inclusion	100	14	2.5

Eq. (1) the dominant dependance for the available SNR is, however, $\propto a^6$. The SNR obtained in Table I were measured by exciting the transducers with a step voltage of 280 V exponentially decaying in 13 ns. By using rf-carrier excitation (20 W at 50 Ω), an improvement in the SNR of 10 dB can be obtained, albeit at the cost of axial resolution [4]. As a rule of thumb, we can say that at present we can detect spherical defects as small as 30 μm present within a depth of 3 mm in ceramic components with a SNR of 6 dB by high-frequency ultrasound.

C-Scans

The high-frequency UT-system also comprises of a scanning system enabling one to obtain C-scan images with a step-resolution of 10 μm [7]. After rectification the portion of interest from the A-scans is cut out by a gate and fed into an integrating amplifier. Its output is then digitized and the digital value displayed on the screen of a monitor as a color level. The number of color levels is at present 16. We use the integrator because all defect echoes within the gate contribute to the color level displayed. If we, however, use a peak detector only the largest echo within the gate would contribute to the color level. In order to demonstrate the capability of our system, C-scan images of a flat-bottom holes of 50 μm, 100 and 200 μm diameter present at a depth of 11 mm in a fine-grained steel were taken (grain size a few micrometers). The PVDF-transducer with 25 MHz center frequency was excited by a step-voltage as described above. In this case the 200 μm flat bottom hole could be imaged with a SNR of 17 dB (Fig. 2a). In the case where the transducer was excited by a rf-burst of 25 MHz center-frequency containing 2 cycles and with a peak-to-peak voltage of ~ 100 V, the SNR was 22 dB. The lateral width of the image in both cases was 350 μm (3 dB-width). In the case of the 100 μm flat bottom hole, the SNR obtained was 9.5 dB and 15.5 dB for broadband and narrowband excitation, respectively. The 3 dB-width of the scans were 500 μm and 350 μm, respectively. The values for the 50 μm flat bottom hole were more difficult to measure. The SNR was of the order of 10 dB for both excitation mechanisms. Since the 50 μm flat bottom hole was only 350 μm long extending from a 500 μm prebore, it was difficult to clearly separate the signals from both reflectors in an A-scan. To this end a transducer with a higher center frequency would be necessary.

Our findings were further corroborated by imaging a soldering layer through 3 mm thick ceramic (Fig. 3). Fig. 3a displays the image obtained with broadband excitation and Fig. 3b displays the one obtained with narrow-band excitation. In this case, the center frequency of the transducer was 20 MHz and the improvement in the SNR of the features displayed (probably disbonding) 5 dB. More striking is, however, the improvement in lateral resolution as can be seen directly from the images.

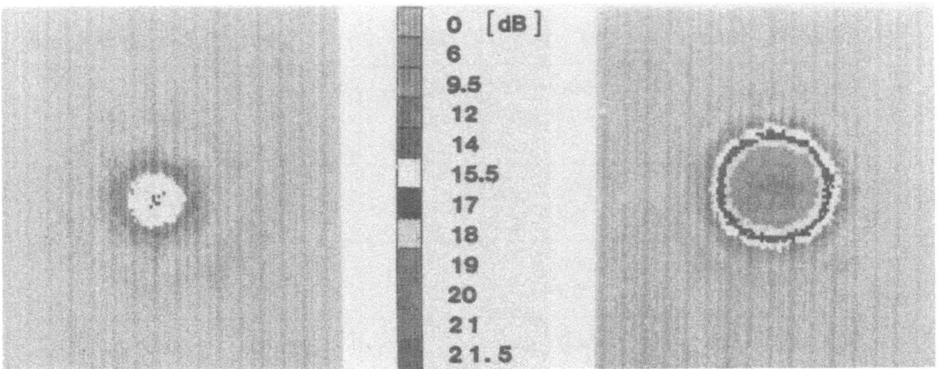

Figure 2a   Figure 2b

C-scan image of a 200 μm diameter flat-bottom hole drilled into a fine-grained steel. The depth of the flat-bottom hole was 11 mm, a) broadband excitation: SNR 17 dB   b) narrowband excitation: SNR 22 dB. Scan width 1.25 × 1.25 mm².

These findings can be easily explained. The Fourier-Spectrum of the broadband pulse and the tone-burst excitation of the transducer are displayed in Figs. 4a and 4b, respectively. Additionally, the insertion loss $Il(\omega)$ of the transducer is also displayed in Fig. 5. The sound-field intensity I_s radiating from the transducer is given by:

$$I_s \propto Il(\omega) \cdot P(\omega) \cdot \exp(-\alpha(\omega)x) \qquad (2)$$

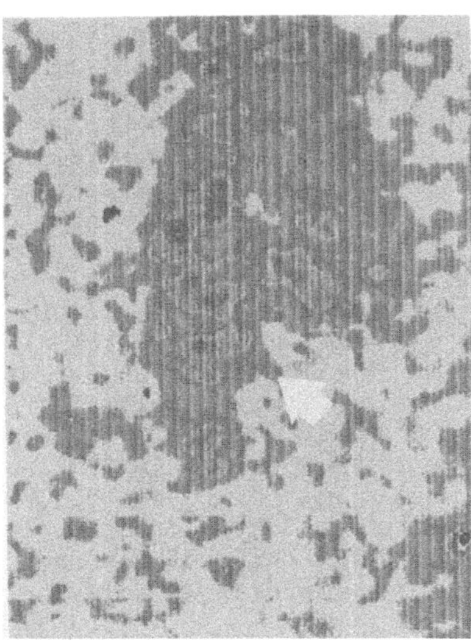

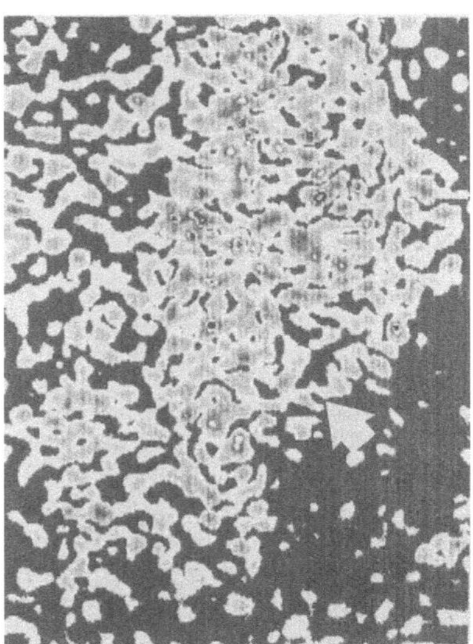

Figure 3a   Figure 3b

C-scan image of a 0.2 mm thick soldering layer. The soldering layer bonded a 3 mm thick SiC ceramic plate onto a metal block. In a) the C-scan image was obtained by exciting the transducer with a step-pulse of 280 V decaying exponentially and in b) with a sine-burst of 2 cycles with a center frequency of 20 MHz and an amplitude of 30 V. The arrow indicates the area of disbonding. Scan-width was 12 x 19 mm².

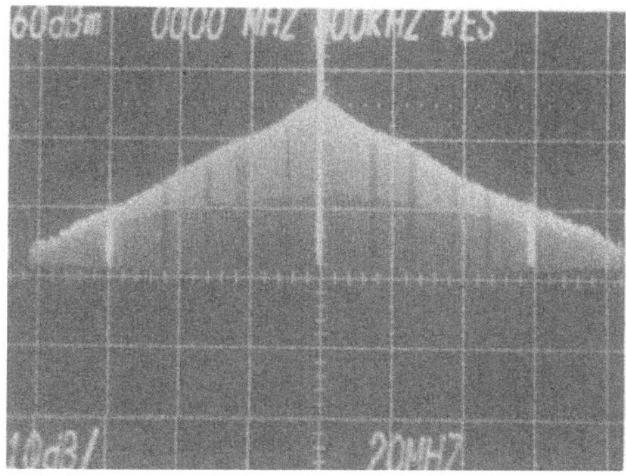

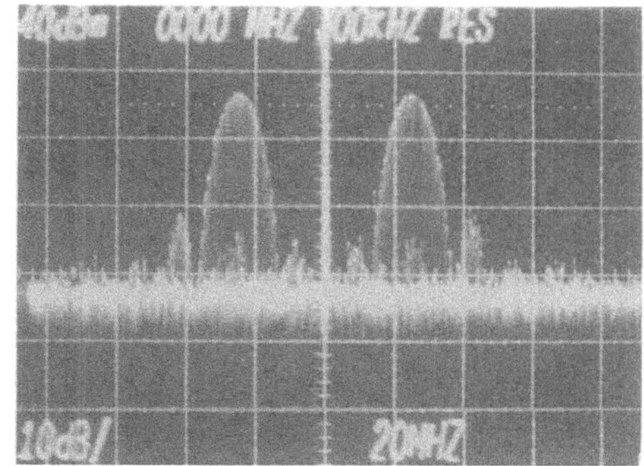

Figure 4a

Measured Fourier spectrum of exciting broadband pulse. The maximum at zero frequency corresponds to a power of 0.95 mW as calibrated with an external signal-generator. In a) and b) the vertical scale is logarithmic with 10 dB/division and the horizontal scale is 20 MHz/division.

Figure 4b

Measured Fourier Spectrum of sine burst with two oscillations. Center frequency is 25 MHz. The value of the maximum at 25 MHz corresponds to a power of 6.3 mW.

Here, $P(\omega)$ is the power spectrum of the exciting pulse, and $\alpha(\omega)$ is the frequency-dependent ultrasonic attenuation. In the case of tone-burst excitation, there is more power at the frequency of the minimal insertion loss of the transducer in comparison with broadband excitation. From Figs. 4a and 4b one can discern a difference of approximately 11 dB at the frequency of 25 MHz. Therefore, $Il(\omega)P(\omega)$ is larger rendering a larger back-reflected signal from a given defect. In turn this leads to a larger dynamic range in the C-scan image as observed.

The 3dB-width d_{sc} of the focal spot in the component under examination is given by the following expression [8]

$$d_{sc} = 1.03\ \lambda_2 Z/D_L \tag{3}$$

Here, λ_2 is the ultrasonic wavelength in water. Due to the frequency dependent absorption $\alpha(\omega)$ and the Fourier spectrum for broadband excitation as displayed in Fig. 4a, always the lower frequency components relative to the minimal value of $Il(\omega)$ (Fig. 5) dominate in the amplitude levels digitized for image formation. According to Eq. (3), these components have a larger d_{sc}, and therefore the lateral resolution is reduced. The above arguments always apply to C-scan images depending on the frequency response of the rf-chain and the defect imaged when broadband excitation and detection schemes are used.

We used our system to inspect components made out of high-strength ceramics, diffusion-bonded ink-jet printers, soldering joints between ceramic and metal plates, and fittings made out of an aluminum alloy. All the features discussed above could be observed in the C-scan images of defects such as pores, inclusions, cracks and debonding.

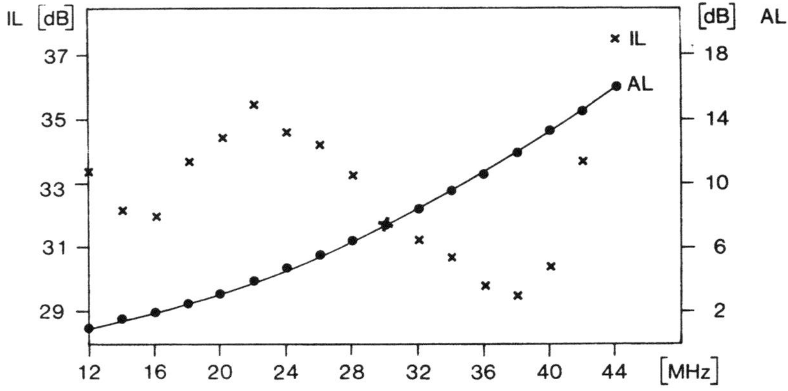

Figure 5. Insertion loss Il(ω) of employed PVDF-transducer as a function
of frequency as measured using a well-known technique [9].

SYNTHETIC APERTURE

In conventional C-scan images of defects one uses the peak amplitude
of the signal scattered from a defect within a certain preset gate width.
All other information is lost and in particular the defect is not reconstructed using the physical principles underlying scattering. Using Synthetic Aperture Focussing Techniques (SAFT) and corresponding algorithms
for reconstruction of the defect more information can be obtained. The
basic idea of the SAFT is to measure the complete soundfield Φ_s scattered
by a defect on a closed surface around this defect [10]. During reconstruction this soundfield is calculated back into that region where the
scattering occurred, employing mostly scalar wave propagation theory
[11]. Ideally, this will be non-zero only on the surface of the scatterer, thus rendering an image of the defect. With the advent of transient
recorders with a digitization rate of at least 400 MHz, it is possible to
combine SAFT with UT-systems with frequencies above 20 MHz. The reconstructed image of a segment of the rotor axis of a turbo-charger made of
sintered SiC (ϕ 10 mm) is shown as an example in Fig. 6. A crack extending almost half the circumference can be seen. The signals at the center
of the axis are probably artefacts due to the reconstruction algorithm. A
50 MHz center frequency PVDF-transducer was used for insonification in a
water tank. The component was rotated by 360° and 900 single A-scans were
digitized at 200 MHz rate and stored in a PC for reconstruction. Due to
absorption of the high frequency components, the dominant amplitude
received was found to be at 30 MHz.

FUTURE RESEARCH WORK

In the further research work emphasis will be laid on fast data-
acquiring and handling, and a scanning system using a precision robot.

ACKNOWLEDGEMENT

        This research work was started with a grant from the European Community (Material Technology Programme). Part of this work was financed by the German Ministry of Research and Technology within the German Material Research Programme. We also thank Krautkrämer-Branson Köln-Hürth, FRG, for close collaboration. The measurements using synthetic apertures techniques were made together with Dr. V. Schmitz from our institute.

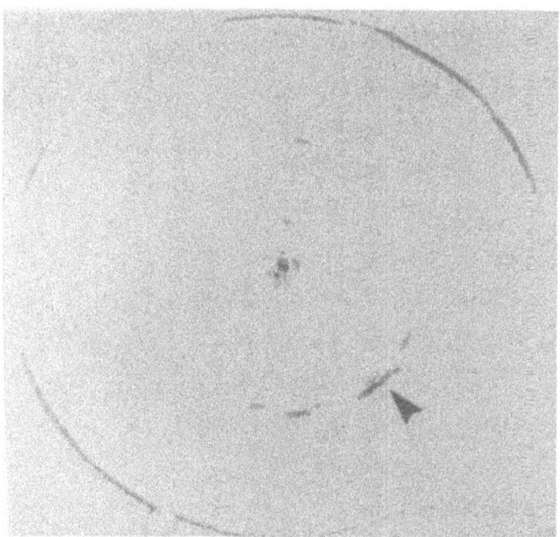

Figure 6. Reconstructed image of a segment of a turbo-charger axis (φ 10 mm) made of sintered SiC. The center-frequency of the PVDF-transducer was 50 MHz. A crack extending almost half the circumference can be seen (arrow).

REFERENCES

[1]   Evans, A.G., Proc. NATO-ASI Nitrogen Ceramics (1981)
[2]   Ermolov I.N., Non-Destr. Test. 5, 87 (1972)
[3]   Hirsekorn, S., IzfP-Report Nr 780157-TW, unpublished (1978)
[4]   Pangraz, S., and Arnold, W., Ferroelectr. 93, 251 (1989)
[5]   Manufactured by Krautkrämer-Branson, D-5000 Köln-Hürth
[6]   Simon, H., Diploma Thesis, University of Saarbrücken, Department of Electrotechnique and Fraunhofer-Institute for Nondestructive Testing (1989), unpublished
[7]   Manufactured by Cogent Inc. Ascot, Berks., England
[8]   R.S. Gilmore, K.C. Tam, J.D. Young, and D.R. Howard, Phil. Trans. Roy. Soc. London A320, 215 (1986)
[9]   Bömmel, H.E. and Dransfeld, K., Phys. Rev. 117, 1245 (1960)
[10] Müller, W., Schmitz, V., Schäfer, G., in: Proc. 3rd German-Japanese Seminar Research Struct. Strength and NDE in Nucl. Eng. (1985)
[11] Langenberg, K.J., in: "Tomography and Inverse Problems", Malvern Phys. Series, Ed. Sabatier, P.C., Adam Hilger Bristol (1987), p. 126

ACOUSTIC MICROSCOPY OF CERAMIC BEARING BALLS

C-H. Chou and B. T. Khuri-Yakub

Edward L. Ginzton Laboratory
W. W. Hansen Laboratories of Physics
Stanford University, Stanford, CA 94305-4085

INTRODUCTION

Ceramic bearing balls have great potential for replacing steel bearing balls in most applications because of their lower weight, larger strength at high temperatures, and abundance of raw materials. However, ceramic materials are brittle, and the advantages of ceramic parts can be lost if small surface cracks and bulk defects are present in part. This work will report on a method we developed to detect small sub-micron surface cracks in ceramic bearing balls. We present a theory to calculate the scattering from these small "trenches" or cracks, and we will present an amplitude and phase measuring acoustic microscope capable of detecting these defects. Finally, we will present a signal processing scheme that enhances the contrast in images obtained on curved surfaces.

THEORY

In order to predict the scanning acoustic microscope's ability for detecting and imaging narrow surface depressions, we calculate the scattering from trenches using a theory similar to the one dealing with the imaging of trenches with a confocal scanning optical microscope [1,2].

As shown in Fig. 1, when an acoustic beam illuminates the surface trench (depression) with width L and depth h, the reflected beam can be considered as the sum of a homogeneous wave (i.e., the wave reflected from a perfect plane boundary) and a scattered wave (caused by the depression). That is:

$$V_r = V_{hom} + V_s \tag{1}$$

where V_r, V_{hom}, and V_s are the particle velocities in the z-direction of reflected, homogeneous and scattered waves, respectively.

The boundary conditions at $z = 0$ are:

$$V_{tr} = V_{inc} + V_r = V_{inc} + V_{hom} + V_s \tag{2}$$

$$\partial V_{tr}/\partial z = \partial V_{inc}/\partial z + \partial V_r/\partial z = \partial V_{inc}/\partial z + \partial V_{hom}/\partial z + \partial V_s/\partial z \tag{3}$$

where V_{inc} is the particle velocity in the z-direction of the incident wave and V_{tr} is that of the wave inside of the trench.

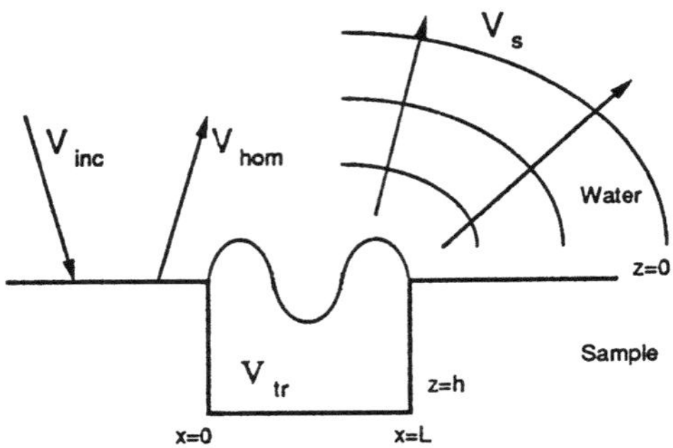

Fig. 1.   Trench structure for scattering simulation.

For the samples with much higher acoustic impedance (such as Si_3N_4) than the working couplant (water), at the boundary, we have:

$$V_{hom} = -V_{inc} \tag{4}$$

$$\partial V_{hom} / \partial z = \partial V_{inc} / \partial z \tag{5}$$

Therefore, Eqs. (2) and (3) can be simplified as:

$$V_{tr} = V_s \tag{6}$$

$$\partial V_{tr} / \partial z = 2\partial V_{inc} / \partial z + \partial V_s / \partial z \tag{7}$$

In order to solve V_s, we decompose V_s at $z = 0$ into spatial frequency components, and then decompose each component into trench modes:

$$V_s = \sum_{n=1}^{N} \int \int A_n(p,q)\sin(n\pi x/L)e^{-jkpx}e^{-jkqy}dpdq \tag{8}$$

where k is the wave number. To satisfy Eq. (6), we can write:

$$V_{tr} = \sum_{n=1}^{N} \iint A_n(p,q)\sin(n\pi x / L)[\sin(\gamma_n(z+h)) / (\sin\gamma_n h)]e^{-jkpx}e^{-jkqy}dpdq \tag{9}$$

where:

$$\gamma_n = k\sqrt{1-p^2-q^2} \tag{10}$$

$\partial V_{inc}/\partial z$ can also be decomposed as:

$$\partial V_{inc} / \partial z = \sum_{n=1}^{N} \iint D_n(p,q)\sin(n\pi x / L)e^{-jkpx}e^{-jkqy}dpdq \tag{11}$$

As in Ref. 2, $A_n(p,q)$ can be solved from the following equation:

$$\sum_{m=1}^{N} H_{mn}(q)A_n(p,q) = D_m(p,q) \tag{12}$$

where $H_{mn}(q) = (1/2)\gamma_n \cot(\gamma_n h)\delta_{mn} - \Gamma_{mn}(q)$.

$$\Gamma_{mn}(q) = \frac{jk^2}{\pi L} \int_{-\infty}^{\infty} \frac{(m\pi/L)(n\pi/L)\sqrt{1-p^2-q^2}}{\left[k^2p^2-(m\pi/L)^2\right]\left[k^2p^2-(n\pi/L)^2\right]}\left[e^{-j(kpL-m\pi)}-1\right]\left[e^{-j(kpL+n\pi)}-1\right]dp$$

(13)

The received signal by the transducer of the SAM is:

$$v = v_s + v_h$$

$$= 2\int_{s_1} V_s^* \frac{\partial V_{inc}}{\partial z}ds + 2\int_{s_1} V_h^* \frac{\partial V_{inc}}{\partial z}ds$$

(14)

where $*$ denotes a complex conjugate and s_1 is the sample plane $z = 0$. Since the incident wave at $z = 0$ depends on the focal location (x_0, z_0) of the acoustic beam, v is a function of x_0 and z_0 .

Figure 2 shows a result of the theoretical calculations as a line scan across a trench, 5 µm in width and 1 µm in depth. The transducer is operated at a frequency of 118 MHz and the lens has an F-number of 0.8 . The focus of the lens is placed at the surface of the bearing ball. A maximum phase shift of 17.2 degrees, and a maximum relative amplitude change of 2.5% are predicted by the theoretical calculation for the trench under consideration.

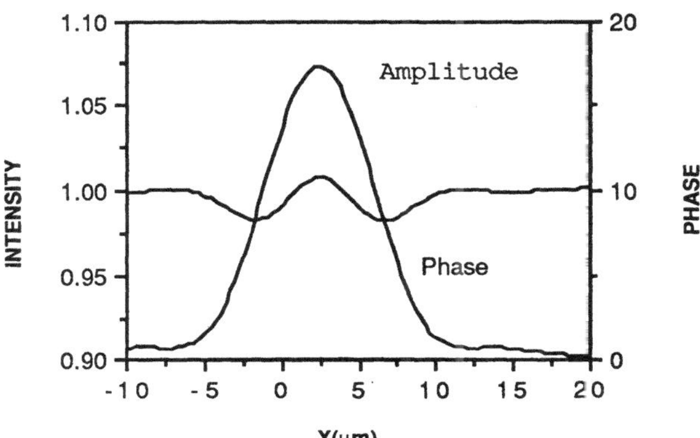

Fig. 2.   Intensity and phase of line scan of a 5 µm (width) by 1 µm (depth) trench. Operating frequency = 118 MHz ; F-number of lens = 0.8 focused on top.

AMPLITUDE AND PHASE MEASUREMENTS

Figure 3 shows a block diagram of the amplitude-phase measurement system which operates in the range of 1-200 MHz . The principle of operation of the system is detailed in our previous paper [3]. Due to vibration of the x-y scanning stage, the noise levels are ±7.5° for phase and 4% for an amplitude measurement at an operating frequency of 118 MHz . According to the theoretical prediction, our microscope is capable of detecting depressions equivalent to, or larger than 5 µm x 1 µm in the phase image. If the noise, which is due to mechanical vibrations, is reduced, sub-micron defects would be easily detected in this system.

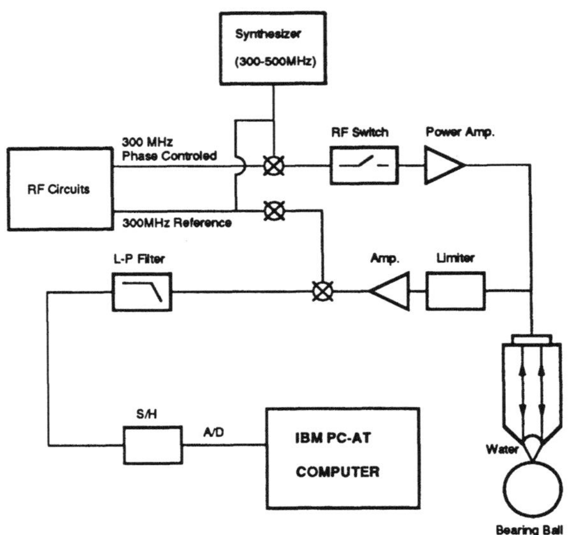

Fig. 3.   Block diagram of the amplitude and phase measurement system.

In the case of the bearing ball inspection, we encounter large phase variations and many amplitude fringes caused by the spherical shape. These large variations add to the small changes of amplitude and phase due to defects and limit the inspection ability of the microscope. Consequently, we developed a signal processing technique that allows us to remove the effects of the spherical shape of the bearing ball.

DATA PROCESSING

In order to overcome the difficulties caused by the shape of the bearing ball, it is necessary to apply proper data processing techniques. We first take the phase and amplitude images of the sample under inspection (i.e., object images). We also take an image of a reference sample which is the same object, but without a defect. The reference object could be the same or another bearing ball that is known to have no defects. This type of process is viable for bearing balls which are usually made to tolerances that are a fraction of a micron. Of course, the object and reference images are taken under the same working condition, including the distance from the lens to the sample. The phase images must be carefully unwrapped. Finally, we subtract the reference image from the corresponding object image. Ideally, the spherical features are removed and the defects are enhanced in the processed images.

EXPERIMENTAL RESULTS

We have inspected several Si_3N_4 bearing balls. As an example, Figs. 4-7 show the amplitude and phase images of a ball with a 16 mm diameter. Figures 4 and 5 are the amplitude images of the original object and the processed data, respectively. Figures 6 and 7 are the phase images. The transducer was operated at 118 MHz with the lens of 0.8 F-number. The acoustic beam was focused on the top of the ball. The fields of view are all   1.6 mm x 1.6 mm .

From the figures shown, the processed phase image shows the small defects which we were not able to see before the data processing. The processed amplitude image also shows the defects much better than the original one. The remaining fringes are due to the imperfect relocation of the x-y scanner. The bearing balls were provided to us by J. Hannoosh of CERBEC Corp.

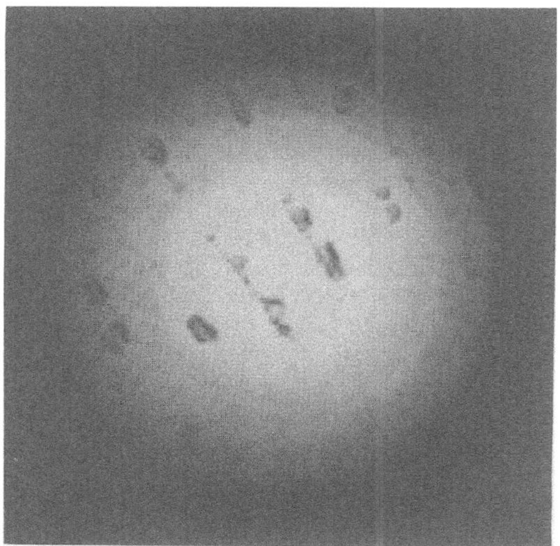

Fig. 4. Amplitude image of Si_3N_4 bearing balls. F-number of lens $= 0.8$ of diameter 16 mm (original); operating frequency $= 118$ MHz ; field of view $=$ 1.6 mm x 1.6 mm .

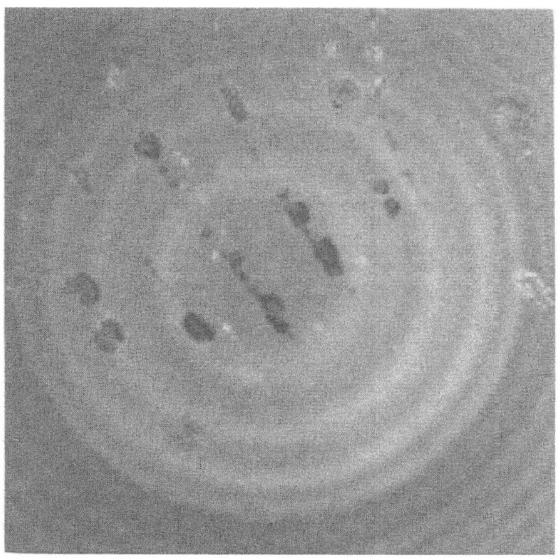

Fig. 5. Processed amplitude image from Fig. 4.

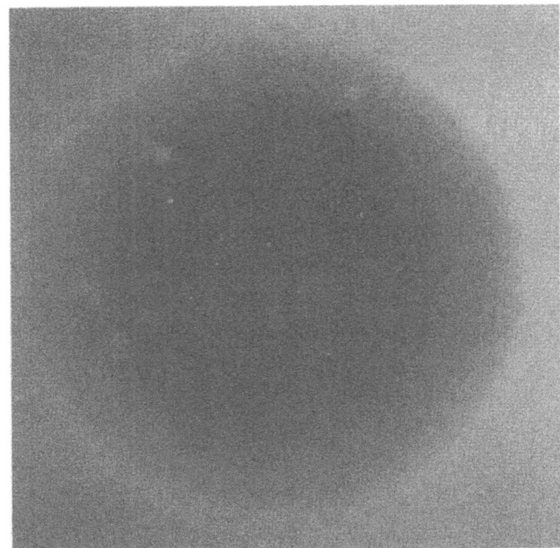

Fig. 6.   Phase image of the same object as in Fig. 4 (original).

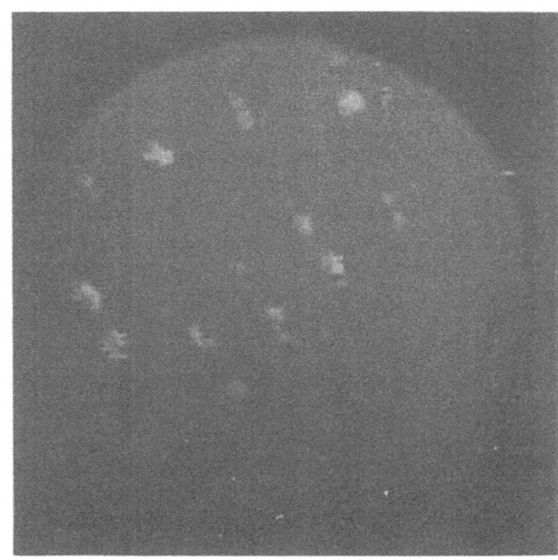

Fig. 7.   Processed phase image from Fig. 6.

CONCLUSION

We have applied data processing to enhance bearing ball defect inspection. Due to the vibration of the x-y scan system, the defect detection limit is 5 μm (w) x 1 μm (d) or higher. With a rotational scan system, we can use differential phase measurements which allow us to improve the defect inspection ability to the size of 1 μm x 1 μm [4].

ACKNOWLEDGMENT

This work was supported by the Department of Energy on Contract No. DE-FG03-84ER45157.

REFERENCES

1. Hobbs, P. C. D., "Heterodyne Interferometry with a Scanning Optical Microscope," Chapter 6, Ph.D. Dissertation, Stanford University, Stanford, CA (August 1987).
2. Kino, G. S., et al, "Confocal Microscopy of Trenches," to be submitted to Optics.
3. Parent,P., Chou, C-H., and Khuri-Yakub, B. T., "Ball Bearing Inspection with an Acoustic Microscope," Proc. IEEE Ultrasonics Symp., Ed: B. R. McAvoy (Institute of Electrical & Electronics Engineers, Inc., New York, 1988).
4. Chou, C-H., Parent, P., and Khuri-Yakub, B. T., "A SAM Bearing Ball Inspection System," Review of Progress in Quantitative Nondestructive Evaluation, Eds: D. O. Thompson and D. E. Chimenti (Plenum Press, New York, 1989).

EVALUATION OF SKIN BIOPSY SAMPLES USING ACOUSTICAL MICROSCOPY AND COMPARISON WITH CONVENTIONAL PATHOLOGICAL STUDIES AND LIGHT MICROSCOPY

R.J. Barr*, L.B. Shaw**, P.A. Ross*, and J.P. Jones**

*Dept. of Dermatology / ** Dept. of Radiological Sciences
University of California Irvine
Irvine, CA 92717

INTRODUCTION

The present study examines representative disorders of the skin using a commercially available scanning acoustical microscope, and seeks to determine whether or not the acoustical images (in general, made from unstained samples) have sufficient detail to render an effective diagnosis. In all cases the acoustical images were compared to results obtained from a conventional pathological analysis.

The idea of creating magnified views using sound waves was first proposed by Sokolov in 1936 (1). However, the first functional devices were not developed until the 1970's when Korpel and Kessler introduced the Scanning Laser Acoustical Micrscope (SLAM) (2) and Lemons and Quate introduced the Scanning Acoustical Microscope (SAM) (3). These devices and their applications have been well described in the literature (4-8). The instrument used in the present study, an Olympus UH-3 Scanning Acoustic Microscope, is essentially a pulse-echo device capable of making C-mode images, focused at selected depths within the sample. The basic design of the device is illustrated in Figure 1. Here a piezoelectric transducer generates an ultrasonic pulse which is focused by a sapphire lens on the specimen. A coupling liquid (usually water) carries the sound wave between the lens and the specimen. Moving the lens along the Z-axis (that is, closer to or further away from the specimen) moves the focal point of the ultrasonic pulse to various depths within the specimen. The reflected wave from the focal zone is collected by the same lens, received by the same transducer, and converted to an electrical intensity for display as a single pixel element on a CRT. By mechanically scanning the specimen in the horizontal (x - y) plane in a raster fashion, a two-dimensional image at the focal depth of the speciment is created. This C-mode image is produced within 6 to 10 seconds of scanning time. Although the present study only utilized a 600MHz transducer, others are available ranging between 30MHz and 1GHz.

MATERIALS AND METHODS

An Olympus UH-3 Scanning Acoustic Microscope with a 600MHz transducer was used to examine a variety of tissue specimens from human skin. The 600MHz transducer was selected because images in this frequency range seemed comparable with optical image of standard H and E sections.

The 600MHz lens provides a surface resolution of 1.7 μm, magnification
is dependent on scanning width and varies from 48 to 950 power on standard polaroid prints.

Cutaneous tissue specimens representing eleven different neoplastic
and inflammatory disorders were fixed in 10% neutral buffered formalin,
embedded in paraffin, sectioned at 6 μm, placed on glass slides, then
deparaffinized. Two adjacent sections were put on each slide, and one,
the control, was stained with hematoxylin-eosin andexamined with conventional light microscopy. The other (unstained) section was examined
with the UH-3 acoustical microscope.

RESULTS

Figures 2 thru 12 show comparisons between the standard H and E
sections and images made with the acoustical microscope for eleven
different skin disorders. The "a" numbered figures at the top of the
page are conventional optical microscope images of the H and E sections
and represent the gold standard for pathological analysis. The "b"
numbered figures at the bottom of the page are acoustical microscope
images of <u>unstained</u> sections immediately adjacent to the above H and E
sections. Let us now consider each of the figures in turn.

Figures 2a and 2b are Basal Cell Carcinoma. The acoustical image
shows characteristic islands of basaloid cells within the dermis focally
communicating with the overlaying epidermis. The pattern as well as
the cytological features are consistent with a basal cell carcinoma even
though some cytological details are somewhat obscured.

Figures 3a and 3b are a wart (Verruca Vulgaris). The low-profile
of this lesion is easily appreciated on the acoustical image allowing
for a definitive diagnosis.

Figures 4a and 4b are Lichen Planus. This lesion is characterized
by a thickened epidermis and a band-like inflammatory infiltrate at the
dermoepidermal junction and within the upper dermis. The low
power pattern can be appreciated in the acoustical image. Because
the precise identification of inflammatory cells is not possible at
this power level, other lichenoid dermatoses cannot be excluded.

Figures 5a and 5b are Actinic Keratosis. The acoustical image
shows a disruption of the normal epidermal architecture characterized
by "atypical" cells in the lower one-third of the epidermis. The resolution is not, however, great enough to pick up subtle atypical or
dyplastic features.

Figures 6a and 6b are Nevus. The acoustical image exhibits characteristic nesting of nevus cells within the dermis. No obvious cytological
atypia is identified, but subtle cytological atypia or dysplasis could
be missed due to lack of resolution.

Figures 7a and 7b are Pemphigus Vulgaris. The acoustical image
exhibits acantholysis (separation of squamous keratinocytes) within the
upper epidermis. This is similar to what is seen in the H and E stained
slide and allows for a specific diagnosis.

Figures 8a and 8b are Eczematous Dermatitis. A definitive diagnosis
of an eczematous (spongiotic) dermatitis can be made due to the fact that
spongiosis (intercellular edema) is easily identified. Eczematous dermatitis applies to a variety of lesions, all characterized by a similar
change.

Figures 9a and 9b are Seborrheic Keratosis. The acoustical images
showing a papillated surface and numerous cystic structures is diagnostic.

Figures 10a and 10b are Lichen Simplex Chronicus. The acoustical
image exhibits the characteristic thickening of the stratum corneum and
expansion of the granular cell layer. In addition, the epidermis shows
characteristic hyperplasia and the dermis appears to be scarred. Taken
together, these features are diagnostic of the pathology in question.

Figures 11a and 11b are malignant melanoma. The acoustical image
demonstrates the lower power profile consistent with the pathology,
namely the haphzard arrangement of cells within both the epidermis and
the dermis. The cells are so bizarre that this resolution is adequate
for diagnosis.

Figures 12a and 12b are Bullous Pemphigoid. The accustical image
clearly demonstrates the characteristic subepidermal blister.

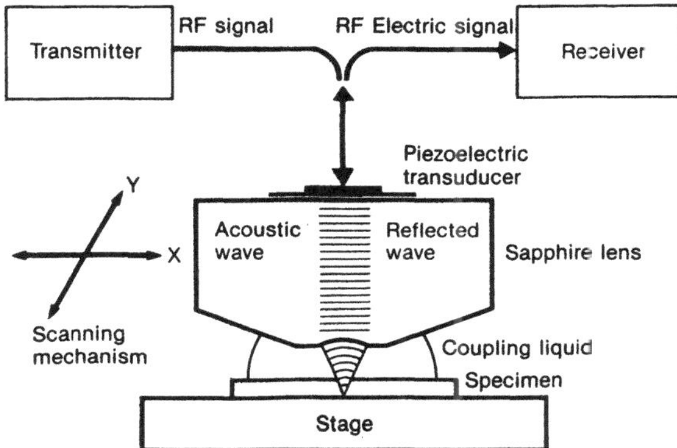

Figure 1. Diagram showing the operation of
the Scanning Acoustical Microscope.

CONCLUSIONS

Cutaneous tissue specimens representing eleven different neoplastic
and inflammatory disorders were examined. Acoustical images of unstained
sections were compared with conventional light microscopic images of ad-
joining sections stained with hematoxylin-eosin. In all cases, the
acoustical images were sufficient for a diagnosis. Although individual
cells could be visualized in the acoustical images, cytological atypia
was difficult to appreciate because nuclear details were poorly defined.
In addition, for inflammatory disorders the composition of the infiltrate
was difficult to determine.

Acoustical microscopy would seem to offer a number of advantages
over conventional pathological analysis including the feasibility, at
least in principle, of making an in situ or in vivo diagnosis.

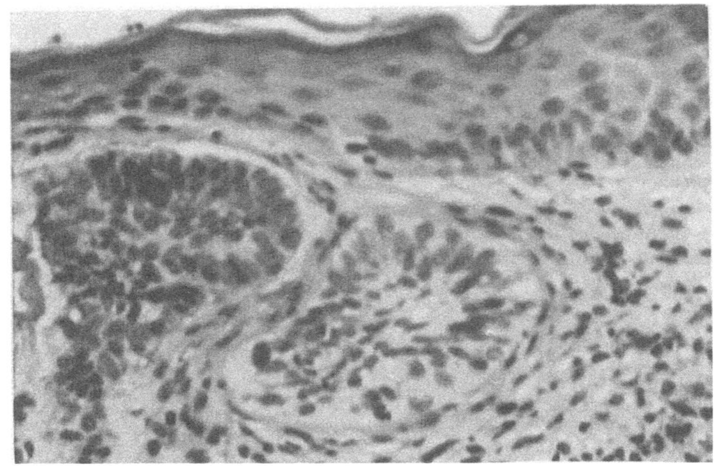

Figure 2a. Basal Cell Carcinoma. Optical microscope image
of standard H and E section.

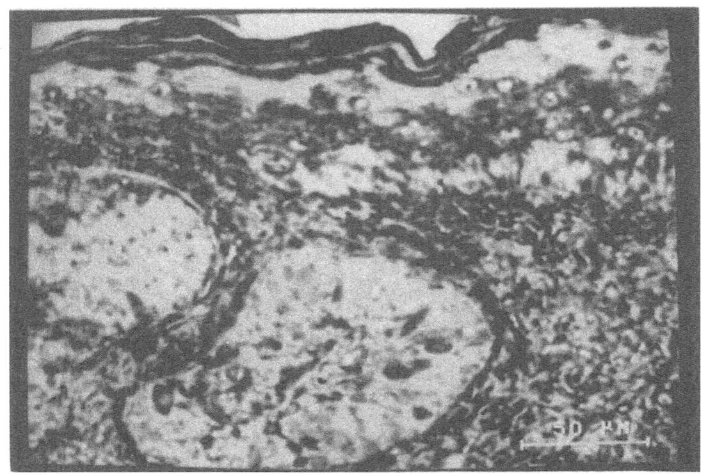

Figure 2b. Basal Cell Carcinoma. Acoustical microscope
image (unstained).

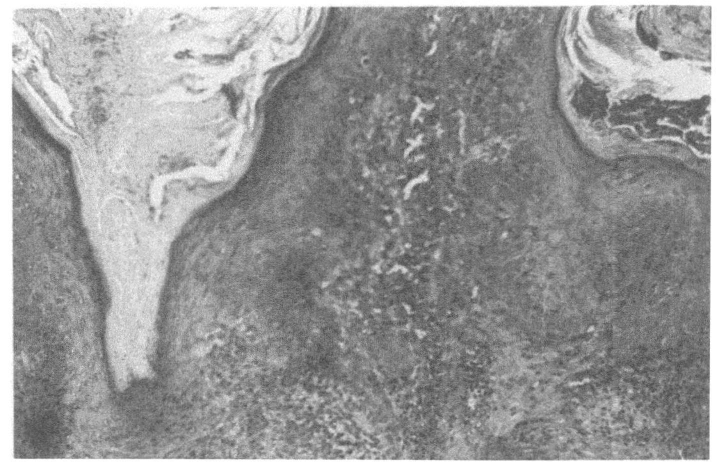

Figure 3a. Wart. Optical microscope image of standard
H and E section.

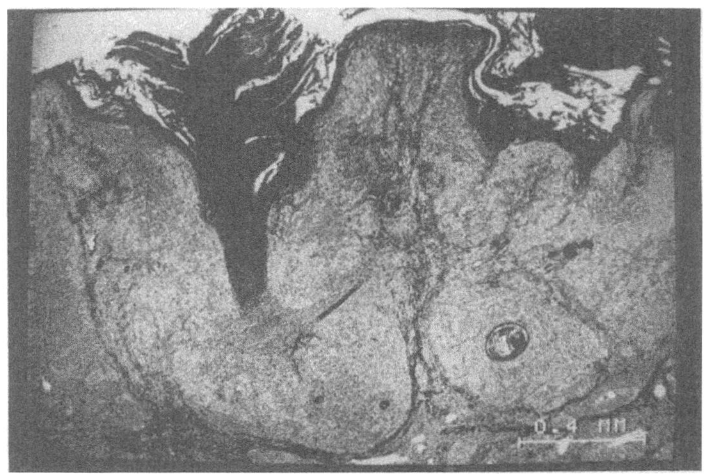

Figure 3b. Wart. Acoustical microscopy image (unstained).

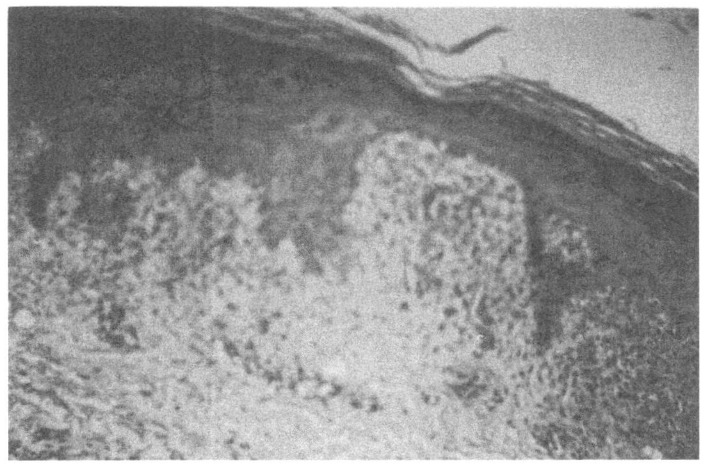

Figure 4a. Lichen Planus. Optical microscopy image of standard H and E section.

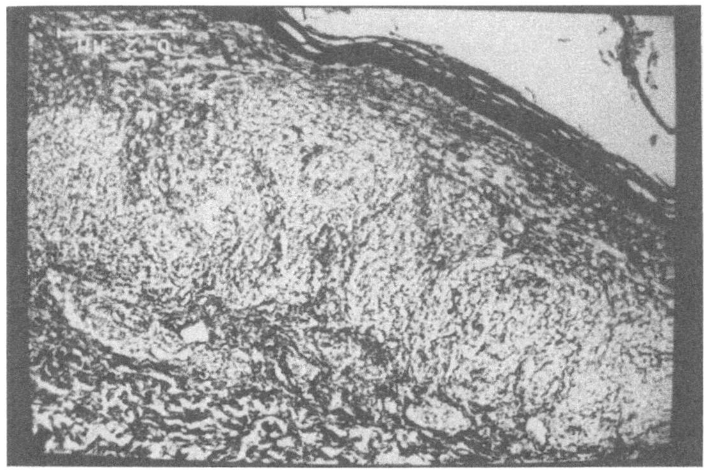

Figure 4b. Lichen Planus. Acoustical microscopy image (unstained).

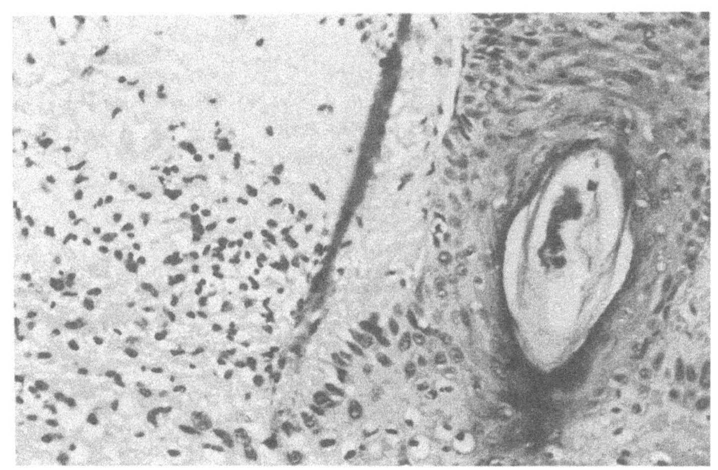

Figure 5a. Actinic Keratosis. Optical microscopy image
of standard H and E section.

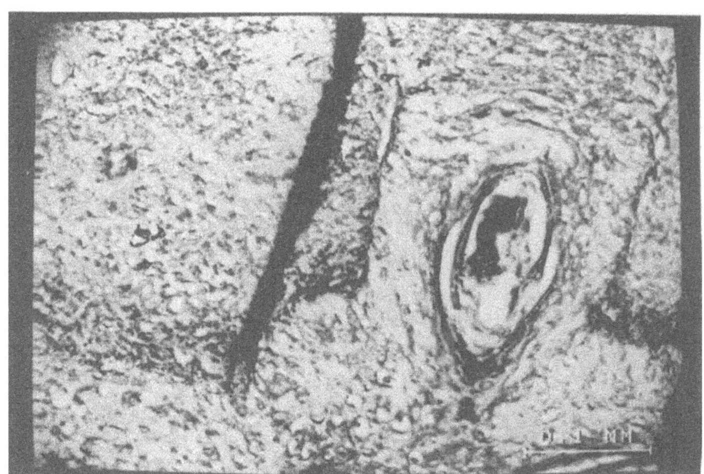

Figure 5b. Actinic Keratosis. Acoustical microscopy image
(unstained).

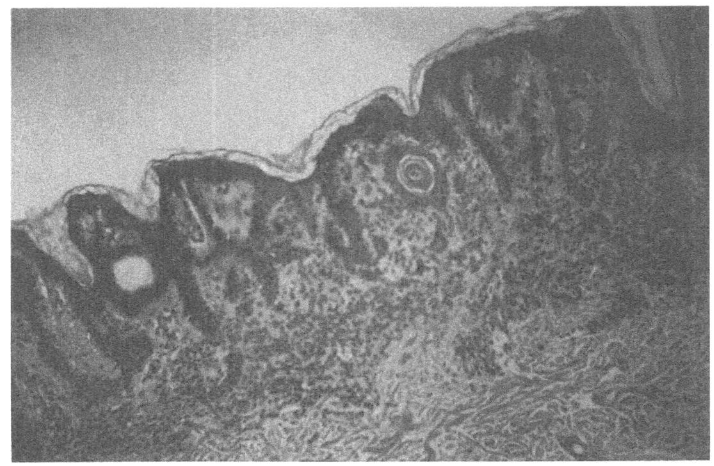

Figure 6a.   Nevus.   Optical microscopy image of H and E
section.

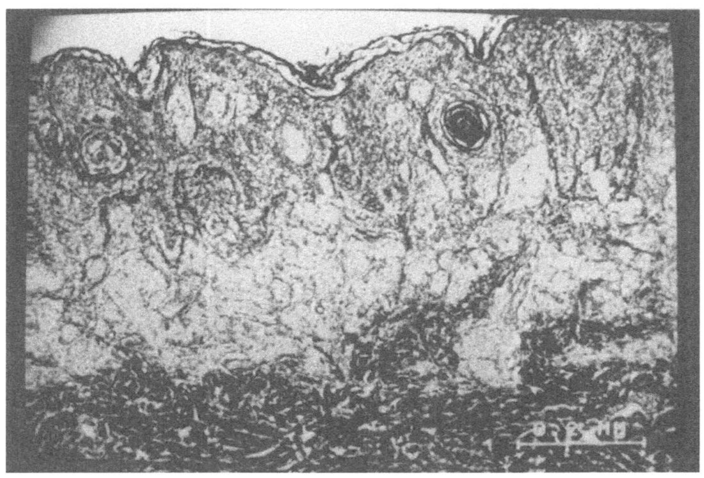

Figure 6b. Nevus. Acoustical microscopy image (unstained).

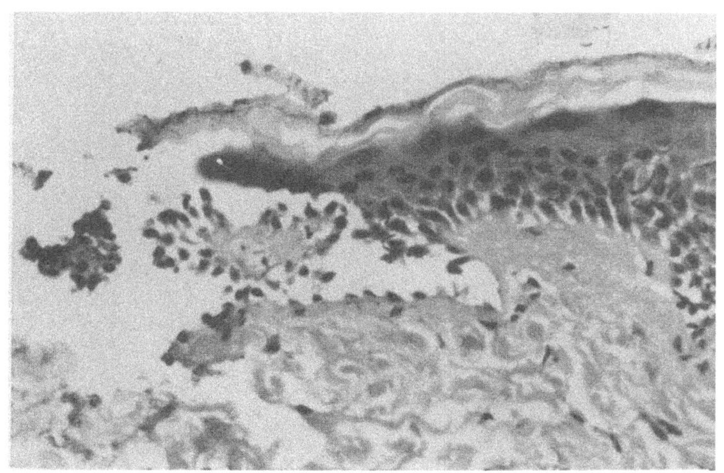

Figure 7a. Pemphigus Vulgaris. Optical microscopy image
of standard H and E section.

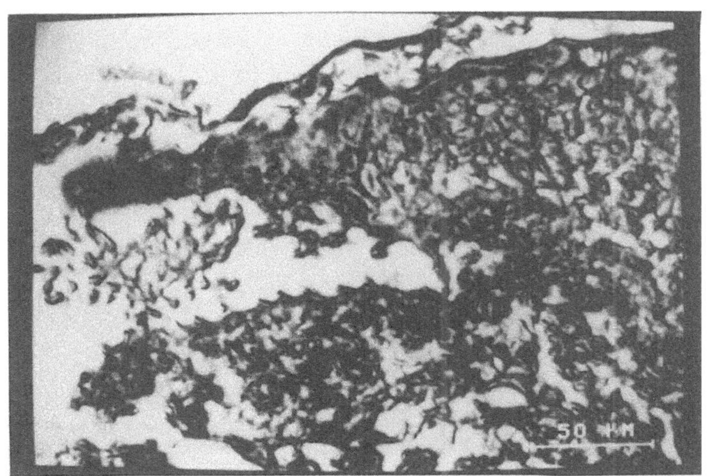

Figure 7b. Pemphigus Vulgaris. Acoustical microscopy
image (unstained).

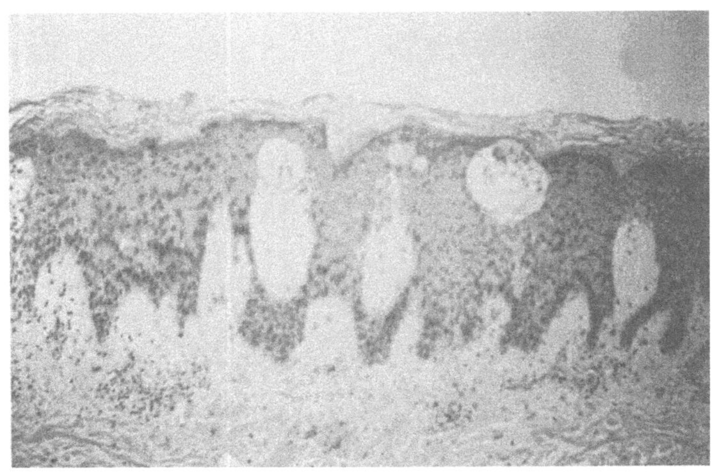

Figure 8a. Eczematous Dermatitis. Optical microscopy image
of standard H and E section.

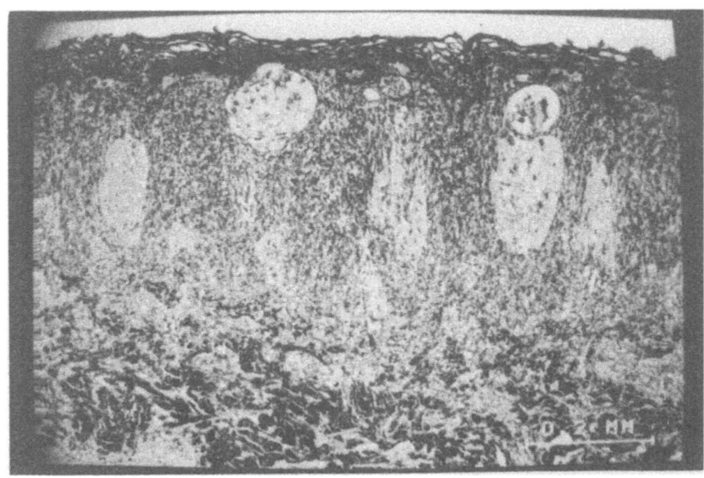

Figure 8b. Eczematous Dermatitis. Accoustical microscopy
image (unstained)

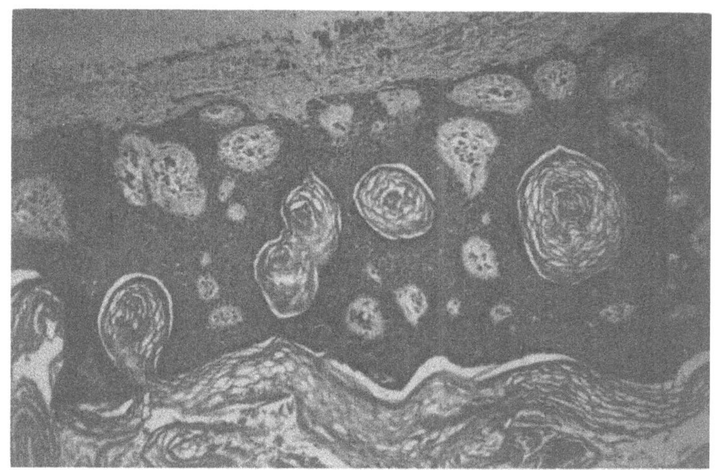

Figure 9a. Seborrheic Keratosis. Optical microscopy image
of standard H and E section.

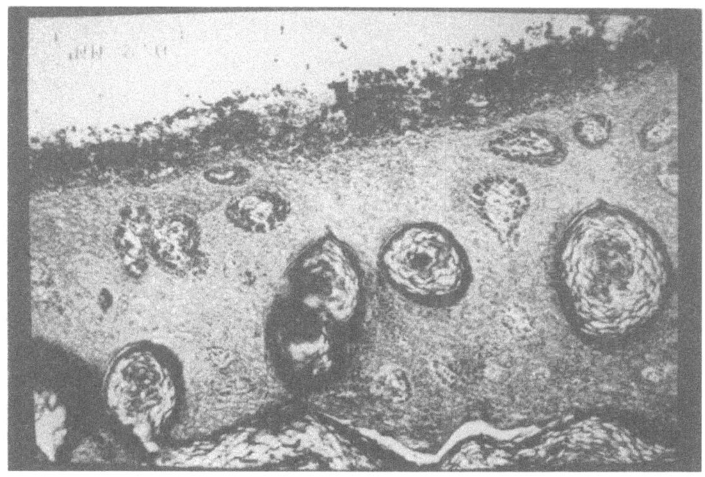

Figure 9b. Seborrheic Keratosis. Acoustical microscopy
image (unstained).

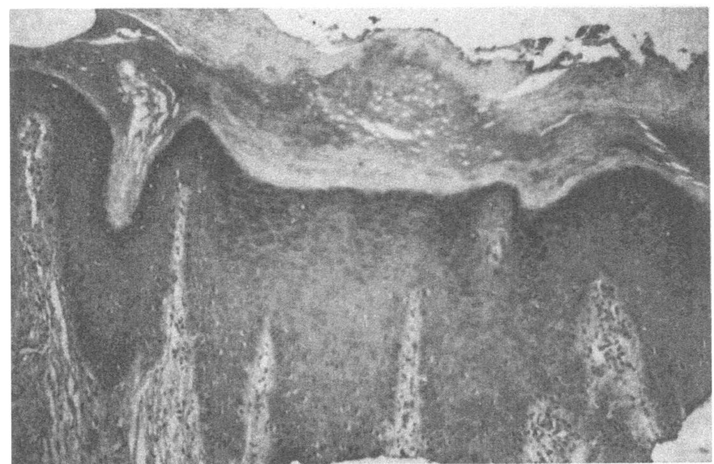

Figure 10a. Lichen Simplex Chronicus. Optical microscopy
image of standard H and E section.

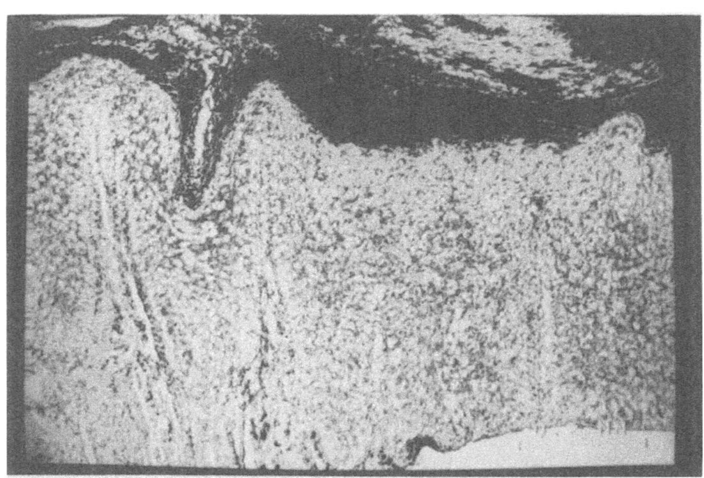

Figure 10b. Lichen Simplex Chronicus. Acoustical microscopy image (unstained).

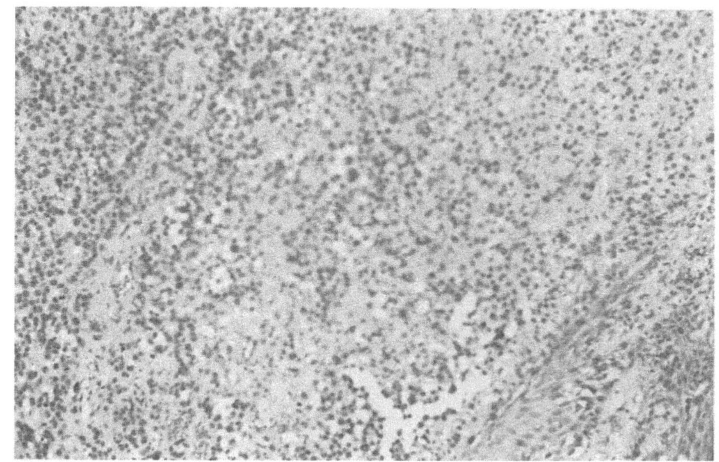

Figure 11a. Malignant Melanoma. Optical microscopy image of standard H and E section.

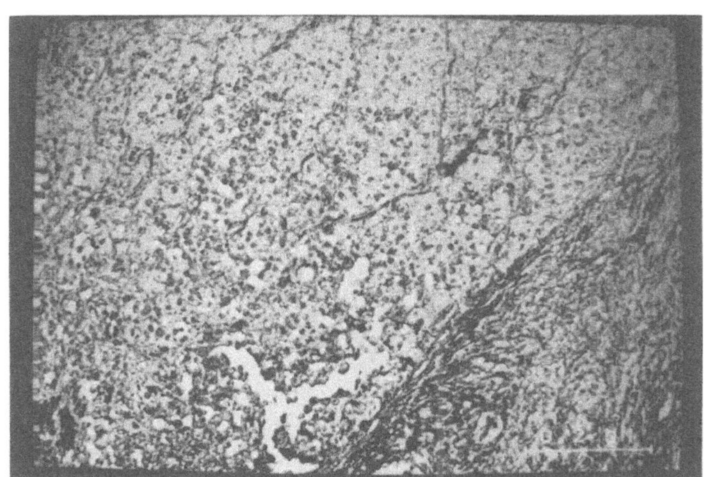

Figure 11b. Malignant Melanoma. Acoustical microscopy image (unstained).

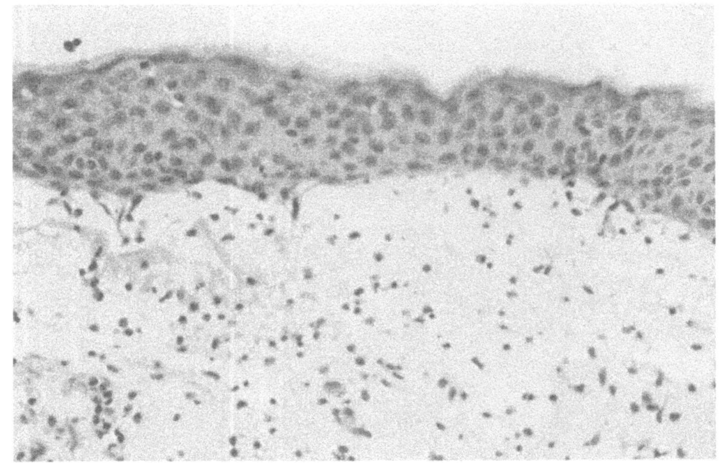

Figure 12a. Bullous Pemphigoid. Optical microscopy image of standard H and E section.

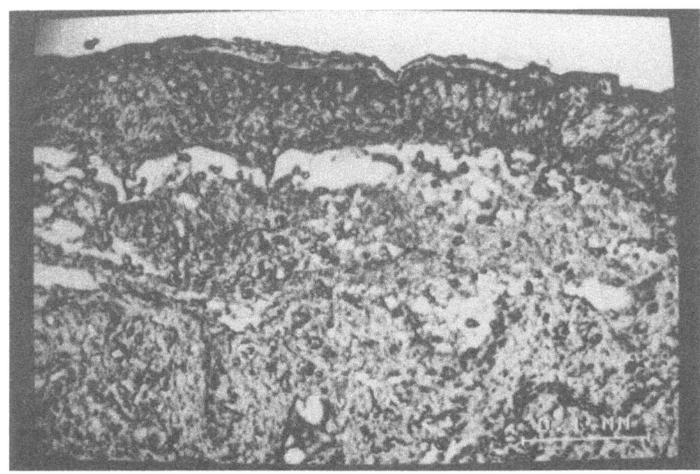

Figure 12b. Bullous Pemphigoid. Acoustical microscopy image (unstained).

REFERENCES

(1) S. Y. Sokolov, USSR Patent No. 49 (August 31, 1936); British Patent No. 477 139 (1937); and US Patent No. 21 64 125 (1939).

(2) A. Korpel, L.M. Kessler, and P.R. Palermo, "Accustic Microscope Operating at 100MHz", _Nature_, Vol. 232, No. 53C6, pp. 110-111 (July 9, 1971).

(3) R.A. Lemons and C.F. Quate, "Acoustic Microscope -- Scanning Version", _Applied Physics Letters, 24_, pp.163-165 (1974).

(4) C.F. Quate, "The Acoustic Microscope", _Scientific American, 241_, pg. 62 (1979).

(5) C.F. Quate, A. Atalar, and H.K. Wickramasinghe, "Acoustic Microscopy with Mechanical Scanning -- A Review", _Proc. IEEE, 67 (8)_, 1092-1114 (1979).

(6) L.W. Kessler, "Acoustic Microscopy Commentary: SLAM and SAM", _IEEE Trans. Sonics and Ultrasonics, SU-32_ (2), pp. 136-137 (1985).

(7) J.E. Olerud, W. O'Brien, et al., "Ultrasonic Assessment of Skin and Wounds with the Scanning Laser Acoustic Microscope", _J. Investigative Dermatology, 88_, pp. 615-623 (1987).

(8) C.M.W. Daft and G.A.D. Briggs, "The Elastic Microstructures of Various Tissues", _J. Acoust. Soc. Am., 85_ (1), pp. 416-422 (1989).

OBSERVATION OF STABLE CRACK GROWTH IN AL_2O_3-CERAMICS BY ACOUSTIC

MICROSCOPY AND ACOUSTIC EMISSION

A. Quinten, C. Sklarczyk, and W. Arnold

Fraunhofer-Institute for Non-Destructive Testing
Bldg. 37, University,
D-6600 Saarbrücken 11, FRG

INTRODUCTION

When cracks propagate upon loading in certain ceramics, one observes
an increase in crack-resistance with increasing crack length (R-curve
behavior). This phenomenon is controversially discussed presently [1,2].
The dissipation of energy, in one class of models, is explained by the
assumption that microcracks are present at the crack tip having certain
extension (process-zone) [1]. In the other models, the increase in crack-
resistance is explained by the interaction of the crack interfaces due to
friction between serrated crack walls [3]. From practical point of view,
it is necessary to understand these phenomena because they may possibly
enable one to improve the fracture toughness of high-strength ceramics.
In this paper we report on measurements of the elastic properties at the
crack-tip of a stably extended crack using Scanning Acoustic Microscopy
(SAM). Acoustic Emission (AE) was also monitored during crack-growth. We
also discuss phenomena observed at the crack-walls.

EXPERIMENTS

Acoustic Microscopy

An experiment was designed which allows one to monitor stable crack-
growth in a Scanning Acoustic Microscope [4]. The aim is to distinguish
between the two different models mentioned above. One should keep in mind
that the SAM is particularly suitable for detecting micro-cracks [5] even
if the crack-opening is far below the resolution-limit of the microscope
because they form an elastic discontinuity interrupting the flow of
ultrasonic energy. Our samples are notched three-point bending specimens
made of Al_2O_3 with varying grain-sizes (between 10-20 µm). The samples
were stressed in the SAM by a suitable loading device at a velocity of 1
µm/min, so that stable crack propagation occurred in small controlled
steps (~ 100 µm large) starting from the notch. The total length of the
crack observed was about 2 mm. V(z)-measurements were carried out at the
crack-tip in order to measure the Rayleigh-velocity with high spatial
resolution, here ~ 10 µm. A decrease of up to 5% in the velocity could
indeed be observed relative to reference points a few mm away from the
crack (Fig. 1). This corresponds to a decrease in the surface elastic

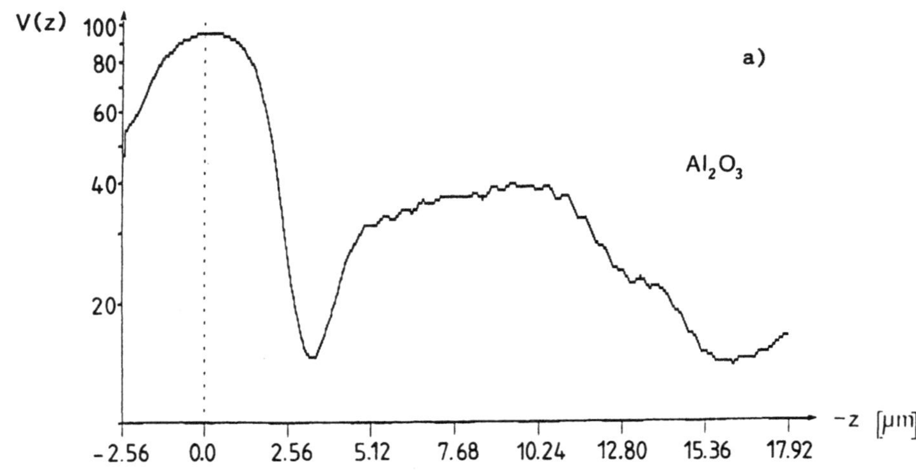

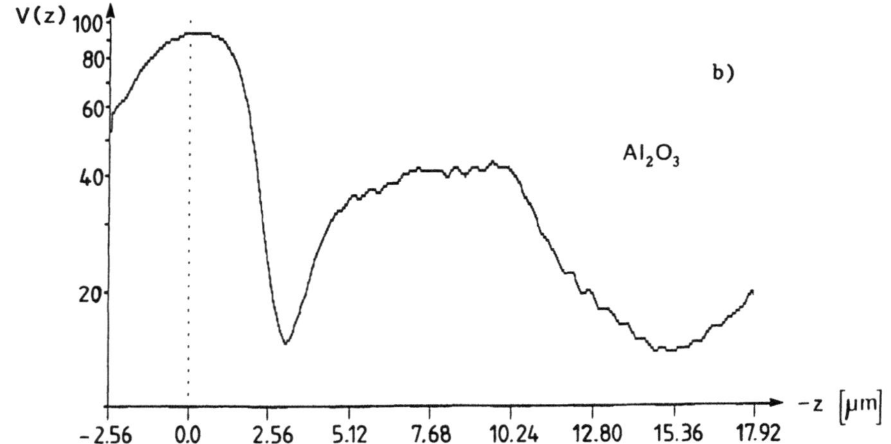

Figure 1. V(z)-Kurve at the crack tip of an Alumina sample (N28). The frequency employed was 1.4 GHz and the water droplet was heated to 60 °C in order to reduce the attenuation in the water.
a) Unloaded condition. Periodicity (distance between the minima) δz = 12.8 µm and corresponding surface wave verlocity v_R = 5.33 µm/µs. b) Loaded condition just before stable crack growth at 15 N. δz is now 12.2 µm corresponding to a surface velocity v_R = 5.21 mm/µs. Repeating the measurements with the B40 - material yielded a larger effect [4].

modulus of 10% roughly in agreement with theoretical predictions [6].
This result was further corroborated by velocity measurements near the
crack tip with longitudinal waves at a frequency of 100 MHz. These measurements yielded a decrease in the velocity by ~ 2%. This smaller value
is probably due to the larger probing area of ~ 100 μm [4].

Quite interestingly, it was also possible to observe effects which
can be interpreted in terms of interaction of serrated crack-walls [2].
Fig. 2 displays the acoustic image of a stably propagated crack in Al_2O_3-
ceramic (B40 – material) in its whole length. It may be noted that although the image looks like an optical micrograph, the contrast mechanism
in SAM reveals changes in the acoustic impedance! After loading and unloading, various locations behind the crack tip were imaged with the SAM.
One can acoustically see crack-closure, opening of new crack-path after
unloading, and bridging-grains (see Figs. 2, [4]) similar to the observations made with optical microscopy [7]. This means that the crack-walls
close at least down to the nm-scale because only then acoustic energy is
transmitted [8]. It should be noted that B40 material shows a transcrys-
talline fracture in contrast to N28 material which exhibits an intercrys-
talline fracture behavior with strongly damaged crack walls. In our opi-
nion this enhances the friction of the crack-walls and therefore its
contribution to the energy dissipation. In turn, this should lead to a
smaller reduction of the elastic moduli caused by the process-zone, in
agreement with our observations.

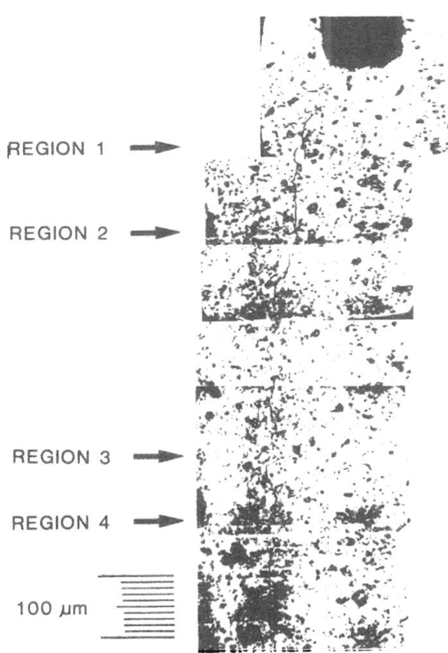

Figure 2. Image of a crack obtained with SAM at a frequency of 1 GHz. The
crack was propagated by slowly loading the sample in a three-
point bending device mounted in the microscope. The crack propagated in small steps of approximately 100 μm length.

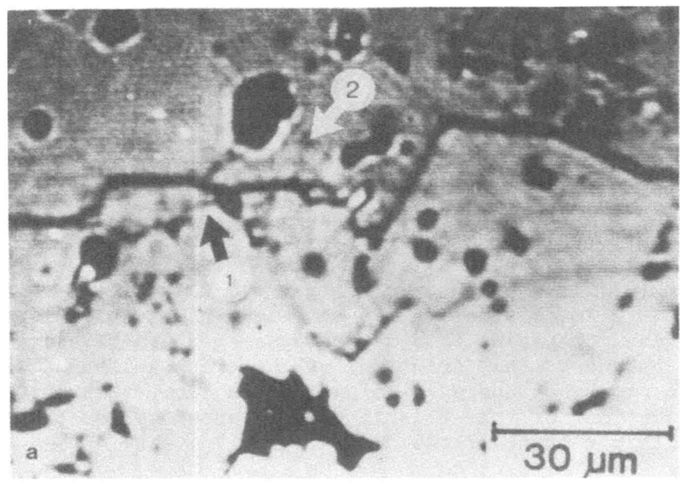

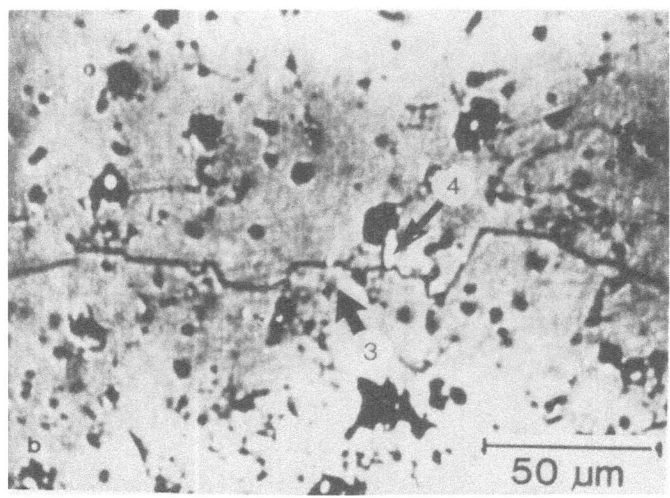

Figure 3. SAM image of a stably propagated crack in a notched three-point bending specimen at two different stages (times). In a) the arrow 1 indicates a side-crack which is no longer observed in b) after unloading the sample (arrow 3). At the loacation indicated by arrow 2 (Fig. 3a), a new crack appears after unloading the sample (arrow 4, Fig. 3b). The frequency employed was 1 GHz and the defocus z was z = -3 µm. The two images were taken with different magnifications in region of Fig. 2. Similar effects were observed in the other regions indicated in figure 2.

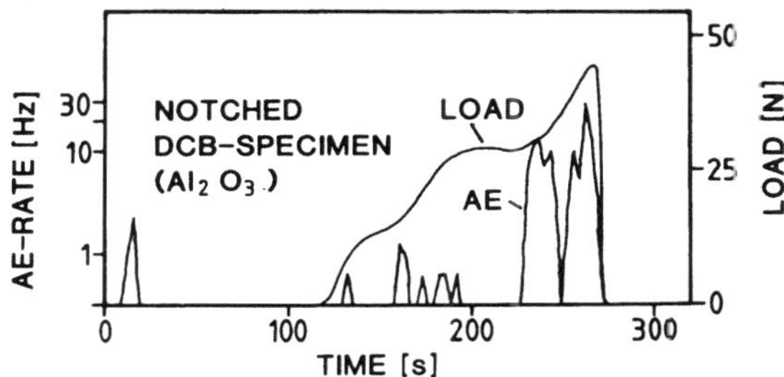

Figure 4. Acoustic emission signals obtained from a notched DCB-specimen made of Al_2O_3. AE signals could be observed although no crack growth was visible after examination with a SEM. The loading of the sample was carried out manually.

Acoustic Emission

Parallel to these measurements, acoustic emission studies were carried out. Double-cantilever beam specimens were loaded, and it was possible to observe AE-signals as shown in Fig. 4 in the frequency range 100-500 kHz. After this, the samples were examined with a scanning electron mivroscope in order to check whether any crack formation from the notch had occurred. In some of the specimen examined, no crack growth larger than 10 μm was visible. Therefore, we ascribed the observed AE signals to the formation of microcracks. It was also possible to localize the origin of the signals to be from the tip of the notch with an accuracy of ± 2 mm. The AE results were further corroborated by small-angle x-ray scattering studies carried out on the same specimen. These studies revealed the presence of scattering centers at the notch tip which were interpreted as micro-cracks [9].

ACKNOWLEDGEMENT

We thank the German Science Foundation for financial support.

REFERENCES

[1] Buresch, F. E., Materialprüfung 29, 261 (1987)
[2] Knehans R. and Steinbrech, R., Fortschritts. Ber. Deutsch. Keram. Gesellsch. 1, 59 (1985)
[3] Knehans R. and Steinbrech R., J. Mat. Sci. Let:., 1, 327 (1982)
[4] Quinten, A., and Arnold, W., Mat. Sci. Eng., to be published (1989)
[5] Briggs, G. A. D., "An Introduction to Scanning Acoustic Microscopy", Oxford University Press (1985)
[6] Babilon, E., Buresch, F. E., Kleist, G., and Nickel, H., Proc. Europ. Conf., Fract. Mech. (Budapest 1988), to be published
[7] Swanson, P. L., Fairbanks, C. J., Lawn, B. R., Mai, Yiu-Wing, and B. J. Hockey, J. Am. Ceram. Soc. 70, 279 (1987)
[8] Buck, O. , Morris, W. L., and Richardson J. M.; Appl. Phys. Lett. 33, 371 (1978)
[9] Babilon, E., private communication

DIRECT MEASUREMENTS OF THE SAW VELOCITY AND ATTENUATION USING CONTINUOUS WAVE REFLECTION SCANNING ACOUSTIC MICROSCOPE (SAMCRUW)

A. Kulik, G. Gremaud, S. Sathish

Ecole Polytechnique Fédérale de Lausanne, Institut
de Génie Atomique CH-1015 Lausanne (Switzerland)

INTRODUCTION

Scanning Acoustic Microscopy (SAM) has already several applications in Material Science. However interpretation of SAM images is not straightforward, since contrast can have several origins :

- Topography of the surface
- Local Acoustic Impedance and Bulk Acoustic Waves propagation constants
- Surface Acoustic Wave (SAW) propagation constants

Therefore possibility of imaging separately different physical properties can enlarge domain of SAM applications. Several attempts were made using different approaches :

i) phase sensitive SAM which allows separation of the surface topography of the sample, actually limited to low frequency range [1],

ii) nonlinear parameters (B/A) imaging for which main interest lies in biological applications,

iii) short pulse (time resolving, broadband) SAM [2,3] which is probably the most promising, however limited to low frequency range, or allowing localized measurements but no imaging.

Short pulse SAM electronics and lens transducers must be able to transmit and receive sufficiently short ultrasonic pulses, allowing time separation of specular and SAW reflections [2], or separation of reflections from the two different areas of the sample which have to be resolved [3].

We will discuss here applicability of the short pulse SAM, keeping in mind its two main drawbacks :

- poor Signal to Noise Radio (SNR) related with necessary
  wide bandwith of receivers,
- difficulties in studying dispersive media, where layered
  structures are one of the most common examples.

We propose short pulse measuring technique, which can be
applied for imaging of the SAW propagation constants in
nondispersive media. Our method is unusual in realisation :
instead of using directly pulse echo electronics, we use
Continuous Wave (CW) sweep and calculate equivalent time domain
response using Inverse Fourier Transform (IFFT). Advantage
consists in much more better SNR with trade off measuring
speed. Proposed method can be applied as well for classical
pulse echo equipment. Method will be compared with Continuous
Wave V(z) (CW V(z)) measurements made at constant frequency.

MEASURING EQUIPMENT

We use Continuous Wave Reflection Scanning Acoustic
Microscope (SAMCRUW) described in details in [4]. Instrument is
composed from commercially available parts as : acoustic lens,
x y z (1, 1, 0.1 µm/step) sample movement, Vector Network
Analyzer with time domain capability and controlling computer.
Reflection coefficient s_{11} of the system lens-water-sample can
be measured as a function of :

- frequency (**F**),
- distance **z** between sample and lens,
- position **x y** on the sample

EQUIVALENT SHORT PULSE MEASURING METHOD

Measuring the s_{11} reflection as a function of the
frequency **F** of the ultrasonic wave allows to obtain the time
domain equivalent short pulse response by calculating the
Inverse Fourier Transform (IFFT) of $s_{11}(\mathbf{F})$ (Fig. 1).

<u>Short pulse method used for nondispersive media</u>

Separation of the specularly reflected pulse and the
delayed through SAW propagation can be observed near the top of
Fig. 2 (lens close to the sample). By increasing the distance z
between sample and lens, one decreases the time separation of
the pulses which start to overlap, developing classical V(**z**)
curve near **z** = 0, where focal point is on the surface.

Using two properly choosen focal distances $\mathbf{z}_o$, $\mathbf{z}_1$ (fig. 2)
and measuring time shifts (Δt_d, Δt_r) and amplitudes of directly
reflected pulses (A_{do}, A_d) and SAW pulses (A_{ro}, A_r) one can
calculate propagation constants for SAW :

The sound velocity in coupling fluid is given by:

$$V_W = \frac{2\,\Delta z}{\Delta t_d} \qquad [m|s] \qquad\qquad (1)$$

and the SAW velocity :

$$V_r = \frac{2\,\Delta z}{\sqrt{(\Delta t_d)^2 - (\Delta t_r)^2}} \quad [\text{m/s}]$$

(2)

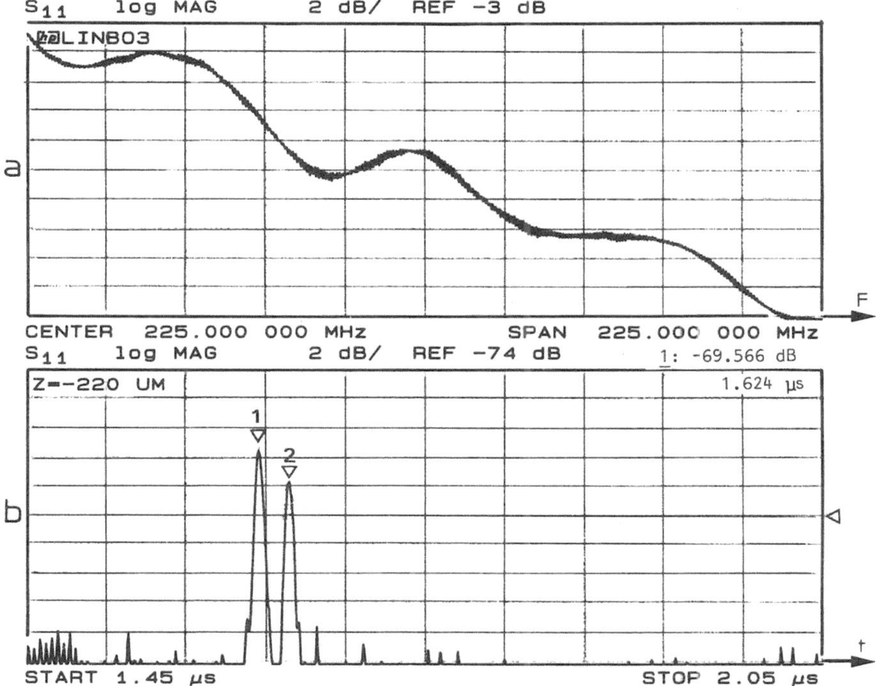

Fig. 1. a) Frequency response $s_{11}(F)$ of the system : acoustic
lens, water, LiNbO₃ sample ($z = - 220$ μm).
b) Equivalent short pulse response calculated using
Inverse Fourier Transform.

The SAW attenuation can also by calculated by :

$$\alpha_R = \frac{1}{(\Delta t_d)^2 - (\Delta t_r)^2} \, (\Delta t_r \ln \frac{Ar}{Ar_o} - (\Delta t_d)^2 \, \alpha_w) \quad [\text{Neper/s}]$$

(3)

where α_w is the attenuation in coupling fluid expressed in
Neper/s.
SAW attenuation is the sum of leaky losses and intrinsic
SAW attenuation of the sample and they can be separated [2].
Therefore a measurement procedure can be the following :

- make two consecutive scans at two different defocuses
  which differs by Δ_z

- measure and store 4 values for each pixel : pulse shifts
  Δt_d and Δt_r; amplitudes A_r and A_{ro}.

- display calculated SAW velocity (2) or attenuation (3)

Short pulse method used for dispersive media

Result from Fig. 3 shows that short pulse method cannot be
used. Several propagating modes and their dispersion leads to
several "smeared" pulses where neither amplitude nor time shift
can be determined.

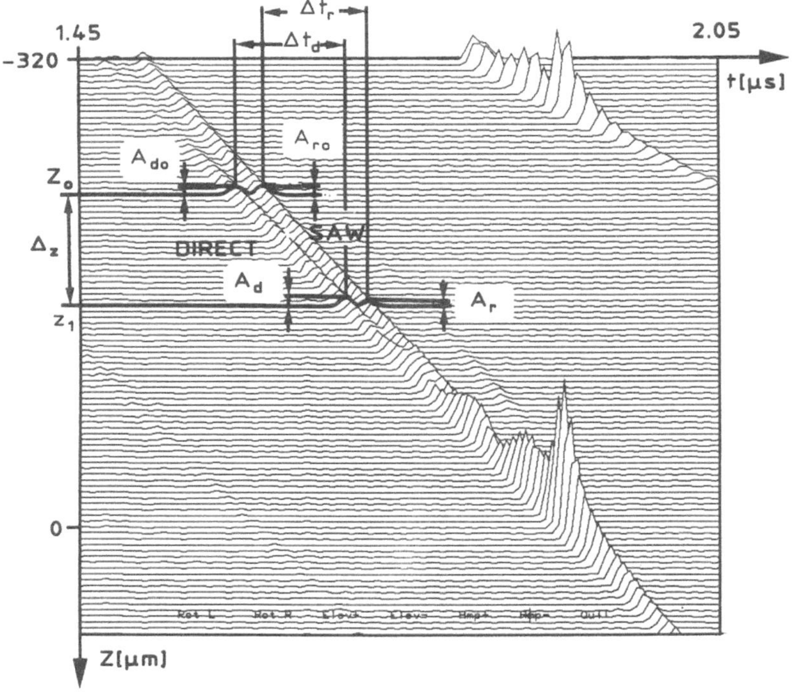

Fig. 2. Short pulse response of LiNbO$_3$ sample (calculated from
frequency (**F**) response of the system as at Fig. 1b) as
a function of distance **z** (sample-lens). Central frequency is 225 MHz, sweep span is 225 MHz.

CONTINUOUS WAVE V(z) MEASURING METHOD

Measuring reflection coefficient **s**$_{11}$ of the system as a
function of distance **z** between lens and sample one obtains CW
V(z) curve (fig. 4a).As CW V(**z**) is measured as a function of
z(t) by scanning **z** with a translation speed $\dot{z}$, one can calculate
the Fourier Transform of the CW V(**z**) curve. One obtains the
spatial frequency spectrum represented in Fig. 4b, where two
peaks are present as a function of f_z, related with the
specular reflection (f_{z1}) and the SAW reflection (f_{z2}).

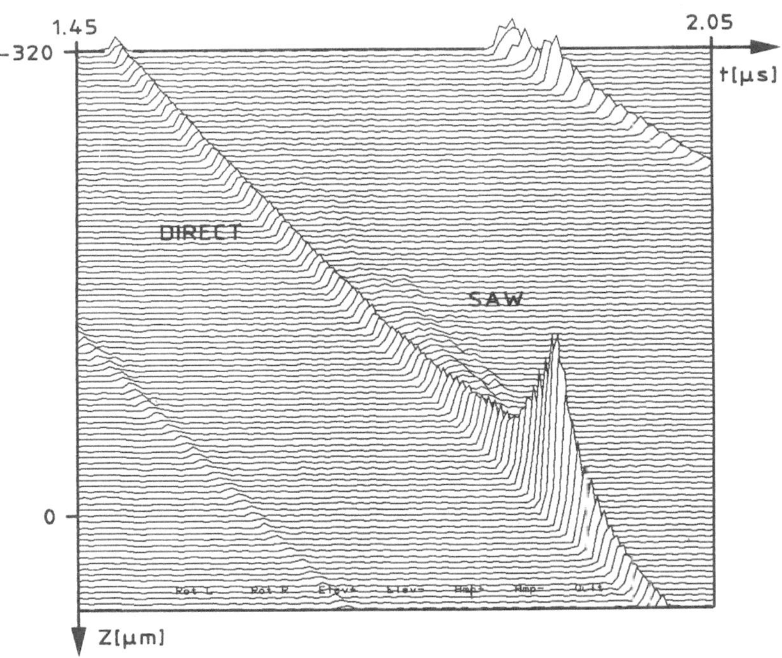

Fig. 3. Short pulse response of a LiNbO$_3$ sample (the same type of substrate as used in Fig. 2) coated with a 1.2 μm thick layer of Gold. SAW pulses are smeared out.

Sound velocity in coupling fluid can be obtained from (Fig. 4b):

$$V_W = \frac{2\,\dot{z}\,F}{f_{z1}} \qquad (4)$$

where $\dot{z}$ is the translation speed, F the working frequency and f_{z1} the spatial frequency of the water peak in the Fourier Transform domain.

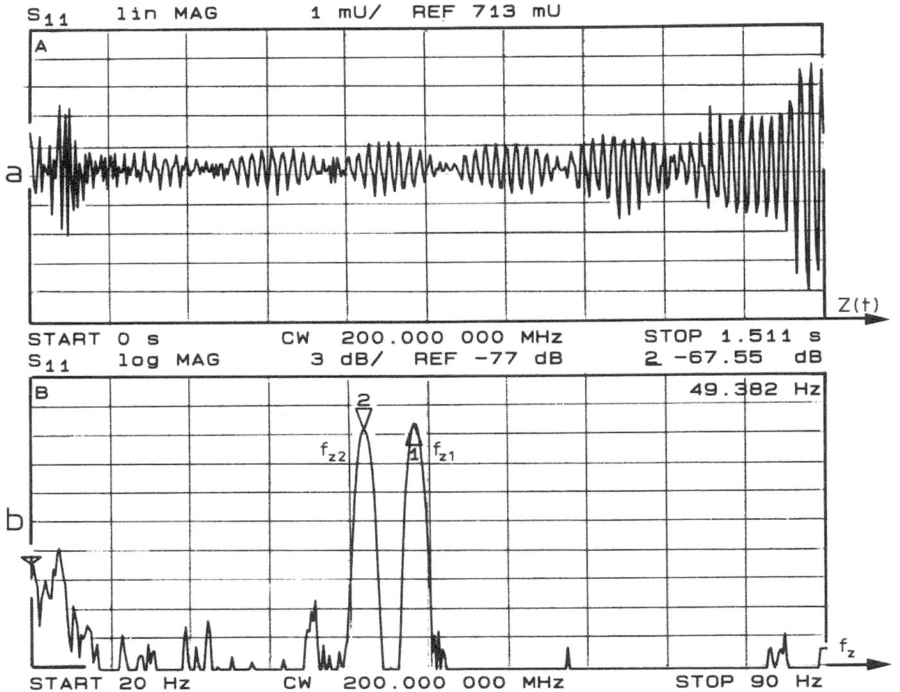

Fig. 4. a) Continuous Wave V(z) curve for LiNbO$_3$ sample (the same as in Fig. 1) measured at F = 200 MHz (Translation speed $\dot{z}$ = 200.8 µm/s)).
b) Spatial frequency spectrum obtained by Fourier Transform of CW V(z) curve.

SAW velocity is given by :

$$V_R = \frac{V_W}{\sqrt{1 - \left(\dfrac{f_{z2}}{f_{z1}}\right)^2}} \qquad (5)$$

where f_{z2} is the spatial frequency of the SAW peak in the Fourier Transform domain (Fig. 4b).

SAW attenuation can be also calculated from the amplitudes
of the f_{z1} and f_{z2} peaks (Fig. 4b).

CW V(z) used for nondispersive media

Making Fourier Transform of CW V(**z**) (as Fig. 4b) and
representing them as a function of the ultrasonic frequency **F**,
one obtains the diagram presented on Fig. 5. Amplitude of the
FFT of CW V(**z**) curves was used to modulate the brightness of
Fig. 5 where black correspond to highest amplitude. Image was
further diffused using Floyd-Steinberg algorithm what explains
"grainy" structure of the figure. The two peaks related with
the specular reflection and the SAW reflection respectively are
responsible for two straight lines as a function of the
ultrasonic frequency **F**, which means that the SAW velocity does
not depend on the frequency and, as a consequence, that this
particular medium is nondispersive for the Rayleigh waves.

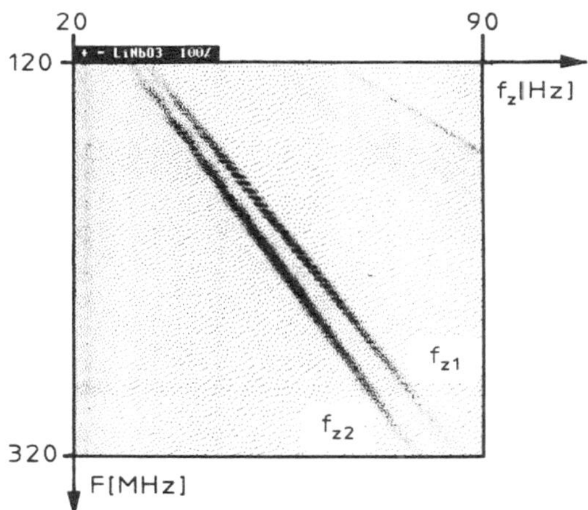

Fig. 5. Fourier Transform of CW V(**z**) as a function of
ultrasonic frequency **F** measured in LiNbO$_3$. The
substrate is the same as used in Fig. 1. Straight SAW
line (f_{z2}) shows that SAW velocity does not depend on
frequency (nondispersive medium).

CW V(z) used for dispersive media

CW V(**z**) was measured in a wide frequency range (120-320 MHz)
for the LiNbO$_3$ sample coated with 1.2 μm Gold layer (Fig. 6).
At least two SAW modes can be distinguished, both presenting a
strong dispersion, seen by the curvature of the two lines as a
function of the frequency **F** (SAW$_1$ and SAW$_2$).

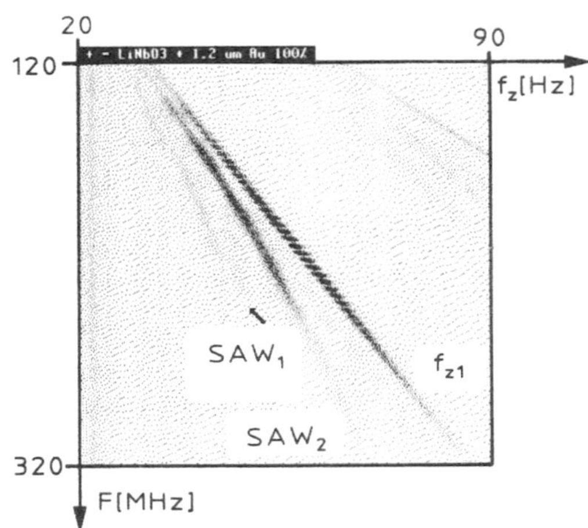

Fig. 6. Fourier Transform of CW V(z) as a function of working frequency (**F**) measured in LiNbO$_3$ substrate coated with a 1.2 µm Gold layer.

The CW V(**z**) measurements allows us to separate and measure different SAW modes as well as their dispersion. Result presented in Fig. 6 help us to understand messy short pulse measurement of the same sample (Fig. 3) where the wide bandwidth SAW pulses are smeared out, making short pulse method almost useless. In fact, different spectral components of the SAW pulse have different propagation velocities, what in time domain give pulse widening with redoubling tendency due to presence of two different SAW modes.

CONCLUSIONS

Construction of SAMCRUW allowed us to use two different methods with the following advantages and disadvantages :

<u>Advantages of the short pulse method</u>

- is fast, therefore potentially can be implemented for imaging SAW velocity and attenuation
- it can give additional informations when studying low SAW velocity materials (e.g. biological), as topography of the sample or thickness of the layers

<u>Disadvantages of the short pulse method</u>

- poor SNR related with wide bandwith (can be improved working in frequency domain as presented here, with trade-off measuring speed)
- wide spectral content disallowing measurements of dispersive media

<u>Advantages of CW V(**z**) method</u>

- excellent SNR since because narrow bandwidth receivers
  are used
- working frequency is perfectly known; by repeating
  measurements at different frequencies (**F**), the
  dispersion relation can be studied at any frequency
  which lies inside transducer's bandwidth
- sound velocity in coupling fluid is measured at the same
  time (4). This feature can be important for use in
  cryogenic SAM [5].

<u>Disadvantages of CW V(**z**) method</u>

- slow, not adequate for imaging
- it needs high sensitivity equipment which is
  commercially available, but expensive.

ACKNOWLEDGEMENTS

This work was partially supported by the Swiss National Science Foundation.

REFERENCES

1. B.T. Khuri-Yakub, P. Reinholdtsen, C.H. Chou, P. Parent and C. Cinbis, Amplitude and Phase Acoustic Microscopy and its application to QNDE, <u>in</u> : "Acoustical Imaging," vol. 17, 173 (1988), ed. Shimizu, N. Chubachi and J. Kushibiki.

2. R.S. Gilmore, R.A. Hewes, L.J. Thomas III and J.D. Young, Broadband Acoustic microscopy : Scanned images with Amplitude and velocity information, <u>in</u>: "Acoustical Imaging," Vol. 17, 97 (1988), ed. H. Shimizu, N. Chubachi and J. Kushibiki.

3. A.M. Sinton, G.A.D- Briggs and Y. Tsukahara, Time resolved Acoustic Microscopy of polymer coatings, <u>in</u>: "Acoustical Imaging," vol. 17, 87 (1988), ed. H. Shimizu, N. Chubachi and J. Kushibiki.

4. A. Kulik, G. Gremaud and S. Sathish, Continuous Wave Reflection Scanning Acoustic Microscope (SAMCRUW), <u>in</u>: "Acoustical Imaging," vol. 17, 71 (1988), ed. H. Shimizu, N. Chubachi and J. Kushibiki.

5. J.O. Fossum and J.D.N. Cheeke, Acoustic Microscopy applied to physical acoustics, <u>Rev. Sci. Inst</u>., 57: 636 (1986).

ELECTRONIC SCANNING IN ACOUSTIC MICROSCOPY USING A WEDGE TRANSDUCER

L. Germain and J. D. N. Cheeke

Centre de Recherche en Microélectronique (CERMUS)
Département de physique, Université de Sherbrooke
Sherbrooke, Québec, CANADA   J1K 2R1

Abstract

A new method for electronic scanning in one direction in an acoustic microscope is investigated. The principle consists in using a wedge piezoelectric transducer as the source for a line focusing device, such as a cylindrical lens. As the exciting RF frequency is varied, the transducer resonates at different positions along its length. This provides a simple method for varying the position of the focal spot in one direction. This electronic scan is combined with a mechanical scan in the orthogonal direction to give a hybrid electronic-mechanical scanning instrument. Design considerations for optimal performance of the transducer assembly is discussed. A theoretical simulation of the wedge transducer is also developed and compared to experimental results.

Introduction

There is an increasing need for non-destructive inspection in production lines to achieve greater reliability of components such as integrated circuits. Ultrasonic imaging systems have proven to be useful in detecting internal defects in components but the time taken by the mechanical system to scan the object is usually too long to be practical for use in production lines.

In order to accelerate the imaging process, the fast axis mechanical scanning system has to be replaced by an electronically controlled scanning system which can be much faster in principle. In this work, a simple approach of doing such an electronic scanning of an acoustic beam for fast acoustic imaging is presented. The technique makes use of a wedge piezoelectric transducer as a position dependant resonant transducer which can be easily scanned by sweeping the frequency of the excitation signal. Fast acoustic images can be obtained by combining a wedge transducer and a line focusing system such as a cylindrical lens for fast scanning on one axis, with a unidirectional mechanical scanning system in the orthogonal direction.

The potential of this approach has been demonstrated in previous work.[1,2] A wedge transducer bonded on an aluminum cylindrical lens was used to obtain images of surface and subsurface holes in a test specimen. However quite a low resolution was obtained in the frequency scanning axis and a closer look to the resolution problem was needed.

In the following sections, the performance of the wedge transducer system is investigated. Design considerations for an optimal wedge transducer are first discussed. A simple staircase model of the wedge transducer is then used to evaluate the best operating conditions for maximum resolution. Experimental measurements for two transducer assembly configurations are then compared with the calculated results.

DESIGN CONSIDERATIONS

Fig. 1 shows the typical configuration of a wedge transducer. It consists of a tapered piezoelectric plate (such as LiNbO3) used in the thickness resonant compressional mode. In the following, the central, lower and higher frequencies are noted F_c, F_1 and F_2 respectively, and the corresponding thicknesses are T_c, T_1 and T_2. The angle between both faces of the transducer is called θ.

One of the main considerations in designing a transducer for an imaging system is to obtained the best possible resolution. With the wedge transducer, where the beam displacement is done by frequency scanning, this is obviously obtained by maximizing the frequency range supported by the transducer along its length. However, due to the odd harmonic resonance capability of compressional transducer, the higher frequency end of the transducer must be kept at a value below the third harmonic of the low frequency end. Failure to do so would produce two separate beams at different positions on the transducer, and a severe degradation of the image would be observed. The bandwidth of the wedge transducer is then limited to $F_2 = 3\,F_1$.

In principle, the wedge transducer can also be designed to be used in one of its odd harmonics. For transducers with a given center frequency F_c, using a higher harmonic leads to a thicker transducer and a narrower frequency range. Using simple geometrical considerations and avoiding any harmonic overlap over the transducer length, the following general relations for an optimal transducer designed for use at an harmonic h can be easily obtained:

$$F_1 = \frac{h + 1}{h + 2}\,F_c\,, \qquad\qquad F_2 = \frac{h + 1}{h}\,F_c\,, \qquad (1)$$

$$T_1 = \frac{h\,(h + 2)}{(h + 1)}\left[\frac{2\,v}{F_c}\right]\,, \qquad\qquad T_2 = \frac{h^2}{(h + 1)}\left[\frac{2\,v}{F_c}\right]\,. \qquad (2)$$

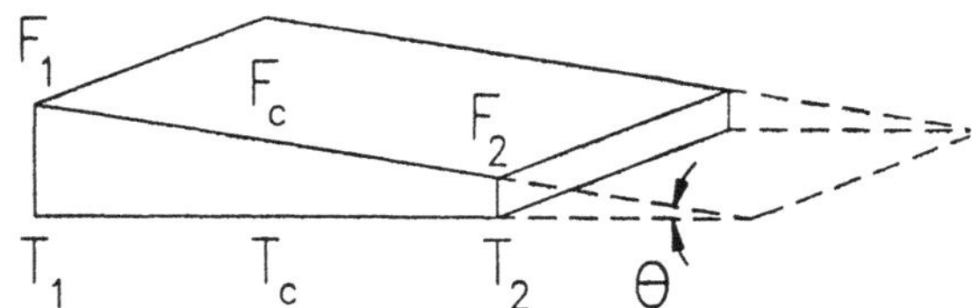

Fig. 1. Typical configuration of a wedge transducer.

238

Table 1. Typical values for an optimized LiNbO$_3$ wedge transducer at F_c = 40 MHz, for various harmonics

h	F_1 MHz	F_2 MHz	ff_1 MHz	ff_2 MHz	T_1 mm	T_2 mm	θ deg
1	26.7	80.0	26.7	80.0	0.138	0.046	1.15
3	32.0	53.3	10.7	17.8	0.345	0.207	1.72
5	34.3	48.0	6.9	9.6	0.537	0.383	1.91
7	35.6	45.7	5.1	6.5	0.725	0.564	2.00

(note: ff_1 and ff_2 are the fundamental frequency limits of the transducer.)

Here, v is the bulk compressional sound velocity in the piezoelectric
material. The results are all expressed as a function of the frequency
F_c at the center of the transducer.

Table 1 lists some typical design values for a 4.6 mm long LiNbO3
transducer with a center frequency of 40 MHz and for various harmonics.
The values listed are the usable frequency bandwidth, the fundamental
frequency bandwidth of the transducer, the thicknesses at each extremity
and the angle between both surfaces.

Staircase Model of a Wedge Transducer

The wedge transducer has been studied earlier, mainly for its use as a
wide band transducer.[3-5] Barthé et al.[4] used a staircase model to
calculate the input impedance and pressure profile of a tapered transducer
excited by a broad band pulse. Their results showed a good agreement with
experimental measurements.

Here, a similar staircase approach is used to estimate the resonance
width and position on a wedge transducer excited at a single frequency.
The staircase model consists essentially in dividing the transducer in N
smaller elements, each of them being considered as flat, with a thickness

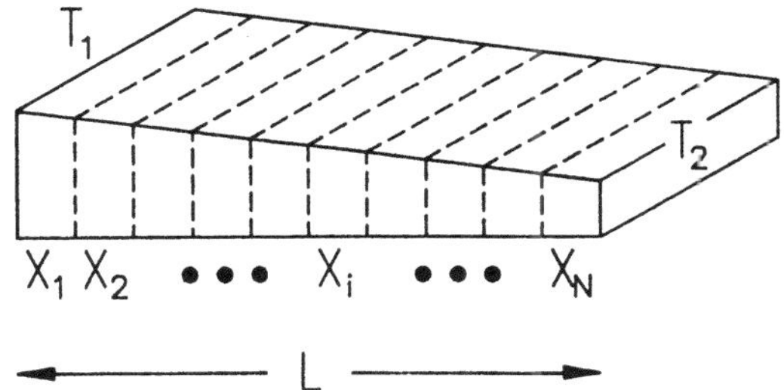

Fig. 2. Division of the wedge transducer as used in the staircase model.

equal to that of its center position on the transducer. The emitted power
is calculated independently for each element, using a standard transducer
theory. The emission pattern for the full transducer is then obtained by
plotting the contribution of each element as a function of their position
on the transducer.

The typical configuration for the staircase model is showed in Fig. 2.
The center position X_i of the element i of a transducer of length L,
divided in N pieces, is given by

$$X_i = (i - .5)\ (L\ /\ N) \tag{3}$$

and the corresponding thickness t_i is

$$t_i = T_1 + (T_2 - T_1)\ X_i\ /\ L\ . \tag{4}$$

The acoustic power emitted by a standard air-backed transducer into a
material of acoustic impedance Z is given by[6]

$$P = 4\ K^2\ C_0\ f_0\ \frac{Z_t}{Z\ M_0(f)}\ v_0^2 \tag{5}$$

where K^2 is the electromechanical coupling constant, C_0 is the capacitance
of the transducer, f_0 is the thickness resonant frequency, Z_t is the
acoustic impedance of the transducer material, and v_0 is the amplitude of
the electric signal applied to electrodes. $M_0(f)$ is the resonance form
factor given by

$$M_0(f) = \frac{\left[\cos\ \pi\ \dfrac{f}{f_0}\right]^2 + \left[\dfrac{Z_t}{Z}\ \sin\ \pi\ \dfrac{f}{f_0}\right]^2}{\left[\sin\ \dfrac{\pi}{2}\ \dfrac{f}{f_0}\right]^4} \tag{6}$$

where f is the applied signal frequency.

The relative power emitted from the element i on the wedge can be
rewritten from (5) as

$$P_i(f) = C\ \frac{Z_t}{Z}\ \frac{1}{t_i^2\ M_{0i}(f)} \tag{7}$$

where $C = (2K^2\varepsilon^s A v v_0^2)$ is a constant for a given transducer arrangement (A
is the area of an element, ε^s is the constant strain dielectric constant of
the piezoelectric material), and $M_{0i}(f)$ is given by (6) where f_0 is
replaced by the resonant frequency of the element $f_i = (2v/t_i)$.

Using (3), (4) and (7), the transducer response at a given frequency can now be calculated. Fig. 3 shows typical curves obtained at various frequencies, for a third harmonic optimized LiNbO3 transducer emitting in water. The beam shift with frequency can be clearly seen. The displacement of the beam is not linear with the frequency since the resonant frequency varies as the inverse of the thickness. This will have to be accounted for in the imaging system to avoid image distortion.

The effect of some parameters over the beam width on the transducer are now calculated using the above model. Fig. 4 shows the effect of using different harmonics h in the design of a transducer at a given center frequency F_c. The parameters of Table 1 were used for the calculations, with emission in water. A noticeable gain in resolution is observed between first and third harmonic while the gain is much more modest for the higher ones. The gain in resolution is directly correlated to the angle between the two faces of the wedge transducer, as can be seen by the angle values of Table 1. Using a third harmonic transducer in the case of Table 1 seems to be a good choice due to the convenient thicknesses, modest frequency range (just below an octave) and almost optimum resolution. A third harmonic transducer will hence be used in all the following discussions.

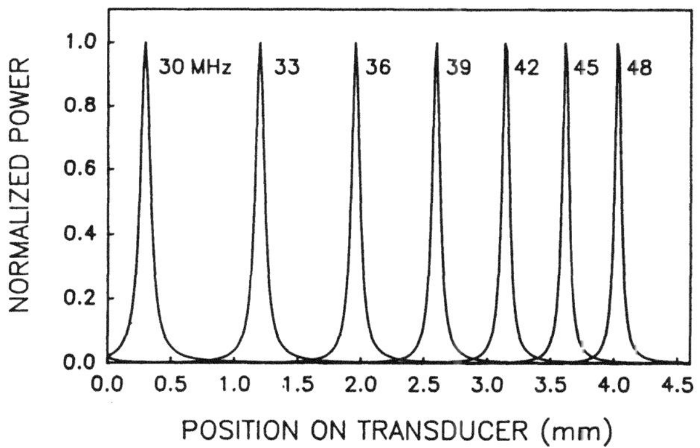

Fig. 3. Resonance peak positions for a typical wedge transducer excited at various frequencies.

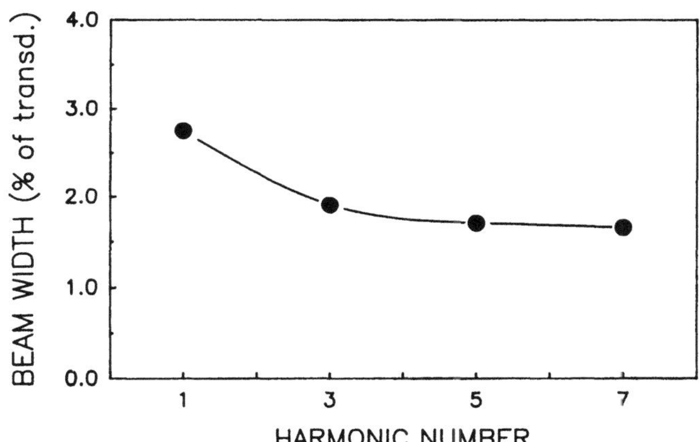

Fig. 4. Resonance width as a function of the harmonic used in the design of the transducer.

Table 2. Resolution in pixels for a third harmonic LiNbO$_3$ wedge transducer emitting into various materials.

Material	Z / Z_t	Pixels
Ethanol	0.0278	78.5
Water	0.0437	51.5
Epoxy	0.0951	24.5
Lead	0.229	10.3
Aluminium	0.508	4.4

The acoustic impedance of the material (Z) into which the wedge transducer emits is also an important parameter. It is well known that an acoustically matched transducer presents a wide frequency bandwidth. In the case of the wedge transducer, matching the acoustic impedance will enlarge the resonance width, which is clearly not wanted here. Fig. 5 illustrates the beam shape (as a percentage of the total transducer length) for an excitation at the center frequency F_c and for various acoustic impedance ratios. Best results are obtained for emission in very low acoustic impedance media such as liquids. This is clearly seen in Table 2 where resolution values in pixels (defined here as the transducer length divided by the half power width of the resonance at the center frequency) are listed for the third harmonic transducer of Table 1, with emission in different materials. We can see the relatively low resolution value for the case of aluminum. This explains the low resolution obtained in the preliminary images[2] (a resolution of about 7 pixels was measured). The development of an imaging system with direct emission in water seems to be much more promising.

EXPERIMENTAL RESULTS

Experimental measurements are done using LiNbO3 wedged transducers with center frequency near 40 MHz and optimized for third harmonic operation. The transducers are 4.6 mm long and 4.0 mm wide, and the thickness ranges from 0.21 to 0.37 mm. The lower face is fully electroded while the upper has a 2 mm wide electrode strip along the center.

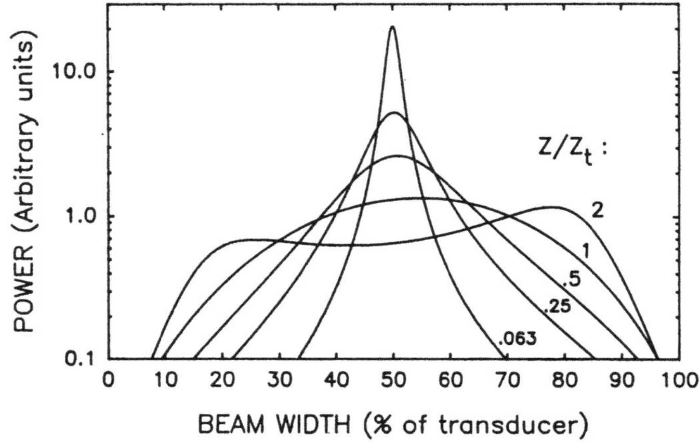

Fig. 5. Beam shape emitted from a wedge transducer for various acoustic impedance ratios.

Two experimental arrangements are investigated. In the first case, a wedge transducer is held up only along its sides where there is no upper electrode, in order to have direct emission in water. In the second configuration, the transducer is epoxy bonded on top of an aluminum cylindrical acoustic lens, the axis of the cavity being along the length of the transducer. This cavity is used to concentrate the scanned beam along the focal line in order to increase the lateral resolution. This transducer-lens assembly is similar to that used in previous work[2].

Fig. 6 shows the insertion loss as a function of frequency for the transducer-lens arrangement. The first harmonic band is clearly seen between 10 and 17 MHz. The third harmonic band, identified by the arrows, extends from about 31 to 48, which is in good agreement with the design parameters of Table 1. The fifth harmonic begins just to the right at about 50 MHz.

The resolution and position of the acoustic beam emitted from the transducer were measured in both arrangements for many frequencies within the third harmonic range. These measurements were done by sending short tone bursts of a known frequency to the transducer while scanning a sharp edge under it, along its length. The returned echo amplitude was recorded and plotted as a function of the edge position. The resolution is then simply obtained by measuring the steepness of the curve. The results obtained for both configuration are presented in Fig. 7.

We first observe in these figures that the measured beam translation with frequency is very similar to that calculated in fig. 3. We also remark a higher resolution for the case of emission in water compared to aluminum. This is again in good agreement with the theory.

Fig. 8 shows simulations of the response to an edge calculated with the staircase model, for both cases of Fig. 7. (These curves were obtained by integrating and normalizing the corresponding resonance curves, as those of Fig. 3). Comparing the simulation to the measurements, we first observe that the measured resolution in water is slightly lower than that calculated. This may be due to a small misalignment of the reflecting edge compared to the line of emission of the transducer which would produce an artifact widening of the step. Some loading of the transducer due to the holder configuration could also be responsible for a slightly larger beam width. In the case of emission in aluminum, on the other hand, we measure

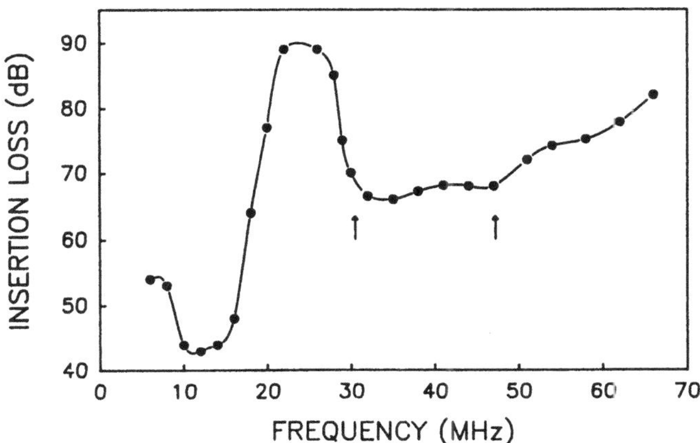

Fig. 6. Insertion loss as a function of frequency for the wedge transducer - aluminum lens arrangement.

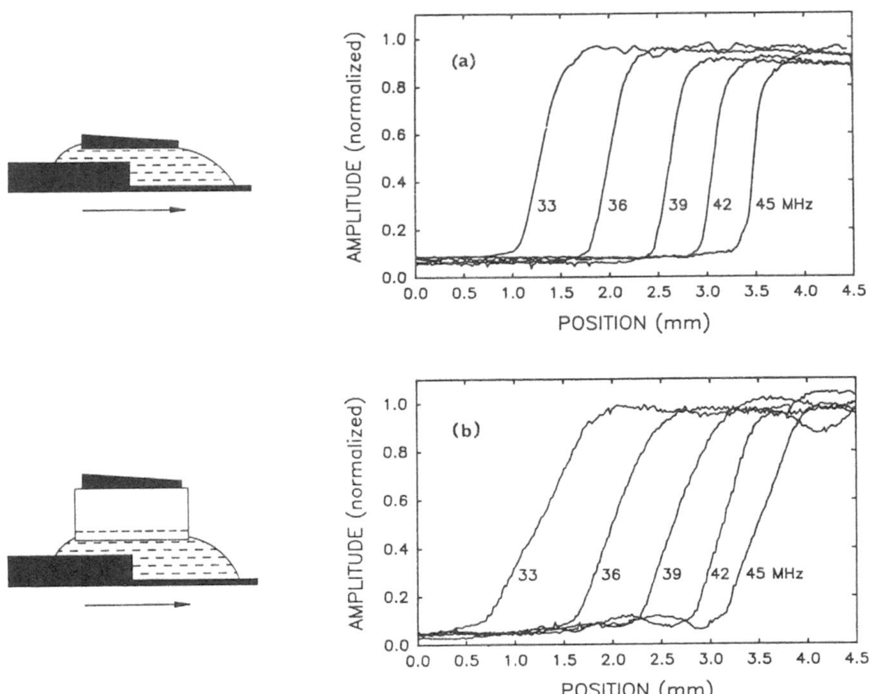

Fig. 7. Edge detection with the wedge transducer at various frequencies.
(a) emission in water, (b) transducer - lens arrangement.

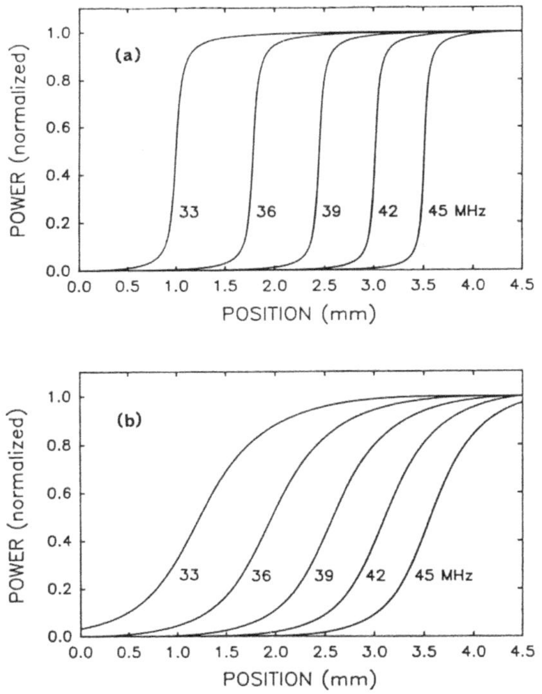

Fig. 8. Step detection as predicted from the staircase model.
(a) emission in water, (b) transducer - lens arrangement.

a better resolution than that calculated. This may be explained by the presence of the epoxy bonding layer that lowers the acoustic impedance seen by the transducer.

The curves of Fig. 7 gives beam widths of the order of 250 μm for water and 600 μm with the aluminum lens. These values are still much larger than what can be obtain with a normal diffraction limited spherical acoustic lens at the same frequency ($\simeq$ 40 μm). However, in cases where very fast imaging is needed and modest resolution is acceptable, the wedge transducer method should be a good solution.

Kubota et al.[7] have recently introduced another method of electronic scanning for fast imaging. Their system use a complex switching circuitry to scan the emitting signal over a transducer array. The resolution obtained with their technique is quite comparable to that obtained here with direct emission in water. However, the wedge transducer system has the advantage of being much more simple to construct and to use. With a rapid VCO source, the beam position can be moved as fast as needed over the transducer length and the imaging time is only limited by transit time of the acoustic pulse between the transducer and the object. A fully optimized system could produces 64 by 64 pixels images at a rate of a few images per second. This would be quite good for direct acoustic inspection in a mass production line.

ACKNOWLEDGEMENT

This work was supported by the Conseil National de Recherche en Sciences et en Génie du Canada (CRSNG). We would like to thank S. Arsenault for his help in the programming of the acquisition system and the Valpey Fisher Co. (Hopkinton, MA) for providing the wedge transducers.

REFERENCES

[1] J.D.N. Cheeke, J.-O. Fossum and N. René, "Line Focus Acoustic Lens with Wedge Transducer Source", in Acoustical Imaging, vol. **17** (Ed. by H. Shimizu, N. Chubachi and J. Kushibiki, Plenum, 1989), 159-166.

[2] J.D.N. Cheeke and L. Germain, Review of Progress in Quantitative NDE, vol. 10, (ed. D.O. Thompson, Plenum Press), Brunswick, Maine, july 1989 (to be published).

[3] Y. Tomikawa, H. Yamada and M. Onoe, "Wide Band Ultrasonic Transducer Using Tapered Piezoelectric Ceramics for Non-Destructive Inspection", Proc. 4th Symp. Ultrasonic Electronics, Jap. J. Appl. Phys., **23**, suppl. 23-1, 113-115 (1984).

[4] P.G. Barthé and P.J. Benkeser, "A staircase model of tapered piezoelectric transducers", Proc. 1987 IEEE Ultrasonic Symposium, p. 697-700 (1987).

[5] P.G. Barthé and P.J. Benkeser, "Analysis of the mechanically-uncoupled electric resonance in tapered-thickness piezoelectric transducers", Proc. 1988 IEEE Ultrasonic Symposium, 717-720 (1988).

[6] E. Dieulesaint and D. Royer, "Ondes élastiques dans les solides", (Masson et Cie, Paris, 1974), chap. 7, pp. 250-274.

[7] J. Kubota, H. Okada, Y. Musha, Y. Takishita, A. Iwasaki, and S. Sasaki, "Electronic scanning of 25MHz ultrasound for imaging IC packages", Proc. 1988 IEEE Ultrasonic Symposium, 767-770 (1988).

HOLOGRAPHIC SCANNING LASER ACOUSTIC MICROSCOPY AND APPLICATIONS

A. C. Wey, and L. W. Kessler

Sonoscan, Inc.
530 E. Green Street
Bensenville, Illinois

ABSTRACT

Scanning Laser Acoustic Microscopy (SLAM) has become an important nondestructive testing (NDT) technique because of its capability of real-time flaw detection and high resolution for imaging the detailed structure of internal flaws. One the one hand, SLAM operates in a transmission mode and can interrogate the entire sample volume in real time. On the other hand, however, the micrograph produced by SLAM is essentially a two dimensional (2-D) shadowgraphic view of 3-D object. Because of diffraction, the resultant images are often unfocused. To overcome this problem we developed a new holographic imaging system which incorporated holographic image reconstruction technique into a modified SLAM. In this paper, we describe the modification of SLAM for holographic data acquisition and the process of holographic image reconstruction. The applications of HOLOSLAM for nondestructive testing and characterization of microelectronic components, biomedical tissues and structural materials are presented.

INTRODUCTION

Acoustic microscopy is the name given to high frequency, 10 MHz to 3 GHz ultrasonic visualization. The scanning laser acoustic microscopy (SLAM) is an important branch of acoustic microscopy which uses ultrasound in the frequency range of 10 to 200 MHz to produce high resolution ultrasonic images.[1,2] In contrast to other visual observation techniques, SLAM provides direct access to the structural elastic properties of solid materials and biological tissues. By using this technique, valuable insight can be gained into mechanisms responsible for the changes of elastic architecture over areas tens of microns in diameter.

SLAM's principle of operation has been presented and discussed extensively in the literature.[3,4] The technique is based on transmitting ultrasound through the specimen and obtaining a dynamic ripple caused by the ultrasound on a solid mirror-like free surface above the specimen. The resulting image depicts the acoustic transmission properties of the insonified specimen and is obtained by employing a scanning laser beam as a point-by-point detector of ultrasound wave. With laser beam scanning technology, the images are produced in real-time, that is, at conventional TV rates of 30 frames per second.

The variations in ultrasound transmission are displayed on a TV monitor where the bright regions correspond to defect-free areas of high transmission through the sample, whereas, the darker areas correspond to regions of higher ultrasonic attenuation attributed to defects or changes in elastic properties because of scattering, reflection or absorption of ultrasound. Therefore, SLAM allows us to actually see inside objects and locate defects which are not evident at the surface.

SLAM operates in a transmission mode. On the one hand, it can interrogate the entire thickness of the sample in real time. On the other hand, however, SLAM micrograph is a composite shadowgraphic image which contains both in-focus and out-of-focus information. Because of diffraction the resultant images are often unfocused and difficult to comprehend especially when test specimens have substantial thickness. To overcome this difficulty we have modified a conventional SLAM and developed a holographic image processing technique which enables SLAM to focus object in the same fashion as an optical microscope.

In this paper we describe the conversion of a SLAM to a holographic imaging system (HOLOSLAM). The modification of SLAM to acquire data for holographic image reconstruction and the system transfer functions which are essential in processing the acquired data are discussed. The applications of holographic SLAM in nondestructive evaluation of microelectronic components, biomedical tissues and structural materials are presented.

DATA ACQUISITION FOR HOLOSLAM

In a practical SLAM, a low power laser beam scans the dynamic surface caused by ultrasound wave. A knife-edge technique is employed to visualize the surface perturbation[5]. Fig. 1 shows the knife-edge technique used in SLAM for acoustic signal detection. The laser beam intensity is detected by a knife-edge and photodiode combination whose output is an electronic signal carrying the spatial information of the ultrasound wavefield. The intensity of the scanning laser can provide an amplitude distribution of the ultrasound wavefield. However, to perform holographic reconstruction, phase information must be preserved.

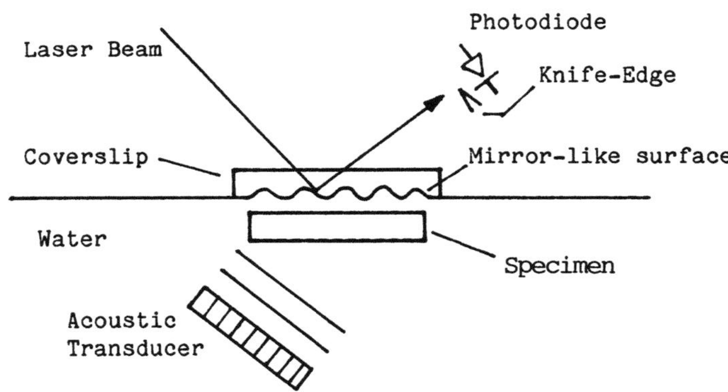

Fig. 1 Knife-edge technique for acoustic signal detection in SLAM.

A quadrature receiver was designed for HOLOSLAM to obtain both the amplitude and phase information for holographic reconstruction. Quadrature detection involves multiplying the signal with two coherent electronic references[6]. One reference has its phase shifted by 90 degrees with respect to the other. The schematic diagram of quadrature receiver is

shown in Fig. 2. The two outputs represent the real and imaginary parts of
the complex amplitude of the ultrasound, respectively, and are given by[6]

$$y_1(x) = -K2\pi f_x A(x)\sin(2\pi f_x x - \psi(x)) \qquad (1)$$

$$y_2(x) = K2\pi f_x A(x)\cos(2\pi f_x x - \psi(x)) \qquad (2)$$

where f_x is spatial frequency of the surface ripple and K a constant which
depends upon the electronic circuit of the system. The amplitude $A(x)$ and
phase $\psi(x)$ can be obtained by directly solving equations (1) and (2).

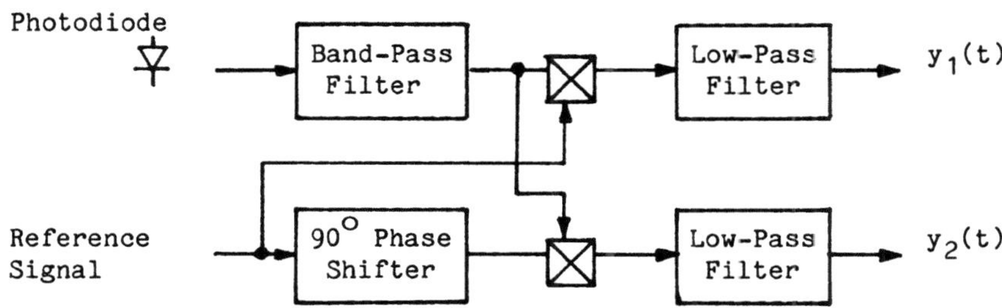

Fig. 2 Block diagram of the electronic circuit for quadrature
        detection.

HOLOGRAPHIC IMAGE RECONSTRUCTION

In SLAM, an ultrasonic plane wave propagates through a test object.
Consider an object consisting largely of homogeneous material but with a
particular internal plane having a distribution of inhomogenieties. The
distribution may consist of defects or a pattern of different materials
within the specimen. It is the objective of holographic image reconstruction to image this particular subsurface plane nondestructively.

The SLAM imaging process can be modeled as a linear system[7] as shown
in Fig. 3. The input to the system is the wavefield distribution at the
particular plane of interest. The system contains two linear filters.
One corresponds to the wave propagation through the homogeneous medium.
The other characterizes the response of the knife-edge detection. The
transfer functions representing wave propagation and knife-edge detection
can be found in the reference[7]. With these transfer functions, it is
possible to design inverse filters for the image reconstruction.

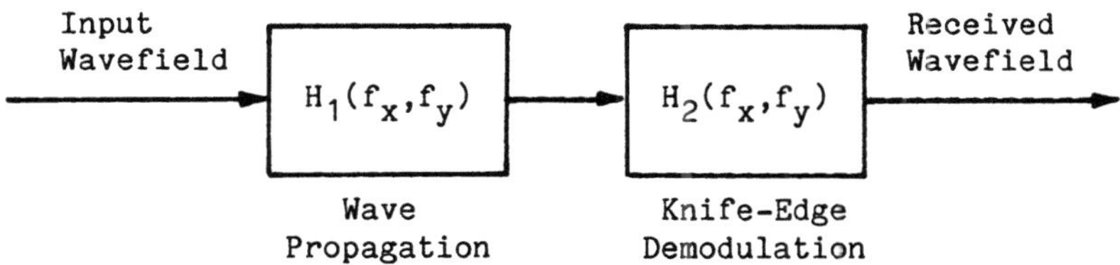

Fig. 3 A linear system model of SLAM imaging process.

The block diagram of the image reconstruction is shown in Fig. 4. The
inverse filter for $H_2(f_x,f_y)$ is designed as follows[7]:

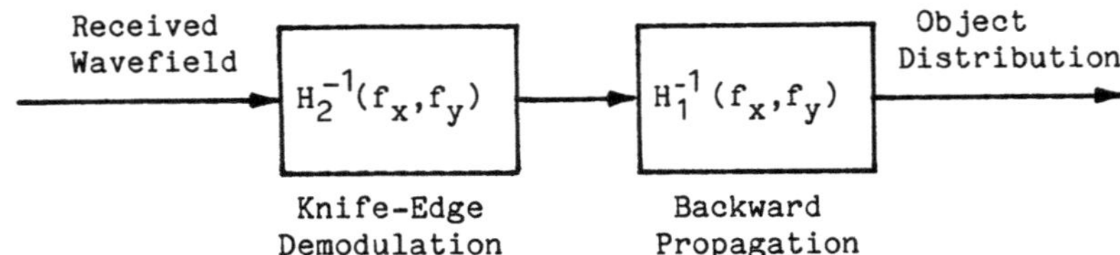

Fig. 4 Inverse filtering model of image reconstruction.

$$H_2^{-1}(f_x,f_y) = \begin{cases} \dfrac{H_{max}\exp\left(\dfrac{\pi^2 r_o^2}{2}(f_x^2+f_y^2)\right)}{j\,\mathrm{erf}\left(\dfrac{\pi\,r_o\,f_x}{\sqrt{2}}\right)}, & \text{if } \left|H_2(f_x,f_y)\right| \geq H_s \\[4mm] 1 & \text{otherwise} \end{cases} \qquad (3)$$

where H_s is the threshold chosen in such a way that after inverse filtering the main spectrum of the signal is recovered and the noise amplification does not degrade the image reconstruction[7]. H_{max} is the amplitude of the maximum response for $H_2(f_x,f_y)$. Here we assume that the laser beam scans in the x-direction and has a Gaussian intensity profile and an effective beam radius of r_o.

The diffraction experienced by wavefield propagation can be corrected by computing the corresponding backward propagation. The corresponding inverse filter to accomplish this is

$$H_1^{-1}(f_x,f_y) = \begin{cases} \exp\left(-j2\pi z\left(\dfrac{1}{\lambda^2} - f_x^2 - f_y^2\right)^{\frac{1}{2}}\right), & f_x^2 + f_y^2 \leq \dfrac{1}{\lambda^2} \\[4mm] 0, & \text{otherwise} \end{cases} \qquad (4)$$

where λ is the wavelength of the ultrasound within the homogeneous medium, and z the distance between the object plane and receiving plane. With this inverse filtering technique, the corresponding object distribution at a specific depth can be reconstructed from the received wavefield.

The holographic SLAM contains a standard SLAM with modified electronic circuit for quadrature detection and an IBM Personal Computer AT compatible with a 10 MHz 32 bit co-processor board. The control program was written in the C programming language and runs on the 32-bit co-processor which acts as the main computer using the MS-DOS computer as an I/O controller. The outputs of quadrature receiver are digitized by a video-rate frame grabber. The data acquisition algorithm extracts phase and amplitude data from the digitized outputs of quadrature receiver and then uses them to reconstruct a complex wavefield. The backpropagation algorithm digitally back-propagates the wavefield to a specified depth in a matter of seconds.

APPLICATIONS AND EXPERIMENTAL RESULTS

It is expected that the power of SLAM could be greatly enhanced through the use of holographic image reconstruction to yield more precise information about the internal structure of a given sample.

Nondestructive Testing of Microelectronic Components

Holographic image reconstruction enables HOLOSLAM to focus object in the same way as an optical microscope. Therefore, for example, it can be used to produce in-focus acoustic images during the examination of plastic leaded chip carriers (PLCC's). Fig. 5.a shows a conventional SLAM acoustic micrograph of a defective PLCC. The ultrasonic frequency used to produce this image was 10 MHz. This test device has disbond between the die and molding compound in the area of the die paddle, where appears dark in SLAM image. Fig. 5.b shows a reconstructed image of the same device. The image was restored at a depth where the die was bonded to the paddle. It reveals the detail of the bonding quality. So, HOLOSLAM can be used to quantitatively and nondestructively measure the degree of disbond.

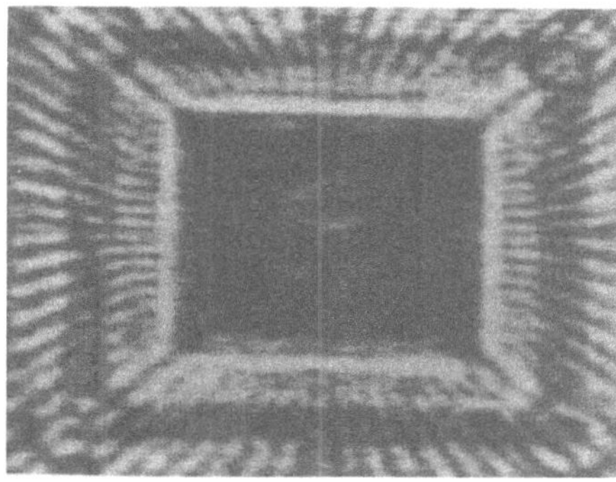

(a) Typical 10 MHz SLAM image   (b) Reconstructed image

Fig. 5   (a) A 10 MHz SLAM image of a defective PLCC device, and
         (b) reconstructed image at the plane where the die is
         bonded to the paddle. The field of view is 30 x 28 mm.

Biomedical Imaging

Optical visualization of thick biological tissues is very difficult since the specimens are optically opaque. In contrast, acoustic images on elasticity and density at the microscopic level can be easily obtained by SLAM which offers a high degree of differentiation for hard and soft tissues. For instance, acoustic micrographs of an unstained, 10th-day mouse embryo, are produced and shown in Fig. 6. The frequency of ultrasound used in SLAM was 99.8 MHz and the field of view was 2.7 mm by 2.5 mm.

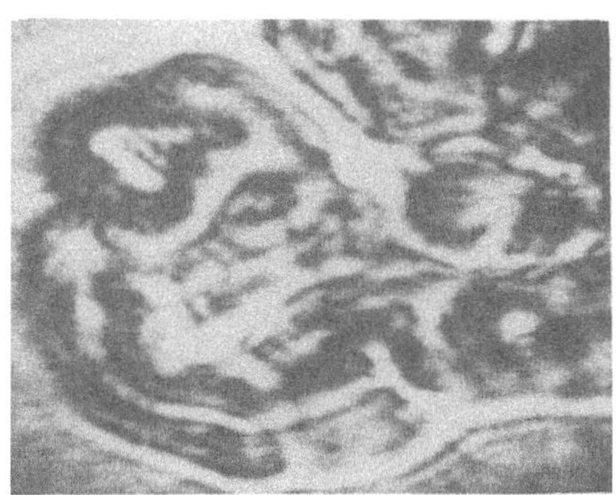

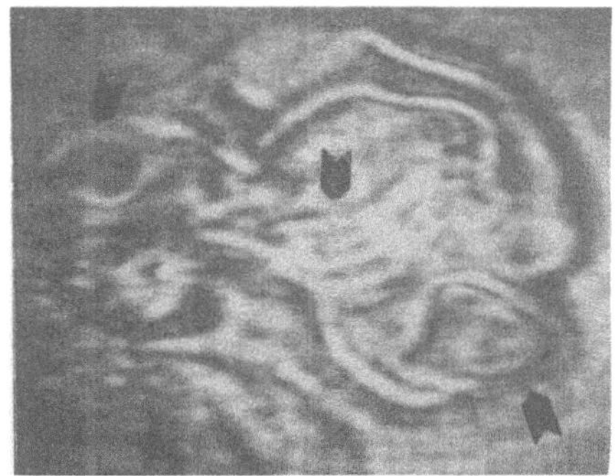

(a) Typical 100 MHz SLAM image   (b) Reconstructed image

Fig. 6   (a) 100 MHz SLAM image of a 10-day mouse embryo, and (b)
         reconstructed image at 1 mm depth. The field of view is
         2.7 mm by 2.5 mm.

The typical gestation period for a mouse is approximately 18 days. By
the tenth day of gestation the mouse is almost completely developed. In
this figure, the head of the mouse embryo is clearly differentiated. To
the left is the top of the embryo's head and at the bottom is the back of
the head. The telencephalon where the front part of the brain will be
developing is identified and so is the optic cup where mouse's eye will be
developing. The heart and basalar arteries which feed blood to the head
are also shown in the micrograph.

The HOLOSLAM micrograph containing the reconstructed image of Fig 6.a
is shown in Fig. 6.b for comparative purpose. This image is reconstructed
at a depth of 1 mm below the coverslip. Some structural details of the
mouse's embryo are brought into focus in this image. The telencephalon,
basalar arteries and fore limb bud are now clearly shown.

3-D Localization of Defects in Ceramics

We also demonstrated the power of HOLOSLAM by performing 3-D
localization of flaws in silicone nitride specimen. The thickness of test
specimen was 5 mm. The ultrasound source used in experiments operated at
100.38 MHz and the field of view was 2 mm by 1.87 mm. The specimen was

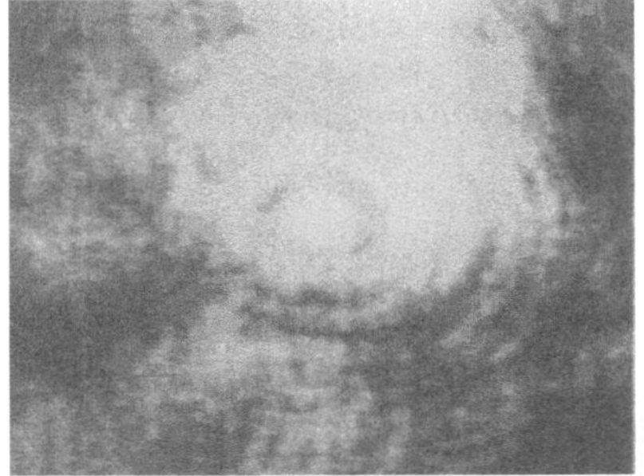

(a) Typical 100 MHz SLAM image of internal flaw.

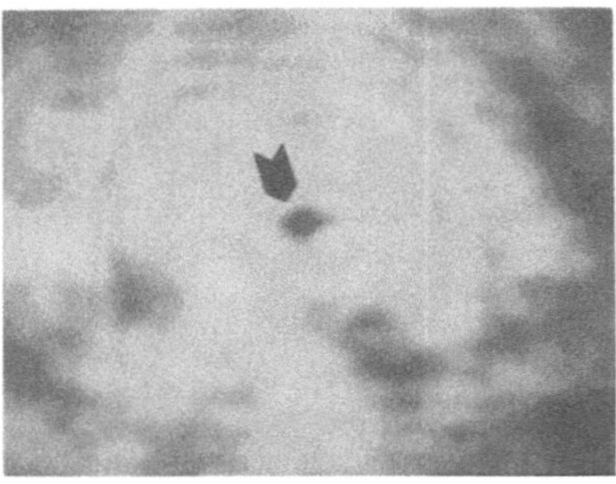

(b) HOLOSLAM reconstructed image
at a depth of 0.45 mm.

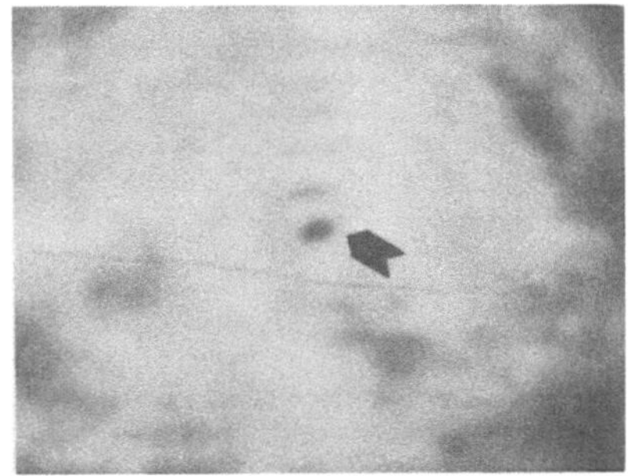

(c) reconstructed image at
0.68 mm deep.

Fig. 7.   (a) A SLAM micrograph showing diffraction pattern of
internal flaws in Si_3N_4 sample, while HOLOSLAM images
showing in (b) a 100 um size void at 0.45 mm deep and
(c) a 80 um void 0.68 mm deep. The frequency used was
100 MHz with a 2 x 1.87 mm field of view

252

placed 0.5 mm below the coverslip to avoid damage to coverslip and the gap
was filled with water. The compressional wave velocity inside the sample
was 10,500 m/sec and the velocity of ultrasound in water was 1,500 m/sec.
Figure 7.a shows a standard SLAM amplitude image which displays diffraction
patterns of flaws below the sample surface. One can easily find the flaw
by locating the diffraction pattern in SLAM image, but may have difficulty
to determine the actual shape, size and depth of the flaws.

Fig. 7.b and 7.c show two reconstructed images at 0.95 mm and 1.18 mm
depth, respectively, produced by HOLOSLAM for the same defect shown in Fig.
7.a. The solid black round spot in Fig. 7.b, as pointed by an arrow,
indicates a void of 100 um in diameter at a depth of 0.45 mm below the top
surface of the sample, or 0.95 mm below the coverslip. Fig. 7.c reveals
another 80 um size void in the sample 0.68 mm deep. It is interesting to
learn that there are two voids, instead of one, behind the diffraction
patterns as shown in Fig. 7.a. These voids are lined up vertically so that
their diffraction patterns overlap to each other at the receiving plane.

SUMMARY

We have reviewed briefly the principles of scanning laser acoustic
microscopy and discussed the required modification of a conventional SLAM
to become holographic SLAM. Inverse-filtering techniques used in the
digital holographic image process are described. Our experimental results
show that the wavefield diffraction due to propagation in regular SLAM can
be corrected by holographic image reconstruction. Applications of HOLOSLAM
in NDT of microelectronic components, biomedical imaging and 3-D localization of flaws in structural materials are presented.

ACKNOWLEDGMENT

We would like to thank professor Hua Lee and his student at the
University of Illinois, Urbana-Champaign for their valuable help in the
development of computer algorithms and also Professor O'Brien and his
student in conducting experiments of biomedical imaging. This work was
supported by National Science Foundation under Grant ISI-8604227.

REFERENCE

1. L. W. Kessler and D. E. Yuhas, "Acoustic Microscopy-1979," Proc. IEEE,
    vol. 67, no. 4, p. 526, 1979.
2. L. W. Kessler, "Acoustic Microscopy Commentary: SLAM and SAM," IEEE
    Trans. Sonics Ultrason., vol. SU-22, p. 136, 1985.
3. L. W. Kessler and D. E. Yuhas, "Principles and Analytical Capabilities
    of the Scanning Laser Acoustic Microscopy," Scan. Elec. Microscopy,
    vol. 1, p. 555, 1978.
4. L. W. Kessler and M. G. Oravecz, "SLAM Analysis of Advanced Materials
    for Internal Defects and Discontinuities," Proc. of NDT/E of Adv.
    Mater. and Comp., p.173, 1986.
5. R. L. Whitman, A. Korpel, "Probing of Acoustic Surface Perturbations
    by Coherent Light," Applied Optics, vol. 8, no. 8, p. 1567, 1969.
6. Z. Lin, H. Lee, G. Wade, M. Oravecz and L. W. Kessler, "Data
    Acquisition in Tomographic Acoustic Microscopy," Proc. IEEE Ultrasonic
    Symp., p. 627, 1983.
7. G. Wade and A. Meyyappan, "Scanning Tomographic Acoustic Microscopy:
    Principles and Recent Developments," SPIE vol. 768, p. 267, 1987.

ACOUSTIC IMAGES OBSERVED BY DIRECTIONAL PFB MICROSCOPE

N. Chubachi, J. Kushibiki, T. Sannomiya,
I. Naruge, K. Saito, and S. Watanabe

Department of Electrical Engineering
Faculty of Engineering, Tohoku University
Aramaki-Aza-Aoba, Sendai 980, Japan

INTRODUCTION

Directional acoustic microscopy has been expected to play an important role in the development of ultrasonic microspectroscopy (UMS) for use in material research[1]. A cylindrical lens was introduced to the scanning acoustic microscope system to develop a directional microscope for the study of anisotropic materials[2]. This microscope has been called "a line-focus-beam (LFB) microscope", and quantitative measurement methods for material analyses have been successfully established for material analyses using this microscope[3].

The LFB microscope, however, does not suit for two dimensional imaging because the spatial resolution along the cylindrical lens-axis is not considered. Since 1983, spherical lenses combined with directional transducers have been cultivated for application in point-focus-beam (PFB) microscopes to obtain directional images with higher resolution [4,5,6]. Recently, a spherical lens with an oblong transducer has been introduced into the quantitative measurement system developed for the LFB microscope to realize a new directional PFB microscope system with a high resolution while preserving the measurement accuracy [7,8]. It has been revealed that anisotropic materials affect remarkably the accuracy of angular dependence measurements of LSAW velocity.

In this paper, directional images observed from polycrystalline samples such as Mn-Zn ferrites and Ti metals are demonstrated. Contrast variations of the directional images with respect to the propagation direction of LSAW are discussed in relation to the V(z) curves measured with the directional PFB lens.

DIRECTIONAL PFB LENS WITH OBLONG TRANSDUCER

An oblong transducer has been introduced to give directionality to a conventional PFB lens-transducer system. In order to investigate the imaging pattern observed with this directional PFB microscope, the beam steering effect of LSAWs appearing on anisotropic materials has been taken into accounts to design the oblong transducer combined with the PFB lens system as pointed out in Ref. 7. As a typical oblong transducer, a rectangular transducer with dimensions of 1.73 mm x 0.5 mm was fabricated by ZnO piezoelectric film with 11 μm thick deposited on a flat end surface of a PFB sapphire lens with the dimensions as depicted in Fig. 1. In addition, a chalcogenide glass film was coated on the spherical lens surface to give a matching layer to water.

IMAGES AND CONTRAST VARIATIONS

Acoustic images observed with the directional acoustic microscope developed here are able to present new information associated with the anisotropic properties on surfaces of anisotropic materials, while those observed with the conventional SAM system show only the mean properties around the acoustic lens axis.

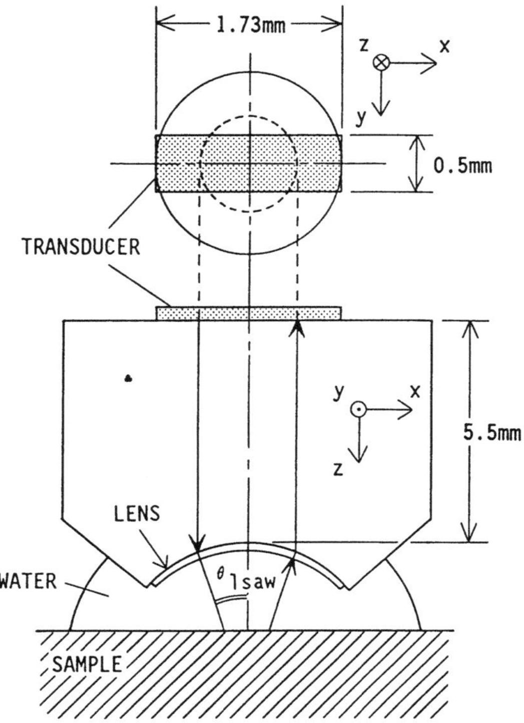

Fig.1   Configuration of directional PFB acoustic
        lens with rectangular transducer.

Figures 2 and 3 show directional acoustic images observed on a polycrystalline sample of Mn-Zn ferrite with an average grain size of 100 μm at three different defocus distances of z=-20μm, -30μm, and -40μm, for the LSAW propagations along the x-axis and y-axis, respectively. These images have been taken at a frequency of 225 MHz. In these pictures, it is clear that the contrast for every grain varies in complicated manners according to the defocus distance. These contrast variations can be explained by the V(z) curve analysis performed for each grain surface.

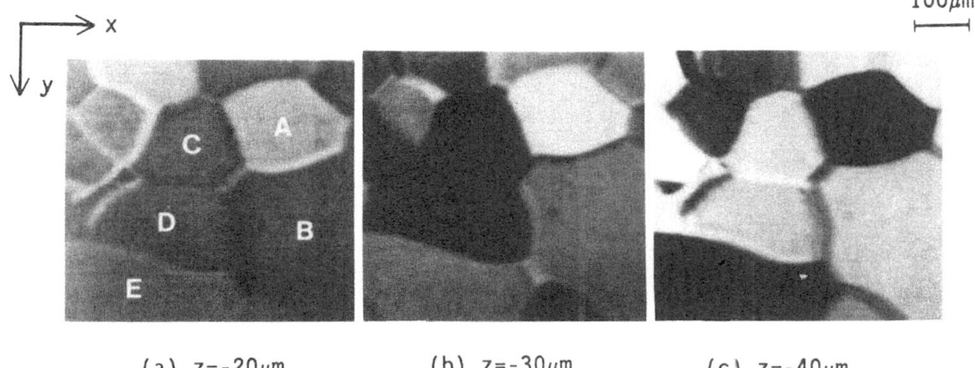

(a) z=-20μm   (b) z=-30μm   (c) z=-40μm

Fig.2   Directional acoustic images of Mn-Zn ferrite
         with average grain size of 100 μm observed at
         three different defocus distances for LSAW
         propagation along the x-axis (f=225MHz).

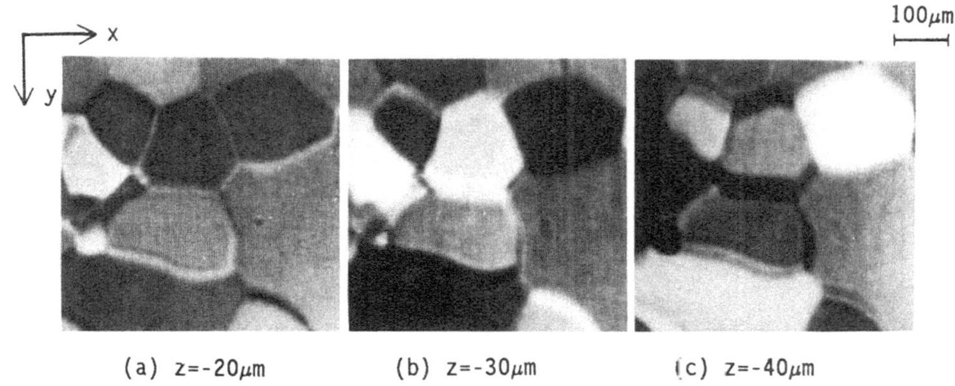

(a) z=-20μm   (b) z=-30μm   (c) z=-40μm

Fig.3   Directional acoustic images of Mn-Zn ferrite
         with average grain size of 100 μm observed at
         three different defocus distances for LSAW
         propagation along the y-axis (f=225MHz).

Taking the typical grains of A and C marked in Fig.2, for example, two
$V(z)$ curves have been measured for the LSAW propagations along x-axis for
the grains as shown in Fig.4. Comparing the two curves, it is easily understood that the output voltage differences between the two curves at the
every defocus distance of z=-20μm, z=-30μm, and z=-40μm in Fig.4 directly
correspond to the contrast variations appearing in Fig.2(a), (b), and (c),
respectively. For example, the contrast inversions for grain A and grain C
occurred in the two pictures of (b) and (c) are well explained by the change
of output voltages at z=-30μm and at -40μm for the curves of grain A and
grain C.

Analyzing the directional acoustic images shown in Fig.2 and Fig.3, the
anisotropic properties of every grain can be investigated. In Fig.5, two
$V(z)$ curves measured for the grain A are shown for the different directions
of LSAW propagations. Contrast inversion appearing between Fig.2 and Fig.3
can be also explained with the measured $V(z)$ curves in Fig.5.

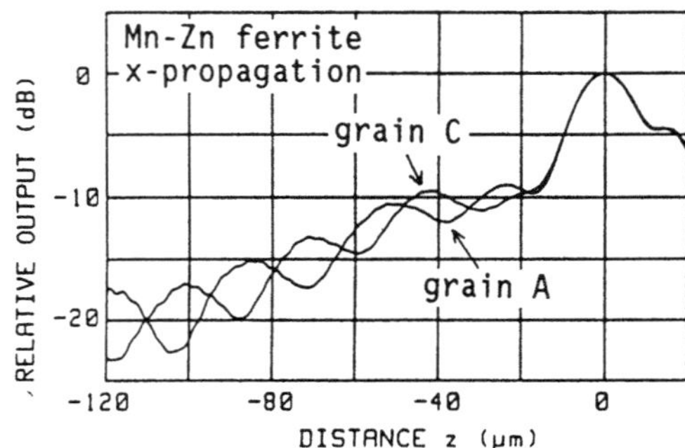

Fig.4   V(z) curves measured for grain A
        and grain C for LSAW propagation
        along the x-axis.

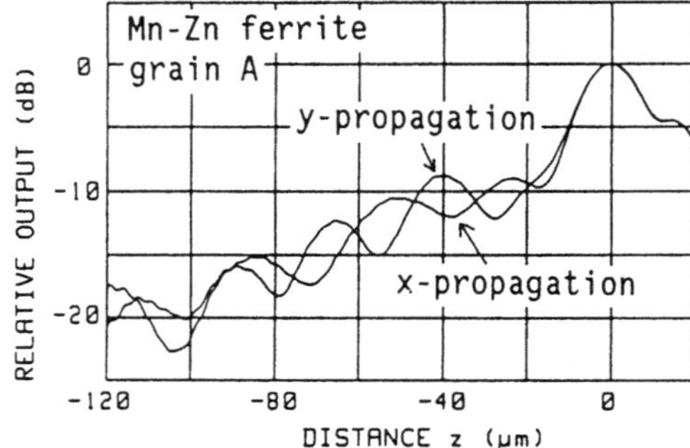

Fig.5   V(z) curves measured for grain A
        for different directions of LSAW
        propagations.

ANALYSES OF ANISOTROPIC GRAIN SURFACE

Quantitative measurements can be made with the directional acoustic microscope by using the V(z) curve analysis equipment developed previously[3]. The beam steering effects due to the elastic anisotropy have been suppressed by limiting the region of defocusing distance for the V(z) curve analysis from z=-20 to -80μm. Figures 6 and 7 show the measured LSAW velocities as a function of propagation directions for grains A, B, C, D, and E as marked in Fig.2. It has been revealed with this analysis that the x and y-axis directions taken in this study are actually the directions of 5° and 95° in this figure, respectively. Experiments on single crystals of Mn-Zn ferrite have been made with the LFB acoustic microscope to identify the orientations of every grain surface. It has been revealed that the surfaces of grain A and grain B correspond to a (110) plane and a (111) plane, respectively, and the surface of grain E is estimated as a (110) plane with a little inclination angle.

Similar experiments have been performed with a sample of Ti polycrystalline metal. Figure 8 shows the directional acoustic images observed on the sample surface for different propagation directions of LSAWs at 225 MHz. The contrast variations can also be explained by the V(z) curve analysis in the same way as described above.

258

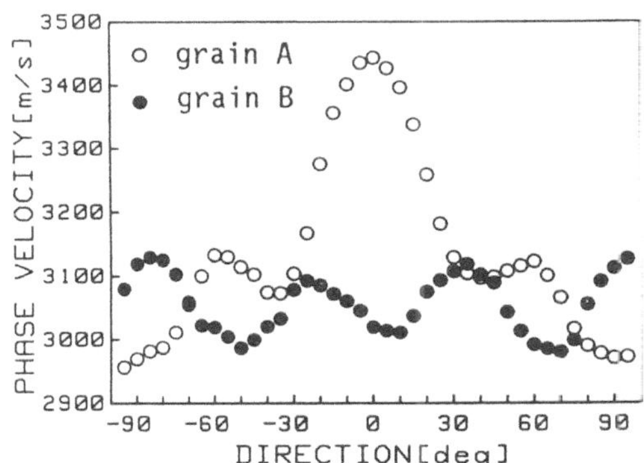

Fig.6   Measured results of LSAW velocity
         as a function of propagation
         direction for grain A and grain B.

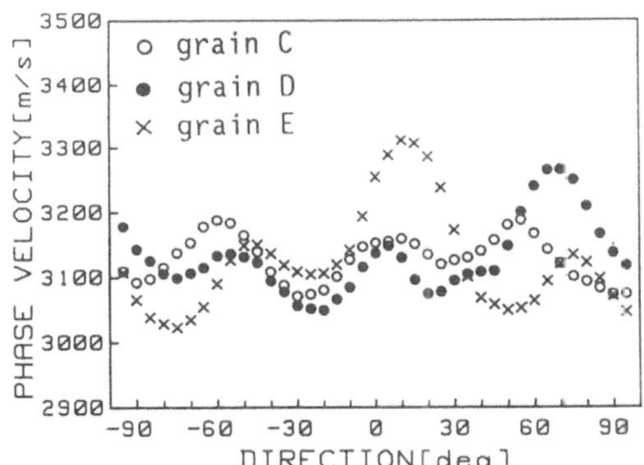

Fig.7   Measured results of LSAW velocity as
         a function of propagation direction
         for grains C, D and E.

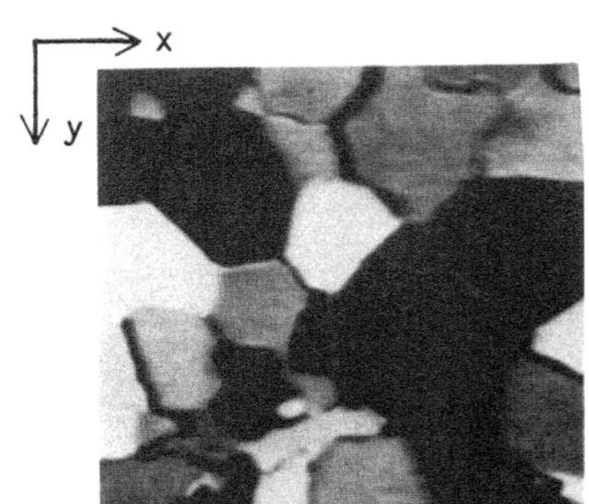
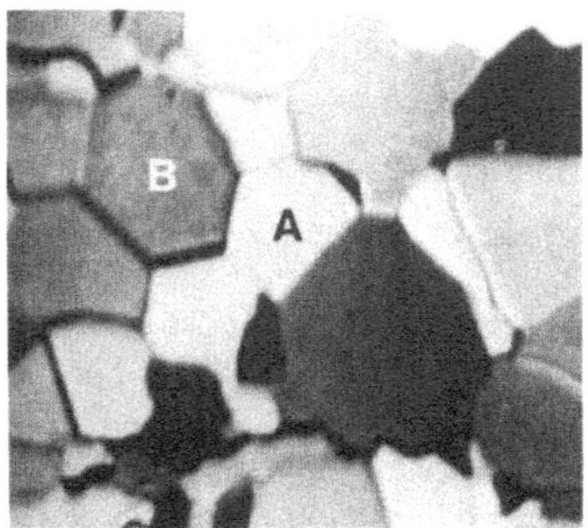

(a) x-propagation   (b) y-propagation

Fig.8   Directional acoustic images of Ti polycrystalline
        metal observed at a defocus distance of -25μm for
        LSAW propagations along the x-axis (a) and y-axis
        (b) (f=225MHz).

CONCLUSION

A PFB lens with a simple rectangular transducer has been successfully introduced to the acoustic microscope system to obtain directional acoustic images with a high resolution. Experiments on polycrystalline samples of Mn-Zn ferrite and Ti metal have been performed at a frequency of 225 MHz, and several directional images have been demonstrated at three defocus distances for the two LSAW propagation directions of x and y. The contrast variations of directional images have been explained by the V(z) curve analysis. The anisotropic properties have been satisfactorily measured with this directional acoustic microscope in each grain area. The directional acoustic microscope with high spatial and angular resolutions developed here is expected to be preferably applied in material research as a useful UMS system.

REFERENCES

1.   N. Chubachi, Ultrasonic micro-spectroscopy via Rayleigh waves, in "Rayleigh-Wave Theory and Application", E. A. Ash and E. G. S. Paige, eds., Springer-Verlag, Berlin (1985).
2.   J. Kushibiki, A. Ohkubo, and N. Chubachi, Linearly focused acoustic beams for acoustic microscopy, Electron. Lett. 17:520 (1981).
3.   J. Kushibiki and N. Chubachi, Material characterization by line-focus-beam acoustic microscope, IEEE Trans. SU-32:189 (1985).
4.   J. A. Hildebrand and L. K. Lam, Directional acoustic microscope for observation of elastic anisotropy, Appl. Phys. Lett. 42(5):413 (1983).
5.   N. Chubachi and T. Sannomiya, Acoustic interference microscope with electrical mixing method for reflection mode, IEEE Ultrason. Symp. Proc. :604 (1984).
6.   D. A. Davis and H. L. Bertoni, Bow-tie transducers for measurement of anisotropic materials in acoustic microscopy, IEEE Ultrason. Symp. Proc. :735 (1986).
7.   E. Tejima, J. Kushibiki, and N. Chubachi, Anisotropy measurements of materials with directional acoustic microscope, Proc. Acoust. Soc. Jpn. :691 (1989) [in Japanese].
8.   J. Kushibiki, N. Chubachi, and E. Tejima, Quantitative evaluation of materials by directional acoustic microscope, Ultrasonics International 89 Conf. Proc. :736 (1989).

INITIAL PHASE ESTIMATION AND TOMOGRAPHIC RECONSTRUCTION

FOR MULTIPLE-FREQUENCY ACOUSTIC MICROSCOPY

Richard Y. Chiao and Hua Lee

Department of Electrical and Computer Engineering
University of Illinois at Urbana-Champaign
Urbana, Illinois 61801, USA

Abstract - In multiple-frequency tomographic acoustic microscopy, projections acquired with different illumination acoustic frequencies are combined to form the tomographic reconstruction. A major source of degradation in the reconstructed image is the presence of an unknown phase offset in each projection. This unknown phase term cannot be determined directly in the space domain because of the loss of evanescent waves during propagation. In this paper, we present a simple technique to estimate the relative phase difference between projections in the spatial-frequency domain. The estimation procedure uses the overlapping region between two spectral distributions to estimate the relative phase offset. Experimental results showing the tomographic reconstruction of a honeycomb test grid using four multiple-frequency projections are presented to demonstrate the resolution enhancement capability of this technique. In addition, this approach can be applied to multiple-angle microscopic tomography.

Introduction

The conventional Scanning Laser Acoustic Microscope (SLAM) is designed to operate in the intensity-mapping mode to produce images of thin specimens. To perform high-resolution subsurface imaging for thick objects, we have modified the data acquisition system of SLAM with a quadrature receiver to detect the complex wavefield distribution [1]. As a result, the SLAM system is now capable of holographic and tomographic subsurface imaging [2]. The improvement of the imaging capability is due mainly to the detection of the phase information of the resultant wavefield. This also implies that the resolution of holographic and tomographic microscopy depends on the accuracy of the phase information, and therefore the phase error becomes a dominant parameter of image degradation.

Recently, a data acquisition phase error has been identified in holographic acoustic microscopy [3,4]. To resolve this problem, a single-sideband estimation and correction algorithm has been developed for the removal of the phase error. In tomographic microscopy, an additional degradation factor exists due to the uncertainty of the relative phase difference of the projections. This unknown initial phase term does not introduce image degradation to holographic imaging, and the degradation occurs only in the tomographic superposition process.

In this paper, we present a simple technique to estimate the relative phase difference of the projections. This estimation process operates on the spectral distributions at the overlapping area in the spatial-frequency domain. Multiple-frequency tomographic reconstruction will be presented to demonstrate the resolution enhancement capability of this technique. In addition, this approach can be applied to multiple-angle microscopic tomography.

Data Acquisition and Holographic Reconstruction

Figure 1 shows the SLAM data acquisition system, where the object plane is located at z_0, the incident wavefield is denoted by $u(x,y,z)$, and the transmitted wavefield is denoted by $v(x,y,z)$. The incident wavefield is a 100 MHz plane wave. After being modulated by the specimen, the transmitted wavefield causes dynamic surface ripples on the gold-plated coverslip [5]. A laser beam is raster scanned over the reflective surface to pick up the wavefield information. At each point in the raster scan, the laser beam is angularly deflected by the surface ripples. The instantaneous angular deflection is converted to an intensity signal by positioning a knife-edge to block part of the laser beam [6]. At this point the spatial wavefield distribution is encoded in a temporal intensity-modulated laser signal. A photodetector placed after the knife-edge converts the laser light signal into an electrical signal, which is then fed into the quadrature receiver [1]. The output of the quadrature receiver is digitized for computer processing.

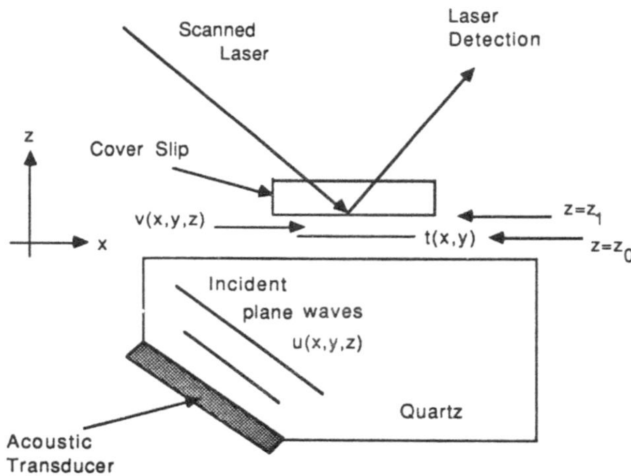

Fig. 1 SLAM data acquisition system.

The insonification wave can be written as

$$u(x,y,z) = u_0 \exp(j\, 2\pi(f_{xi}x + f_{yi}y + f_{zi}z)), \qquad (1)$$

where u_0 is the amplitude of the plane wave, λ is the acoustic wavelength, and $f_{xi}, f_{yi}, f_{zi} = \sqrt{1/\lambda^2 - f_{xi}^2 - f_{yi}^2}$ are the spatial frequencies of the plane wave. At the object plane, $z = z_0$, the wavefield is modulated by the object transmittance, $t(x,y)$, and the transmitted wavefield is given by

$$v(x,y,z_0) = u(x,y,z_0)\, t(x,y). \tag{2}$$

In the Fourier domain, Eq. (2) can be written as

$$V(f_x,f_y;z_0) = u_0 \exp(j2\pi f_{zi} z_0)\, T(f_x - f_{xi}, f_y - f_{yi}). \tag{3}$$

Thus the transmitted wavefield is proportional to the object transmittance function modulated by f_{xi} and f_{yi}. The wavefield $v(x,y,z_1)$ detected at the receiving plane is related to $v(x,y,z_0)$ by the the relation

$$V(f_x,f_y;z_1) = V(f_x,f_y;z_0)\, H(f_x,f_y;z_1-z_0), \tag{4}$$

where $V(f_x,f_y)$ is the Fourier transform of $v(x,y)$, and H is the propagation transfer function given by

$$H(f_x,f_y;z_1-z_0) = \begin{cases} exp\,[\,j2\pi(z_1-z_0)\sqrt{1/\lambda^2 - f_x{}^2 - f_y{}^2}\,], & f_x{}^2 + f_y{}^2 < \dfrac{1}{\lambda^2} \\ 0, & otherwise. \end{cases} \tag{5}$$

The propagation transfer function is a low-pass function because of the loss of evanescent waves.

Given the received wavefield, $v(x,y,z_1)$, the object transmittance, $t(x,y)$, can be reconstructed in two steps. First the scattered wavefield, $v(x,y,z_0)$, is estimated by matched filtering

$$\hat{V}(f_x,f_y;z_0) = V(f_x,f_y;z_1)\, H^*(f_x,f_y;z_1-z_0). \tag{6}$$

This step is also commonly known as back-propagation. Then $\hat{v}(x,y,z_0)$ is demodulated by shifting the frequency index to form an estimate of the object transmittance

$$P(f_x,f_y) = \hat{V}(f_x+f_{xi}, f_y+f_{yi}, z_0). \tag{7}$$

Figure 2 shows the block diagram for the holographic reconstruction algorithm.

The back-propagation operator is constructed by ignoring the evanescent waves, thus the reconstructed transmittance is a low-pass version of the actual transmittance

$$P(f_x,f_y) = T(f_x,f_y)\exp(j2\pi f_{zi} z_0), \quad (f_x+f_{xi})^2 + (f_y+f_{yi})^2 < \dfrac{1}{\lambda^2}. \tag{8}$$

Figure 3 shows the frequency domain coverage of a projection after back-propagation and demodulation. As shown in Fig. 3, the circular passband is centered at $(-f_{xi},-f_{yi})$ with radius $\dfrac{1}{\lambda}$. In tomographic processing, several projections with different f_{xi}, f_{yi}, f_{zi} or λ are combined to form the tomographic reconstruction.

The constant phase offset due to the term containing f_{zi} does not degrade the holographic reconstruction. However, for tomographic reconstruction this phase offset may be different for each projection, thus creating a constant relative phase offset between the projections. A relative phase offset between the tomographic projections results in severe degradation in the tomographic reconstruction. The above phase term is one of the contributions to the unknown initial phase in each projection. In the next section, a method will be presented to estimate this phase offset.

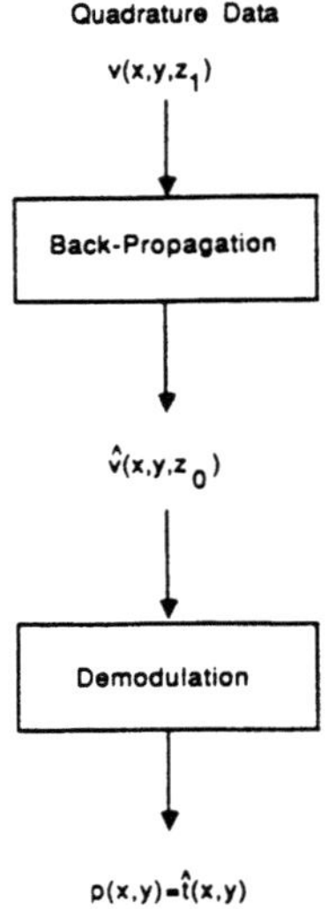

Fig. 2 Holographic reconstruction algorithm.

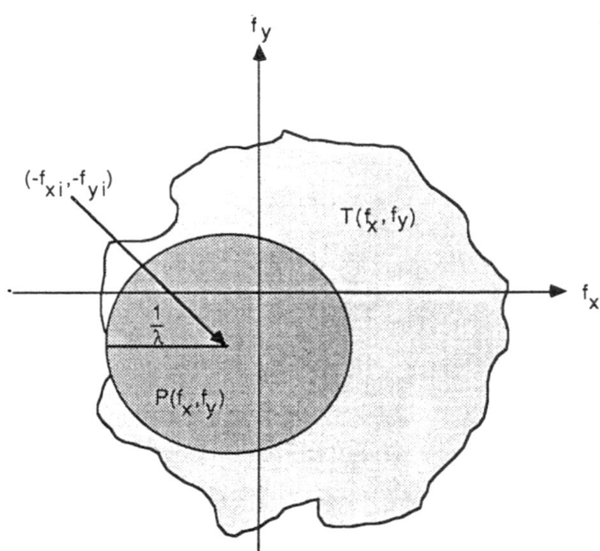

Fig. 3 Passband of a projection.

Initial Phase Estimation

A major cause of degradation in the tomographic reconstruction is the presence of an unknown phase offset between the projections. This unknown phase offset arises from two sources of phase uncertainty in the data acquisition process. One source is due to the uncertainty of the phase of the insonification plane wave at the object plane as mentioned in the previous section. The other source of uncertainty is due to the temperature fluctuations in the acoustic medium, which results in fluctuations in the acoustic propagation velocity. For large propagation distances, the velocity fluctuation contributes to the phase offsets in the detected wavefields.

To estimate the relative phase offset in tomographic projections, Equation (8) is first rewritten as

$$
P_k(f_x,f_y) = \begin{cases} u_0\, exp\,(j\,\Omega_k)\, T(f_x,f_y), & (f_x+f_{xi}^{(k)})^2 + (f_y+f_{yi}^{(k)})^2 < \dfrac{1}{\lambda_k^2} \\ 0, & otherwise, \end{cases} \tag{9}
$$

where Ω_k is the unknown initial phase to be estimated, and the subscript (k) denotes the different projections. Notice that the circular passband is different for each projection. For multiple-frequency projections, the passbands are centered at different locations and have different radius. For multiple-angle projections, the passbands are centered at different locations but have the same radius. It is in the overlapping region of the passbands of the projections that the relative phase offset is estimated. To estimate the phase offset between two projections m and n, we write from Eq. (9)

$$
P_m(f_x,f_y) = exp\,(j\,(\Omega_m-\Omega_n))\,P_n(f_x,f_y), \tag{10}
$$

which is valid for f_x and f_y in the overlapping region of the passbands of $P_m(f_x,f_y)$ and $P_n(f_x,f_y)$. Figure 4 shows the overlapping region denoted by A. From Eq. (10), the least-squares estimate of $exp\,(j\,(\Omega_m-\Omega_n))$ can be computed as

$$
exp\,(j\,(\Omega_m-\Omega_n)) = \frac{\iint\limits_A P_n^*(f_x,f_y)P_m(f_x,f_y)\,df_x df_y}{\iint\limits_A |P_n(f_x,f_y)|^2\,df_x df_y}. \tag{11}
$$

Thus, the spatial frequencies in the overlapping region between two spectral distributions are used to estimate the relative phase offset. Using Eq. (10) in the denominator of Eq. (11), we obtain

$$
exp\,(j\,(\Omega_m-\Omega_n)) = \frac{\int\limits_{-\infty}^{\infty}\!\!\int P_n(f_x,f_y)P_m(f_x,f_y)\,df_x df_y}{\int\limits_{-\infty}^{\infty}\!\!\int |P_n(f_x,f_y)|\,|P_m(f_x,f_y)|\,df_x df_y}. \tag{12}
$$

By using Eq. (12), it is unnecessary to explicitly compute the overlapping region A, and the integration needs only be performed over an area that includes the overlapping region.

Application to Multiple-Frequency Tomography

In multiple-frequency tomography, projections acquired using different acoustic frequencies are combined to form the tomographic reconstruction. The spatial frequencies for the insonification plane wave are given by

$$f_{xi}{}^{(k)} = \frac{\sin(\theta)}{\lambda_k} \tag{13-a}$$

$$f_{yi}{}^{(k)} = 0 \tag{13-b}$$

$$f_{zi}{}^{(k)} = \frac{\cos(\theta)}{\lambda_k}, \tag{13-c}$$

where θ is as shown in Fig. 1. Given a set of N projections $v_k(x,y,z_1)$, $k=1,2,...,N$ the subsurface plane $t(x,y)$ can be reconstructed using the back-and-forth propagation algorithm. The reconstruction equation for the back-and-forth propagation algorithm is given by [7,8]

$$f(x,y) = \frac{\sum\limits_{k=1}^{N} u_k^*(x,y,z_0)v_k(x,y,z_0)}{\sum\limits_{k=1}^{N} |u_k(x,y,z_0)|^2}. \tag{14}$$

The transmitted wavefield at the object plane, $v(x,y,z_0)$, can be obtained by back-propagating the received wavefield to to object plane using Eq. (6). However, the incident wavefield at the object plane, $u(x,y,z_0)$, cannot be obtained from the received data. Instead, we use the plane wave model and substitute Eq. (1) into Eq. (14) to obtain

$$f(x,y) = \frac{1}{Nu_0} \sum\limits_{k=1}^{N} \hat{v}(x,y,z_0) \, exp\,(-j\,2\pi(f_{xi}x + f_{yi}y + f_{zi}z)). \tag{15}$$

$$= \frac{1}{Nu_0} \sum\limits_{k=1}^{N} p(x,y).$$

Ideally, we have

$$f(x,y) = \frac{1}{Nu_0} \sum\limits_{k=1}^{N} t_{LP}{}^{(k)}(x,y), \tag{16}$$

where the Fourier transform of $t_{LP}{}^{(k)}$ is given by

$$T_{LP}{}^{(k)}(f_x,f_y) = \begin{cases} T(f_x,f_y), & (f_x + f_{xi}{}^{(k)})^2 + (f_y + f_{yi}{}^{(k)})^2 < \dfrac{1}{\lambda_k^2} \\ 0, & \textit{otherwise.} \end{cases} \tag{17}$$

However, because of the presence of the initial phase

$$f(x,y) = \frac{1}{Nu_0} \sum\limits_{k=1}^{N} t_{LP}{}^{(k)}(x,y) \, exp\,(j\,\Omega_k). \tag{18}$$

The unknown phase term $exp\,(j\,\Omega_k)$ causes severe degradation in the superposition process. Using the algorithm described in the last section, the phase term can be estimated and compensated. The block diagram for the multiple-frequency reconstruction algorithm is shown in Fig. 5.

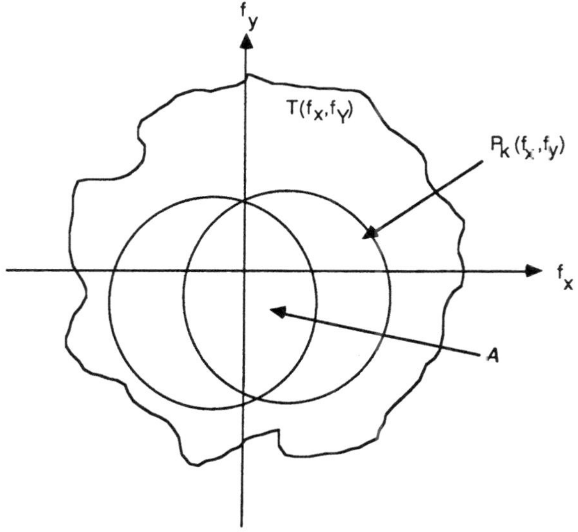

Fig. 4 Overlapping region in the spectral domain for two projections.

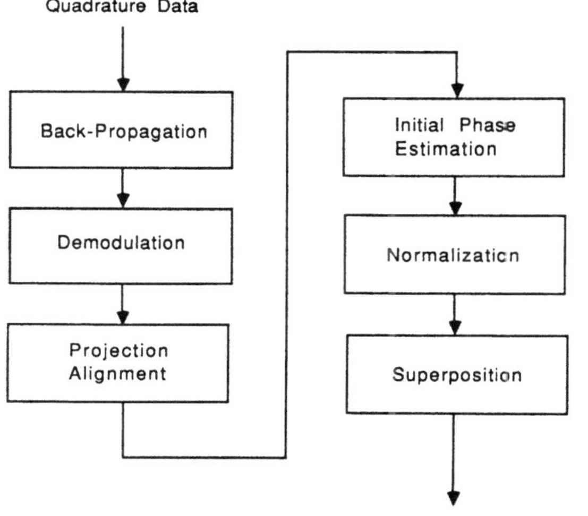

Fig. 5 Multiple-frequency tomographic reconstruction algorithm.

Experimental Results

Figures 6(a)-(d) show the magnitude images of four multiple-frequency projections. The field of view is 2 mm by 2 mm. The specimen is a honeycomb test grid located 0.5 mm beneath the detection plane ($z_1 - z_0 = 0.5$ mm). The insonification frequencies used are 97 MHz, 99 MHz, 101 MHz, and 103 MHz, respectively. Figures 7(a)-(d) show the magnitude images of the back-propagated and demodulated wavefields. Figure 8(a) shows the tomo-

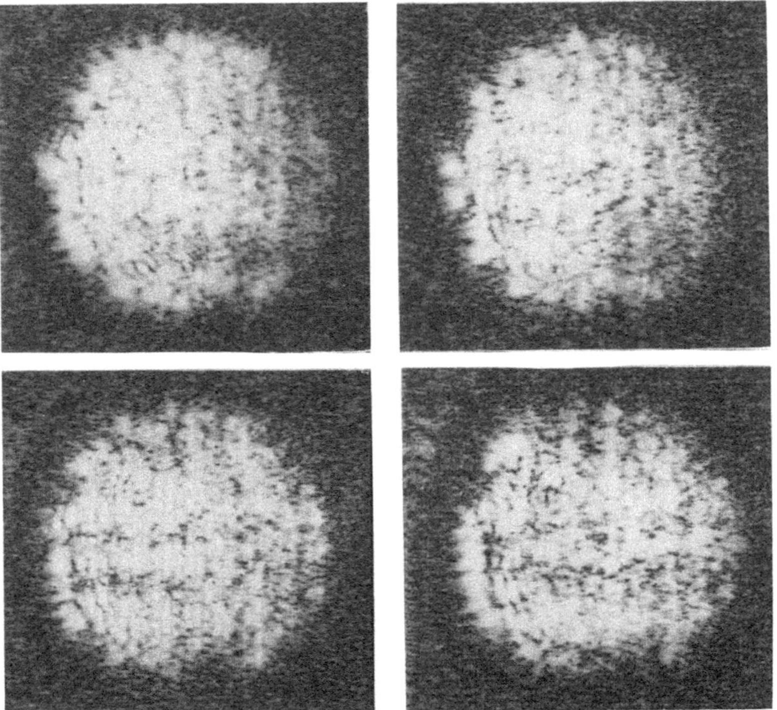

Fig. 6 Magnitude of received wavefields.

graphic superposition of the four images of Fig. 7 without initial phase correction, and Fig. 8(b) shows the tomographic reconstruction using corrected projections. As can be seen, tomographic superposition without the initial phase correction is worse than the the individual projections. With the initial phase correction, the reconstruction shows significant improvement.

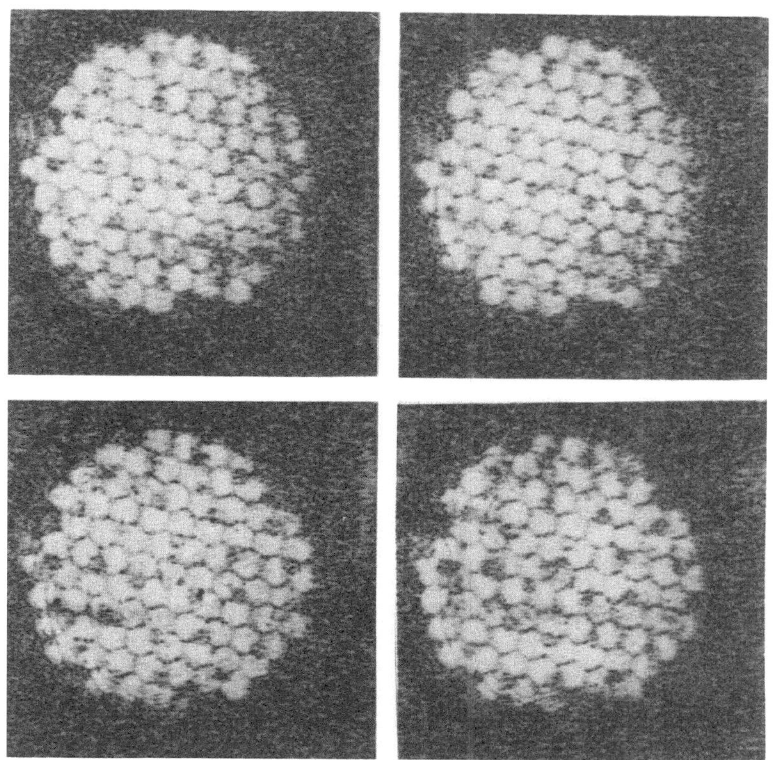

Fig. 7 Back-propagated and demodulated wavefields.

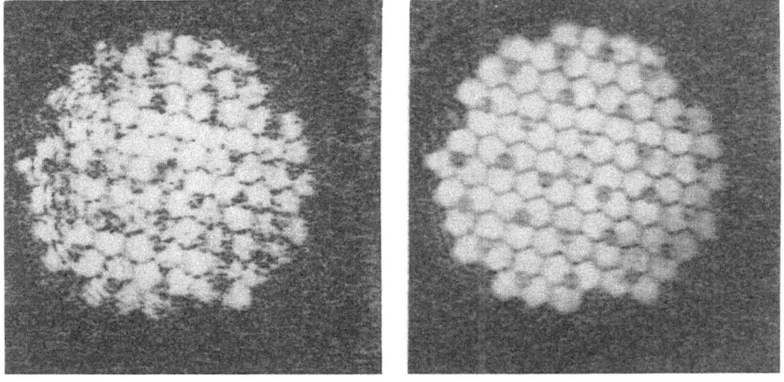

Fig. 8 (a) Without initial phase correction. (b) With correction.

Application to Multiple-Angle Tomography

In multiple-angle tomography, the frequency of the acoustic wave is kept constant while the insonification angle is varied for the different projections. The spatial frequencies for the insonification plane wave are given by

$$f_{xi}(k) = \frac{sin\,(\theta_k)cos\,(\phi_k)}{\lambda} \tag{19-a}$$

$$f_{yi}(k) = \frac{sin\,(\theta_k)sin(\phi_k)}{\lambda} \tag{19-b}$$

$$f_{zi}(k) = \frac{cos\,(\theta_k)}{\lambda}, \tag{19-c}$$

where θ and ϕ are the spherical angles of the propagation vector. Eq. (15) can again be used to combine the multiple-angle projections for tomographic reconstruction.

Two methods of obtaining multiple-angle projections have been studies [8]. One method is the linear scan method in which ϕ is held constant and θ is varied. The other method is the rotational scan method in which θ is held constant while ϕ is varied. Both methods have their respective advantages in terms of the resolution in the reconstructed image, however, only the rotational scan method can be practically implemented by rotating the specimen. While rotating the specimen circumvents the difficulty of rotating the acoustic transducer, it creates the need to rotate and align the received projections. The rotation axis for the received projections needs to be identical to that used to rotate the specimen, and the rotation angle is negative that used to rotate the specimen. Because the tomographic projections are not simply related by a rotation (otherwise no new information would be gained in each projection), estimating the rotation axis and angle from the projections is not a simple task. Without the projections properly aligned, the initial phase estimation cannot be performed. Fig. 9 shows the block diagram to implement the tomographic reconstruction algorithm for multiple-angle data.

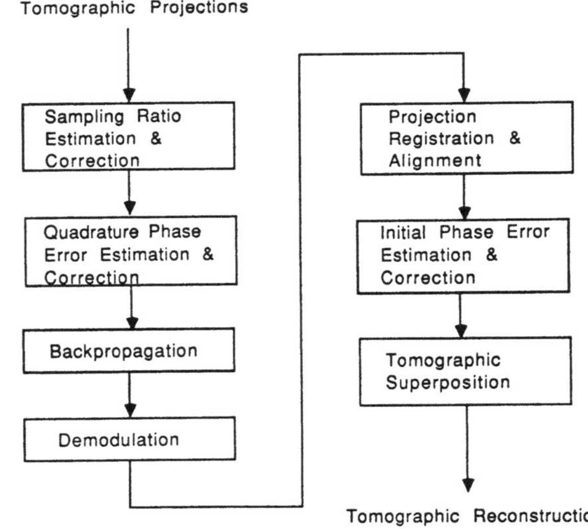

Fig. 9 Multiple-angle tomographic reconstruction algorithm.

Conclusion

We presented a technique to estimate the initial phase difference between projections for multiple-frequency and multiple-angle tomographic acoustic microscopy. The relative phase difference between the projections is a significant degradation factor in the tomographic superposition process. Because of the loss of evanescent components during propagation, the estimation cannot easily be performed in the space domain. Instead, we use a simple method which operates in the overlapping region of the projection spectral distributions. Finally, a multiple-frequency tomographic reconstruction using four projections is used to demonstrate the resolution enhancement capability of the technique.

Acknowledgment

This research is supported by the National Science Foundation under Grant EET-8451484.

References

[1]   Hua Lee and Carlos Ricci, "Modification of the scanning laser acoustic microscope for holographic and tomographic imaging," *Appl. Phys. Lett.*, vol. 49, no. 20, pp. 1336-1338, November 1986.

[2]   Z. C. Lin, H. Lee, G. Wade, M. G. Oravecz, and L. W. Kessler, "Holographic image reconstruction in scanning laser acoustic microscopy," *IEEE Trans. Ultrason. Ferroelec. Freq. Contr.*, vol. UFFC-34, pp. 293-300, May 1987.

[3]   Hua Lee and Richard Chiao, "Data acquisition phase error estimation and correction in tomographic acoustic microscopy," *Acoustical Imaging,* vol. 17, H. Shimizu, N. Chubachi and J. Kushibiki, Eds. New York: Plenum Press, 1989, pp. 131 - 141.

[4]   R. Y. Chiao, H. Lee, and G. Wade, "Image restoration and wave-field error removal in holographic acoustic microscopy," *Proceedings of 1989 IEEE International Conference on Acoustics,* Speech and Signal Processing, pp. 1508 - 1511, 1989.

[5]   R. L. Whitman and A. Korpel, "Probing of acoustic surface perturbations by coherent light," *Applied Optics,* vol. 8, no. 8, pp. 1567-1576, August 1969.

[6]   L. W. Kessler, P. R. Palermo, and A. Korpel, "Recent developments with the scanning laser acoustic microscope," *Acoustical Imaging,* vol. 5, P.S. Green, Ed. New York: Plenum, 1974, pp.15-23

[7]   Hua Lee, Carl Schueler, Gail Flesher, and Glen Wade, "Ultrasonic planar scanned tomography," *Acoustical Imaging,* vol. 11, J. Powers Ed. New York: Plenum, 1982, pp. 309-323.

[8]   Z. C. Lin, H. Lee, and G. Wade, "Scanning tomographic acoustic microscopy: A review," *IEEE Trans. Sonics Ultrason.,* vol. SU-32, pp. 168-180, March 1985.

DECONVOLUTION OF EXPLORATION SEISMIC DATA

Enders A. Robinson

College of Engineering and Applied Sciences
The University of Tulsa
Tulsa, Oklahoma 74104

INTRODUCTION

A layered-earth seismic model is subdivided into two subsystems. The upper subsystem can have any sequence of reflection coefficients but the lower subsystem has a sequence of reflection coefficients which are small in magnitude and have the characteristics of random white noise. It is shown that if an arbitrary wavelet is the input to the lower lithologic section, the same wavelet convolved with the white sequence of reflection coefficients will be the reflected output. That is, a white sedimentary system passes a wavelet in reflection as a linear time-invariant filter with impulse response given by the reflection coefficients. Thus, the small white lithologic section acts as an ideal reflecting window, producing perfect primary reflections with no multiple reflections and no transmission losses. The upper subsystem produces a minimum-delay multiple-reflection waveform, called the multiple wave train or the multiple wavelet. Therefore, the received seismic signal within the time gate corresponding to the lower subsystem is given by the convolution of the multiple wavelet with the white reflection coefficients of the lower subsystem. This is the linear time-invariant seismic model used in predictive deconvolution. This model explains why time-invariant deconvolution filters can be used within various time gates on a received seismic signal, which at first appearance might look like a continually time-varying phenomenon.

THE GOUPILLAUD MODEL

Because a received seismic signal is digitized at equal increments of seismic traveltime, it is convenient to model the Earth with layers defined by the same equal increments of traveltime. Such a model is the classic Goupillaud (1961) layer-cake earth model. There are k interfaces with *reflection coefficients* c_1, c_2, ..., c_k. The sequence of reflection coefficients is called the *reflectivity function*, or simply the *reflectivity*. The magnitude of each reflection coefficient is less than one. Interface 1 represents the surface of the Earth, but for mathematical simplicity, the source and receiver are not located on the surface. Instead both source and receiver are located on the datum, namely interface 0 (with $c_0 = 0$), which is above the surface of the Earth. Interface 0 is a fictitious interface and has no physical existence. Layer 1 is defined as the fictitious layer between the datum (interface 0) and the surface of the Earth (interface 1). Interfaces 2, 3, ..., k lie under the surface of the Earth in that order. The first stratum (between interfaces 1 and 2) is called layer 2, the next stratum (between interfaces 2 and 3) is called layer 3, and so on, until we come to the last stratum (between interfaces $k - 1$ and k). Above the surface

```
      Air
----------------  Datum (with reflection coefficient $c_0 = 0$)
  Air (layer 1)    Surface or interface 1 (with reflection coefficient $c_1$)
----------------
    Layer 2        Interface 2 (with reflection coefficient $c_2$)
----------------
    Layer 3        Interface 3 (with reflection coefficient $c_3$)
----------------
    Layer 4
----------------  Interface 4 (with reflection coefficient $c_4$)
```

Fig. 1. The layered-earth model.

(interface 1) is the air, and below the bottom interface (interface k) is homogeneous basement rock; see Figure 1. The two-way traveltime in each layer is taken to be one time unit. For more complex models, see Lindseth (1971), Matsuoka and Ulrych (1984), Mendel (1977), Webster (1978), and Ziolkowski (1984).

The system has one input (a downgoing plane wave at normal incidence to interface 1) and two outputs (a downgoing plane wave emitted from the bottom interface and an upgoing plane wave emitted from the surface). Because geophysicists and electrical engineers have different conventions with respect to the z-transform, in this paper the letter s is used instead of the letter z, and the term "generating function" is used instead of "z-transform". Accordingly, s represents a unit delay operator. If a small letter such as x represents a signal, then usually the corresponding capital letter X is used for the generating function.

The source of energy is an impulsive downgoing unit spike occurring at time zero. If the total number of interfaces in the system is k, then the discrete time signal $r_{k1}, r_{k2}, r_{k3}, \ldots$ (i.e., r_{kn} for time index $n = 1, 2, 3, \ldots$) denotes the *reflection response* to this impulsive unit source, and is the upgoing *received seismic signal* recorded at the receiver on the datum. The generating function of the received seismic signal is

$$R_k(s) = r_{k\,1}s + r_{k\,2}s^2 + r_{k\,3}s^3 + \cdots.$$

(Note: The subscript k indicates the layered system has k interfaces.) The discrete-time signal $t_{k,(k/2)}, t_{k,(k/2)+1}, t_{k,(k/2)+2}, \ldots$ (i.e., t_{kn} for time index $n = (k/2), (k/2) + 1, (k/2) + 2, \ldots$) denotes the transmission response to the impulsive unit source, and is the downgoing wave train occurring just below interface k. The generating function of the transmission response is

$$T_k(s) = t_{k,\,(k/2)}s^{k/2} + t_{k,\,(k/2)+1}s^{(k/2)+1} + t_{k,\,(k/2)+2}s^{(k/2)+2} + \cdots.$$

Note that the first nonzero amplitude of the reflection response r_{kn} is at $n = 1$, because it takes one time unit for the round trip between the datum (interface 0) and the surface of the Earth (interface 1). Also the first nonzero amplitude of the transmission response t_{kn} is at $n = (k/2)$, because it takes $k/2$ time units for the one-way trip through the k-layers.

THE REFLECTION AND TRANSMISSION RESPONSES

Let us now specify the initial conditions (Figure 2). Assume that the source is a downgoing unit spike initiated at time 0 at the top of layer 1, so

$$D_1(s) = 0$$

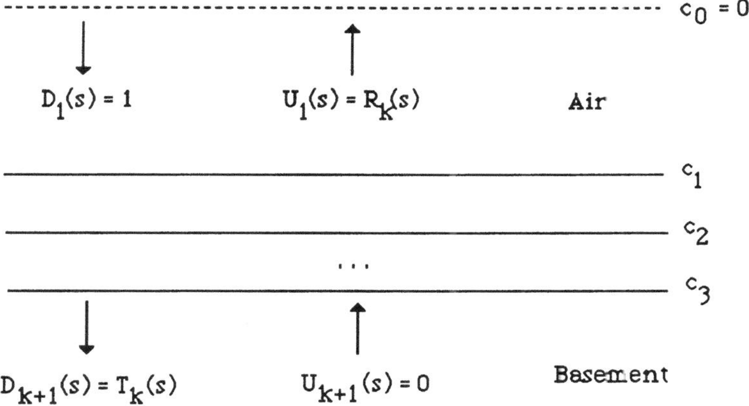

Fig. 2. Definition of the reflection response R_k and the transmission response T_k for a k interface system.

Moreover, assume that there can be no upgoing wave motion in the basement, so

$$U_{k+1}(s) = 0$$

The reflection seismogram measured in layer 1 is the reflection response. The generating function of the reflection response is thus

$$R_k(s) = U_1(s).$$

The transmission seismogrm is the trace that would be measured in the basement, by a seismometer placed in a deep well in a vertical seismic profiling (VSP) survey (Levin and Lynn, 1959). The generating function of the transmission response is thus

$$T_k(s) = D_{k+1}(s).$$

Concerning notation, we are using R_k and T_k as the symbols for the reflection and transmission responses for k interfaces. This is a different use for the subscript k, as usually k indicates the layer in which the wave motion occurs. Since $R_k = U_1$, the received seismic signal R_k occurs in layer 1, and since $T_k = D_{k+1}$ the signal T_k occurs in the basement.

The fundamental set of polynomials P and Q are generated recursively from the reflection coefficients by the equations

$$P_1(s) = 1$$

$$Q_1(s) = -c_1 s$$

$$P_i(s) = P_{i-1}(s) - c_i s^i Q_{i-1}(s^{-1})$$

$$Q_i(s) = Q_{i-1}(s) - c_i s^i P_{i-1}(s^{-1})$$

for $i = 2, 3, ..., k$ (Robinson, 1967, 1984). By computing the first few of these polynomials, one sees that their coefficients are nonlinear combinations of the reflection coefficients.

The *transmission coefficient* across interface i is given by the positive root

$$\tau_i = \sqrt{1 - c_i^2} \, .$$

Thus the *one-way transmission factor* is given by the product of transmission coefficients

$$\sigma_k = \tau_1 \tau_2 \cdots \tau_k \, .$$

The transmission factor σ_k is a number in the range from 0 to 1. It represents the transmission loss through the k layers.

In terms of the polynomials, the *reflection response* (or *received seismic signal*) is

$$R_k = \frac{-Q_k}{P_k}$$

We write the minus sign with the Q_k, because $-Q_k$ represents a physical entity. The expression for the *transmission response* is

$$T_k = \frac{\sigma_k \, s^{0.5k}}{P_k}$$

This equation states that the transmission response is made up of the direct pulse from the shot to (and through) interface k and the pulses of the attached multiple reflections. The direct pulse has amplitude σ_k (i.e., the downward transmission factor) and is delayed by the traveltime $0.5k$ downward through the k layers (as given by the delay factor $s^{0.5k}$). The attached multiple reflections are generated by the feedback system given by $1/P_k(s)$. For these derivations, see Robinson (1984).

Because physically the transmission response is a causal transient time function, it follows that $P_k(s)$ is minimum delay. However, the minimum-delay property of $P_k(s)$ can also be established mathematically by using the fact that each reflection coefficient $c_1, c_2, \ldots, c_k$ is less than one in magnitude.

SMALL REFLECTION COEFFICIENTS

Interface i between two rock layers is characterized by the reflection coefficient c_i (more specifically, a local reflection coefficient). The reflection coefficient is a real number and can be positive or negative, but its magnitude must be less than one. Generally, the magnitudes of the reflection coefficients encountered are much less than one. If they tend to cluster around the mean value of zero, they are considered *small* reflection coefficients. The concept of smallness is made more precise as follows.

The two-way transmission factor through k interfaces is

$$\sigma_k^2 = \tau_1^2 \tau_2^2 \cdots \tau_k^2 = (1 - c_1^2) \cdots (1 - c_k^2) \, .$$

The factor σ_k^2 is necessarily less than or equal to one. It is said that the reflection coefficients $c_1, c_2, \ldots, c_k$ are small provided that σ_k^2 is equal to one or is nearly equal to one. The logarithm of σ_k^2 is

$$\log \sigma_k^2 = \log \prod_{i=1}^{k} (1 - c_i^2) = \sum_{i=1}^{k} \log (1 - c_i^2).$$

If the magnitude of c_i is small, then c_i^2 is very small, so the logarithm has the approximation $\log (1 - c_i^2) = -c_i^2$. Thus,

$$\log \sigma_k^2 = -\sum_{i=1}^{k} c_i^2 = -\gamma_0,$$

where γ_0 is defined as the zero-shift autocorrelation coefficient of the reflection coefficient series.

Small reflection coefficients are important because an essential mathematical simplification occurs in the case of small reflection coefficients. The simplification is expressed as follows. If a layered system of k interfaces has reflection coefficients $c_1, c_2, ..., c_k$ which are small in magnitude, then the fundamental polynomials are approximately given by the expressions

$$P_k(s) = 1 + \gamma_1 s + \gamma_2 s^2 + \cdots + \gamma_{k-1} s^{k-1},$$

and

$$Q_k(s) = -c_1 s - c_2 s^2 - \cdots - c_k s^k,$$

where γ_i is the autocorrelation coefficient

$$\gamma_i = \sum_{j=1}^{k-i} c_{j+i} c_j \qquad \text{for } i = 1, 2, \cdots, k - 1$$

of the reflection coefficient sequence. For small reflection coefficients, the only two mathematical entities that enter into the layered-earth model are the reflection coefficient sequence itself, and the autocorrelation of the reflection coefficient sequence. In the case of small reflection coefficients, the layered-earth model does not require any higher-order function of the reflection coefficients.

SMALL WHITE REFLECTION COEFFICIENTS

If the reflection coefficient sequence is white noise, then its autocorrelation coefficients (except the one for lag zero) are approximately zero. Thus, when the lithologic section has small white reflection coefficients, $P_k(s) = 1$ and $Q_k(s) = -C(s)$ where $C(s)$ is the generating function of the reflection coefficient series. The transmission response and reflection response of a white lithologic section are, respectively,

$$T_k(s) = \sigma_k s^{0.5k}, \qquad R_k(s) = C(s) = c_1 s + c_2 s^2 + \cdots + c_k s^k.$$

A white lithologic section merely produces a scalar attenuation σ_k and a delay of $0.5k$ on transmission, and reproduces the reflection coefficient series $c_1, c_2, ..., c_k$ (with no transmission losses and no multiple reflections) on reflection. That is, a white lithologic section produces a reflection response consisting of only primary reflectison with no transmission losses and no multiple reflections. Because the transmission response represents an impulse response, it follows that if an arbitrary wavelet is the input to a white lithologic section, the same wavelet multiplied by the scalar σ_k will be the transmitted output. Therefore, a white lithologic section passes a wavelet in transmission with no change in shape. Also, if an arbitrary wavelet is the input to a white lithologic section, the same wavelet convolved with the white reflectivity function is the reflected output. Thus a white lithologic section passes a wavelet in reflection as a linear time-invariant filter with impulse response $c_1, c_2, ..., c_k$. The time-invariant seismic convolutional model is based upon this central result. A small white lithologic section acts as an ideal window, producing perfect transmission and perfect primary reflections. Randomness gives perfection, or near perfection, as far as seismic characteristics are concerned.

THE CONVOLUTIONAL MODEL

In order to derive the *convolutional model*, a layered system of N interfaces is subdivided into an upper subsystem made up of the first k interfaces and a lower subsystem made up of the remaining $N-k$ interfaces. Assume that the lower subsystem is a small white lithologic section (i.e., a layered system with small white reflection coefficients). The lower subsystem acts as free space with respect to everything except the primary reflections. Because of the small white hypothesis, transmission effects and multiple reflections are completely absent in the reflection response of the lower subsystem.

The reflection response (or received seismic signal) R_N is made of the physical reflection response of R_k of the upper subsystem plus the primary reflections from the lower subsystem. The primary reflection from interface i in this lower portion has the following down-and-up travel path. The pulse travels down from the source through the upper subsystem (transmission reponse T_k), it bounces from interface i (reflection factor $c_i s^{i-k}$), and then travels back up through the upper system (transmission response T_k). This primary with its attached multiple train has generating function

$$T_k^2 c_i s^{i-k} = \left[\frac{\sigma_k s^{0.5k}}{P_k} \right]^2 c_i s^{i-k} = c_i \left(\frac{\sigma_k^2}{P_k^2} \right) s^i .$$

where we have used the expression for T_k as given by $\sigma^k s^{0.5k}/P_k$. The multiple wave train m_n (where the subscript n is the time index) is defined as the wave train with generating function

$$M(s) = \sum_{n=0}^{\infty} m_n s^n = \sigma_k^2/P_k^2(s) .$$

In terms of $M(s)$ the generating function of the primary with its attached multiple train is $c_i M(s) s^i$. The factor s_i represents the arrival time i of the multiple train m_n, and the reflection coefficient c_i is the weighting factor. Thus, in the time domain, this primary with its attached multiple is $c_i m_{n-i}$. Because the polynomial P_k is minimum-delay, it follows that $M(s)$ is also minimum-delay (i.e., a casual linear, time-invariant filter with minimum phase spectrum). For further treatment of multiple reflections, see Backus and Simmons (1984).

To get the total reflection response, we add the reflection response r_{kn} of the upper subsystem to all these primaries (with their attached multiples) from the lower subsystem; that is

$$r_{Nn} = r_{kn} + \sum_{i=k+1}^{N} c_i m_{n-i},$$

o r

$$R_N = R_k + \sum_{i=k+1}^{N} c_i M(s) s^i = R_k + M(s)\left(\sum_{i=k+1}^{N} c_i s^i\right).$$

If r_{kn} is regarded as a noise term, this model represents a linear time-invariant convolutional model of the received seismic signal r_{Nn} for $n > k$. This is the *convolutional model*. By dropping the additive noise term,

$$r_{Nn} = \sum_{i=k+1}^{N} c_i m_{n-i}, \quad \text{for } n = k+1, k+2, \ldots,$$

o r

$$R_N(s) = C(s)M(s),$$

w h e r e

$$C(s) = \sum_{i=k+1}^{N} c_i s^i, \quad M(s) = \frac{\sigma_k^2}{P_k^2(s)}.$$

In other words, within the time gate $k < n \leq N$ the time-invariant convolutional model is

(reflection response) = (reflection coefficients) * (multiples)

which is

$$r_{Nn} = c_n * m_n,$$

where the asterisk denotes convolution.

As shown, within the time gate $k < n \leq N$ corresponding to a small white lithologic section, the seismic reflection response (or received seismic signal) is given by the time-invariant convolution model. That is, within the time gate the received seismic signal is the convolution of the white reflectivity $c_{k+1}, c_{k+2}, \ldots,$ c_{k+n} with the minimum-delay multiple wave train $m_0, m_1, m_2, \ldots$ (Silvia and Robinson, 1978).

CONCLUSIONS

This paper concentrates on the specification of a mathematical model of the received seismic signal (the so-called *seismic field trace*). As is well-known, the field trace can be described by a linear time-varying system. However, for deconvolution (Robinson, 1954), the entire field trace is not dealt with as a single unit, but is subdivided into time sections with each section defined as that portion of the trace between two time limits (the so-called *time gate*). Corresponding to each time gate on the trace there is a certain vertical section of rock layers within the Earth, such that the primary reflections from these layers all arrive within the time gate. There is a one-to-one correspondence between vertical sections cut through the rock layers within the Earth and time gates on the received seismic trace, the portion of trace within each of the gates is deconvolved separately, and then the results are blended to make up the final output trace. Therefore, in practice the portion of the field trace within a specified time gate and the corresponding portion of the subsurface rock layers are always specified.

Under the hypothesis that a section of the lithologic column of the subsurface of the Earth can be characterized by a reflectivity function with small white reflection coefficients, then within the corresponding time gate the seismic field trace can be described by a linear time-invariant convolutional model. The model states that within the time gate, the field trace is equal to the convolution of the white reflectivity with a minimum-delay wavelet. This is the model required for predictive deconvolution, both spike deconvolution and gapped deconvolution.

ACKNOWLEDGEMENT

I want to express my sincere thanks to Rosa Jackson for typing this paper.

REFERENCES

Backus, M. M., and Simmons, J. L., 1984, Multiple Reflections as an Additive Noise Limitation in Seismic Reflection Work, <u>Proc. Inst. Elect. Electron. Eng.</u>, 72:1370-1384.

Goupillaud, P. L., 1961, An Approach to Inverse Filtering of Near-Surface Layer Effects from Seismic Records, <u>Geophysics</u>, 26:754-760.

Levin, F. K. and Lynn, R. D., 1959, Deep-Hole Geophone Studies, <u>Geophysics</u>, 23:662-671.

Lindseth, R. O., 1971, Recent Advances in Digital Processing of Geophysical Data - A Review, <u>Soc. Explor. Geophys.</u>, Tulsa, Oklahoma.

Matsuoka, T. and Ulrych, T. J., 1984, Phase Estimation Using the Bispectrum, <u>Proc. Inst. Elec. and Electron. Eng.</u>, 72:1403-1411.

Mendel, J. M., 1977, White Noise Estimators for Seismic Data Processing in Oil Exploration, <u>Inst. Elect. Electron. Eng. Trans. on Automatic Control</u>, AC-22:694-706.

Robinson, E. A., 1954, Predictive Decomposition of Time Series with Application to Seismic Exploration, Ph.D. Dissertation, M.I.T, Reprintered in <u>Geophysics</u>, 1967, 32:418-484.

Robinson, E. A., 1967, Multichannel Time Series Analysis with Digital Computer Programs, Second Edition, 1983: Prentice Hall, Englewood Cliffs, New Jersey, 485 pp.

Robinson, E. A., 1984, Seismic Inversion and Deconvolution, Pergamon Press, New York, 361 pp.

Silvia, M. T., and Robinson, E. A., 1978, Deconvolution of Geophysical Time Series in the Exploration of Oil and Natural Gas, Elsevier Science Publishing Co., 274 pp.

Webster, G. M., 1978, Deconvolution, <u>Soc. Explor. Geophys.</u>, Tulsa, Oklahoma.

Ziolkowski, A., 1984, Deconvolution, Prentice Hall, Englewood Cliffs, New Jersey.

TOMOGRAPHIC RECONSTRUCTION FOR IMAGING EOR PROCESSES IN HYDROCARBON RESERVOIRS

J.H. Justice, A.A. Vassiliou

Mobil Research and Development Corporation
Dallas Research Laboratory
P.O. Box 819047
Dallas, Texas 75381-9047

ABSTRACT

Clastic reservoirs saturated with heavy oils have been observed to exhibit a marked relationship between velocity of propagation of acoustic waves and temperature of the oil saturated sediments. This observation forms the basis for a method of monitoring the changes which occur in the reservoir when thermal enhanced oil recovery (EOR) procedures are used. New developments in crossborehole tomographic imaging allow accurate images of the reservoir to be obtained, clearly delineating the zones affected by thermal EOR. Using both compressional and shear components of the wavefield (representing distinct coupled modes of propagation in an elastic medium), a variety of physical parameters such as incompressibility (bulk modulus), shear modulus, and Poisson ratio can be tomographically reconstructed.

I. INTRODUCTION

Tomographic imaging techniques have found numerous applications in many fields and are now being extended successfully for use in geophysical imaging applications. Perhaps the first major use of this technology in geophysical imaging has been for the purpose of monitoring enhanced oil recovery operations in heavy oil reservoirs. Oil which is too viscous to flow at normal reservoir temperatures and pressures can often be successfully produced when heated to several hundred degrees Celsius. This is often accomplished by injecting superheated steam or by in-situ combustion of the oil, in which some of the oil is consumed to provide the energy necessary to produce the remainder. It was observed several years ago that heavy oil sands can exhibit a marked decrease in the velocity of propagation of acoustic waves and a significant increase in acoustic attenuation when heated. This observation led to the first successful attempts to apply tomographic imaging to monitoring the movement of the heat front in a thermal enhanced oil recovery (EOR) project in heavy oil reservoirs. By computing time-lapse images of a section of the reservoir during the EOR process, we may reasonably assume that the changes are induced by the EOR process, everything else being constant.

Because variations in velocity of acoustic wave propagation due to heating of heavy oil sands can be on the order of several tens of percent, the acoustic raypaths cannot be assumed to be straight. In addition, sharp contrasts in acoustic velocity at lithologic boundaries generally refract acoustic energy resulting in strongly deviated raypaths and focusing of energy. Any attempt to apply standard straight-ray tomographic imaging in these environments usually results in an uninterpretable image. As a result, it has been necessary to develop robust and reliable algorithms for diffraction tomography for this application.

Seismic tomographic data for reservoir imaging is usually collected between two boreholes penetrating the reservoir, or between boreholes and the surface (in the case of very shallow reservoirs). The producing reservoir can be a very noisy environment and data is usually corrupted by a variety of acoustic and electrical noise. Resonant waves induced in the borehole by the acoustic source (tube waves) can carry particularly large amounts of energy and can obscure events of interest in the data. As a result, considerable effort is required to detect the desired signal in noisy data. The most robust form of tomographic reconstruction for geophysical imaging uses first arrival traveltime data. When velocity contrasts are large enough, later arrivals should be included in the reconstruction process for accurate image reconstruction.

Raytracing is a tempting tool to use in diffraction tomography, but experience shows that it is inadequate when small diffractors with a correlation length less than a wavelength are present and velocity contrasts exceed perhaps ten percent of background. These conditions are not uncommon in practice. As a result, only diffraction tomography based on the full wave equation (acoustic, elastic, or poroelastic, as appropriate) should be used. This is particularly true since an incorrectly reconstructed image may be worse that no image at all.

Accurate tomographic images of reservoirs carry an enormous amount of useful information which is simply not available otherwise. These images provide the only capability which we have for direct imaging of the reservoir, and which extends more than a few inches from the borehole. In a typical application, images are constructed over many hundreds of feet across the reservoir. The information acquired can tell us about the geologic structure of the reservoir, including information about lithology and faulting. It can help us to make decisions about the placement of new development wells, and can help us to evaluate the effectiveness of an EOR program. It provides significant information valuable for understanding the reservoir, its structure and dynamics, and can help us to build more accurate reservoir models. Finally, the economic impact of more efficient EOR programs and more effective recovery can be very substantial. This becomes obvious when we realize that hundreds of billions of barrels of heavy oil can be found in oil fields in North America alone.

II. TOMOGRAPHIC RECONSTRUCTION FOR RESERVOIR IMAGING

Although there has been much research devoted to transform methods (straight ray) and inverse scattering formulations for tomographic reconstruction, none of this research has found much application in geophysical tomographic image reconstruction to date. We shall focus our attention here on an algebraic approach to diffraction tomography using the full wave equation (in whatever form is appropriate for the problem). As pointed out earlier, algebraic formulations based on ray tracing are fairly simple to construct, but should not be used in reservoir imaging applications.

In algebraic reconstruction, the unknown velocity field to be reconstructed is represented fully by a finite set of parameters (which can be chosen in many different ways). The objective, of course, is to use the observed data to determine these parameters, thereby reconstructing the unknown velocity field. The reconstruction problem is nonlinear and is solved by linearizing it and iterating from some initial model, in order to converge to the solution.

Let P denote the parameter set to be reconstructed. The (traveltime) data, T, is a nonlinear function of the parameter set, P

$$T = F(P) \tag{1}$$

where the nonlinear function, F, is determined by the appropriate wave equation.

Given an initial estimate, P_0 of the parameter set, we may compute an updated estimate from the first order expansion

$$T = F(P_0) + J(P_0) \, \Delta P \tag{2}$$

where T is the set of observed traveltimes and J is the appropriate Jacobian matrix determined by the choice of F. The updated estimate of the parameter set is found by solving equation (2) for ΔP and then computing the update

$$P_1 = P_0 + \Delta P \tag{3}$$

This process is repeated until the residual error is in the range of errors in the observations.

Equation (2) can be solved in a variety of ways, using an error criterion which can be selected to fit the characteristics of the problem. For example, the L_2 norm can be minimized using Singular Value Decomposition (SVD) or, for large problems, a conjugate gradient method would be preferred. It is also possible to use partial singular value decomposition [3]. We have also used the total least squares criterion [2] with encouraging results.

It has been pointed out that data are usually noisy and are limited in accuracy due to rather severe bandwidth limitations on acoustic waves propagating in a highly attenuating medium, which may exhibit relatively low Q. While the L_2 norm can serve well to fit data sets with limited accuracy, it often displays undesirable characteristics in the presence of noise or outliers in the data. The L_1 norm is preferable in this situation.

Various implementations of the L_1 norm are available, including various Linear Programming algorithms as well as reweighted least-squares [6]. The Karmarkar and Simplex algorithms tend to find extreme values in the solution and require almost exact bounds on the parameter space [7]. It is difficult, in practice, to assign these bounds with the required accuracy. Even when this can be done, the extreme values taken by the solution tend not to be physically meaningful. The implementation of the L_p norm using reweighted least squares, with p near 1 provides a satisfactory solution to the problem and results in reconstructions which are relatively insensitive to outliers in the data.

III. WAVE EQUATION VS. RAYTRACING IN DIFFRACTION
TOMOGRAPHY

As pointed out earlier, it is tempting to implement tomographic reconstruction algorithms for geophysical imaging using ray tracing as an approximate substitute for the wave equation, and this approach has been reported by various authors [1]. We may recall that ray theory is based on a high frequency approximation to the solution of the acoustic wave equation. In geophysical imaging, frequencies of several kiloHertz may be successfully transmitted over short distances in some types of rocks. In the typical hydrocarbon reservoir environment, it is far more common to observe a useable bandwidth of perhaps several hundred Hertz. This is particularly true in the highly attenuating environment of most heavy oil reservoirs which are being thermally stimulated (which, we recall, also reduces the velocity of propagation). Typically, in these environments, dominant wave lengths will be on the order of 5 meters, more or less. As a result, it is not difficult for a velocity anomaly to act as a diffractor of seismic energy, which cannot be correctly modeled by ray tracing.

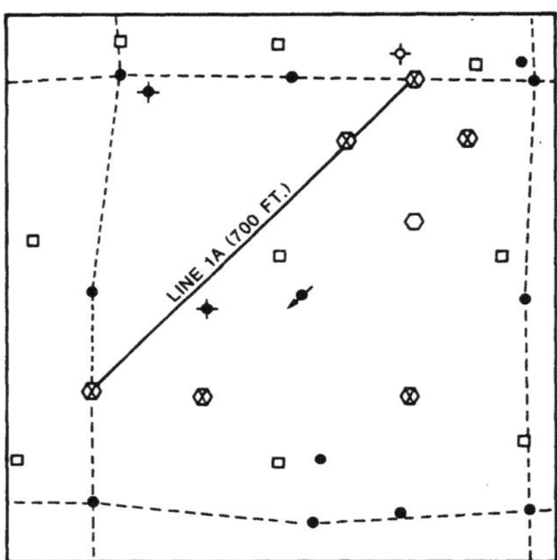

Figure 1

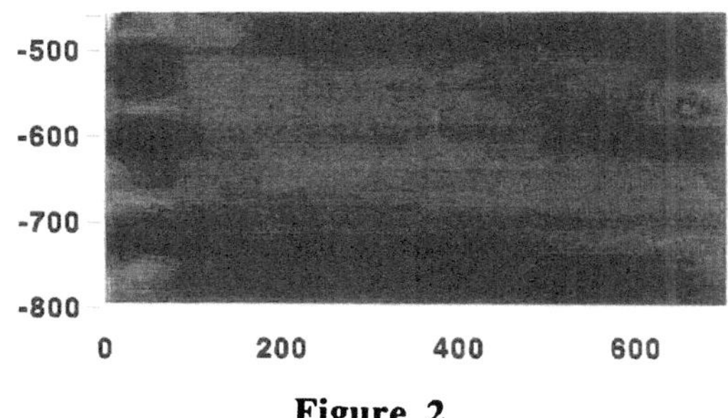

Figure 2

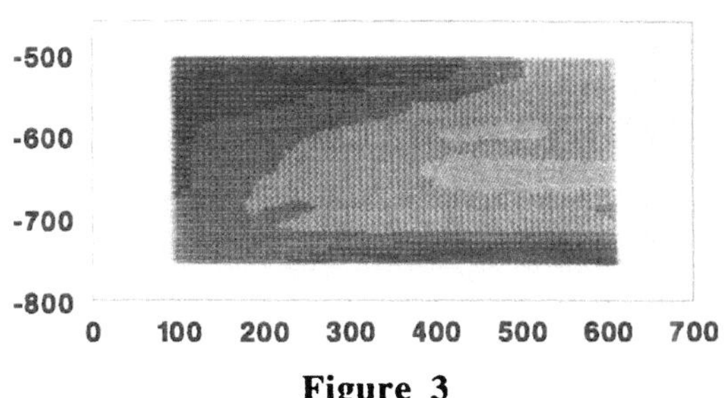

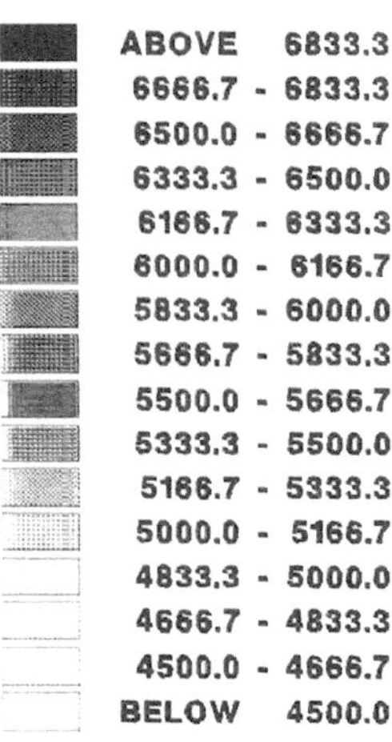

Figure 3

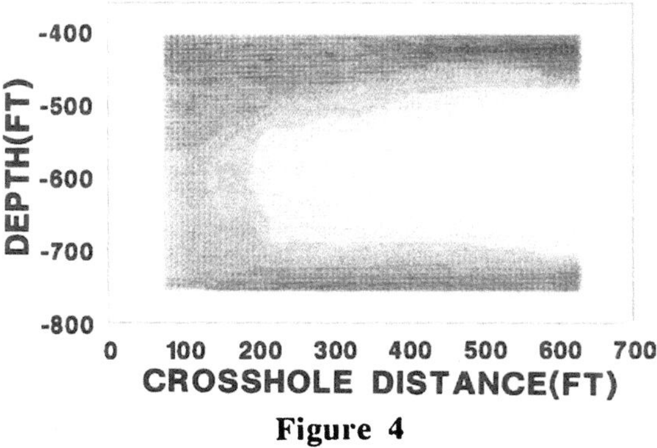

Figure 4

Another way to look at this is in terms of in-phase first arrivals. Because of Huygens principle, raypaths which are "close" to geodesics (true minimum time raypaths) will represent valid propagation paths for energy which will arrive essentially in phase with the true first arrival. If small diffractors are present, then these paths cannot be found of even approximated by traditional raytracing, in general. Let us look at the implications of this statement in terms of the reconstruction (inversion process)

In tomographic reconstruction, the observed data must be 'back-propagated' in some way to the complete set of parameters which gave rise to it. If we reconstruct an image using ray tracing, then only the parameters which contribute to the ray (or rays) connecting source and receiver will be influenced by the corresponding data. According to our comments above, this data should have been distributed over a much larger set of parameters than those associated with a single (or multiple) ray. The only exception to this statement would occur in case we were reconstructing only a very (spatially) sparse parameter set, which is so sparse that small diffractors could not be imaged anyway. In either case, the reconstruction will be incorrect.

We may now make this concept much more precise by looking at the Jacobian in equation (2). The non-zero components in this Jacobian show clearly which parameters are contributing to each element in the data. What we have just said is that the Jacobian associated with a ray-trace reconstruction will be much more sparse than a Jacobian associated with a wave equation inversion. We believe that the inability of a ray-trace reconstruction to correctly account for diffraction effects is important enough in high resolution geophysical imaging of the kind described here, that only wave equation reconstruction should be used.

IV. COMPUTING ISSUES AND CONFIDENCE BOUNDS

The algebraic approach to tomographic reconstruction described in this paper is inherently massively parallel in computer implementation. When coupled with a massively parallel implementation of the wave equation, the full power of a parallel computing architecture can be brought to bear on the reconstruction problem. Our experience has shown that N parallel processors can lead to an almost N-fold increase in computation speed when this approach is used.

The Jacobian matrix in equation (2) carries significant information about the reconstruction process and this information can be tapped to tell us something about the reliability (confidence level) of each parameter in the final reconstruction. What the Jacobian matrix is telling us is the sensitivity of the observed data to each parameter in the reconstruction. As we converge on the final solution, the sensitivity of each datum to each parameter can be checked. When all the data are extremely sensitive to perturbations of a single parameter, we may place a high confidence on the final value of that parameter. On the other hand, if the data are relatively insensitive to perturbations in a parameter, then we may place little confidence on the accuracy of that parameter. Perhaps not surprisingly, it turns out that this measure of confidence correlates very well with the geometry of the survey and the zones where energy is concentrated due to focusing. There are a number of ways in which the concept explained here can be quantized, and these details will be outlined in later publications. An example of a color-coded 'confidence' plot for an actual seismic survey is shown in the examples. Use of this concept can be extremely useful in interpreting geophysical tomograms.

V. EXAMPLES

In 1987, a tomographic crossborehole survey was taken in a thermal EOR project just after the project had started. The position of the wells used in the survey as well as the nearby injection well are shown in Figure 1. The initial reconstruction (Figure 2) shows little variation in the compressional velocity field when the first survey was taken. The

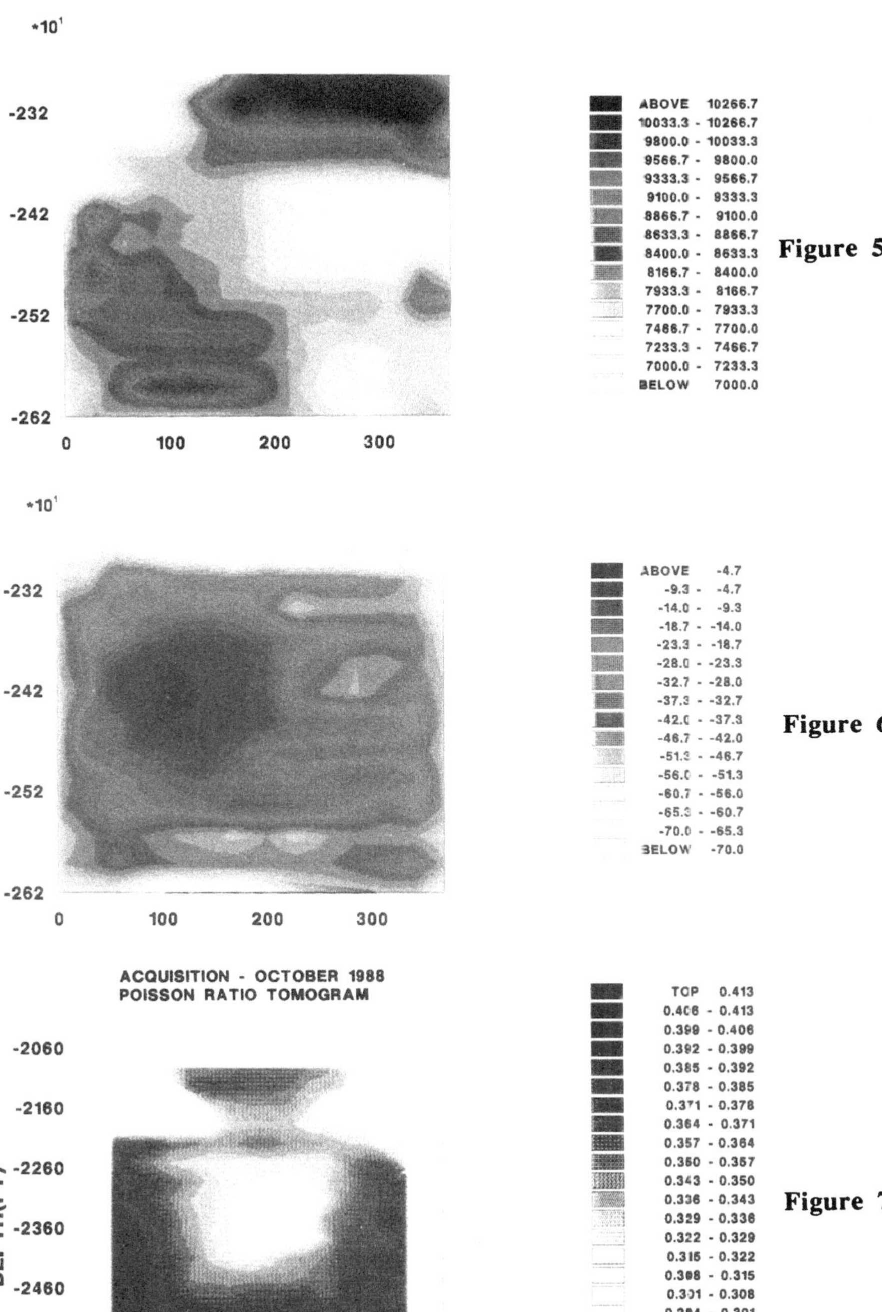

Figure 5

Figure 6

Figure 7

287

second survey was completed six months later, and the effects of the injection fluid are beginning to be clearly imaged as a decrease in the acoustic velocity field (Figure 3). The fluid is seen to be moving toward the well at the right end of the line (as confirmed by field monitoring of production). The third survey was completed after another six month interval and the result is shown in Figure 4. The effects are very clearly visible, with a strong decrease in the acoustic velocity field visible near the well at the right end of the line where the fluid appears to be concentrated.

Figure 5 shows a tomogram taken through an in-situ burn project. The heated/burning zones are clearly visible as lower velocity zones. A confidence plot was made of this tomogram as discussed earlier and is shown in Figure 6. The darkest grays represent areas where the reliability of the reconstruction is highest. These areas correlate nicely with zones where energy was focused by the variable velocity field.

We conclude with an example showing the reconstruction of a physical parameter other than acoustic wave velocity. To compute this tomogram, both compressional and shear wave tomograms were reconstructed for a cross section of a hydrocarbon reservoir. Both tomograms indicate a channel-type structure in cross section. The compressional and shear wave tomograms were combined [5] to compute the Poisson Ratio, which is related to the mechanical properties of the rock. The result is shown in Figure 7. The Poisson ratios computed are in the right range for this reservoir, and the channel structure stands out clearly as a type of rock with different mechanical properties than the surrounding material.

VI. CONCLUSIONS

We have considered a number of issues related to tomographic image reconstruction for geophysical imaging. Experience with many tomographic seismic surveys carried out in various kinds of reservoirs has led to a good understanding of the process of diffraction tomography and the development of practical algorithms which are in use today for imaging these previously poorly understood environments. Knowledge which can be gained from this kind of surveillance will add significantly to our ability to maximize our recoverable reserves at a lower cost per barrel for enhanced recovery, due to increased efficiency and our improved ability to monitor EOR processes.

REFERENCES

1. Dines, K.A., and Lytle, R.J., 1979, Computerized Geophysical Tomography, Proc. IEEE, 67:1065.

2. Golub, G.H. and Van Loan, C.F., 1980, An Analysis of the Total Least Squares Problem, SIAM J. Num. Anal., 17:883.

3. Justice, J.H., Vassiliou, A.A., and Nguyen, D.T., 1988 Geophysical Diffraction Tomography, Signal Processing, 15:227 .

4. Justice, J.H., Vassiliou, A.A., Singh, S., Logel, J.D., Hansen, P.A., Hall, B.R., Hutt, P.R., and Solanki, J.J., 1989, Acoustic Tomography for Monitoring Enhanced Oil Recovery, Leading Edge, 10:12.

5. Justice, J.H., and Vassiliou, A.A., to appear 1990, Diffraction Tomography for Geophysical Monitoring of Hydrocarbon Reservoirs, Proc. IEEE.

6. Scales, J., Gersztenkorn, A., and Treitel, S., 1988, Fast l_p solution of large, sparse, linear systems, J. Comp. Phys., 75:314.

7. Vassiliou, A.A., Justice, J.H., and Guinzy, N.J., 1989, Alternative Error Measures for Non-Diffraction and Diffraction Tomography, Acoustical Imaging, 17:627.

GEOPHYSICAL DIFFRACTION TOMOGRAPHY

Benli Gu AND Ying Ji

Biomedical Engineering Department
Southeast University
Nanjing, P.R.China

INTRODUCTION

People now are paying much more attention to the geophysical tomography technique, which is a very useful tool used to explore and reconstruct subsurface structures. The methods used in the past in geophysics, are usually not unique. Because of the uniqueness of tomography technique in math, many scientists have been attracted to make it useful in geophysics. Although in actual geophysical situations, the projection are uncompleted, therefore the results are still not unique. However we can do better now than before, if we use tomography technique.

Up to the present stage, what has been applied practically in geophysics is the tomography reconstruction according to the ray theory.Many scientists, such as Dine [1], have made much contribution in applying tomography to geophysics using straight ray reconstruction or ray tracing iterative reconstruction.

But, as we all know, ray method is only usable when the wavelength of exploration wave is much shorter than that of the measured object. What is more, only one point of the record curve is used in seismic exploration, i.e.,the information of arrival time, not the information of all waveform. So, it is inevitable that the inversion of wave equations using all the wave field records become an attractive aim, especially because the wave length of actual seismic wave (its low frequency element) are really comparable to the size of the objects measured. In 1978, R. K. Mueller [2,3] put forward the diffraction tomography method. In 1982, A. J. Devaney [4] raised the filtered back-propagation algorithm. Many scientists, such as S.X.Pan, Glen Wade, D. Nahamoo, A. C. Kak et al.[5-10] improved the theory and algorithm. In 1984, again Devaney [11], applied diffraction tomography to geophysics, so to make great contributions to the theory and algorithm. But still there exists a gap between the theory and application. R.S.Wu, Tien-chen, Toksoz, R.Gerhard[12-14], did some laboratory tests, and this made a great progress in practical using. Their work is admired by all of us. Yet many problems still remain unsolved in practical using.

Considering the laboratory experiment research, our paper discusses problems of 2.5 dimension, signal preprocessing, the relation between resolution and the length of wells, the results curve and images of computer simulation and the image reconstructed using the experimental data.

THE PROBLEM OF 2.5 DIMENSION

The diffraction tomography theory is founded on the basis of two dimension model. However, in geophysical exploration, what we meet are problems of three-dimension. This include the source and the receiver are point, not line, and objects imaged are three-dimension, not cylinder. We still can not solve the problem of three dimension wave equations inversion completely at present stage. In laboratory, objects can be made like cylinders, sources and receivers are used in the form of points. This is the so-called 2.5 dimension problem. Apparently, this model is much more similar to that of the actual situation of seismic exploration compared with two dimension problem.

Though we only have inversion formula of two dimension, it is mentioned in paper [13,14] that scattering field data of 2.5 dimension can be used with the inversion formula of two dimension so as to reconstruct the image of objects in far field approximation. But no deep discussion and comparison have been made. Bleistein [15] derived the method to solve Born inversion formula of 2.5 dimension problem with asymptotic ray theory. But high frequency approximation does not exist in diffraction tomography, and the field is not always much far away. In experiments, the length of cylinder is limited and the probes (or transducers) may be have directivity along the axis of cylinder, the background medium attenuates the amplitude of wave, the distance between object and transducer is different. We don't know clearly the influences produced by these factors to the scattering field data and the inversion results. So it is necessary to make an estimate of the influence of these factors before experiment.

In medium without the acoustical source, the frequency of monofrequeucy acoustical wave is ω, the scattering field U_s at receive point r_g can be expressed as

$$U_s(r_g) = - k_0^2 \int_V O(r')U(r')G(r',r_g)dr'$$

(1)

where $U(r)$ is total field, G is the Green's function, $O(r) = \dfrac{C_0^2}{C^2(r)} - 1$, expresses the anomalous velocity distribution. Assume

$$U(r)=U_i(r)+U_s(r)$$

and $U(r')$ in (1) is replaced by $U_i(r')$ based on the first order approximation, we obtain

$$U_s(r_s,r_g) = - k_0^2 \int_V O(r')U(r_s,r')G(r',r_g)dr'$$

(2)

here, r_s expresses the location of transmission source. Two-dimension Green's function and two-dimension incident field are

$$G_{2D}(r',r) = \frac{i}{4} H_0^{(1)}(k_0R)$$

(3)

its far field approximation is

$$G_{2D}(r',r) = \frac{i}{4} \left(\frac{2}{\pi k_0 R}\right)^{\frac{1}{2}} e^{i(k_0R - \frac{\pi}{4})}$$

(4)

three-dimension Green's function and three-dimension incident field are

$$G_{3D}(r',r) = \frac{e^{ik_0R}}{4\pi R}$$

(5)

The far field approximation of two kinds of formula is similar in form, i. e. amplitude changes with $1/\sqrt{R}$ and $1/R$ respectively, phase difference is $\pi/4$.

Fig. 1 shows the situation of well-to-well in geophysical tomography. From Fig. 1 , we can see

2-D: $\quad R = \sqrt{(x - x')^2 + (y - y')^2}$ $\qquad\qquad$ (6)

3-D: $\quad R = \sqrt{(x - x')^2 + (y - y')^2 + z'^2}$ $\qquad\qquad$ (7)

only when z' is small, can two kinds of R be consistent, otherwise the difference is quite great, so it is impossible to estimate the difference of the features between them from the formal similarity.

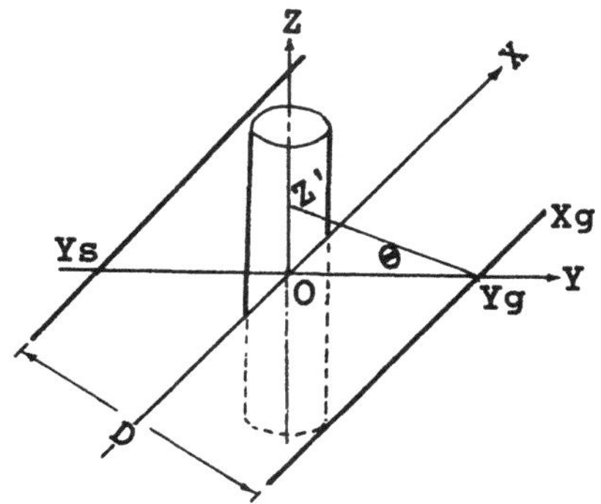

Fig. 1. well-to-well geophysical diffraction tomography in 2.5-D

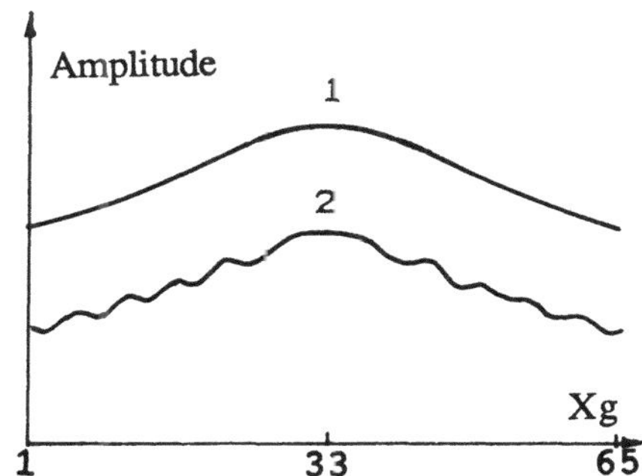

Fig. 2. The amplitude of scattering field versus the position of receiver, X_g for model 1 (point anomaly locate in center). 1) 2-D 2) 2.5-D

As Fig. 1 shown, by inserting (3) or (5) to (2), the 2-D or 3-D scattering field changed with X_g can be calculated. In 3-D formula (7), let $z'=D \tan(\theta)$ (see Fig. 1) $-90° < \theta < 90°$, so, the unlimited integral calculus can be changed into limited integral calculus along the direction of

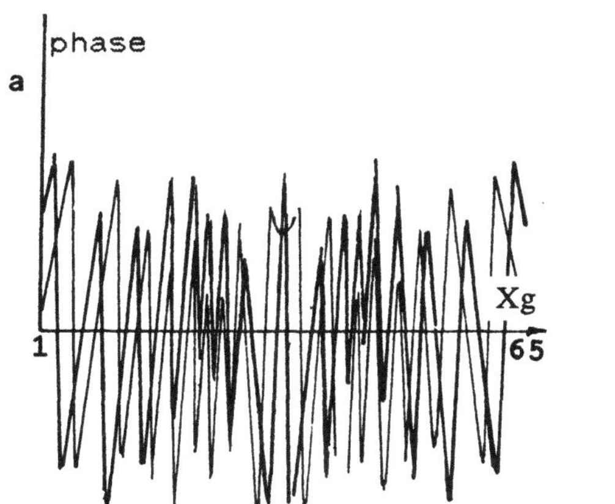

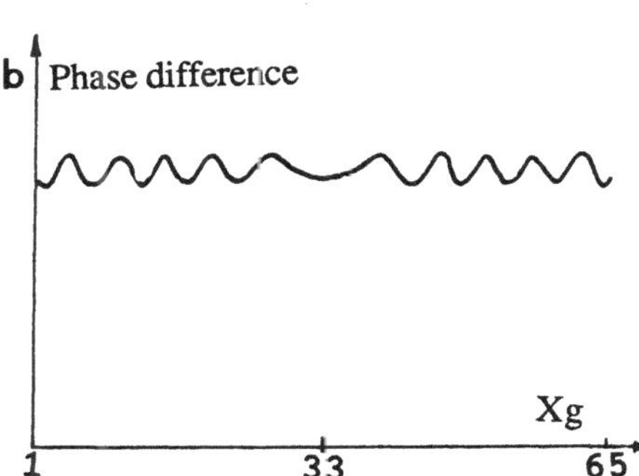

Fig.3.The phase of scattering field versus the position of receiver X_g,for model 1.
a). phase curves of 2-D and 2.5-D.
b). different phase between 2-d and 2.5-D.

θ. Actually, because of the limited length of cylinder, medium attenuation and the directivity of the probe and Us decreasing while R(z') increased, an integral calculus of only a limited range is needed. The calculus results have been researched in the range of $\theta \leq \pm 72°$.

Model 1 is a line object in the "o" point in Fig. 1. The well distance is 16 wavelength. The well depth is 32 wavelength. The sample interval is 0.5 wavelength. Fig. 2 shows the amplitude changing curve of the 2.5-D scattering field with X_g compared to 2-D situation when the integral range is $\theta = \pm 72°$. We can see that the trend of variation of amplitude distribution are similar except the numerical value becomes smaller and the range of variation has a little difference. Their phases are shown in Fig.3(a). The difference of two phases versus X_g is shown in Fig.3(b). We can see that the phase difference is approximately a constant. Its average value is $0.974 \, \pi/4$. Fig.4(a) shows the point expansion function (ambiguity function) reconstructed by 2-D inversion formula using the 2.5-D scattering field data.

We once calculated the amplitude, phase curve in different integral calculus range of z direction, such as $\theta = \pm 17.5°, \pm 26.4°, \pm 52.73°$ and $\pm 72°$, we find that the amplitude curves had a similar trend (in form). But the curvature is different. The curves of phase differences between 2-D and 2.5-D are smooth. Average value is near $\pi/4$. When the integral calculus range $|\theta| > 35°$, their point expansion function are similar with Fig. 4a. We have simulated a probe which have about $\pm 50°$ angle along the z direction. The point expansion function is also quite similar with Fig.4a. In the situations of $17.5°$ and $26.4°$, the main difference between their point expansion functions obtained and figure 4(a) is that the negative peaks are deeper, but the main peak extent doesn't have many differences.

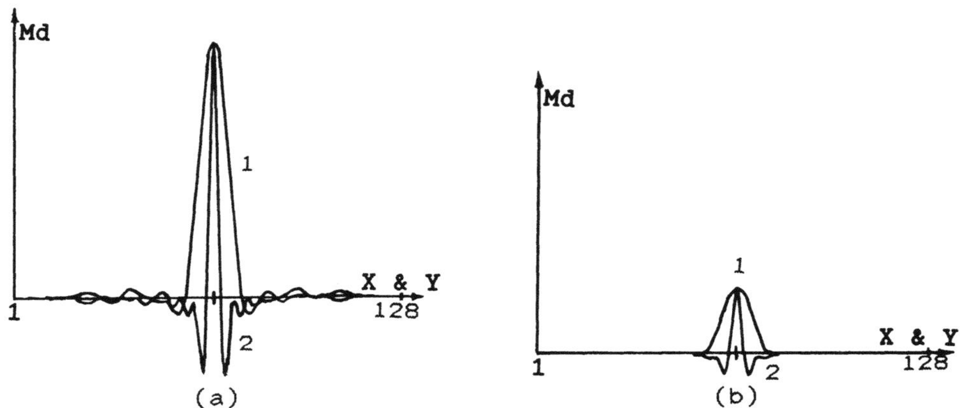

Fig.4. The point expansion function of model 1. D=16λ, H=32λ, $\theta= \pm 72°$.
1). X-direction 2). Y-direction a). $k=k_0$ b). $k=k_0(1- j0.01)$

Amplitude and phase curves are again calculated in different distances between two wells, as well as the point expansion function, when the distances are 8,12,16,20,24,28,32 wavelength. We find that when the well distances are greater than 20 wavelength, their phase difference (compared to the 2-D phase) is changed and have a great undulation. But if the ratio of the well depth H to the well distance D is kept constant, satisfactory point expansion function can still be obtained. If the ratio H to D is decreased, then the point expansion function becomes unsatisfactory as we have predicted.

Also, we have calculated the point expansion function(PEF) with the attenuation of the background medium. When $k=k_0 (1-j0.01)$, well distance is 16 wavelength, well depth is 32 wavelength, the PEF obtained is shown in Fig.4b. We can see the unsatisfactory PEF. When the well distance increased, the PEF becomes shorter and wider, indicating that no good image can be obtained.

From the calculation above we know that the scattering field data of 2.5-D, usually can be used for reconstructing diffraction tomography image by 2-D inversion formula. The influences of the integral calculus range along the z direction are not so great. Of course, in actual model, if the model only have a limited length, the scattering field will change because of the effect of cutting off. But It's out of our discussion.

THE EXTRACTION OF SCATTERING FIELD AND THE DIRECTIVITY OF PROBE

The scattering field U_s are used in the reconstruction processing but we can only measured the total field in the experiment. so the incident field must be removed. Though the incident field can be calculated theoretically, because of the directivity of probe, and the influence of the fixed equipment and surroundings, it is more suitable to use the real incident field i. e. measuring the total field when having objects firstly, then measuring the incident field without the objects using the same scan procedure.

Assume the transmitting signal is S(t), its Fourier transform is S(ω).Without objects, the incident field measured is

$$U_i(\omega, r_s, r_g) = S(\omega)D^2(\omega, \theta_1)G(\omega, r_s, r_g) \tag{8}$$

where $D(\omega, \theta)$ is the direction coefficient of the probe in X-Y plane. See Fig. 5. With objects, the total field measured at the receiver position is

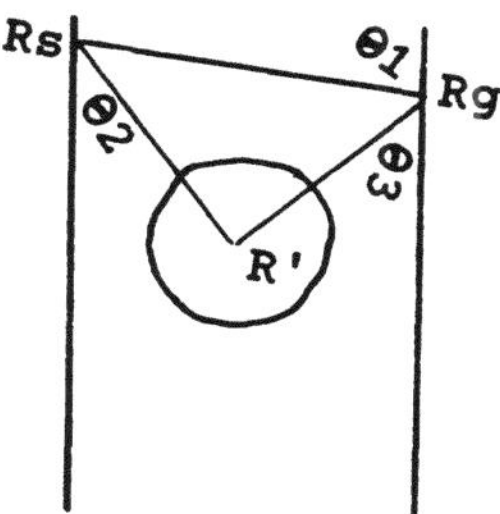

Fig.5. The directional angle for incident field and scattering field, respectively, in diffraction tomography

$$U_t(r_s, r_g) = U_i - k_0^2 \int_v S(\omega)D(\omega, \theta_2)G(r_s, r')O(r')D(\omega, \theta_3)G(r', r_g)dr' \tag{9}$$

(8) (9) divided by $S(\omega)D^2(\omega, \theta_1)$, we obtain

$$\widehat{U}_s(r_s, r_g) = \frac{U_t(r_s, r_g)}{S(\omega)D^2(\omega, \theta_1)} - G(r_s, r_g) \tag{10}$$

From (9) (10), we obtain the equation

$$\widehat{U}_s(r_s, r_g) = - k_0^2 \int_v G(r_s, r')O(r')G(r', r_g)\frac{D(\omega, \theta_2)D(\omega, \theta_3)}{D^2(\omega, \theta_1)}dr' \tag{11}$$

If the probe has weak directivity, then the direction coefficient can be neglected. Then we can reconstruct the object successfully. Otherwise, it is difficult to inverse the object.

According above analysis and because in geophysical application, the source is really no directivity, So in our experiment, we don't use the ordinary transducer of circle plane. There are two kinds of probe which are acceptable:

1). Sphere probe, no directivity at X,Y,Z, it is analogous to the actual seismic source and receiver which are no directivity. That is the typical 2.5-D problem.

2). Cylinder probe, no directivity at X,Y, but Z direction, it has a range of small angle (for example ±7.5°in z direction), so it is similar with cylindrical wave. That is the 2-D problem.

THE WTW DISTANCE RESOLUTION AND THE COMPUTER STIMULATION STUDY

It is necessary to do some computer stimulations before doing the experiment. In the computer stimulations (2) is the forward formula and the inversion formula is

$$\tilde{U}_s(\alpha_x,k_x) = \iint U_s(r_g,r_s)e^{i(k_x x_s - \alpha_s x_g)}dx_g dx_s \tag{12}$$

$$\tilde{U}_s(\alpha_x,k_x) = \frac{k_0^2}{4\gamma_s\gamma_g}e^{i(\gamma_g y_g - \gamma_s y_s)}\tilde{O}(\alpha_x - k_x, \gamma_g - \gamma_s) \tag{13}$$

Where α_x and k_x are the wave numbers along the receiver line and the source line, $\gamma_g = \sqrt{k_0^2 - \alpha_x^2}$, $\gamma_s = \sqrt{k_0^2 - k_x^2}$. From (12), we obtain $\tilde{U}_s$, then insert to (13). For each α_x, k_x, we can calculate γ_g, γ_s, then we obtain $\tilde{O}(\alpha_x - k_x, \gamma_g - \gamma_s)$, assume

$$u = \alpha_x - k_x$$
$$v = \gamma_g - \gamma_s \tag{14}$$

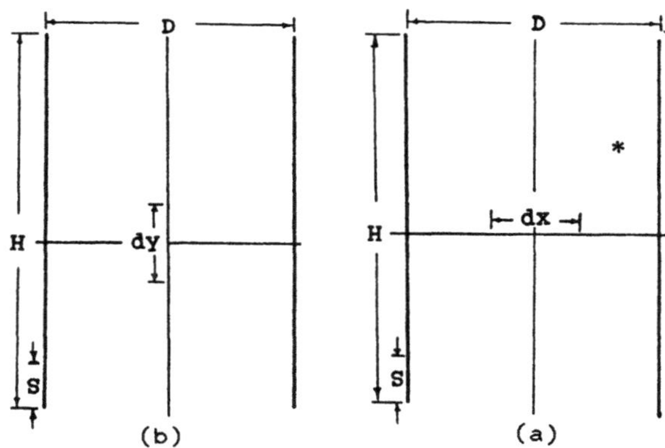

Fig. 6.The experimental geometry for calculating the distance resolution.
a). X-direction   b). Y-direction

Due to the fact that in an experiment measurement, $\tilde{O}(u,v)$ of these sampled data will not uniformly spaced, but only the rectangular grids suitable for image reconstruction, we use interpolation to obtain $\tilde{O}(u,v)$ at each grid.Then take the inverse two-dimension FFT, and obtain $O(x,y)$.

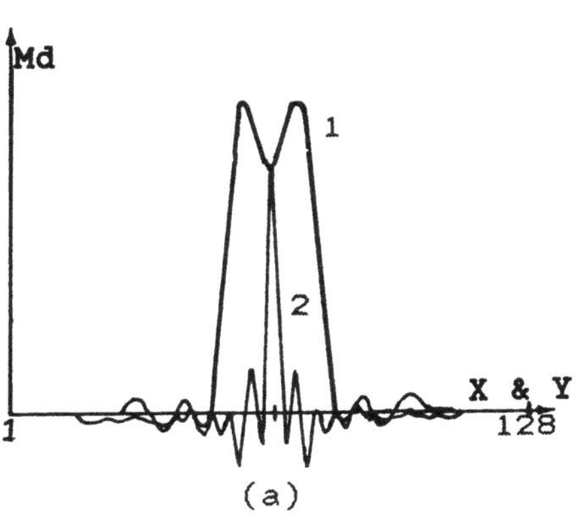
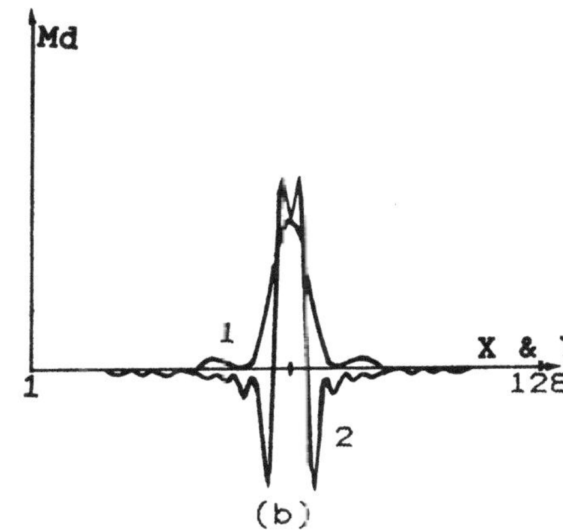

Fig.7. The distribution of image function from two point scatters
a).arrange two points in X-direction b).arrange two points in Y-direction
1).distribution in X-direction   1).distribution in X-direction
2).distribution in Y-direction   2).distribution in Y-direction

In practical WTW tomography, the resolution must be estimated, then the reconstructed image can be evaluated and explained correctly. We have done computer stimulation to study the image reconstructed by using scattering field from two point scatters in different well length.

Fig. 6 shows the experiment geometry, for $D=16\lambda$, $H=32\lambda$, $s=0.5\lambda$, we use 2.5-D scattering field data to do the reconstruction. In situation (a), we obtain the distribution of image function of Fig. 7a in different direction. In situation (b), we obtain 7b.

If the center concave of distribution of the image function is 0.707 Mdmax, we define the distances of two scattering point and as the perpendicular distance resolution Rsy and parallel distance resolution Rsx, respectively.

In WTW tomography, because of lacking of the projections of perpendicular direction, Rsx is greater than Rsy obviously.

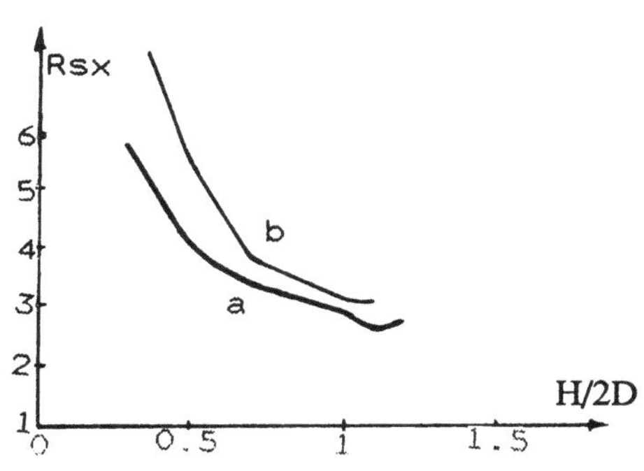
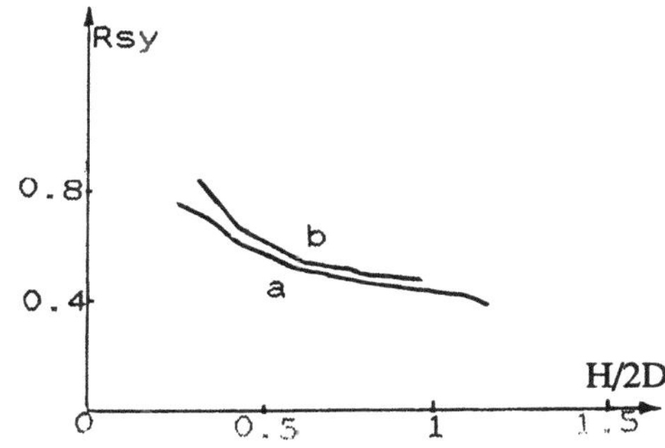

Fig. 8 The distance resolution in
Y-direction R_{sy}. a). 2-D b). 2.5D

Fig. 9 The distance resolution in
X-direction R_{sx}. a). 2-D b). 2.5-D

Fig. 8 shows the changing of Rsy versus the depth of the well. The down curve is the results reconstructed using two dimensional scattering field, and the up curve is the results of using 2.5-D scattering field $(-72° \sim +72°)$. It indicates that the R_{sy} in 2.5-D is slightly greater than that in 2-D, but not far off. Fig. 9 shows the distance resolution of X-direction. It is 3-4 times than that of Y-direction.

When the depth of the well is less than the distance between the wells.The resolution ecomes worse and the image will be unsatisfactory. Fig. 10 shows the curve of distribution of ᵢmage function when the depth of well is 16 wavelength. From above results, we conclude that the image is unsatisfactory when the ratio of the depth H to the distance of wells D is less than 1, the resolution and the image will be good when the ratio of H/D is about 2-2.2.

In order to show the changing of resolution with the position of the scattering object, we move the two scattering point to the upright, its center situated in the * in Fig. 6b. In this situation we find that although the resolution is not so good, the two scattering point still can be distinguished. Here, the diffraction tomography over ray tomography is shows its advantage.

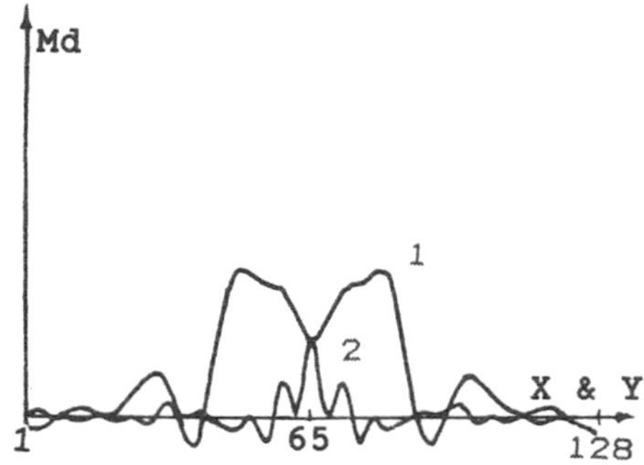

Fig. 10. The distribution of the image function. H=14λ, D=16λ
1) X-direction   2) Y-direction

Fig. 11 is the reconstructed image when the cylinder located in the center, the distance between two wells is 16λ, the depth of each well is 32λ Fig. 12 is the image which the cylinder located in the * in Fig. 6b. The diameter of model is 3.375 cm. The frequency is 100 KHz. Because the width of the point expansion function in X-direction is large, the image in such direction becomes wider, so the image becomes an ellipse.

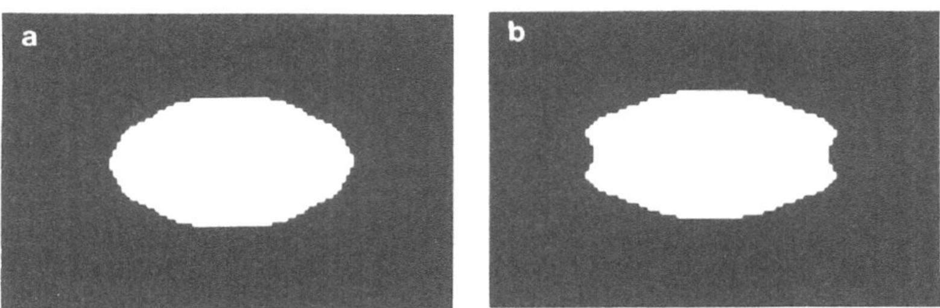

Fig. 11. The reconstructed image of a cylinder located in the center, computer simulation
H = 32λ , D = 16λ , f = 100 KHz, diameter of model is 3.375 cm. a) 2-D   b) 2.5-D

EXPERIMANTAL SYSTEM AND RESULTS

Fig. 13 shows the block diagram of the measurement system. The volume of tank is 83x 130x60 cm^3 . The surroundings and the bottom of tank are covered with the noise elimination rubber in order to get rid of the influence of multireflection of wave. The motion of the transducers is controlled by the step motor, the minimum step is 0.00625 mm. The transmitting

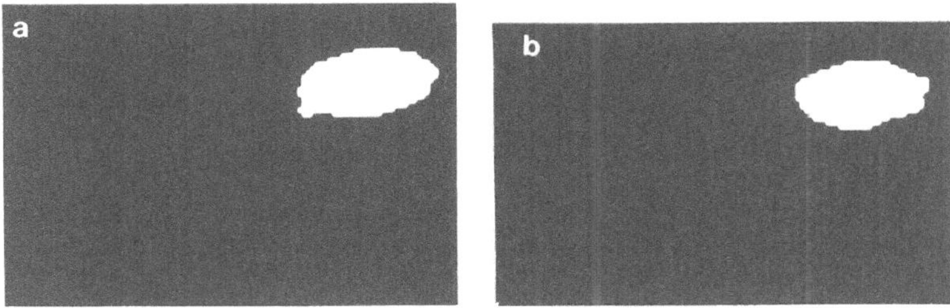

Fig. 12 The reconstructed image of a cylinder located in the upright corner, computer simulation
H = 32λ, D = 16λ , f = 100 KHz, diameter of model is 3.375 cm   a) 2-D   b) 2.5-D

signal is generated by MFS-2A pulse generator, and is amplified by ZHL-32A amplifier, then input to the transmitting probe. After pre-amplified and filtered, the receiving waveform is sampled by HP-1980B Oscilloscope measurement System and stored in the disk of IBM-PC. A window is used to extract the useful signal and then transformed by Fourier transformation. According to formula(13), we reconstruct the object using the scattering field data. The inversion was performed in the IBM PS-2-80.

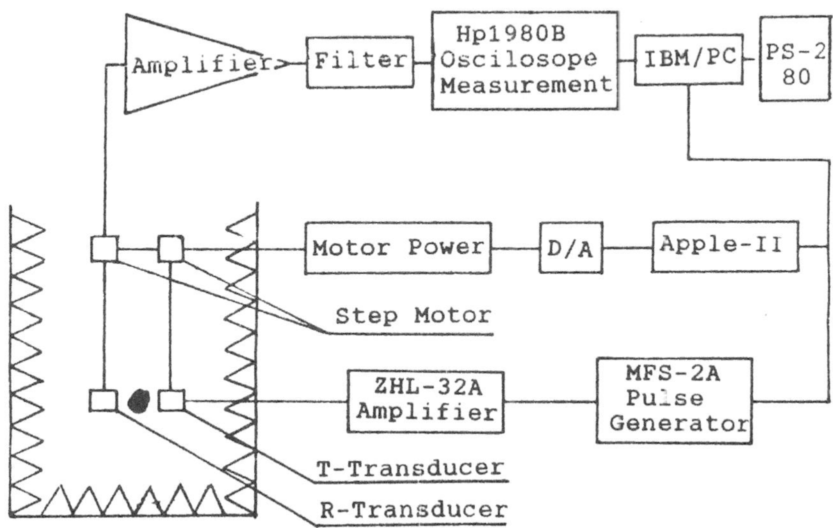

Fig. 13. The block diagram of experimental system

The object is a cylinder which is made of agar, the diameter is 3.5 cm. The frequency is 100 KHz. The distance between two wells is 13 cm. The length of each well is 18 cm. Fig.14. shows the image reconstructed.

Fig. 14. The reconstructed image of a cylinder located in the center with the experimental data of 2.5-D, H = 18 cm , D = 13 cm , f = 100 KHz , the dia meter of model is 3.5 cm.

ACKNOWLEDGMENTS

We are grateful to Prof. Wei Yu for her supporting this research. Thanks to R. S. Wu for his useful discussion. And also thanks to Qiao, Li, Lu,et al. for their help.

REFERENCE

1. Denies, K. A. and Lytle, R. J. , Computerized geophysical tomography, Proc. IEEE, 67:1065-1073(1979).
2. R. K. Mueller, M. Kaveh and R. D. Iverson, A new approach to acoustic tomography using diffraction techniques, Acoustical Imaging, 8:615-628(1978).
3. R. K. Mueller, M. Kaveh and G. Wade, Reconstructive tomography and Application to Ultrasonics, Proc. IEEE, 67:567-587(1979).
4. Devaney A. J. , A filtered back-propagation algorithm of diffraction tomography, Ultrasonic Imaging. 4:336-350(1982).
5. S. X. Pan and A. C. Kak, A computational study of reconstruction algorithm for diffraction tomography: Interpolation versus filtered backpropagation, IEEE Trans. on ASSP, ASSP-31:1262-1275(1983).
6. S. K. Kenue and J. F. Greenleaf, Limited Angle multi-frequency Diffraction tomography, IEEE Trans. on SU, SU-29:215-217(1982).
7. D. Nahamoo, S. X. Pan and A. C. Kak, Synthetic aperture diffraction tomography and its interpolation-free computer implementation, IEEE Trans.on SU, SU-31:218-229(1984).
8. H. L. Xue and Y. Wei, A fast reconstruction Algorithm for diffraction tomography, 16th International Symposium on Acoustic Imaging, June:10-12 (1987).
9. Jian Yu Lu, A computational study for synthetic aperture diffraction tomography-Interpolation Versus Interpolation free, 16th International Symposium on Acoustical Imaging, June:10 12(1987).
10. Slaney M. , Kak A. C. and Larsen L. , Limitation with first-order diffraction tomography, IEEE Trans. on MIT, MIT-32:860-874(1984).
11. Devaney A. J. , Geophysical Diffraction Tomography, IEEE Trans, GE-22:3-13(1984).
12. Wu R. S. and Toksoz M. N. , Diffraction Tomography and multisource holography applied to seismic imaging, Geophysics, 52:11-25(1987).
13. Tien-When Lo et al., Ultrasonic laboratory tests of geophysical tomography reconstruction, Geophysics, 7:947-956(1988).
14. R. Gerhard Pratt et al., The application of diffraction tomography to cross-hole seismic data, Geophysics, 10:1284-1294(1988).
15. Norman Bleistein et al., Two and one-half dimensional Born inversion with an arbitrary reference, Geophysics, 1:26-36(1987).

IMAGING OF THE NONLINEAR ACOUSTIC PARAMETER B/A

Charles A. Cain

Dept. of Electrical Engineering and Computer Science
University of Michigan
Ann Arbor, MI 48109

ABSTRACT

The nonlinear acoustic interaction between a single frequency sinusoid and a broadband pump waveform propagating in the opposite direction produces phase changes in the probe proportional to the nonlinear parameter B/A in the spatial region of interaction. The instantaneous phase change along the received probe can be expressed as the convolution of the pump waveform with the spatial distribution of B/A along the propagation path over which the pump and reflected probe interact. In theory, the phase modulated sinusoidal probe can be processed (phase detection and deconvolution) to produce an "A-mode" representation of B/A. If the pump is an intense impulse and the probe a swept-frequency sinusoid, then the pump interacts with the probe at each spatial position along the propagation path at a unique frequency. Thus, the phase modulation, proportional to $\frac{B}{A}(x)$ along the propagation path, can be extracted directly from the phase spectrum obtained from the Fourier transform of the swept-frequency probe. This is somewhat analogous to the mapping of space to frequency which forms the basis for magnetic resonance imaging.

If the impulsive pump is replaced by another swept-frequency sinusoid, then the phase change in the probe due to B/A at a particular point along the propagation path will be spread out for the duration of the pump along the probe. Passage of the received signal through an appropriate matched filter restores spatial coherence to the phase information in the probe so that it can be processed as if the pump were a broadbanded impulse. This approach suggests a means of approaching the design of effective pump waveforms which can resolve a wide range of spatial frequencies in $\frac{B}{A}(x)$.

I. Introduction

The nonlinear parameter B/A arises from the nonlinearity in the equation of state of a medium in terms of the pressure and density. In a nonlinear medium the sound speed is a function of the pressure amplitude. For an isentropic system, sound speed c is related to pressure by

$$\frac{\partial c}{\partial p} = \frac{1}{2\rho_0 c_0} \frac{B}{A} \tag{1}$$

where p is pressure and ρ_0 and c_0 are values of density and sound speed about some static pressure p_0 [1]. An expression of the instantaneous phase shift in a reflected sinusoidal probe due to nonlinear interaction with an arbitrary pump waveform propagating in the opposite direction has been derived from Eq. (1) [2]. This instantaneous phase shift $\Delta\phi(\tau)$ is proportional to the convolution of the pump waveform $p(x,t)$ with the spatial distribution of $\frac{B}{A}(x)$ of the nonlinear parameter B/A along the propagation path of the pump and interacting probe. Thus

$$\Delta\phi(\tau) = -\pi f \int_{0}^{\frac{c_0\tau}{2}} (\rho_0 c_0^3)^{-1} \, p(\frac{c_0\tau}{2} - x) \, \frac{B}{A}(x) dx \qquad (2)$$

where f is the probe frequency. The sensitivity is directly proportional to the pump pulse width, the probe frequency, and the peak pressure of the pump pulse. Because of the convolution operation and the finite width of the pump pulse, $\Delta\phi(\tau)$ is a filtered (convolved) version of $\frac{B}{A}(x)$. Further signal processing (deconvolution) is necessary to obtain an undistorted version of the B/A profile. In frequency domain terms, Eq. (2) implies that the pump waveform must be sufficiently broadbanded to supply spectral components with wavelengths to match the spatial frequencies of $\frac{B}{A}(x)$. Considerations which influence the choice of both pump and probe waveforms, and the processing of the received reflections for information leading to an estimate for $\frac{B}{A}(x)$, will be discussed in the conference presentation.

That the mechanical property of biological materials represented by B/A may be useful in tissue characterization has been verified by a number of investigations [3,4,5]. Measurements have shown significant differences in B/A among different normal tissues and between normal and malignant human tissues [5]. These results may have considerable significance if practical techniques can be developed for imaging B/A in human body.

Ichida et al. [6] produced the first images of biological materials using a transmission mode technique based on a low frequency pump wave propagating perpendicularly to a higher frequency probe signal. This technique was modified with the use of impulsive pump waveforms allowing for the generation of real time B/A images [7]. Nakagawa et al. [8] proposed a quite different method based on the nonlinear production of sum and difference frequencies after the parametric array theory of Westervelt [9].

II. Results

In what follows, the B/A imaging scheme outlined in [2] will be analyzed by computer simulations. The simulation algorithms were discussed in [10]. Briefly, the simulation takes into account both finite amplitude distortion and absorption of acoustic energy. The algorithm interacts an infinitesimal amplitude probe with a collinear counter-propagating pump pulse. The situation is shown in Fig. 1. The discretized waveforms propagate simultaneously, where the local velocity of each wavelet is determined by the local values of pressure, media density, infinitesimal sound speed, and acoustic nonlinearity parameter B/A. Since absorption is frequency dependent, the attenuation function is periodically applied to the fast Fourier transform of the pump pulse, and the attenuated waveform is reconstructed via inverse fast Fourier transform. The modulated phase of a single frequency sinusoid is extracted using a simulated narrow-band FM demodulator. The smearing effect of the finite width pump waveform is removed through the deconvolution of the recovered phase with the initial spatial profile of the pump waveform [10]. This simulation algorithm, an adapatation from [11], will be described in depth in a future communication.

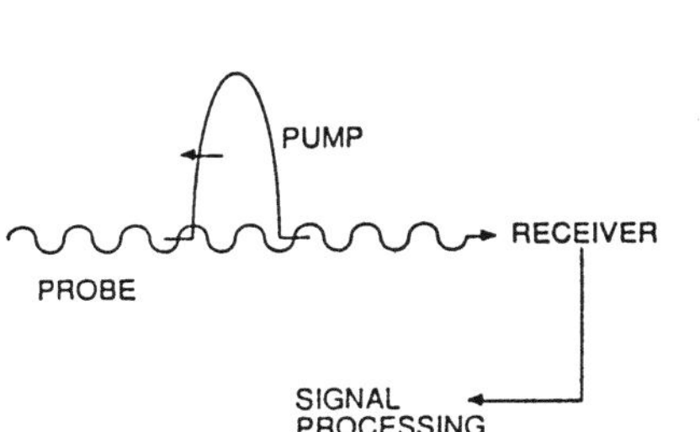

Figure 1.   Diagram of the imaging arrangement.

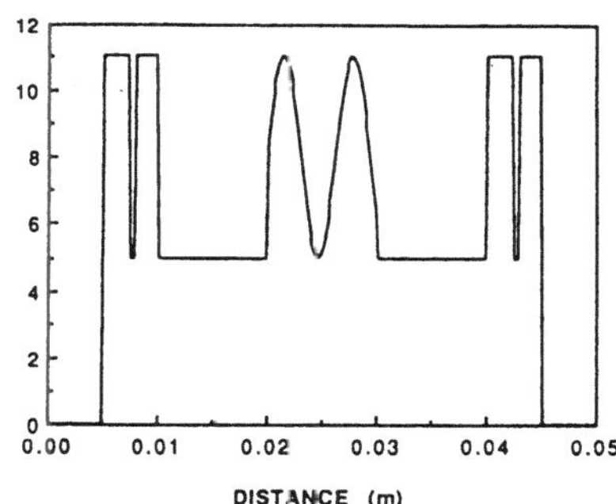

Figure 2.   Profile of the acoustic nonlinearity parameter.

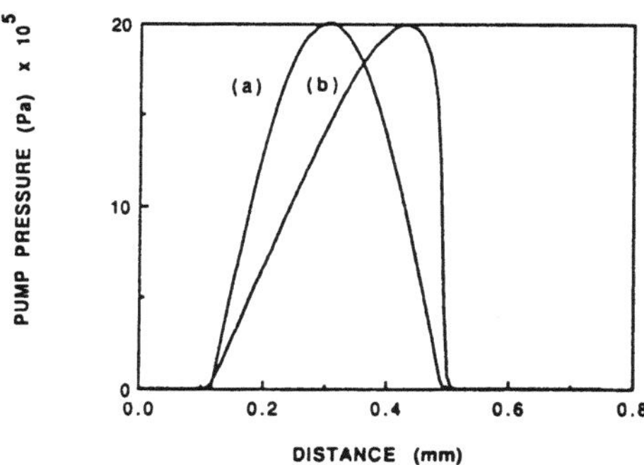

Figure 3.   a) Spatial profile of an impulsive pump pulse before entering the medium. b) Spatial profile of the impulsive pump pulse at maximum depth.

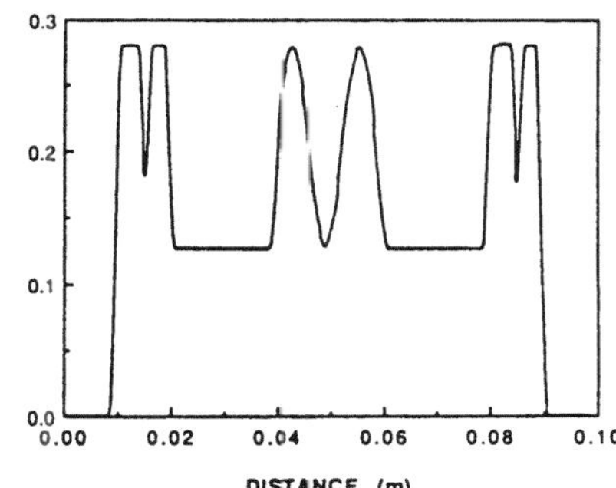

Figure 4.   The instantaneous phase of a 2MHz sinusoidal probe signal interacted with an impulsive pump pulse.

A profile of the B/A acoustic nonlinearity parameter useful for imaging simulations is shown in Fig. 2 where the transmitter/receiver plane is located at 0.5cm and the reflector plane is at 4.5cm. The total width of the simulated medium is four centimeters. This B/A profile contains discontinuities in order to allow the study of the resolving capabilities of this simulated imaging system. The main features of this B/A profile are repeated at both ends so that the effects of absorption and nonlinear distortion of the pump pulse on the recovered images can be studied.

The simulation was first carried out in a low attenuation medium (water). An impulsive pump is approximated with the positive half cycle of a 2MHz sinusoidal. The peak pressure of this signal is 20 atmospheres. This pump pulse before entering the media and after propagating through it are depicted in Fig. 3. The major change in this pulse is its progress towards a sawtooth waveform shape. Since the attenuation in water is negligible the peak pressure has changed only slightly.

The probe signal is a single frequency (SF) sinusoidal signal with a frequency of 2MHz. This waveform is propagated into the medium, reflected, and when it reaches the launch point the pump pulse is propagated into medium. The effective spatial length of the probe is then twice the length of the medium. The probe waveform is received and simultaneously a narrow-band FM demodulator detects the instantaneous phase of the probe. This phase function carries the information on the convolution of the pump pulse with the B/A profile. This recovered phase is depicted in Fig. 4. The maximum phase deviation is 0.3 degrees. The low sensitivity of the detected phase was discussed on in a numerical example along with possible approaches to alleviate the problem [2]. This phase profile matches the underlying B/A profile very closely as predicted by Eq. (3); this instantaneous phase is the convolution of the impulsive pump with the B/A profile which reproduces the profile of the acoustic nonlinearity parameter. The deconvolved image of the B/A parameter is shown in Fig. 5. As expected, the broadband nature of this pump waveform has reproduced an estimate very similar to the underlying acoustic nonlinearity parameter profile.

To see the effects of attenuation, a medium much more attenuative than tissue was assumed. The impulsive pump is chosen as the positive half cycle of a sinusoid with spatial extent of 0.2cm and peak pressure of 10 atmospheres. The attenuation function for this medium was assumed to be 8dB/MHz/cm. The pump pulse is depicted in Fig. 6a and the pulse after propagation through this medium is displayed in Fig. 6b. The peak pressure has been attenuated from 10 atmospheres to 3 atmospheres.

The probe signal is a single frequency sinusoidal with a temporal frequency of 2MHz. It is propagated into the medium and reflected. When it reaches the transmitter the pump pulse is introduced into the medium. The probe data after nonlinear interaction with the pump pulse is processed by a narrow-band FM demodulator. The extracted phase is displayed in Fig. 7. The maximum phase deviation is 0.7 degrees. The sensitivity decreases for recovered phase representing the deeper regions of the medium under study. This phenomenon could be explained as follows: as the pump wave propagates its amplitude is reduced because of energy loss due to absorption of acoustic energy. Since the phase sensitivity is directly proportional to this amplitude, it is also reduced. The result of the deconvolution of this phase information with the initial spatial profile of the pump pulse is provided in Fig. 8. The general features of this reconstruction match the underlying B/A profile, but there is a general decrease in the magnitude of this function. This could be compensated for by a variable delay gain which adjusts the amplitude compensation factor depending on the time of arrival. This function depends directly on the rate of energy loss of the pump pulse and hence is dependent on the finite-amplitude distortion and the attenuation properties of the medium.

III.   <u>Summary</u>

Only sample simulations are given. More will be presented and discussed in the conference presentation. From these simulations one can conclude that high resolution imaging of B/A is indeed theoretically possible and that the convolution relation of Eq. (2) is basically correct. However, the simulations also show that sensitivity is quite low

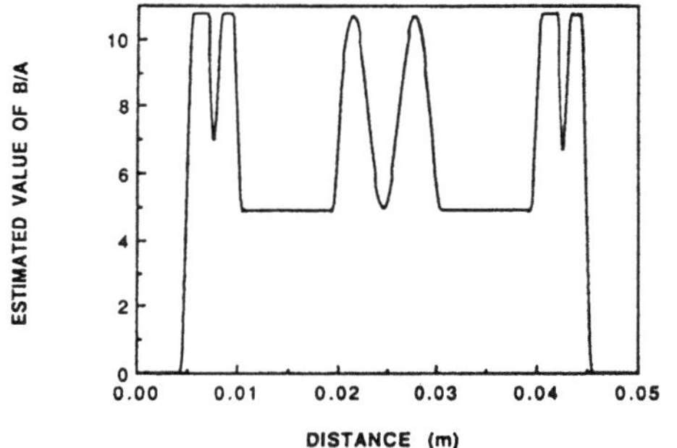

Figure 5.   The deconvolved estimate of the nonlinearity parameter.

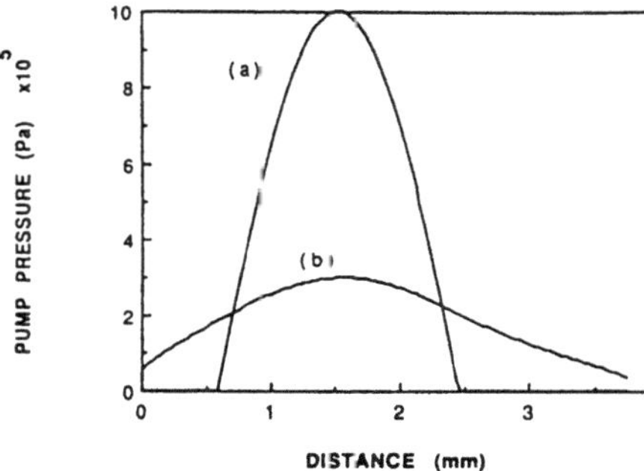

Figure 6.   a) The spatial profile of an impulsive pump entering a medium with an attenuation of 8dB/MHz/cm. b) The impulsive pump after propagating 5cm. The medium is about 10 times more attenuative than tissue.

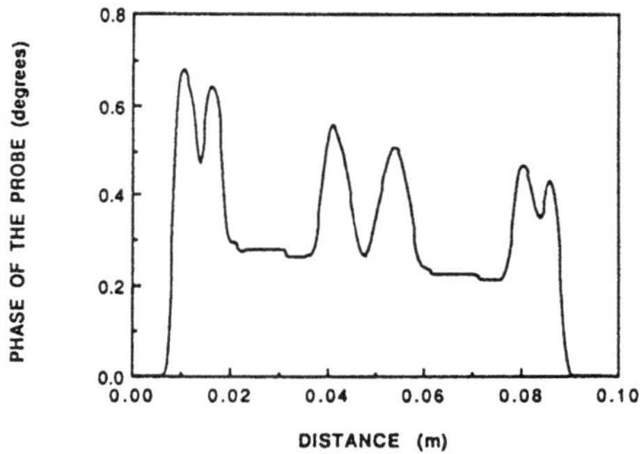

Figure 7.   The detected phase of the 2MHz coherent probe signal interacted with the impulsive pump in a tissue like medium.

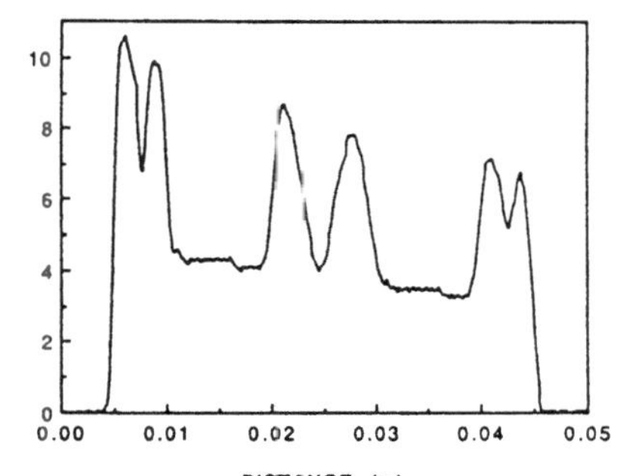

Figure 8.   The deconvolved estimate of the acoustic nonlinearity parameter in a tissue like medium.

which may require either high peak pressure pump pulses or broad-banded "chirp" pulses spread out in time. Problems of diffraction, reflection, and a reverberant environment must also be addressed in a practical imaging system. These were not taken into account in the simulations.

ACKNOWLEDGEMENTS

This work was funded by grant CA 44938 from the National Institutes of Health, grant ECS 870001 from the National Center for Supercomputing Applications (NCSA) at the University of Illinois, and an award from Hitachi Central Research Laboratory, Tokyo, Japan.

REFERENCES

1. R. T. Beyer and S. V. Letcher, "Nonlinear Acoustics," in Physical Ultrasonics, edited by H. S. W. Massey and K. A. Bruecner (Academic, New York, 1979), Vol. 7, pp. 202-230.
2. C. A. Cain, "Ultrasonic Reflection Mode Imaging of the Nonlinear Parameter B/A: I. A. Theoretical Basis," J. Acoust. Soc. Am., Vol. 80, pp. 28-32, 1986.
3. W. K. Law, L. A. Frizzell, and F. Dunn, "Ultrasonic Determination of the Nonlinearity Parameter B/A for Biological Media," J. Acoust. Soc. Am., Vol. 69, pp. 1210-1212, 1980.
4. F. Dunn, W. K. Law, and L. A. Frizzell, "Nonlinear Ultrasonic Wave Propagation in Biological Materials," IEEE Ultrasonics Symposium (IEEE, New York, 1981), pp. 527-532.
5. C. M. Sehgal, R. C. Bahn, and J. F. Greenleaf, "Measurement of the Acoustic Nonlinearity Parameter B/A in Human Tissues by a Thermodynamic Method," J. Acoust. Soc. Am., Vol. 76, pp. 1023-1029, 1984.
6. N. Ichida, T. Sato, and M. Linzer, "Imaging the Nonlinear Ultrasonic Parameter of a Medium," Ultrasonic Imaging, Vol. 5, pp. 295-299, 1983.
7. N. Ichida, T. Sato, H. Miwa, and K. Murakami, "Real-Time Nonlinear Parameter Tomography Using Impulsive Pumping Waves," IEEE Trans. Sonics and Ultrasonics, Vol. SU-31, pp. 635-641, 1984.
8. Y. Nakagawa, M. Nakagawa, M. Yoneyma, and M. Kikuchi, "Nonlinear Parameter Imaging Computer Tomography by Parameter Array," IEEE Ultrasonics Symposium, Dallas, 1984, in press.
9. P. J. Westervelt, "Parametric Acoustic Array," J. Acoust. Soc. Am., Vol. 35, pp. 534-537, 1963.
10. H. Houshmand, R. J. McGough, E. Ebbini, H. Lee, and C. A. Cain, "Ultrasonic Transmission Mode Imaging of the Nonlinear Parameter B/A: A Simulation Study," IEEE Ultrasonics Symposium, Chicago, 1988.
11. L. B. Orenstein and D. T. Blackstock, "The Rise Time of N Waves Produced by Sparks," ARL-TR-82 51, (ADA 120817),1982.

LEAST SQUARES ESTIMATION AND IMAGING USING PHASE AND FREQUENCY MEASUREMENTS OF THE NONLINEAR PARAMETER B/A

Daehoon Kim, James A. Nicholson, James F. Greenleaf, and Chandra M. Sehgal

Biodynamics Research Unit, Department of Physiology and Biophysics, Mayo Clinic/Foundation, Rochester, MN 55905

INTRODUCTION

There is growing interest in acoustic nonlinearity since Muir and Carstensen[1] and Carstensen et al.[2] questioned the assumption of linear propagation of ultrasound in biological media. Nonlinearity of the pressure-density relationship of a medium is measured quantitatively in terms of the B/A parameter, which is the ratio of the coefficients of quadratic and linear terms of a Taylor series used to express an equation of the state of a medium in terms of pressure and density. Because the parameter B/A can be used to characterize tissues, interest in measuring B/A has increased significantly.[3,4,5,6,7]

There are two procedures for determining B/A: the finite-amplitude procedure and the thermodynamic procedure. Advantages and disadvantages of these two procedures are reviewed in Ref. 4.

In this paper we describe a finite-amplitude measurement method for measuring B/A and an improved imaging method for the nonlinear parameter by estimating shift in phase and frequency (SIPAF) of a high frequency acoustic probe wave by a low frequency acoustic pump wave. The proposed measurement method involves transmission of frequencies for f_1 and $2f_1$, and measurements of the pressure amplitudes of f_1 and $2f_1$ as a function of distance in the far field. These measurements provide the values of attenuation coefficients enabling calculation of B/A without using extrapolation and approximation procedures required for traditional finite amplitude methods.

The proposed measurement method has several attractive features. It does not require measurements of acoustic pressure amplitude of the fundamental sinusoidal wave at the source. It also does not involve the assumption that $(\alpha_1 - \alpha_2/2)x$ is small, where α_1 and α_2 are attenuation coefficients of fundamental and second harmonics, and x is the source to measurement distance. Additionally, the proposed method provides a means of measuring the frequency dependence of the attenuation coefficient.

Until recently, very few efforts have been made to use nonlinear parameter in medical imaging.[6,8,9,10] The early B/A images of biological tissue were obtained by Ichida et al.[8,9] Ichida's method is based on measuring phase shift of a high frequency acoustic probe wave due to a low frequency acoustic pump wave. This change in phase, measured by a phase demodulator, provides the real part of the Fourier transform of the distribution of the nonlinear parameter B/A over the wave number of the pump wave.

Like the earlier imaging method, a low frequency pump wave is used to sinusoidally modulate a high frequency probe wave over the region of interest. As a result of this, the phase of the probe wave is modified in proportion to the integral of the product of the nonlinear parameter B/A and the sinusoidally varying pressure of the pump wave.[8,9] This information on the measured probe wave can be considered as a Phase Modulation (PM) signal. This phase change, measured by a phase demodulator, provides the real part of the Fourier transform of the distribution of the nonlinear parameter B/A at the wave number of the pump wave. Ichida's method uses only the real part of the Fourier transform, providing an oscillating reconstruction of the actual magnitude B/A. If we have a PM signal, there is also a Frequency Modulation (FM) signal corresponding to a different modulation wave shape. Thus, the measured probe wave can also be considered as a FM signal. This frequency change can be measured using a frequency demodulator and used to determine the imaginary part of the Fourier transform. The real and the imaginary components of the Fourier transform make it possible to obtain more accurate reconstructions of B/A.

Imaging experiments carried out through computer simulations for one- and two-dimensional distributions of nonlinear parameter show that the method proposed in this paper has better performance than the earlier method. The earlier method provides an oscillating response to the actual magnitude of B/A even though the employed pump wave frequencies are $f_k \neq k/NT$, $k = 1,2...N$. This can lead to an error of the order of 7% in B/A values. Such an oscillating response can be eliminated by using the proposed method with a derived set of wave numbers of the pump wave.

THE LEAST SQUARES ESTIMATION OF THE NONLINEAR PARAMETER B/A

Method

Thuras et al.[11] developed a procedure for calculating fundamental and second harmonic pressure amplitudes as a function of source distance. This procedure is based on the assumptions that the attenuation of the fundamental and the second harmonic amplitude are mutually independent and the rate of change with propagation distance of the second harmonic amplitude is the sum of the changes caused by nonlinear generation of the second harmonic and by its attenuation. Based upon these assumptions, the fundamental pressure amplitude $P_1(x)$ as a function of source distance x, is expressed as[4,12]

$$P_1(x) = P_1(0)\, e^{-\alpha_1 x}, \tag{1}$$

where $P_1(0)$ is the peak acoustic pressure amplitude at the source of the fundamental sinusoidal wave.

The second harmonic pressure amplitude $P_2(x)$ as a function of source distance x, is given by[4]

$$P_2(x) = K\, P_1^{\,2}(0)\, \frac{e^{-\frac{\alpha_2}{2}x} - e^{-2\alpha_1 x}}{2\alpha_1 - \alpha_2}, \tag{2}$$

$$\text{where} \quad K = \frac{(2 + \frac{B}{A})\, f_1}{2\,\rho_o c_o^{\,3}}.$$

In the constant K, ρ_o and c_o are the density and sound speed under atmospheric pressure, respectively, and f_1 is the frequency of the fundamental sinusoidal wave.

One of the popular methods of estimating B/A uses the simplified version of $P_2(x)$ by assuming that $(\alpha_1 - \alpha_2/2)x$ is small. This assumption simplifies Eq. (2) to the form[4,12]

$$P_2(x) = K\, x\, P_1^{\,2}(0)\, e^{(\alpha_1 + \frac{\alpha_2}{2})x}. \tag{3}$$

Using Eq. (3), one can determine B/A by measuring the parameter

$$\frac{P_2(x)}{x\, P_1^{\,2}(0)} \quad \text{as a function of x and extrapolating back to } x = 0$$

However, extrapolation of the second harmonic measurements to zero propagation distance using Eq. (3) cannot be used in B/A determination when the quantity $(\alpha_1 - \alpha_2/2)x$ is not small.[13] Moreover, $P_2(x)$ should be measured outside the near field of the transducer, i.e., the region of a smooth-intensity decay in order to minimize the errors in extrapolation. The method proposed in this paper circumvents these problems by directly using Eq. (2) for the determination of B/A. Let us consider a homogeneous medium irradiated by frequency f_1. Pressure amplitude of the fundamental frequency, $P_1(x)$ and its second harmonic $P_2(x)$ are measured as a function of distance along the transducer axis. Following these measurements, the same medium is irradiated at frequency $2f_1$ and the corresponding pressure amplitude $q_1(x)$ of the fundamental frequency, $2f_1$, is measured as a function of distance x. Since the attenuation coefficient depends on frequency, $q_1(x)$ can be represented as

$$q_1(x) = q_1(0)\, e^{-\frac{\alpha_2}{2}x}. \tag{4}$$

The measured values of pressure amplitude generate the following set of data for pressure amplitudes

$P_1(x_i)$ for i = 1,2,...N, and

$P_2(x_i)$ for $i = 1,2,...N$, and

$q_1(x_i)$ for $i = 1,2,...N$, and

where x_i represents the distance from the transducer for the i-th measurement.

The source distance x_i should satisfy $x_i \geq N_f$, where N_f is the near field distance of a continuous wave-type transducer. The near field distance is defined as that point on the axis of the transducer separating a region of intense oscillation from a region of a smooth-intensity decay.[11]

By using a least square error fit,[14] each set of pressure value vs. x_i data was fitted to exponential equations of Eqs. (1) and (4) to yield the fitting parameters α_1, α_2, and $P_1(0)$. Substitution of these parameters along with the measured value of $P_2(x_i)$ in Eq. (2) yield the value of B/A.

<u>Experimental Methods and Results</u>

Figure 1 shows the experimental setup consisting of tank and electronic equipment used for determination of B/A. The fundamental frequency for pressure amplitude $P_1(x)$ was selected as 2 MHz using V-306 panametric transducer with 1.2 cm diameter. The fundamental frequency for pressure amplitude $q_1(x)$ was 4 MHz and was generated by using V-310 panametric transducer with 0.6 cm diameter. We used a homemade needle-type receiver with sensitivity 5 KPa/mV to measure the transmitted signal.

The horizontal scanner was used to move the transducer in steps of 1 mm. At each step, the signal was amplified, digitized at 100 MHz, and spectrum analyzed by a Data 6000* to give the pressure amplitude of the wave of each selected frequency. Following this the transducer was moved to the next position and the measurement was repeated. The movement of the transducer and the data collection were carried out under computer control.

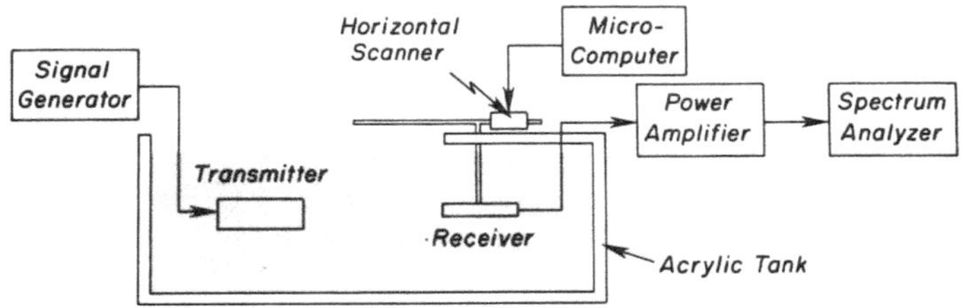

Fig. 1. Test facility used for determination of B/A.

For the case of glycerol, 14 dB of gain was used. The near field distance N_f of 3.8 cm was calculated by using the formula[15]

*Data 6000 is manufactured by Analogic Corporation.

$$N_f = \frac{D^2 f}{4V},\tag{5}$$

where $D = 1.2$ cm is the transducer diameter, $f = 2$ MHz is the transducer frequency, and $V = 1.9 \times 10^5$ cm/sec is the sound speed. The pressure was measured beyond 4 cm to satisfy the condition $x_i \geq N_f$.

The measured pressure amplitude $P_1(x)$ and $q_1(x)$ as a function of source distance are shown by dots in Figs 2a and 2b. The corresponding curve fit of Eqs. (1) and (4) to the measured data are given by solid lines. The attenuation coefficients derived from $P_1(x)$ and $q_1(x)$ by the least square error fitting method yield the values of α_1 and α_2 to be 0.196 Np/cm and 0.856 Np/cm, respectively. The pressure amplitude at the source, viz., $P_1(0)$ is determined to be 3654 KP. Using these values of α_1, α_2, and $P_1(0)$, and taking value of $\rho_o = 1.239$ g/ml[13] and $c_o = 1.9 \times 10^5$ cm/sec,[13] B/A is determined by Eq. (2) to be 8.83. This value compares closely with the value found by Zhu et al.[16] of B/A = 8.96.

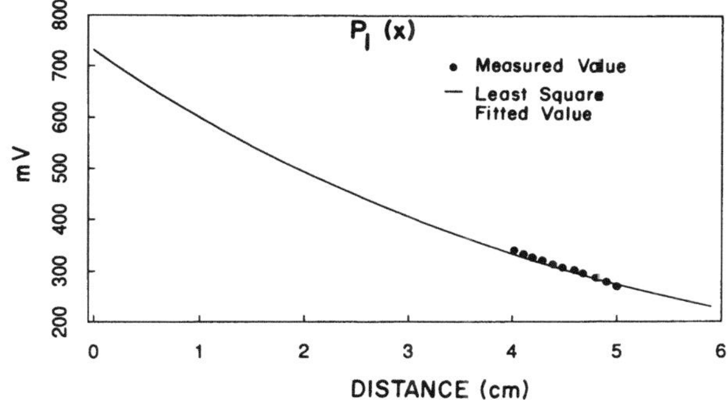

Fig. 2a. Pressure value of $P_1(x)$.

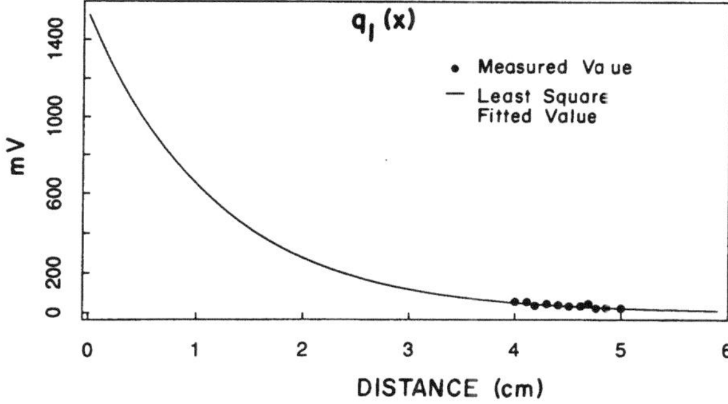

Fig. 2b. Pressure value of $q_1(x)$

The frequency dependence of attenuation coefficient can be represented by a power function of the form

$$\alpha(f) = \alpha_o f^b, \qquad (6)$$

where α_o and b are constants. For most nondispersive viscous fluids, the parameter b is 2. The estimated value of b for glycerol is 2.13. The literature value of b for glycerol is 2.0.[17] Thus, the estimated value of b compares closely with the literature value of b.

NONLINEAR PARAMETER IMAGING USING PHASE AND FREQUENCY MEASUREMENTS

Methods

The nonlinear parameter describes the dependence of ultrasonic velocity on pressure

$$c = c_o + \frac{1}{2\rho_o c_o} \left(\frac{B}{A}\right) \Delta P, \qquad (7)$$

where c is the sound speed, ΔP is the dynamic sound pressure, and c_o and ρ_o are the sound speed and density at atmospheric pressure. The method described in Fig. 3 uses a low frequency high amplitude pump wave, and a relatively low amplitude and high frequency probe wave.[8,9] According to Eq. (7), a change in the sound speed of the probe wave is induced in proportion to pressure variations of the pump wave and the magnitude of B/A. This speed variation produces a cumulative phase shift which is detected at the receiver. It has been shown in Refs. 8 and 9 that under the condition that the acoustic impedance differences along the propagation path are small, the phase perturbation due to the pump wave is given by

$$m_p\left(\frac{1}{\lambda_p}, y, t\right) = \frac{\pi}{\rho_o c_o^2} \frac{|\Delta P|}{\lambda_c} \int_o^L \frac{B}{A}(x,y) \cos\left(\frac{2\pi c_o t}{\lambda_p}\right) dx, \qquad (8)$$

where B/A(x,y) is the spatial distribution of the nonlinear parameter, λ_c is the probe wavelength, λ_p is the pump wavelength, $|\Delta P|$ is the pump pressure amplitude, and L is the length of the path traversed by the probe wave.

Equation (8) implies that the phase shifts of the probe wave, due to the pump wave of the spatial frequency $1/\lambda_p$, is the Fourier cosine transform of the distribution of the nonlinear parameter B/A in the x direction.

One can estimate a set of the phase shifts by changing the wave number of the pump wave. Mechanical scanning of the probe path along the y direction allows to reconstruct a two-dimensional image of the nonlinear parameter. This set of phase shifts, the real part of the Fourier transform, with an arbitrary set of wave numbers of the pump wave, leads to an oscillating response in the reconstructed magnitude of B/A. In this section we propose an improved method of nonlinear parameter imaging which simultaneously measures the real and the imaginary parts of the Fourier transform with a derived set of wave numbers of the pump wave to eliminate the oscillating response.

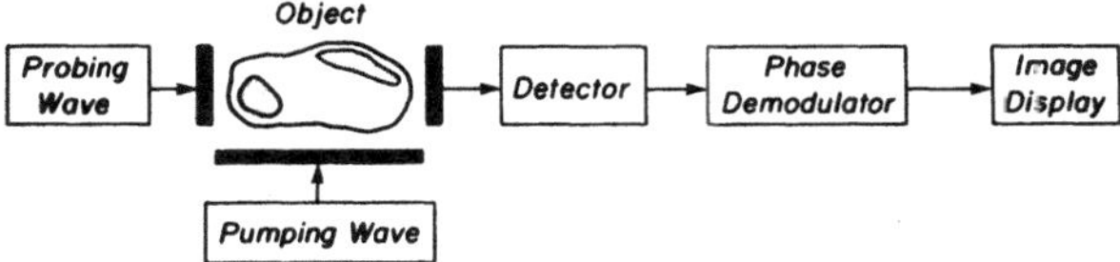

Fig. 3. Imaging system using phase measurements. [Reproduce with permission from D. Kim et al.: Ultrasonic imaging of the nonlinear parameter B/A: Simulation studies to evaluate phase and frequency modulation methods. Ultrasound in Medicine and Biology (In Press)]

PM and FM are special cases of an angle modulated signal.[18] The measured probe wave in Eq. (8) can be considered as a PM signal with modulation wave shape

$$\Delta\phi(\frac{1}{\lambda_p},y,t) = \int_o^L \frac{B}{A}(x,y)\ \cos\ (\frac{2\pi c_o t}{\lambda_p})\,dx. \tag{9}$$

If there is a PM signal modulated by $\Delta\phi(1/\lambda_p,y,t)$, there is also FM on the signal corresponding to a different modulation wave shape which is given by

$$\Delta\Psi(\frac{1}{\lambda_p},y,t) = \int_o^L \frac{B}{A}(x,y)\ \sin\ (\frac{2\pi c_o t}{\lambda_p})\,dx. \tag{10}$$

By approximating the time parameter t in Eq. (9) and Eq. (10) as $\frac{x}{c_o}$, $\Delta\phi(\frac{1}{\lambda_p},y)$ and $\Delta\Psi(\frac{1}{\lambda_p},y)$ represent the real and imaginary parts of a Fourier transform. These factors can be measured by using phase and frequency demodulators as shown in Fig. 4.

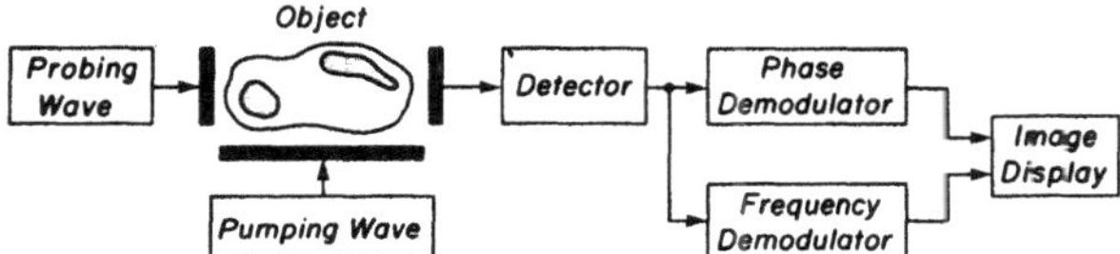

Fig. 4. Imaging system using simultaneous measurements of phase and frequency. [Reproduced with permission from D. Kim et al., Ultrasound in Medicine and Biology (In Press)]

It is necessary to establish a relationship between the continuous and discrete Fourier transforms, mainly because we need to measure discrete values of a Fourier transform of the distribution of the nonlinear parameter to compute the nonlinear parameter in the x direction. It has been shown in Ref. 19 that the discrete Fourier transform $H(k)$ can be expressed in terms of the continuous Fourier transform $H_a(\Omega)$

$$H(k) = \frac{1}{T} H_a\left(2\pi \frac{k}{NT}\right) \qquad k=1,2,\ldots N, \tag{11}$$

where T is the sampling period, and N is the number of samples.

It follows from Eq. (11) that in order to measure discrete values of a Fourier transform of the distribution of B/A for nonlinear parameter imaging, it is necessary to employ the pump wave frequencies described by

$$f_k = \frac{k}{NT} \quad for \quad k = 1,2,\ldots N. \tag{12}$$

Equations (9) and (10) can be rewritten as Eqs. (13) and (14) below in which the condition imposed by Eq. (12) is used, and the time parameter t is approximated as x/c_o.

$$\Delta\phi(f_k,y) = \int_o^L \frac{B}{A}(x,y) \cos(2\pi f_k x)\,dx, \tag{13}$$

$$and \quad \Delta\Psi(f_k,y) = \int_o^L \frac{B}{A}(x,y) \sin(2\pi f_k x)\,dx, \tag{14}$$

respectively, where $\Delta\phi(f_k,y)$ is the phase shift, $\Delta\Psi(f_k,y)$ is the frequency shift, and $k=1,2,\ldots,N$.

Using the discrete Fourier transform pair in Ref. 19, it can be shown that B/A is related to measured phase and frequency in Eqs. (13) and (14) by the following equations;

$$\frac{B}{A}(nT,y) = \sum_{k=1}^{N} \left[\Delta\phi\left(\frac{k}{NT},y\right) \cos\frac{2\pi kn}{N} - \Delta\Psi\left(\frac{k}{NT},y\right) \sin\frac{2\pi kn}{N}\right]$$

$$n=1,2,\ldots N. \tag{15}$$

Equation (15) leads to the important result that B/A is a weighted sum of the phase shift and the frequency shift.

<u>Simulation</u>

Figure 4 shows a block diagram of an imaging system design for measuring phase and frequency modulations.

The length of the path traversed by the probe wave is selected as L = 1 cm. The number of employed pump wave frequencies is N = 100, and the spatial sampling interval T = 0.01 cm. Usually, the wavelength of the bandlimited probe wave is in the order of millimeters (e.g., if the sound speed is 1500 m/s and the frequency of the probe wave is 5 MHz, the wavelength is 0.3 mm). Thus, the spatial resolution of the reconstructed image can be of the order of millimeters. This

312

implies that spatial reconstruction interval $T = 0.01$ cm in this simulation is enough to reconstruct the probed highest spatial frequency of B/A in practical application according to the Nyquist sampling theorem. In this simulation we generate the measured phase and frequency shifts using Eqs. (9) and (10) for a known distribution of B/A. For practical cases, there will be noise in the measurements of phase and frequency shifts. Unbiased measurement noise effects in estimating the nonlinear parameter B/A will be reduced further because the measurements of phase and frequency shifts are heavily averaged in Eq. (15). Thus, we assume that experimental errors are negligible in this simulation.

A two-dimensional phantom image of the nonlinear parameter is shown in Fig. 5. The image of the nonlinear parameter determined by the proposed method is shown in Fig. 6a. The result shown in Fig. 6b is the case when

$$f_k = \frac{k}{NT}$$ using Ichida's method, and Fig. 6C is the case when

$$f_k \neq \frac{k}{NT}$$ using Ichida's method. We also see that Ichida's

method yields symmetric response when pump wave frequencies

are $f_k = \frac{k}{NT}$, $k = 1, 2, \ldots N$.

Fig. 5. The nonlinear parameter phantom image.
[Reproduced with permission from D. Kim et al.,
Ultrasound in Medicine and Biology (In Press)]

CONCLUSION

We present a method for estimating B/A and the frequency dependence of the attenuation coefficient in highly attenuating media using a least squares error fitting method. This method does not use the assumption that $(\alpha_1 - \alpha_2/2)x$ is small. Additionally, it does not require measurement of $P_1(0)$.

Fig. 6a. The nonlinear parameter image using phase and frequency measurements. [Reproduced with permission from D. Kim et al., Ultrasound in Medicine and Biology (In Press)]

Fig. 6b. The nonlinear parameter image using phase measurements ($k_k = k/NT$). [Reproduced with permission from D. Kim et al., Ultrasound in Medicine and Biology (In Press)]

We also propose a method for nonlinear parameter imaging which simultaneously uses the real part and the imaginary part of the Fourier transform of the nonlinear parameter B/A at the wave number of the pump wave. We derive the necessary set of wave numbers of the pump wave which must be employed to accurately image B/A. Simulations of the imaging method show that the proposed method in this paper leads to more accurate imaging of B/A than the earlier method.

Fig. 6c. The nonlinear parameter image using phase measurements ($f_k \neq k/NT$). [Reproduce with permission from D. Kim et al., Ultrasound in Medicine and Biology (In Press)]

ACKNOWLEDGMENTS

The authors thank E. C. Quarve and C. A. Welch for secretarial and graphics. This work was supported in part by grant CA-41324 from the National Institutes of Health and NSF ECS 8310626 from the National Cancer Institute.

REFERENCES

1. T. G. Muir and E. L. Carstensen, Prediction of nonlinear acoustic effects at biological frequencies and intensities, Ultrasound Med. Biol. 6:345-357 (1980).
2. E. L. Carstensen, W. K. Law, N. D. Mckay, and J. B. Muir, Demonstration of nonlinear acoustic effects at biological frequencies and intensities, Ultrasound Med. Biol. 6:359-368 (1980).
3. C. A. Cain, H. Nishiyama, and K. Katakura, On ultrasonic techniques for measurements of the nonlinear parameter B/A in fluid-like media, IEEE Ultrasonics Symposium Proc. pp. 885-888 (1986).
4. "Tissue Characterization with Ultrasound," J. F. Greenleaf, ed., CRC Press, Inc., Boca Raton, FL (1986).
5. C. M. Sehgal, R. C. Bahn, and J. F. Greenleaf, Measurement of the acoustic nonlinearity parameter B/A in human tissues by a thermodynamic method, J. Acoust. Soc. Am. 76(4):1023-1029 (1984).
6. C. M. Sehgal, G. M. Brown, R. C. Bahn, and J. F. Greenleaf, Measurement and use of acoustic nonlinearity and sound speed to estimate composition of excised livers, Ultrasound Med. Biol. 12:865-874 (1986).
7. R. L. Errabolu, C. M. Sehgal, R. C. Bahn, and J. F. Greenleaf, Measurement of ultrasonic nonlinear parameter in excised fat tissues, Ultrasound Med. Biol. 14:137-146 (1988).

8.   N. Ichida, T. Sato, H. Miwa, and K. Murakami, Real-time nonlinear parameter tomography using impulsive pumping waves, <u>IEEE Trans. Sonics Ultrasonics</u> 31(6):635-641 (1984).

9.   N. Ichida, T. Sato, and M. Linzer, Imaging the nonlinear ultrasonic parameter of a medium, <u>Ultrasonic Imaging</u> 5:295-299 (1983).

10. Y. Nakagawa, W. Hon, C. Cai, N. Arnold, and G. Wade, Nonlinear parameter imaging with finite amplitude sound waves, <u>IEEE 1986 Ultrasonics Symposium Proc.</u> 86:901-904 (1986).

11. A. L. Thuras, R. T. Jenkins, and H T. O'Neil, Extraneous frequencies generated in air carrying intense sound waves, <u>J. Acoust. Soc. Am.</u> 6:173-180 (1935).

12. F. Dunn, W. K. Law, and L. A. Frizzell, Nonlinear ultrasonic wave propagation in biological materials, <u>IEEE 1981 Ultrasonic Symposium Proc.</u> pp. 527-532 (1981).

13. W. N. Cobb, Finite amplitude method for the determination of the acoustic nonlinearity parameter B/A, <u>J. Acoust. Soc. Am.</u> 73:1525-1532 (1983).

14. F. B. Hildebrand, "Methods of Applied Mathematics," Prentice Hall, Inc., Englewood Cliffs, NJ (1965).

15. J. L. Rose and B. B. Goldberg, "Basic Physics in Diagnostic Ultrasound," John Wiley and Sons, Inc., Sussex, England (1979).

16. Z. Zhu, M. S. Roos, W. N. Cobb, and K. Jensen, Determination of the acoustic nonlinearity parameter B/A from phase measurements, <u>J. Acoust. Soc. Am.</u> 74:1518-1521 (1983).

17. P. D. Edmonds, "Methods of Experimental Physics," Academic Press, New York (1981).

18. L. W. Couch, II, "Digital and Analog Communication Systems," Macmillan Publishing Co., Inc., New York (1983).

19. A. V. Oppenheim and R. W. Schafer, "Digital Signal Processing," Prentice Hall, Inc., Englewood Cliffs, CA (1975).

ULTRASONIC TIME-OF-FLIGHT TOMOGRAPHY FOR THE NON-INTRUSIVE MEASUREMENT OF FLOW VELOCITY FIELDS

Axel Hauck

Institut für Meß- und Regelungstechnik
Universität Karlsruhe
D-7500 Karlsruhe
Federal Republic of Germany

ABSTRACT

In this paper, ultrasonic time-of-flight tomography is applied for imaging the three-dimensional velocity vector field in a moving fluid. Some fundamentals of tomography for scalar fields are reviewed and its extension to vector fields is discussed. An experimental set-up for an accurate measurement of the transit time based on a phase-locked-loop system is explained and experimental results for various velocity profiles are presented.

INTRODUCTION

Applications of tomographic methods using acoustic transmission measurements are commonly known and have been studied in detail. They allow to reconstruct the spatial distribution of scalar quantities such as the sound velocity or the acoustic absorption in the measurement region. This paper presents tomographic imaging of vector fields. Although vector tomography makes a number of new and interesting non-intrusive measurement methods possible, very little literature can be found in this field. In most publications the used algorithms have not been verified mathematically to be a solution of the reconstruction problem. In this paper, the basic reconstruction formulas for mapping three-dimensional flow velocity fields of moving fluids are derived. The velocity components in the measurement plane are reconstructed using a new method which determines the vector potential of the plane velocity field. To verify this reconstruction algorithm, experimental studies were carried out. An ultrasonic phase-locked-loop system was used, which enabled us to measure sound speed with high resolution. Since this measurement technique requires neither scattering centers nor optically transparent media, it may be applied in many cases where Doppler or optical methods fail.

FUNDAMENTALS OF COMPUTERIZED TOMOGRAPHY

Computerized tomography means reconstruction of pictures from projections taken at different angles. The projections are obtained in the form of line integrals over the measurement region Ω (Fig. 1). For the projections we can write

$$F(\ell,\theta) = \int_{-\sqrt{R^2-\ell^2}}^{+\sqrt{R^2-\ell^2}} f(\ell\cos\theta - s\sin\theta, \ell\sin\theta + s\cos\theta)\,ds \qquad (1)$$

where f denotes the unknown field propery. The totality of projections for all projection angles θ is called the Radon transform $R\{f\}$. Radon has shown /1/ that there exists an inverse transform $R^{-1}\{F\}$ which fulfills the identity

$$f = R^{-1}\{R\{f\}\} . \qquad (2)$$

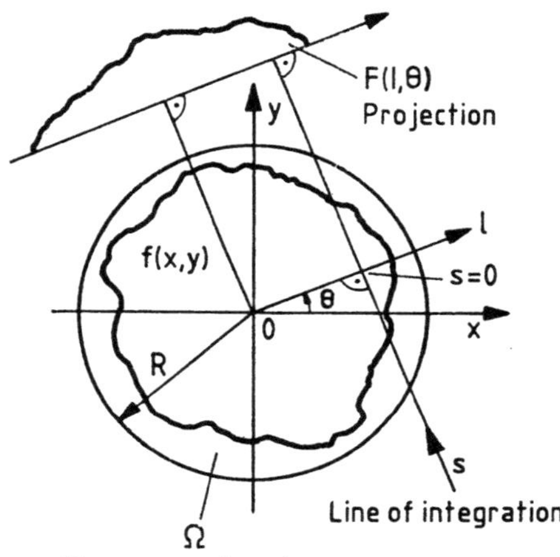

Fig. 1. Measurement geometry and projection
of a scalar function

Radon's inversion formula reads

$$f = \frac{1}{2\pi} \int_0^\pi \int_{-R}^{+R} \frac{1}{x\cos\theta + y\sin\theta - \ell} \frac{\partial F(\ell,\theta)}{\partial \ell}\,d\ell\,d\theta. \qquad (3)$$

The inverse Radon transform is the basis for a variety of reconstruction procedures such as the backprojection or the Fourier inversion method. An explicit derivation of the algorithms is given in /2/. This basic principle of measurement and reconstruction, well known in medical diagnostics, provides a powerful tool for non-intrusive measurement techniques.

For measuring transit times with high resolution, we used a continuous phase-locked-loop (PLL) method which overcomes some problems occurring in transit time measurements of ultrasonic pulses. The measurement principle is based on the following idea. When an acoustic wave is travelling in a moving medium, the total speed of sound Ceff is the superposition of the local sound velocity C and the velocity $\underline{V}$ of the medium:

$$C_{eff} = C + \underline{V}^T \underline{\tau} \; . \tag{4}$$

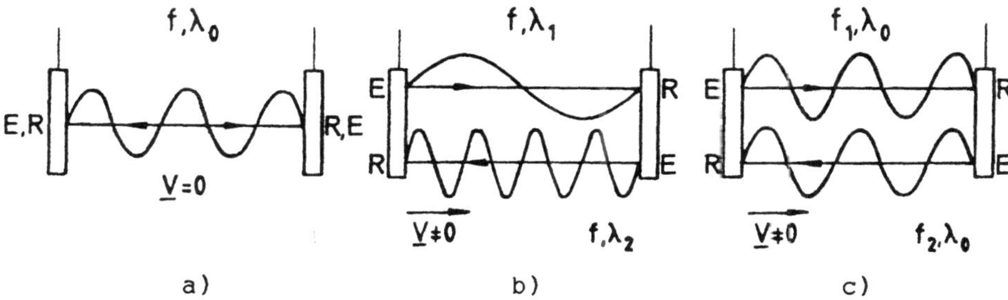

Fig. 2. Phase- and frequency shift in a fluid flow. a) Fluid without motion. b) Phase shift in a moving fluid for a constant frequency. c) Frequency shift for a constant phase. (C = const.)

Assuming a constant frequency f of the continuous wave, a variation of the flow velocity results in a shift of the phase difference between the emitted and the received signal (Fig. 2a,b). In the PLL the variation of the phase difference is measured and the instantaneous frequency is controlled so that the phase shift disappears (Fig. 2c). The mean wavelength λ_0 is therefore constant and independent of the ray direction. The relation between the measured frequency and the transit time of an acoustic wave travelling from a to b can then be expressed by

$$f_{ab \atop ba} = \frac{n}{t_{ab \atop ba}} \approx \frac{n}{\int_a^b \left[\frac{1}{C} \mp \frac{\underline{V}^T \underline{\tau}}{C^2} \right] ds} \qquad \text{where} \quad n = \frac{L}{\lambda_0} \; . \tag{5}$$

L = acoustic path length

$\underline{\tau}$ = tangent vector to the ray

If the variation of sound speed ΔC, and the flow velocity $|\underline{V}|$ are small compared to C, the frequency difference can be approximated by

$$f_{ab} - f_{ba} = \frac{2n}{L^2} \int_a^b \underline{V}^T \underline{\tau} \; ds \tag{6}$$

whereas for the sum we get

$$f_{ab} + f_{ba} = \frac{2n}{L^2} \int_a^b C \, ds. \tag{7}$$

Based on Eq. (7), the distribution of the local sound velocity C can be found by using conventional tomographic methods.

TOMOGRAPHY FOR VECTOR FIELDS

Since the flow velocity is generally a three-dimensional vector field, Eq. (6) has to be evaluated to obtain a component-wise reconstruction /3/.

With $\underline{V}^T = (V_x, V_y, V_z)$ and $\underline{\tau}^T = (-\sin\theta\sin\gamma, \cos\theta\sin\gamma, \cos\gamma)$ we can write

$$f_{a_1 b_2} - f_{b_2 a_1} = \frac{2n}{L^2} \int_{a_1}^{b_2} \left[\left(-V_x \sin\theta + V_y \cos\theta \right) \sin\gamma + V_z \cos\gamma \right] ds. \tag{8}$$

For the projections in the plane A which is normal to the z-axis we have

$$f_{a_1 a_2} - f_{a_2 a_1} = \frac{2n}{L^2} \int_{a_1}^{a_2} \left[-V_x \sin\theta + V_y \cos\theta \right] \sin\gamma \, ds. \tag{9}$$

Under the assumption $\partial \underline{V}/\partial z = 0$, the velocity component in the direction of the z-axis can be separated by the linear combination

$$\left[f_{a_1 b_2} - f_{b_2 a_1} \right] - \left[f_{a_1 a_2} - f_{a_2 a_1} \right] = \frac{2n}{L^2} \cot\gamma \int_{a_1}^{a_2} V_z \, dR. \tag{10}$$

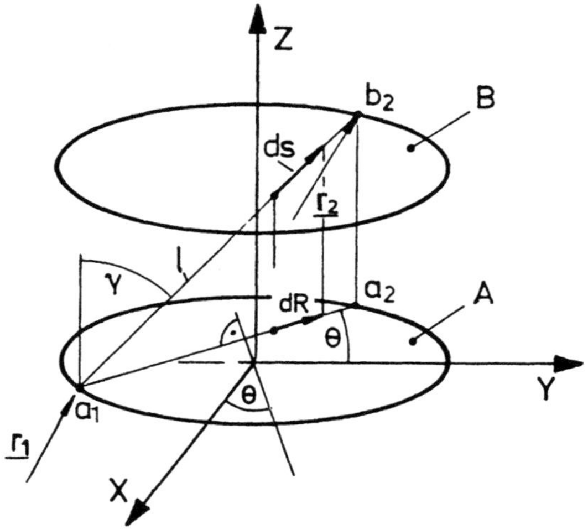

Fig. 3. Transducer arrangement and acoustic path
for a cylindrical bounded pipe flow

Eq. (10) represents the Radon transform of the field component V_z which can be found by using conventional reconstruction algorithms. For the velocity components in the measurement plane the separation turns out to be more complicated. However, reconstruction is also possible.

If the plane velocity field $\underline{V}_A^T = (V_x, V_y)$ is source-free, i.e. div $\underline{V}_A = 0$ we can write

$$\underline{V}_A = \text{curl } W \underline{e}_z \tag{11}$$

Here W denotes the vector potential (stream function) of the field and $\underline{e}_z$ is a unit vector in the direction of the z-axis. Using Eq. (11), Eq. (9) can be replaced by

$$f_{a_1 a_2} - f_{a_2 a_1} = -\frac{2n}{L^2} \int_{a_1}^{a_2} \left[\frac{\partial W}{\partial y} \sin\theta + \frac{\partial W}{\partial x} \cos\theta \right] dR \tag{12}$$

$$= -\frac{2n}{L^2} \frac{\partial}{\partial \ell} \int_{a_1}^{a_2} W \, dR.$$

When comparing (12) with (1) and (3), it is easily seen that the vector potential of the plane velocity field can be reconstructed similarly to the scalar case. The difference is that the derivation with respect to ℓ is discarded in the reconstruction procedure because it is already included in the measurement.

$$W = \frac{1}{2\pi} \int_0^\pi \int_{-R}^{+R} \frac{1}{x\cos\theta + y\sin\theta - \ell} \cdot \frac{-L^2}{2n} \left[f_{a_1 a_2} - f_{a_2 a_1} \right] d\ell \, d\theta. \tag{13}$$

The velocity field finally is

$$\underline{V}_A = \text{curl } W \underline{e}_z . \tag{14}$$

In this evaluation we assumed homogeneous boundary conditions, i.e. the velocity component in the direction of the normal vector $\underline{V}_{An} = 0$ on the boundary of the measurement region. Otherwise, an additional field component will occur which can be expressed either by a harmonic vector potential W_H or by a harmonic scalar potential Q_H

$$\underline{V}_{AH} = \text{curl } W_H \underline{e}_z = \text{grad } Q_H . \tag{15}$$

The reconstruction then turns out to be more complicated. Additional boundary terms will occur in Eq. (12) with the result that the harmonic field $\underline{V}_{AH}$ is reconstructed with only half of its magnitude. This reconstruction problem, arising for example in the case of unbounded flow regions, can be treated by a special algorithm suggested in /4/.

MEASUREMENT SET-UP FOR EXPERIMENTAL FLOW INVESTIGATIONS

In order to verify our theory we carried out some experimental studies on fluid flows (Fig. 4). Two linear ultrasonic arrays each with 36 transducers were placed on the outside of a cylindrical tube with diameter D = 120 mm. To ensure acoustic coupling, both the tube and the arrays were immersed in a water basin. The position of the arrays

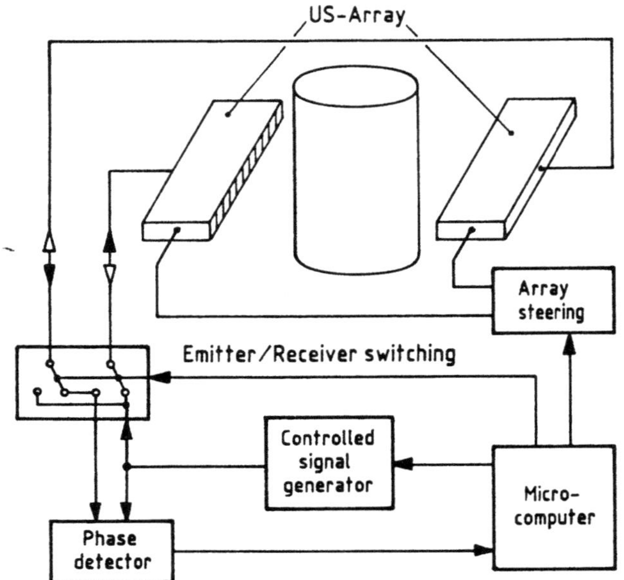

Fig. 4. Block diagram of the phase-locked-loop measurement system

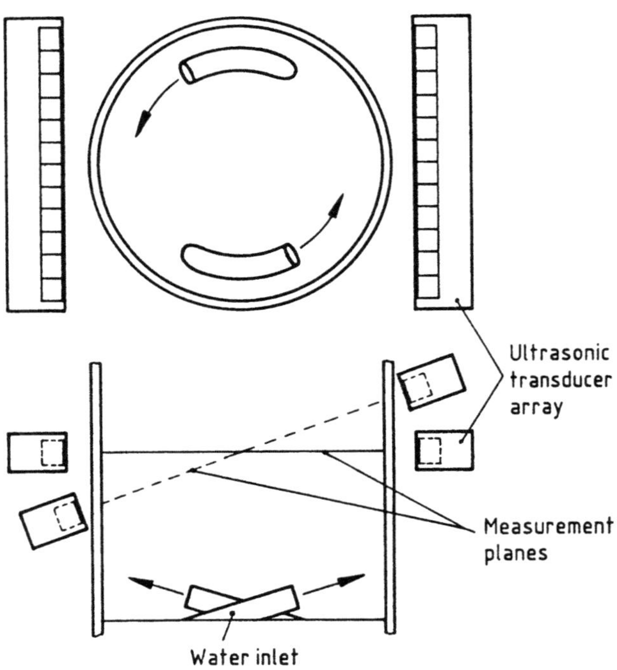

Fig. 5. Experimental arrangement

i.e. projection angle θ and inclination angle γ could be manually adjusted. In a closed loop the frequency of the signal generator is controlled by a microcomputer. Once the PLL locks and the variation of the phase difference between the emitted and the received signal disappears, the frequency is measured and stored for later signal processing. Furthermore, the microcomputer controls the switching of the transducers from emitting to receiving mode and vice versa. The 36 corresponding transducer pairs are then multiplexed until a complete projection is recorded. The investigated flow was a central vortex generated by special water inlets inside the tube (Fig. 5).

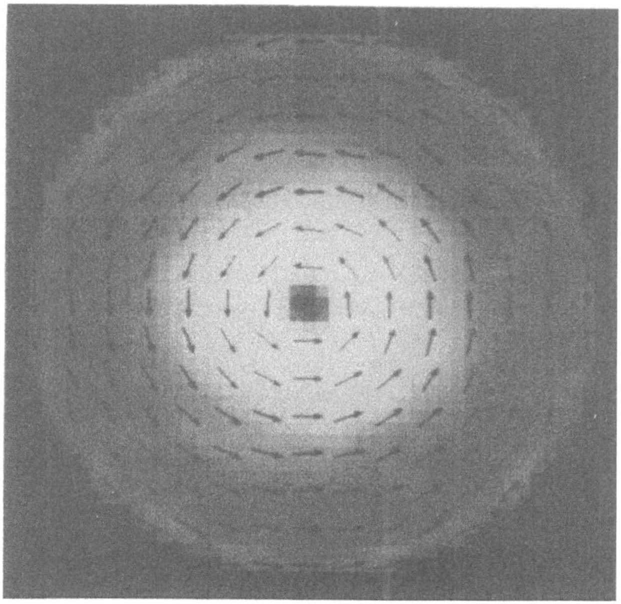

Fig. 6. Reconstruction of the plane velocity field

EXPERIMENTAL RESULTS

The measurements were performed in a plane normal to the tube axis, and at an inclination angle of $\gamma = 70°$. Nine views each with 36 equally spaced parallel rays were recorded. Based on these data a filtered backprojection algorithm for vector fields was evaluated on a rectangle of 45x45 pixels.

In Fig. 6 the magnitude $(V_x^2 + V_y^2)^{1/2}$ and the direction of the fluid velocity component in the measurement plane is shown in the form of a gray scale picture. The flow direction is depicted by the arrows, and the magnitude by local brightness. In Fig. 7, a plot of the magnitude along a central line is compared to the values of a reference measurement provided by a laser doppler anemometer system. It can be seen that both the vortex geometry and the magnitude are reconstructed with good accuracy. The flow velocity in the direction of the tube axis is shown in Fig. 8. This velocity profile has its maximum near the center of the tube. The negative values in the picture indicate local backflow close to the wall.

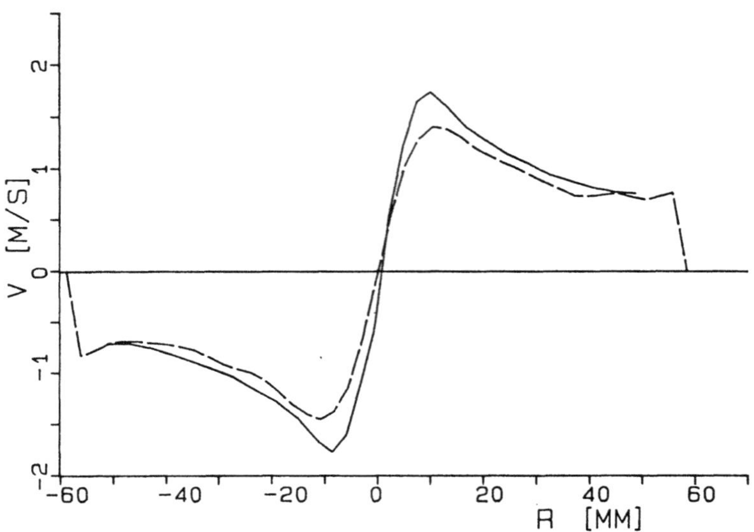

Fig. 7. Plot of the magnitude along a central line
——— LDA measurements
———— tomographic reconstruction

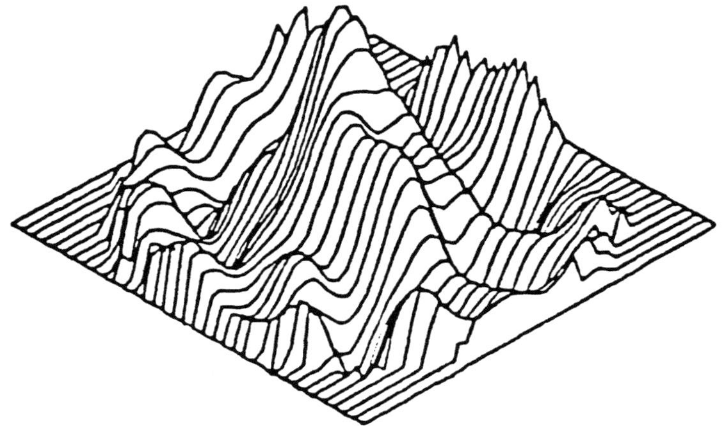

Fig.8. Reconstruction of the velocity in
the direction of the tube axis

CONCLUSION

It has been shown that vector tomography, based on ultrasonic time-of-flight measurements, is applicable to measure flow velocity fields. Since the investigation of two-phase-flows (2 Vol.- % air in water) also gave encouraging results, this measurement principle opens a wide field of potential applications.

REFERENCES

/1/ Radon, J.: Über die Bestimmung von Funktionen durch ihre Integral-werte längs gewisser Mannigfaltigkeiten. Ber. Sächsische Akademie der Wissenschaften 29 (1917), S. 262-277.

/2/ Herman, G.T.: Image reconstruction from projections. Academic Press, London 1980.

/3/ Johnson, S.A.; et. al. Reconstructing three-dimensional fluid velocity vector fields from acoustic transmission measurements. ISA Transactions 16 (1977).

/4/ Braun, H.; Hauck, A.: Tomographic reconstruction of vector fields. To be published in IEEE Transactions on Acoustics, Speech and Signal Processing.

COMPARISON OF ESTIMATION STRATEGIES FOR

THE DETERMINATION OF BLOOD VELOCITY USING ULTRASOUND

Katherine Ferrara and V. Ralph Algazi

Department of Electrical Engineering and Computer Science
University of California, Davis

INTRODUCTION

A number of estimation strategies for the determination of blood velocity have been proposed using narrowband and wideband transmitted signals. Traditional strategies include narrowband "doppler" based estimation techniques, and strategies involving the transmission of a wideband signal with the use of cross correlation for the velocity estimate. We compare the performance of these strategies with two new mixed time-frequency strategies which use both position <u>and</u> phase in order to estimate the blood velocity. The first new mixed estimation strategy utilizes a maximum likelihood estimator which is derived for a wideband periodic transmitted signal. The second new mixed estimation strategy is based upon the concept of cyclic correlation. In each case, results which have been derived previously are cited, and methods which have not been explored are presented in greater detail.

The desirable goal in blood velocity estimation is the ability to map the mean velocity and velocity spread for each small region with a minimum velocity uncertainty, and a maximum energy to noise ratio. In order to maximize the axial spatial resolution, all of these strategies are evaluated with transmitted signals of significant fractional bandwidth. While narrowband doppler estimation can describe the distribution of velocity components along an axial line through a vessel, spatial velocity mapping requires the transmission of a wideband signal. Note that we have utilized a train of short wideband signals in order to maximize the signal to noise ratio and remain within regulatory limits placed upon spatial peak temporal average power in the in-vivo environment.

The comparison of estimator quality is based upon the theoretical local and global accuracy of the estimator evaluated using a statistical model of the blood scattering medium. In each case the local accuracy is determined through the evaluation of the variance of the estimator in the presence of a slowly fluctuating point target, and global accuracy is considered through the evaluation of the expected estimator output in the presence of a general target. In addition, the experimental performance of these estimators has been evaluated using ultrasonic data acquired after transmission of a wideband signal. The experiments support the theoretical analysis and confirm the potential improvement in local and global accuracy using a mixed estimation strategy.

Our paper is organized as follows. The model for the received signal is introduced in section one. The new mixed strategies and their performance are discussed in section two. A cross correlation strategy is discussed in section three, and narrowband estimation is reviewed in section four. A comparative discussion is presented in section five.

1.0 MODEL FOR THE RECEIVED SIGNAL FROM THE BLOOD SCATTERING MEDIUM

Blood consists of a viscous incompressible fluid containing a high concentration of red blood cells which scatter the acoustic wave. The blood cells move through the vessel due to movement of the fluid, at a rate which can be estimated by tracking the movement of a group of cells.

Since the concentration of cells varies randomly through the vessel, the magnitude of the returned signal varies when the group of scatterers being insonified is changed. Between the transmission of one pulse and the next, the scatterers move a small distance within the vessel. The resultant change in axial depth produces a change in the delay of the signal returning to the transducer from each group of scatterers. By transmitting a wideband signal, the magnitude of this change in delay is significant in comparison to the length of the complex envelope, and can be used to estimate scatterer velocity. Using a periodic train of pulses, this change in delay due to axial scatterer velocity becomes a change in the period of the received signal. Traditional narrowband estimators cannot use this information because the change in delay is not easily detected due to the long duration of the transmitted signal.

As discussed in [1], the following model will be used for the received signal from wideband insonification of the blood scattering medium. This model is similar to that used by Atkinson[2] with the generalization of a wideband periodic transmitted signal. For a center frequency w_0, the notation for the received signal, $r(t)=Re\{r\,'(t)e(jw_0t)\}$ will be used to differentiate the complex envelope $r\,'(t)$ from its instantaneous form. Two cylindrical coordinate systems are defined. One system describes the beam position, the second system describes the position of the vessel. The scatterer's initial axial distance from the center of the beam coordinate system is denoted by z. The beam coordinate system is centered at an axial position which corresponds to the center of the sample volume. The scatterer's initial radial(away from the beam axis) position is denoted r. The angular coordinate of the beam is denoted Φ. For the vessel coordinate system $\tilde{z}$, $\tilde{r}$, and $\tilde{\Phi}$ are the cylindrical coordinates. Under the assumption of laminar flow, the velocity profile $v(\tilde{r})$ is solely a function of the initial radial vessel coordinate. The notation d' indicates the temporal delay required for an acoustic pulse to propagate to the sample volume center and return to the transducer.

The deterministic filtered complex envelope is represented by $s'(t)$, where $s'(t)=h_{as}'(t,z)*f\,'(t)$, $h_{as}'(t,z)$ is the filter which models the scattering response and the effect of propagation through tissue, $f\,'(t)$ is the complex envelope of the transmitted signal. The scaling of the velocity in the exponential is given by σ = $2w_0/c$, where c represents the acoustic propagation velocity. The angle between the vessel and the beam axis is represented by α, and T represents the transmitted pulse repetition period.

The model for the received signal consists of the sum over scatterers indexed by i and pulses indexed by k of a product of terms. These terms are the amplitude A_i of the signal from each scatterer, the lateral beam sensitivity $b[r_i(t)]$ of the ith scatterer, the delayed deterministic signal, the phase of the signal from each scatterer, and the doppler shift. Introducing this notation into the sum of products produces the following expression.

$$r'(t)=\Sigma_i\Sigma_k A_i b[r_i(t)]s'[t\text{-}2z_i/c\text{-}kT\{1+2v(\tilde{r}_i)\cos\ \alpha/c\}]\exp[\text{-}j\sigma(z_i+v(\tilde{r}_i)\cos\ \alpha\cdot t)]. \tag{1}$$

For a narrowband transmitted signal, the effect of the scatterer velocity upon the pulse repetition frequency is not significant, and the received signal is approximately given by

$$r'(t)=\Sigma_i\Sigma_k A_i b[r_i(t)]s'[t\text{-}2z_i/c\text{-}kT]\exp[\text{-}j\sigma(z_i+v(\tilde{r}_i)\cos\ \alpha\cdot t)]. \tag{2}$$

2.0 MIXED STRATEGIES

The mixed estimation strategies coherently sum a number of properly delayed pulses, using both the repetition period and doppler shift of the received signal to determine the scatterer velocity. This is shown to produce a very sensitive velocity estimate.

2.1 Wideband MLE

By modelling the spread target and its effect on the received signal, a novel estimation strategy which incorporates the spread nature of the target was derived and described in detail in [3]. Consideration of the range spread nature of the target increases the complexity of the estimator, but also improves the estimator performance. In addition, a simpler estimator based upon the point target assumption was derived and is described here. This strategy will be referred to as the wideband point maximum likelihood estimator. Since this estimator facilitates a direct comparison to alternative strategies, the performance of wideband point MLE will be used as the basis of comparison in this paper. The term wideband MLE will be used to refer to the common properties of both point and range spread estimators. Olinger[4] considered the maximum likelihood estimator for a general wideband signal, but utilized an approximation which included only target fluctuation due to lateral beam sensitivity. This resulted in a complex estimator structure which is suboptimal in the presence of a range or velocity spread target.

The performance of the wideband ML estimators has also been studied in [3], and they were shown to improve local and global accuracy when using a wideband transmitted signal. This improvement is seen in reduced estimator variance, improved spatial mapping of velocity gradients, and reduced subsidiary velocity peaks. In equation 3, using the wideband point MLE, the likelihood $l(v)$ for a discrete velocity value, v, and estimator delay, d, is given by

$$l(v) = |\Sigma_k \int_{-\infty}^{\infty} r'(t)\, s'^*(t-d-kT[1+2v/c])\, \exp[j\omega vt]\, dt\, |^2 \qquad (3)$$

The estimator's local accuracy is evaluated under the assumptions that the energy to noise ratio is large, and the errors are small. The Cramer-Rao bound provides a lower bound on the accuracy of any unbiased estimate. It was shown in [3] that the bound on the variance of the wideband point target MLE is given by equation 4 in the absence of significant frequency dependent attenuation, and considering a constant velocity target. This important expression shows that the variance is inversely proportional to the sum of a term which reflects the mean square transmitted signal bandwidth and a term which reflects the mean square signal duration.

$$\mathrm{var}[\hat{v}-v] \geq [2E_r/N_0\phi^2\{w_0^2\,[E(t^2)-E(t)^2] + [E(w^2)-E(w)^2]\}]^{-1} \qquad (4)$$

with $\phi^2=4\,(\cos\,\alpha)^2/c^2$, $\phi_k=2kT\cos\,\alpha/c$, P the number of transmitted pulses, $\hat{v}$ the estimated velocity, E_r the power in the received signal, and v indicates the scatterer velocity, the terms in 4 are:

$$E[w^2]=\Sigma_k(kT)^2 \int_{-\infty}^{\infty} P\; s'^*(u-d-kT-\phi_k v)\; \partial^2/\partial(\phi_k v)^2[s'(u-d-kT-\phi_k v)]du \qquad (5)$$

$$E[w]^2=|\Sigma_k kT \int_{-\infty}^{\infty} \partial/\partial(\phi_k v)[s'^*(u-d-kT-\phi_k v)]\; s'(u-d-kT-\phi_k v)du\,|^2 \qquad (6)$$

$$E[t^2]= \int_{-\infty}^{\infty} P\;\Sigma_k u^2\; |s'(u-d-kT-\phi_k v)|^2\, du \qquad (7)$$

$$E[t]^2=\{ \int_{-\infty}^{\infty} \Sigma_k t\, |s'(t-d-kT-\phi_k v)|^2\, dt\,\}^2 \qquad (8)$$

This equation for the variance of the wideband point MLE will be compared to the variance of other estimators. Let $\tilde{v}_r = v(\tilde{r})\cos\,\alpha$.

The expected output of the wideband point estimator in the presence of a spread target was derived in [1] and will be compared to the appropriate expressions for the cyclic correlation, cross correlation and narrowband estimators. Assuming the scatterers to be uniformly distributed over the axial range $[c\cdot d'/2-M, c\cdot d'/2+M]$, radial positions 0,R and angular coordinates $0,2\pi$,

$$E\{l(v)\}= \sigma_a^2 W\bar{n}\pi \int_{cd'/2-M}^{cd'/2+M} \int_0^{2\pi} \int_0^R r\,|\Sigma_k \int_{-\infty}^{\infty} b[r(t)]s'[t-2z/c-kT(1+2\tilde{v}_r/c)]$$

$$\cdot\; s'^*[t-d-kT(1+2v/c)]\, \exp(j\omega t[v-\tilde{v}_r])\, dt\,|^2\, dr\, d\Phi\, dz \qquad (9)$$

The derivation follows the theory developed in statistical physics and applied to ultrasound for narrowband signals by Atkinson[2], and by Mo and Cobbold[5]. The first term σ_a^2 represents the variance of the scatterer amplitude. The quantity W is called the packing factor and is a measure of the correlation among scatterers. This quantity and therefore the power in the received signal approach zero for a uniformly dense medium at a very high volume concentration. The notation $\bar{n}$ indicates the average scatterer concentration within a unit volume.[2],[5] Using a wideband transmitted signal, it is possible to evaluate the product of these parameters for small spatial volumes and thus map any change in their magnitude across the vessel.

Numerical evaluation of the height of subsidiary velocity peaks using equation 9 provides a global evaluation of the estimator and shows a significant reduction in comparison with traditional narrowband estimation strategies. For a two cycle rectangular pulse envelope and an eight cycle burst, the ratio of the main peak to first subsidiary peak is already quite low at 0.196. This ratio decreases further as the number of pulses used in the estimate increases, and the effect has been verified using experimental data in [6].

2.2 Cyclic Periodogram

It is shown in [3] that the magnitude of the autocorrelation of the received signal from a sample volume containing a constant velocity target is periodic with a period given by $T(1+2v/c)$, where v represents the velocity of the scatterers. In addition, the phase of the autocorrelation is given by $w_0(k-m)T(2v/c)$, where $(k-m)$ is the difference in pulse indices. Therefore, a velocity estimation strategy using wideband signals which exploits the periodicity of the autocorrelation can be proposed. A desirable property of this estimation strategy is the removal of all assumptions regarding the form of the deterministic portion of the returned signal $s'(t)$. A classical estimator of second order periodicity discussed in Gardner[7] is the cyclic periodogram. Using in this case the complex envelope

$$R_1^{\gamma}(t_1,\tau)=1/T_1 \int_{t_1-(T_1-\tau)/2}^{t_1+(T_1-\tau)/2} r'(t-\tau/2)r'^*(t+\tau/2)\exp[-j2\pi\gamma t]\,dt. \tag{10}$$

The data sample used for this estimate is centered at t_1 and is of length T_1. Note that the $E[r'(t)r'(t+\tau)]=0$ and that the squared modulus of this quantity is used in this comparison as in traditional ambiguity theory for spread targets. Due to the doubly spread nature of the target, the velocity of scatterers must be estimated for discrete spatial regions, and therefore the received signal must be windowed in time along each vector. This window must shift as a function of the pulse index and will be denoted $a_k(t)$. Since the signal is expected to be coherent for a limited time, due to the presence of velocity gradients within a sample volume and the limited transit time of a group of scatterers through the lateral beam width, a temporal window must be used to limit the number of pulses utilized in the estimate. In this case the notation $w(k)$ will be used to indicate a temporal window over the pulse train. Rewriting equation 10 to include these quantities, using the asymmetrical form of τ for convenience, and writing the received signal as the sum of pulse indices $r'(t)=\Sigma_k r'_k(t)$, we have

$$|R_1^{\gamma}|^2=1/T_1^2 \left| \int_{-\infty}^{\infty} \Sigma_k r'_k(t)a_k(t)w(k) \cdot \Sigma_m r'^*_m(t+\tau)a_m(t+\tau)w(m)\exp[-j2\pi\gamma t]dt \right|^2 \tag{11}$$

Using the model for the received signal, it can be shown that this quantity is maximized by $\tau=(m-k)T[1+2v/c]$ and $\gamma=2f_0 v/c$. For these values of τ and γ the scatterer velocity is estimated by maximization of this equation over v. The expected output of the cyclic periodogram is sufficient here for the performance comparison, without the difficult evaluation of the variance.

The expected value of the cyclic correlation estimator in the absence of noise is computed in Appendix A assuming a range spread target. The limits of integration on time are included in the windows $a_k(t)$ and $w(m)$, where the indices k and m are pulse indices. For the radial vessel coordinate the limits over r_1, and r_2 range from $0,R$. For the axial vessel coordinates z_1 and z_2, the limits are $cd'/2-M,cd'/2+M$, where M is the target length which is assumed to be much larger than the axial sample volume size. For the angular vessel coordinates Φ_1 and Φ_2, the limits are $0,2\pi$.

$$E\{|R_1^{\gamma}|^2\} \approx 2[Wn\pi\sigma_a^2/T_1]^2 \iiiint\int r_1 r_2 \; |\Sigma_k \Sigma_m \{ \int a_k(t)w(k)a_m(t+\tau)w(m)$$

$$\cdot \; b[r_1(t)]b[r_2(t+\tau)] \, s'^*(t-2z_1/c-kT[1+2\tilde{v}_{r1}/c])s'(t+\tau-2z_2/c-mT[1+2\tilde{v}_{r2}/c])$$

$$\cdot \; \exp\{-joT[-v(k-m) + (k\tilde{v}_{r1}-m\tilde{v}_{r2})]\} \, dt \; |^2 \quad dr_1 dr_2 dz_1 dz_2 d\Phi_1 d\Phi_2\}. \tag{12}$$

It can be shown that, as expected, this estimator performs best for a constant velocity target. Here, the width of the main velocity lobe depends upon the joint distribution of targets. Thus, in the presence of a velocity spread target the magnitude of the estimator output decreases significantly with a resultant loss in signal to noise ratio.

It is also easily shown that the presence of additive noise produces second order signal products multiplied by second order noise products. This effect lowers the signal to noise ratio.

3.0 TIME SHIFT ESTIMATION

In time shift estimation the estimate is based upon the shift in the rf or baseband signal alone.[8]-[10] Advantages of this method include the improvement in spatial velocity mapping in comparison with narrowband estimation techniques, and the decrease in the illuminated velocity gradient due to the transmission of a wideband signal. The disadvantages of this method include an increase in estimator variance due to the limited scatterer observation time. For a single estimate two pulses are used, and the windowed return from the first pulse is correlated with the windowed return from the second pulse. The delay corresponding to the maximum correlation is used to estimate the target velocity using the expression $\tau = T[2v/c]$.

The time delay estimator proposed by Bonnefous[8] involves the correlation between the rf received signal from the kth transmitted pulse, and the received signal from the next transmitted pulse over a temporal window given by W. This local cross correlation is given by

$$C_k(t,u) = \int_t^{t+W} r_k(t')r_{k+1}(t'+u-\tau)dt' \tag{13}$$

The variance of this estimator was studied by Foster[9] and Embree[10]. It was shown by Embree to be unbiased at high signal to noise levels, and the variance was calculated as

$$var(\tau-\tau_0) = 2N_0/(\beta^2\rho^2 E) \tag{14}$$

where τ is the estimate of the delay, τ_0 is the actual scatterer delay, N_0 is the noise level, β is the RMS bandwidth of the received echoes, ρ is the correlation coefficient and E is the energy of the received signal assumed to be constant for r_k and r_{k+1}. By definition,

$$\rho = \int_{-\infty}^{\infty} r_k(t)r_{k+1}(t)\,dt / E^2 \tag{15}$$

It is the velocity of the blood scatterers which is actually desired, therefore the variance will be rewritten in terms of velocity where $\tau-\tau_0 = 2T\cos\alpha/c(v-v_0)$.

$$var(v-v_0) = c^2[2T^2\beta^2\rho^2(\cos\alpha)^2(E/N_0)]^{-1} = [\phi^2 T^2\beta^2/2(E/N_0)\rho^2]^{-1} \tag{16}$$

Since the pulse to pulse correlation is a decreasing function of the pulse repetition period, the variance reaches its minimum when $T^2\rho^2$ reaches its maximum, and therefore the pulse to pulse delay cannot be increased without limit. Since the total observation time used in the generation of a single estimate will be small in comparison with the mixed estimation strategies, the variance will be significantly larger for the time shift estimator.

An additional problem is the signal to noise ratio of an individual estimate. Since the estimate is based upon two pulses, the energy to noise ratio will be much smaller than estimators which utilize the coherent summation of a large number of pulses.

Finally, since the estimator proposed by Bonnefous does not use the magnitude squared correlation, and does not use an approximate deterministic waveform s() to estimate the delay, the variance of this estimator is larger than the variance of other estimators of delay.

The expected value of the cross correlation estimator is given by

$$E\{C_k(t_0,t_0+\tau)\} = E\{\int_{-\infty}^{\infty} r_k(t+t_0)a(t+t_0)r_{k+1}(t+t_0+\tau)a(t+t_0+\tau)\,dt\} \tag{17}$$

where a(t) represents the temporal window and $0 < t_0 < T$. Using the real envelope s(), this quantity is evaluated in a manner similar to the cyclic correlation estimator shown in Appendix A. The result is given by

$$E\{C_k(t_0,t_0+\tau)\} = \sigma_a^2\,W\bar{n}\pi \int_{cd'/2-M}^{cd'/2+M} \int_0^{2\pi} \int_0^{R'} r\cos[-2w_0\tilde{v}_r(\tau+T)/c+w_0\tau]$$

$$\cdot \int_{-\infty}^{\infty} b[r(t+t_0)]b[r(t+t_0+T+\tau)]a(t+t_0)a(t+t_0+\tau)s(t+t_0-2z/c)s(t+t_0+\tau-2z/c-T[2\tilde{v}_r/c])dtdrd\Phi dz \tag{18}$$

The effect of the lateral beam modulation should be negligible for a single estimate, since the pulse to pulse interval T is constrained to be short with respect to the signal correlation. For this rf correlator, the cosine term will improve the sensitivity of the estimate, at the cost of introducing subsidiary peaks.

For a baseband correlator, using a rectangular pulse envelope, the expected output would show a linear dependence on τ, and therefore the estimator would be less sensitive than strategies which include phase information, but would show a complete absence of subsidiary peaks.

4.0 NARROWBAND MAXIMUM LIKELIHOOD ESTIMATION

In an unbiased situation, the maximum likelihood estimator is the minimum variance estimate. Therefore, although we recognize that other narrowband estimators have been proposed and evaluated, the maximum likelihood estimate of velocity will be used for comparison. The narrowband MLE for a slowly fluctuating point target is a classical structure derived by Van Trees[11].

For maximization over velocity space and a wideband transmitted signal, the bound on the variance of the narrowband MLE is given by

$$\text{var}[\hat{v} - v] \geq \{2c'E_r \; \sigma^2 \; (E[t^2] - E[t]^2)\}^{-1} \tag{19}$$

$$\text{where } \bar{E}[t^2] = P \int_{-\infty}^{\infty} \Sigma_k \; u^2 \; s'(u-kT) \; s'^*(u-kT[1+2v/c]) \; du \tag{20}$$

$$\text{and } c' = 1/N_o(E_r/[E_r + N_o]) \tag{21}$$

and $E[t]^2$ is the mean observation interval squared. The observation time $E[t^2]$ is clearly limited by the axial signal length $s'(\cdot)$, since the product of $s'(u-kT)s'^*(u-kT[1+2v/c])$ equals zero after a number of periods determined by the scatterer's velocity. Therefore the observation window is reduced to a small number of pulses which can be summed coherently, and this number of pulses is determined by the velocity of the scatterers.

The expected value of the narrowband MLE is given by

$$E\{l(v)\} = \sigma_a^2 W\tilde{n}\pi \, | \, \Sigma_k \int_{cd'/2-M}^{cd'/2+M} \int_0^{2\pi} \int_0^{R'} r \int_{-\infty}^{\infty} b[r(t)] \; s'[t-2z/c-kT(1+2\tilde{v}_r/c)]$$

$$\cdot \; s'^*[t-d-kT]\exp(j\sigma t[v-\tilde{v}_r])dt \, |^2 dr \, d\Phi \, dz \tag{22}$$

The presence of the scatterer velocity in the delay of the complex envelope will cause the integral to decrease sharply as a function of pulse index, again demonstrating the limited observation time. This expression also shows that the sensitivity of the mismatched receiver will vary significantly with the actual scatterer velocity. This dependence of the expected estimator output on the velocity produces a biased estimate, with a bias toward lower velocities. In addition, evaluation of equation 22 shows that the subsidiary velocity peaks are of approximately unity height. Using narrowband estimation, aliasing frequently occurs in medical doppler systems regardless of the bandwidth of the transmitted signal.

5.0 DISCUSSION

The discussion will structured in the following manner. The performance comparison of velocity estimators is based upon the variance or local accuracy, and global accuracy through the evaluation of the expected estimator output. In each case the performance of the wideband point MLE is first summarized and used as a basis of comparison for each strategy. Table 1 provides a summary of the behavior of the estimators assuming a wideband transmitted signal in the absence of significant frequency dependent attenuation. The cyclic correlation estimator is discussed separately since its performance is very similar to the wideband point MLE in the absence of a velocity gradient. Finally, estimation of velocity spread is discussed.

	Cross Correlation	NMLE	WMLE
Variance*	$\propto 1/[(E/N)(\text{delay term})]$	$\propto 1/[(E/N)(\text{phase term})]$	$\propto 1/[(E/N)(\text{phase term+delay term})]$
	* (E/N) *and* phase and delay terms depend on *observation interval*		
Observation Interval (OI)	Limited by Correlation Between TWO Pulses	Limited by axial sample volume size.	Optimized for each group of scatterers.
Energy to Noise Ratio	Limited by small OI	Limited by small OI	Maximized by optimal OI and signal power
Bandwidth(BW)	↑ BW : ↓variance	↑ BW : ↑variance	↑ BW : ↓variance
Transmitted Center Freq. (F)	No change in variance without ↑ BW	↑ F : ↓variance (aliasing increases)	↑F : ↓ variance
Bias	unbiased	biased toward low scatterer velocities	unbiased
Subsidiary Peaks	absent	approximately unity height	low-depend on OI and E/N
Estimation of spread	Can be roughly estimated from correlation-noisy	insensitive	Amplitude and width of velocity profile are sensitive estimators

VARIANCE

The denominator of the bound on the variance of the wideband point MLE (equation 4) contains two terms, one which results from the phase information in the received signal, and a second term which results from estimation of the period of the received signal. The first term is proportional to the scaled mean squared observation time of a region of scatterers, normalized by the transducer center frequency. For the wideband MLE the observation interval is limited only by the length of the coherent data segment for each group of scatterers. Using experimental acoustical data it is shown in [6] that the maximum duration of correlated data decreases as the wall of a straight laminar vessel is approached. Therefore the optimal observation interval is <u>not</u> always equal to the transit time through the lateral beam width. Extending the observation window beyond the coherent signal length decreases the signal to noise ratio without improving the performance of the estimator. Also, additional estimates made using subsequent observation intervals can be averaged using a total duration which is less than the period of stationarity of the cardiac dynamics. Therefore, selection of the optimal observation interval for a group of scatterers is critical to the quality of the estimate. Using the wideband MLE the signal processing can be optimized for the mean velocity and velocity gradient in each small spatial area. This term can also be increased by increasing the transducer center frequency.

Term two, which results from the use of the signal delay, is proportional to the mean squared bandwidth of the signal, normalized by a measure of mean squared observation interval. It is significant only for a transmitted signal of substantial fractional bandwidth. Once again the use of the optimal observation interval is important in order to decrease the variance.

The terms discussed above are multiplied by the energy to noise ratio in the denominator of the wideband MLE. This ratio is maximized by the choice of the maximum coherent observation interval for each region of scatterers.

It is shown in [3] that the variance can be reduced further through the use of the wideband range spread MLE.

The denominator of the bound on the variance of the narrowband MLE (equation 19) results from the phase information in the estimator and is again proportional to the scaled mean squared observation time of a region of scatterers, normalized by the transducer center frequency. In comparison to the phase term in the denominator of the wideband point MLE, the observation interval available to the narrowband MLE is typically limited by the axial burst length when the transmitted signal is wideband, or the lateral beam width when the transmitted signal is narrowband. Transmission of a <u>narrowband</u> signal degrades the axial resolution of scatterers. Since the signal length cannot be varied along an axial line, the observation interval cannot be optimized for each axial region. Therefore, the decrease in axial resolution which results from transmission of a narrowband signal does not produce an improvement in velocity resolution for all axial regions.

Using the narrowband MLE, transmission of a <u>wideband</u> signal reduces the maximum observation time to the duration of the axial burst, significantly increasing the variance. In addition, the signal to noise ratio is decreased due to the smaller number of coherently summed signals.

The denominator of the bound on the variance of the cross correlation estimator (equation 16) again results from the delay information in the received signal and is proportional to the mean squared bandwidth of the signal, normalized by a measure of mean squared observation interval. In comparison to the delay term in the denominator of the wideband MLE, the mean squared observation time used for a single estimate using the cross correlation estimator is limited by the correlation of the two pulses used. Since only two pulses are used in the generation of each estimate, the time between the pulses must be minimized in order to maintain a significant correlation between the received pulses. Based upon the experimental data considered in [6], using Sephadex scatterers travelling at 0.16m/s in the center of the straight tube, the pulse to pulse interval used for the cross correlation estimator should be less than 700us in order to prevent spurious peaks in the cross correlation velocity estimate. This delay should be reduced as the vessel wall is approached. Using the wideband MLE, a data segment 7-8 times longer continues to improve estimator performance in the center of the vessel, producing a significant change in estimator sensitivity.

In addition, the use of two pulses in the generation of a single cross correlation estimate significantly reduces the energy to noise level. Although subsequent estimates may be averaged for each strategy, the variance of the cross correlation estimate remains considerably larger than the variance of the wideband MLE.

Thus, in the absence of significant frequency distortion, the bound on the variance of the wideband MLE will be smaller than the variance of the cross correlation estimator because of the larger observation interval, the presence of the two terms in the denominator of equation 4, and the higher energy to noise ratio achieved by the coherent summation of an optimal number of pulses.

EXPECTED ESTIMATOR OUTPUT

In Figure 1, the expected performance of the wideband point MLE (WMLE) is compared to the expected performance of the narrowband MLE and the cross correlation estimator in the presence of a constant velocity range spread target. This figure shows the expected estimator output in the absence of noise for a receiver tuned to a velocity of 1 m/s with the actual scatterer velocity plotted along the bottom axis. Therefore, this represents the ability of the estimator to resolve the target velocity in the absence of noise. In each case the envelope of the transmitted waveform is rectangular, two cycles in duration, and the center frequency of transmission is 5 MHz. The wideband and narrowband MLEs are evaluated based upon an eight pulse train with a pulse repetition interval of 100 us. Therefore the observation times are equivalent for these estimators. The difference in velocity resolution between the estimators is clear.

The positive portion of the cross correlation estimate is shown using two pulses separated by 100 us. Both the width of the mainlobe and height of subsidiary peaks change as a function of the pulse to pulse interval, which must be limited due to the length of the correlated received signal. Increasing the pulse to pulse interval significantly improves the resolution of the estimate, for a constant velocity target.

In addition, it is shown that an observation interval of twenty pulses using the wideband MLE produces a very sensitive estimator. As the number of pulses used in the estimate is increased to the maximum coherent observation interval for a spatial region, the resolution of the wideband MLE improves. The resolution of the narrowband MLE would remain constant as the transmitted pulse train is increased due to the limitation on the observation window imposed by the axial burst length.

Figure 1 also shows the height of the first subsidiary velocity peak of the estimators. The presence of the full height subsidiary peak of the narrowband MLE is as expected. The height of the subsidiary peak is decreased for the wideband point MLE, and decreases further as the observation interval is increased. This reduction in subsidiary peaks for the wideband estimator permits the increase in the transducer center frequency which was shown to reduce the variance.

Figure 2 shows the evaluation of the expected output for a scatterer velocity of 0.2 m/s using thirty five pulses, with a repetition period of 128.5 us. The superior resolution of the wideband estimator continues for the lower velocity. The cross correlation estimator could not be used with an equivalent pulse to pulse duration due to the requirement for signal correlation.

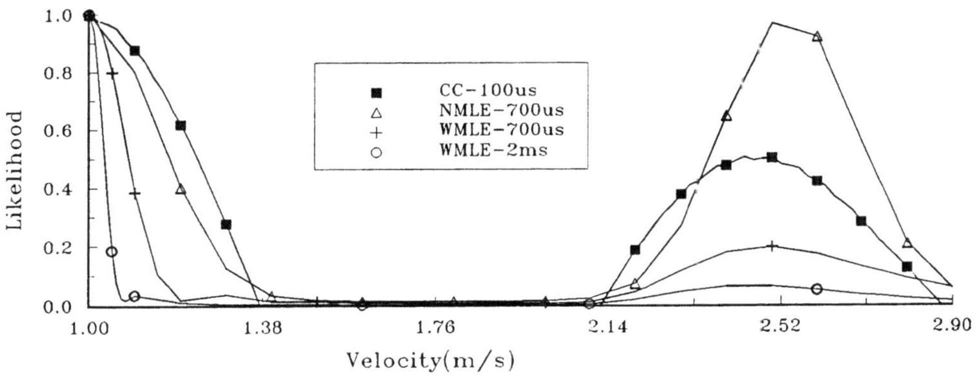

Figure 1- Comparison of Strategies for a Scatterer Velocity of 1 m/s

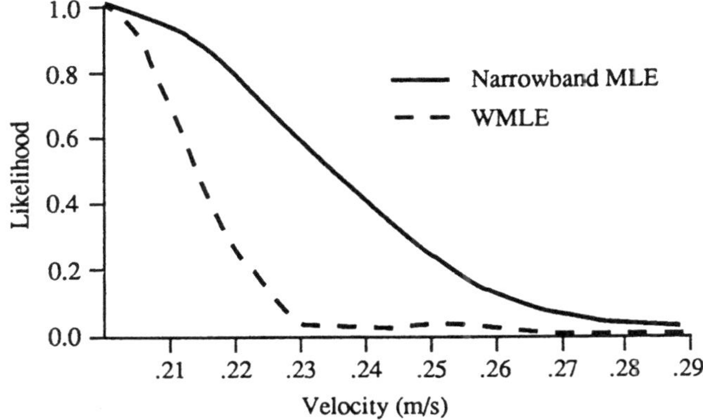

Figure 2- Comparison of Strategies for a Scatterer Velocity of 0.2 m/s and 35 Periods

CYCLIC CORRELATION

Evaluating equation 12 for the atypical case of a constant velocity target, the expected resolution of the cyclic correlation estimator is slightly better than the wideband point MLE. The subsidiary peak for the cyclic periodogram is nearly identical to the wideband MLE in the absence of a velocity gradient. A slight velocity gradient degrades performance and must be displayed as a multidimensional function of the receiver velocity, multiple scatterer positions, and multiple scatterer velocities. Therefore, the expected output of the cyclic correlation estimator is not shown in Fig. 1. Also, in the presence of a spread target, the magnitude of the estimator output decreases significantly with a resultant loss in signal to noise ratio in comparison with the wideband point MLE, and is discussed further in the next section. Evaluation of the expected value in the presence of noise shows additional second order signal-noise products which degrade the signal to noise ratio. It is also noted in using this strategy that the number of computations increase as the square of the number of pulses.

EVALUATION OF VELOCITY SPREAD

In the presence of a velocity spread target, parameters such as mean velocity and velocity spread should be estimated. Using the wideband MLE, the velocity profile is estimated for each spatial region in the vessel. The mean velocity is then estimated for each spatial region using the first moment of the likelihood function, normalized by the zeroth moment. The estimation of velocity spread is based upon the width of the computed velocity profile, and the relative amplitude of the estimator output. Assuming that the zeroth moment is not biased by noise, the wideband point MLE is an unbiased estimate of the expected mean velocity for each spatial position. For the experimental conditions the velocity profile was estimated as a parabolic function of spatial position, and the velocity spread at each position was compared with the amplitude and spread of the wideband MLE using the experimental data in [6]. This evaluation shows that the width and the amplitude of the experimental output at each spatial location vary approximately linearly with the expected velocity spread. Therefore this estimation strategy presents an opportunity to study the spread properties of small spatial volumes of blood.

In comparison, the narrowband MLE in the presence of a wideband transmitted signal produces a biased estimate of mean velocity, since the magnitude of equation 22 decreases for higher velocity targets. Evaluating experimental data, the width of the expected output is insensitive to the presence of velocity gradients due to its limited velocity resolution.

The amplitude of the cyclic periodogram decreases sharply in the presence of a velocity gradient. Equation (A.2) shows that the expected output of the cyclic periodogram is proportional to the square of the signal correlation, and therefore a decrease in the expected estimator output in the presence of a velocity gradient is predicted. This decrease produces a corresponding decrease in the signal to noise ratio, and therefore in the quality of the mean velocity estimate. This feature detracts from its usefulness as a mean frequency estimator. In addition, the width of the expected output of the cyclic periodogram is nearly constant regardless of the degree of velocity spread, and cannot therefore be used as an estimator of velocity spread.

6.0 CONCLUSION

Two traditional strategies for the estimation of scatterer velocity have been compared with two new mixed estimation strategies. The recent research into the temporal correlation of the received signal from blood scatterers has provided a foundation for the development of mixed estimation strategies using delay and phase information [8]-[10].

In the absence of a significant velocity gradient, both mixed estimation strategies provide a sensitive estimate of scatterer velocity, improving performance in several areas. First, small spatial volumes can be studied without a proportionate loss of velocity resolution. The observation window can be optimized for each axial spatial region, improving the bound on the variance and the energy to noise ratio. Using a wideband transmitted signal, the width of the velocity mainlobe using the new mixed schemes is significantly decreased in comparison with the traditional estimation strategies. The height of the subsidiary velocity peaks is significantly reduced in comparison with the narrowband MLE, although they are not absent as is true with a baseband time delay estimator.

The amplitude and width of the mainlobe of the wideband MLE are also sensitive to a change in velocity spread. Combined with the spatial resolution of this estimation strategy, this estimation strategy presents the opportunity for further study of the parameters of small spatial volumes of blood.

REFERENCES

[1] K. W. Ferrara, "Wideband Strategies For Blood Velocity Estimation Using Ultrasound," Ph.D. Dissertation, University of California, Davis, 1989.

[2] P. Atkinson, and J. P. Woodcock, _Doppler Ultrasound and its use in Clinical Measurement_, Academic Press, New York, 1982.

[3] K. Ferrara and V.R. Algazi, "A New Wideband Spread Target Maximum Likelihood Estimator for Blood Velocity, Part One- Theory," IEEE Transactions on Ultrasonics, Ferroelectrics, and Frequency Control, Jan. 1991, 1-16.

[4] M. Olinger, "Ultrasonic Blood Flow Imaging Using Correlation Processing," Ph.D. Dissertation, Michigan State University, 1981.

[5] L. Mo, and R. Cobbold, "A Stochastic Model of the Backscattered Doppler Ultrasound from Blood," IEEE Transactions on Biomedical Engineering, vol. BME-33, no. 1, pp. 20-27, January 1986.

[6] K. Ferrara and V.R. Algazi, "A New Wideband Spread Target Maximum Likelihood Estimator for Blood Velocity, Part Two- Evaluation of Estimators With Experimental Data," IEEE Transactions on Ultrasonics, Ferroelectrics, and Frequency Control, Jan. 1991, 17-26.

[7] W.A.Gardner, _Statistical Spectral Analysis_, Prentice-Hall, Inc. Englewood Cliffs, New Jersey, 1988.

[8] O. Bonnefous, and P. Pesque, "Time Domain Formulation of Pulse-Doppler Ultrasound and Blood Velocity Estimators By Cross Correlation," Ultrasonic Imaging 8, pp. 73-85, 1986.

[9] S. G. Foster, "A Pulsed Ultrasonic Flowmeter Employing Time Domain Methods," Ph.D. Dissertation, University of Illinois at Urbana-Champaign, 1985.

[10] P. M. Embree, "The Accurate Ultrasonic Measurement of the Volume Flow of Blood By Time Domain Correlation," Ph.D. Dissertation, University of Illinois at Urbana-Champaign, 1986.

[11] H. L. Van Trees, _Detection, Estimation and Modulation Theory, Part III_, John Wiley and Sons, New York, 1971.

[12] P. M. Embree, and W. T. Mayo, "Ultrasonic M-Mode RF Display Technique with Application to Flow Visualization," SPIE Vol. 768, International Symposium on Pattern Recognition and Acoustical Imaging, 1987.

APPENDIX A- EXPECTED VALUE OF CYCLIC PERIODOGRAM

The estimator was described in equation 11. The total received signal can be modeled as a Gaussian random process and is therefore completely specified by its mean and autocovariance. It is easily shown that $E[r'(t)] = 0$ because of the uniform distribution of the random phase.

$$E|R_1^\gamma|^2 = \iint_{-\infty}^{\infty} 1/T_1^2 \ E\{r'(t)r'^*(t+\tau)r'^*(u)r'(u+\tau)\}$$

$$\cdot \Sigma_k \Sigma_m \Sigma_d \Sigma_e \ a_k(t)a_m(t+\tau)a_d(u)a_e(u+\tau)\exp[-j2\pi\gamma(t-u)]dtdu \qquad (A.1)$$

Due to the large number of scatterers which contribute to the received signal, the amplitude is assumed to have a Gaussian distribution and zero odd order moments, this can be rewritten as follows.

$$E|R_1^\gamma|^2 = \iint_{-\infty}^{\infty} 1/T_1^2 [E\{r'(t)r'^*(t+\tau)\}E\{r'^*(u)r'(u+\tau)\} + E\{r'(t)r'^*(u)\}E\{r'^*(t+\tau)r'(u+\tau)\}$$

$$+ E\{r'(t)r'(u+\tau)\}E\{r'^*(u)r^*(t+\tau)\}]\Sigma_k a_k(t)\Sigma_m a_m(t+\tau)\Sigma_d a_d(u)\Sigma_e a_e(u+\tau)\exp[-j2\pi\gamma(t-u)]dtdu \quad (A.2)$$

The three terms summed in equation A.2 will be evaluated separately using the model for the received signal. Term one is evaluated in equation (A.3).

$$\iint_{-\infty}^{\infty} E\{r'(t)r'^*(t+\tau)\}E\{r'^*(u)r'(u+\tau)\}\Sigma_k a_k(t)\Sigma_m a_m(t+\tau)\Sigma_d a_d(u)\Sigma_e a_e(u+\tau)\exp[-j2\pi\gamma(t-u)] \, dt \, du =$$

$$= [\sigma_a^2]^2 (Wn\pi)^2 \left| \iiint r\Sigma_k\Sigma_m \left\{ \int b[r(t)] \, b[r(t+\tau)] \ a_k(t)w(k)a_m(t+\tau)w(m) \right. \right.$$

$$\cdot \ s'^*(t-2z/c-kT[1+2v(\tilde{r}) \cdot \cos \ \alpha/c]) \, s'(t+\tau-2z/c-mT[1+2v(\tilde{r}) \cdot \cos \ \alpha/c])$$

$$\cdot \ \exp\text{-}\{j\sigma T(k-m)[-v+v(\tilde{r})\cos \ \alpha]\} \ drd\Phi dz \, dt \left. \left. \vphantom{\int} \right|^2 \right\} \qquad (A.3)$$

The evaluation of term two is shown in equation A.4.

$$\iint_{-\infty}^{\infty} E\{r'(t)r'^*(u)\}E\{r'^*(t+\tau)r'(u+\tau)\}\Sigma_k a_k(t)\Sigma_m a_m(t+\tau)\Sigma_d a_d(u)\Sigma_e a_e(u+\tau)\exp[-j2\pi\gamma(t-u)] \, dt \, du =$$

$$= [\sigma_a^2]^2 (Wn\pi)^2 \iiiint r_1 r_2 \left| \ \Sigma_k\Sigma_m \int b[r_1(t)] \, b[r_2(t+\tau)]a_k(t)w(k)a_m(t+\tau)w(m) \right.$$

$$\cdot \ s'^*(t-2z_1/c-kT[1+2v(\tilde{r}_1)\cos \ \alpha/c]) \, s'(t+\tau-2z_2/c-mT[1+2v(\tilde{r}_2)\cos \ \alpha/c])$$

$$\cdot \ \exp\text{-}[j\sigma T\cos \ \alpha\{v(\tilde{r}_1)k-v(\tilde{r}_2)m\}] \, \exp[j\sigma v(k-m)T] \, dt \ \left. \vphantom{\int} \right|^2 \ dr_1 dr_2 dz_1 dz_2 d\Phi_1 d\Phi_2 \qquad (A.4)$$

The expected value of the integral involving term three is identical to the result for term two. In addition, it is easily shown that the magnitude of equation A.4 will dominate equation A.3 through numerical evaluation. Therefore the expected value of $|R_1^\gamma|^2$ is approximately equal to equation 12.

APPLICATION OF A NEW ALIASING-DEFEATING METHOD

TO PULSED-DOPPLER FLOW IMAGING SYSTEMS

Gabriele Guidi, Piero Tortoli, Francesco Valgimigli

Electronic Engineering Department,
University of Florence via Santa Marta, 3
50139 Firenze, Italy

INTRODUCTION

Although it was introduced in the early 70's,[1] ultrasound pulsed Doppler (PD) technique has only recently become of widespread use in biomedical flowmeters. For many years in fact, the Continuous Wave (CW) technique has been preferred, since it presents less difficulties in positioning the transducer and it does not suffer the *aliasing* problem connected to sampling the Doppler signal at the Pulse Repetition Frequency (PRF). PD analysis was thus limited to the investigation of deep lying vessels where the CW transducers cannot work.

In recent years, however, the rapid growth of electronic technology allowed sophisticated pulsed equipment to be developed with great commercial success. Duplex scanners, for example, where flow in a single sample volume is analysed after selection from a B-mode acoustical image, now represent an unsobstitutable tool in many clinical applications. Two-dimensional (2-D) flow mapping systems[2], capable of providing cross-sectional images of blood velocity in real time, now meet the growing interest of medical people, too. They are based on a *multigate* operation, i.e., on the simultaneous Doppler frequency detection in a number of adjacent sample volumes[3].

In all these cases a pulsed excitation of the transducer is necessary in order to make flow *imaging* possible, i.e. to select the depth at which blood flow is measured. High frequency aliasing generated by an insufficient sampling rate of the Doppler signal often represents a real problem here, since low sample rates can be imposed by system requirements in terms of depth of analysis or of frame rate.

An original technique has been recently introduced,[4] which enables real-time spectral analysis to be extended beyond the Nyquist limits +/-PRF/2. It can be directly applied to Duplex Scanners; the new technique, in fact, does not involve hardware modifications in Doppler electronics, but just a reordering of data provided by a conventional Fast Fourier Transform (FFT) unit.

In this paper, the application of the new approach to a *multigate* Doppler system,[5] is discussed. This system has been so far shown capable of performing real time spectral analysis in a number of sample volumes located along the ultrasound beam axis. Extension of the anti-aliasing technique to this case thus involve the independent application of the new algorithm to spectral data provided for each Doppler cell. Preliminary results obtained *off-line* by using the multigate system in conjunction with a Personal Computer are reported. A discussion about the possible extension of this approach to 2-D flow imaging is also included.

THE ALIASING PROBLEM IN MULTIGATE FLOW IMAGING SYSTEMS

A multigate system employing spectral analysis for Doppler frequency detection, can provide the flow velocity profile across the investigated vessel according to several display formats.

Full information derived by frequency analysis is reported in the format of Fig.1a. Depth is proportional to vertical axis, here, with Doppler frequencies detected within the Nyquist limits +/-PRF/2 distributed along the horizontal axis. For each line of the image, corresponding to one sample volume, spectral power is coded in grey levels, like in conventional Doppler frequency vs. time sonograms.

Fig.1 represents three subsequent velocity profiles obtained *in vitro* by observing at different times the behaviour of a fluid pulsating through a plastic tube. Laminar flow conditions leaded to the parabolic velocity profile detected in Fig.1a. Subsequent flow evolution towards higher velocities involved the turbulence evidenced in Fig.1b where each horizontal line exhibits a noticeable spectral spread. A further velocity increment finally caused a corresponding high frequency aliasing. This is evident in Fig.1c, where Doppler frequency components higher than PRF/2 are shifted left by PRF, so that the true velocity profile appears correspondingly interrupted.

The possibility of resolving this kind of ambiguity through a tracking procedure based on fast spectral analysis, has already been demonstrated. In practice the new approach involves that a variable offset is impressed on the spectrum. The amount of this offset is derived by computation of the instantaneous mean frequency. This parameter, in fact, gives an idea of the position of the spectrum itself and of its tendency towards aliasing. By compensating this tendency with a corresponding offset, the spectrum is forced to be instantaneously centered within a width PRF, and the aliasing effect is avoided in the display.

This procedure has been successfully applied to pulsed Doppler analysis of a single sample volume, even in extreme clinical conditions[6]. Since in the multigate system described in[5], the spectra of several sample volumes are independently provided in real time by an ultrafast processor, the same procedure can be used here, in principle.

In order to test this possibility, the multigate system has been faced to a Personal Computer (PC) capable of acquiring in real time all the Doppler spectra which originated Fig.1. For subsequent spectra referring to the

same sample volume, the tracking procedure has been applied.
The results are shown in Fig.2. As expected, the procedure
has no effects when aliasing is not present (Fig.2a and
Fig.2b). However it allows display ambiguity to be avoided in
situations where Doppler frequencies exceed the Nyquist
limits + PRF/2 (Fig.2c).

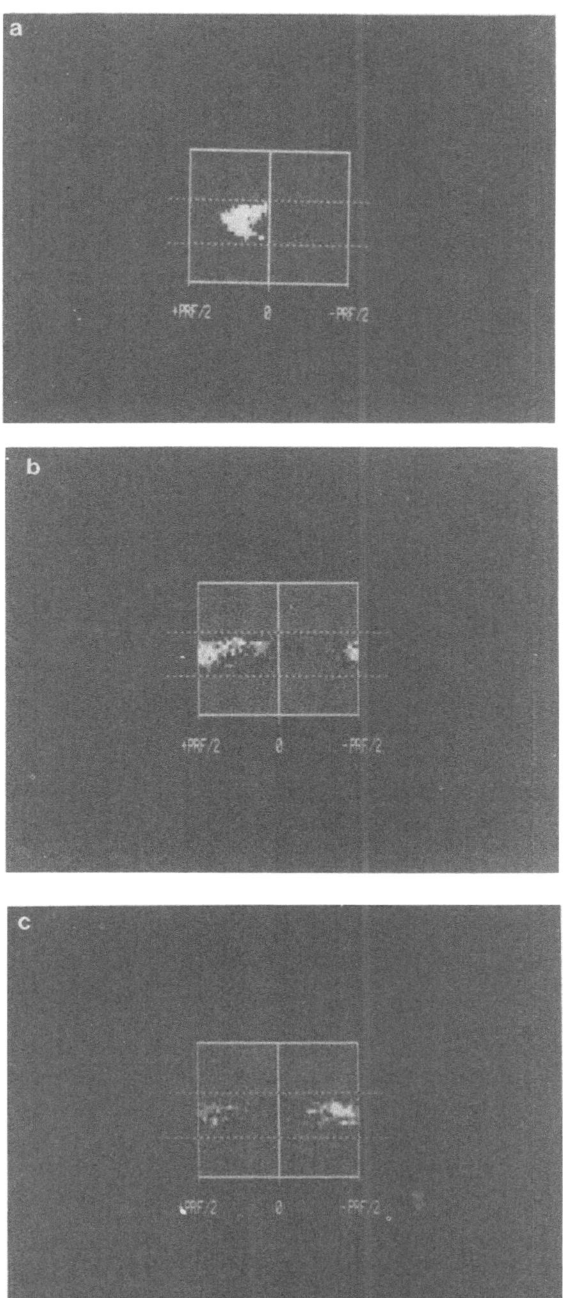

Fig. 1 - Typical spectral flow velocity profiles.

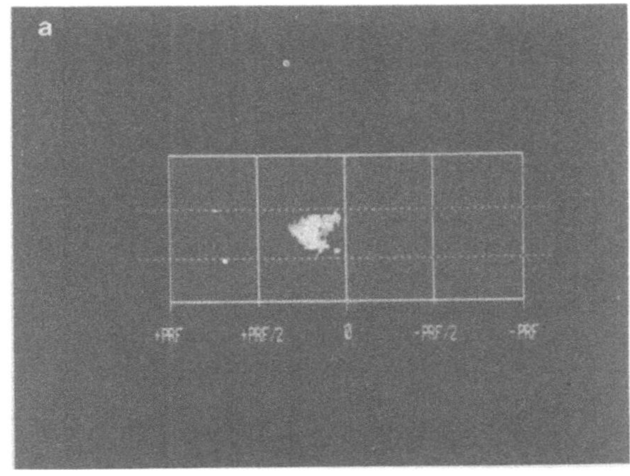

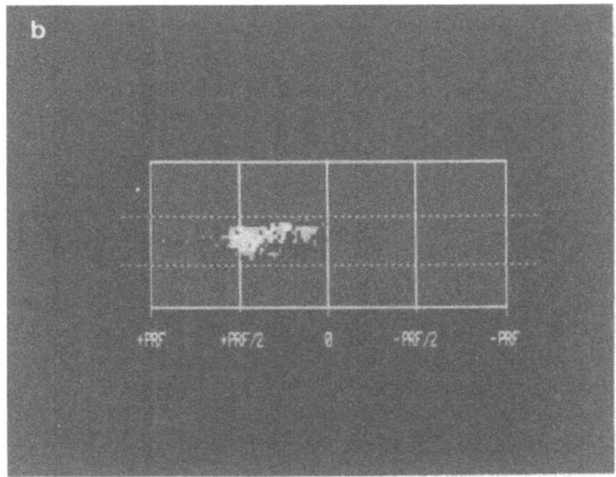

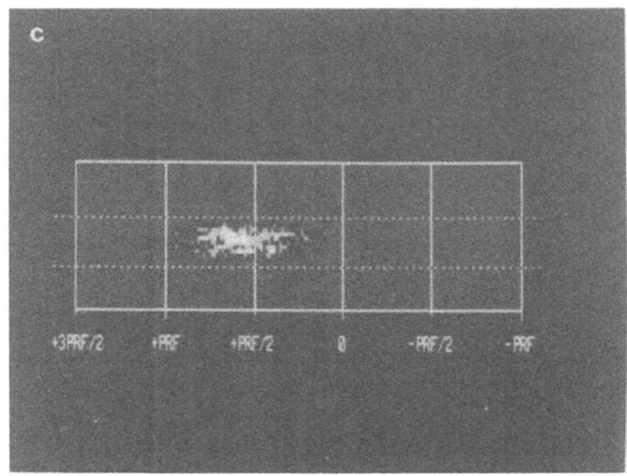

Fig. 2 – Spectral flow velocity profiles obtained by applying the new anti-aliasing procedure.

DISCUSSION

Preliminary *in vitro* results indicate that the recently
introduced anti-aliasing technique can also be applied to
multigate systems employing spectral analysis for Doppler
frequency detection. Since this application just involves
mean frequency computation and spectral data reordering, its
hardware implementation appears feasible, too.

The further application of the method to 2-D flow imaging
can also be considered. Cross sectional flow maps are in fact
obtained by sequentially scanning a number of raster lines:
for each line, the typical operation of a multigate system
detecting the mean velocity at all points in real time is
needed. Preliminary 2-D Doppler images have been
reconstructed off-line by means of our multigate system,
too[7].

A typical feature of Doppler imaging systems, however, is
that in order to obtain a high frame rate, a short time is
available for the investigation of each Doppler line. Besides
a worst signal to noise (S/N) ratio, this fact also involves
a poor frequency resolution, i.e., a large instantaneous
Doppler bandwidth. Moreover, each line is analysed again
after that one frame has been completed. Depending on the
application, this time can be so long that in the meanwhile
flow has suffered a noticeable acceleration.

Low S/N ratio, large instantaneous bandwidths and high
acceleration, together, can represent a severe drawback to
the direct application of the anti-aliasing tracking
procedure[4]. Future studies will be directed to the
investigation of which are the clinical conditions where
these problems compromise the applicability of the new
technique, and how they can be faced.

REFERENCES

1. D. W. Baker, Pulsed Ultrasonic blood flow sensing, <u>IEEE
    Trans. Sonics Ultrason.</u>, vol. SU-17: 170 (1970).
2. Roelandt J., "Color Doppler Flow Imaging", Martinus
    Nijhoff Pub., Dordrecht, (1986).
3. F. D. Mc Leod, Multichannel Pulse Doppler techniques, in
    Cardiovascular Application of Ultrasound, R. S. Renemann
    Ed. Amsterdam: North-Holland/American Elsevier, 85-107,
    (1974).
4. P.Tortoli, A Tracking FFT Processor for Pulsed Doppler
    Analysis Beyond the Nyquist Limit, <u>IEEE Trans. Biomed.
    Eng</u>, BME-36:232-237; (1989).
5. P. Tortoli, F.Andreuccetti, G. Manes, C. Atzeni, Blood
    Flow Images by a SAW-Based Multigate Doppler System,
    <u>IEEE Trans. Ultrason. Ferr. and Freq. Contr.</u>, Vol.35,
    No.5: 545-551, (1988).
6. P. Tortoli, F. Andreuccetti, C. Atzeni, Ultrasound Two-
    Dimensional Flow Mapping using Fast Spectral Analysis,
    <u>in</u> Acoustical Imaging, Vol.16, Ed. Lawrence W. Kessler,
    Plenum Publishing Corp. (1988).
7. P.Tortoli, F.Valgimigli, G.Guidi, P.Pignoli, Clinical
    Evaluation of a New Anti-Aliasing Technique for
    Ultrasound Pulsed Doppler Analysis, Ultrasound in
    Medicine & Biology, in press.

ESTIMATION OF CENTER FREQUENCY AND VARIANCE OF ULTRASONIC DOPPLER SIGNAL BY USING SECOND-ORDER AUTOREGRESSIVE MODEL

Young-Bok Ahn and Song-Bai Park
Dept. of Electrical Engineering
Korea Advanced Institute of Science and Technology
P. O. Box 150, Cheongryang, Seoul, Korea

ABSTRACT

A new estimator using a second-order autoregressive (AR) model, called AR estimator, is proposed to estimate the center frequency and variance of the Doppler signal in the presence of clutter noise. When the center frequency and the variance of the Doppler signal are obtained from the phase and the magnitude, respectively, of a first-order AR model, they become identical to those obtained by the autocorrelation (AC) estimator. In the proposed method, the sampled signal which contains information of both the Doppler signal and the clutter is described by the second-order AR model. The center frequency and the variance of a unidirectional Doppler signal can be estimated from the phase and the magnitude, respectively, of the pole (among two poles) with larger phase. If the blood flows in both directions in a sample volume, and the clutter is rejected to the extent where it no longer obscures the Doppler signal, the proposed method can estimate simultaneously the center frequencies and variances of both the forward and the reverse blood flows. The performances of the proposed estimator are compared with those of the AC estimator by computer simulation and experiments.

INTRODUCTION

Acoustic measurements of blood velocity are based on the Doppler effect. Following the CW and then the pulsed Doppler, the two-dimensional Doppler color flow mapping introduced by Kasai et al. in 1982 is regarded as the third generation Doppler flow meter which provides high resolution imaging of the velocity distribution. Although the autocorrelation (AC) estimator is known as an effective time-domain spectrum estimator unbiased by noise, most of the color flow mapping systems suffer from the small number of available data for AC processing.

As the number N of pulses to be sent out on the same line consecutively with a certain frequency (pulse repetition frequency : PRF) determined by the depth of exploration increases to obtain more data for Doppler processing, the frame rate of 2-D imaging must be sacrificed. Also, the higher the order n of the FIR clutter filter, the better rejection of the clutter. However, n must be restricted to a low value since only $(N-n)$ data will be available from the filter output for estimation of the Doppler frequency. (In real 2-D Doppler systems, N is usually 8 - 12 and n is 1 or 2.)

Clutter refers to the echo received from stationary or slowly moving specular reflectors such as vessel walls and intervening tissue between the probe and the flow volume interrogated. Such an echo is typically 100 times as large as the echo received from blood, and is distinguished by exhibiting a low-frequency Doppler shift. Thus, a high-pass clutter filter is used to attenuate this low-frequency clutter to the extent where it no longer obscures the desired blood flow data. For a given velocity, the Doppler shift obtained from slowly moving tissue will vary according to the Doppler equation with the transmit frequency f_o. In addition, the clutter frequency depends on the position of interest, time, and the patient, and hence the cutoff frequency of the clutter filter is difficult to set beforehand.[1] The artifacts caused by clutter result in poor velocity estimation when it is not perfectly eliminated. In addition, the desired Doppler signal itself is distorted by the clutter filter, causing difficulty particularly in estimating low flow velocity.

This paper proposes a new method using a second-order autoregressive (AR) model, called AR estimator, to estimate the center frequency and the variance of the Doppler signal in the presence of clutter noise, and compares its performance with the AC estimator. When the center frequency and the variance of the Doppler signal are obtained from the phase and the magnitude, respectively, of the pole of a first-order AR model, they become identical to those obtained by the AC estimator. In the proposed method, the sampled signal which contains the information of both the Doppler signal and clutter is described by the second-order AR model with two poles.

AC ESTIMATOR AND FIRST-ORDER AR MODEL

Autocorrelation (AC) Estimator

In the blood flow, the velocity distribution within a resolution cell can be represented by the average (center) frequency and the variance. Letting $P(\omega)$ denote the power spectrum of the complex Doppler signal $z(n)$, the center frequency $\bar{\omega}$ and the variance σ^2 of the power spectrum can be expressed as

$$\bar{\omega} = \frac{\int \omega\, P(\omega)\, d\omega}{\int P(\omega)\, d\omega} \tag{1}$$

and

$$\sigma^2 = \frac{\int (\omega-\omega)^2\, P(\omega)\, d\omega}{\int P(\omega)\, d\omega} . \tag{2}$$

According to the definition of the AC function $R(T)$ for a stationary random process $z(n)$,

$$R(T) = E\{ z(n)z^*(n-1) \}$$

$$= M(T)\, e^{j\phi(T)} \tag{3}$$

where is T is the sampling period. By some manipulations of Eqs. (1) and (2), the center frequency and the variance of the Doppler spectrum for small T can be calculated approximately as

$$\bar{\omega} = \frac{1}{T} \tan^{-1} \left\{ \frac{\mathrm{Im}\,[R(T)]}{\mathrm{Re}\,[R(T)]} \right\} \tag{4}$$

and

$$\sigma^2 = \frac{2}{T^2} \left\{ 1 - \frac{|R(T)|}{R(0)} \right\} . \tag{5}$$

The derivation of the frequency frequency above equations is presented in detail elsewhere,[2] and is not repeated here.

First-Order AR Model

In the periodogram spectrum, the frequency resolution deteriorates when the length of signal samples available is too short. In this case, parametric representation of the spectra by means of AR models can provide much better frequency resolution than the classical periodogram method.[3] The estimate of the AR power spectral density (PSD) is frequently referred to as the maximum entropy method of spectral analysis. If a finite autocorrelation sequence (ACS) $R(0)$, ..., $R(p)$ is assumed to be known, then a question may be posed as to how the remaining unknown autocorrelation coefficients $R(p+1)$, $\cdots$ should be specified in order to guarantee that the entire ACS is positive semidefinite. There are an infinite number of possible extrapolations that will yield a valid ACS. The extrapolation should be made in such a way as to maximize the entropy of the time series characterized by the extrapolated ACS. This time series would then be the most random, in the entropy sense, of all possible series that have the known ACS for lags of 0 to p.

The PSD estimate of the p th order AR model is given as follows :

$$P_p(\omega) = \frac{T\rho_\omega}{\left|1 - \sum\limits_{k=1}^{p} a(k)\, z^{-k}\right|^2} \Bigg|_{z\,=\,e^{j\omega T}} \tag{6}$$

where ρ_ω is the variance of driving white noise and $a(k)$'s are parameters of the AR model.

In the case that order p is 1, the PSD of the AR model is given by

$$P_1(\omega) = \frac{T\rho_\omega}{\left|1 - a(1)\, z^{-1}\right|^2} \Bigg|_{z\,=\,e^{j\omega T}} \tag{7}$$

where

$$a(1) = R(T)/R(0) \tag{8}$$

The power spectrum density $P_1(\omega)$ is determined by the pole $a(1)$, as shown in Fig. 1. The center frequency ω of $P_1(\omega)$ is found from the phase of pole $a(1)$ as follows :

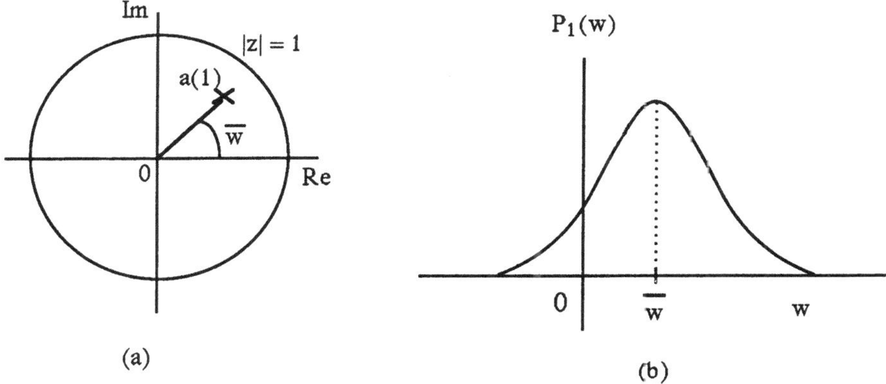

Fig. 1   (a) Pole pattern in the z-plane (b) Power spectrum of $P_1(\omega)$.

$$\omega = \frac{1}{T} \tan^{-1} \left\{ \frac{\text{Im}\ [a\,(1)]}{\text{Re}\ [a\,(1)]} \right\}$$

$$= \frac{1}{T} \tan^{-1} \left\{ \frac{\text{Im}\ [R\,(T)]}{\text{Re}\ [R\,(T)]} \right\}$$

Also, since the variance of $P_1(\omega)$ decreases as the pole $a\,(1)$ approaches the unit circle, we may define

$$\sigma^2 = \alpha \left\{ 1 - |a\,(1)| \right\} \tag{10}$$

Eq. (10) can be made equivalent to Eq. (5) by setting α to $2/T^2$. Then,

$$\sigma^2 = \frac{2}{T^2} \left\{ 1 - \frac{|R\,(T)|}{R\,(0)} \right\} \tag{11}$$

We have shown that the center frequency and the variance obtained by the first-order AR model become identical to those by the autocorrelation estimator.

SECOND ORDER AR ESTIMATOR

By using the second order AR model, which has two poles, it is possible to extract two informations, e.g., the Doppler signal and clutter, from the sampled signal. Because the phase angle of a pole of the PSD is proportional to the center frequency of the PSD as shown above and the center frequency of the Doppler signal is usually larger than that of the clutter, one pole with larger phase can be associated with the Doppler signal, and the other with smaller phase, with the clutter. Thus, the center frequency and the variance of the Doppler signal can be estimated from the phase and magnitude, respectively, of the Doppler pole.

When the order p is 2 in Eq. (6), the PSD of the AR model is given by

$$P_2(\omega) = \frac{T\rho_\omega}{|1 - a\,(1)\,z^{-1} - a\,(2)\,z^{-2}|^2} \Big|_{z\ =\ e^{j\omega T}} \tag{12}$$

The AR parameters $a\,(1)$ and $a\,(2)$ are obtained by the Burg method (given in Appendix),[3] which needs $3Np - p^2$ operations. The two poles of $P_2(\omega)$, which are always within the unit circle,[3] are easily calculated from the following formula:

$$p_{1,2} = \frac{a\,(1) \pm \sqrt{a^2(1) + 4a\,(2)}}{2} \tag{13}$$

If $|\angle p_1| < |\angle p_2|$, p_1 can be considered as the pole due to the clutter, and p_2 due to the Doppler signal. Therefore, the center frequency and the variance of the Doppler signal can be obtained from the phase and magnitude, respectively, of the pole p_2, as in the case of the first-order AR model.

$$\bar{\omega} = \frac{1}{T} \tan^{-1} \left\{ \frac{\text{Im}\ [p_2]}{\text{Re}\ [p_2]} \right\} \tag{14}$$

$$\sigma^2 = \frac{2}{T^2} \left\{ 1 - |p_2| \right\} \tag{15}$$

Before giving into detailed simulation by the the second-order AR model using realistic Doppler signals, we will give the estimation results using a simple Doppler signal model, namely the Gaussian spectrum as given by Eq. (16). We assume two Doppler signals separated by 0.3 PRF but having the same variance and the same peak, as shown by curve A in Fig. 2 (a). The inverse Fourier transform of this spectrum is then processed through the second-order AR estimator to get curve B in the same figure. We note that the AR estimator gives very good estimation of the center frequencies of the original spectrum although the variances differ somewhat. Fig. 2 (b) shows good tracking of the center frequencies when they are varied over a wide range, whereas the AC estimator can not detect the two center frequencies separately but gives only the average of the two center frequencies. Since the spectrum having the lower center frequency can be considered as the clutter, the influence of the clutter on the center frequency estimation for the Doppler signal is lessened in the AR estimator.

When the blood flows in both directions within a sample volume, the proposed method can estimate the center frequencies and the variances of both of the forward and the reverse flows, simultaneously. If the clutter is eliminated, one pole corresponds to the forward flow, and the other to the reverse flow. Since the AR estimator requires less computation time than the FFT estimator, the former method can be used to detect both flows simultaneously in multi-gate or 2-D Doppler systems. The direction of the flow is determined by the position of the pole p_2. If it is located in the first or second (in the third or fourth) quadrant of the z-plane, the blood flows in the forward (reverse) direction.

COMPUTER SIMULATIONS AND EXPERIMENT

Modeling the Doppler Signal

In all simulations to be presented below the Doppler signals were synthesized following the process described in,[4]. The synthesis begins in the frequency domain assuming a Gaussian-shaped PSD with a center frequency $\bar{f}$ and a standard deviation $\bar{\sigma}$

$$g(f) = \frac{1}{\bar{\sigma}} e^{-\frac{(f - \bar{f})^2}{2\bar{\sigma}^2}} \tag{16}$$

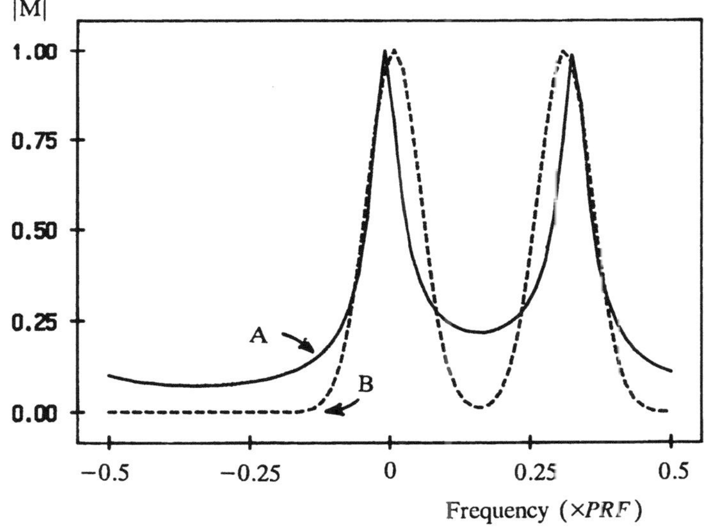

Fig. 2 (a)

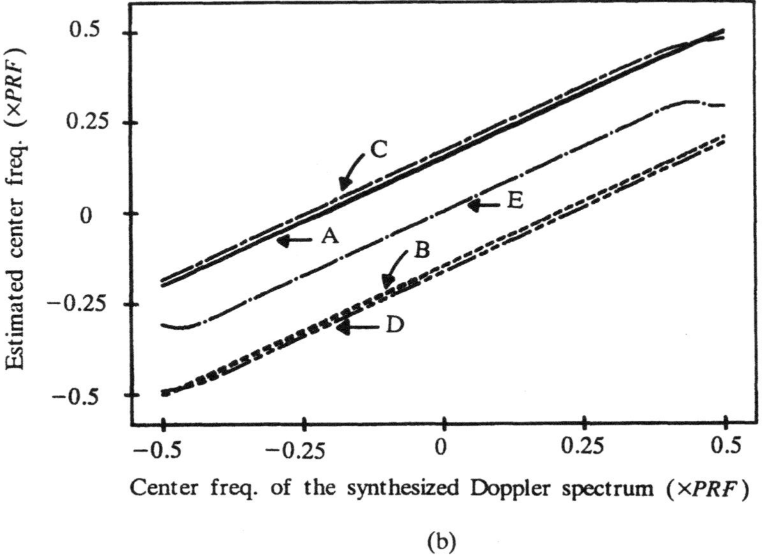

(b)

Fig. 2   (a) Spectral match between original spectrum (dotted) and 2nd order AR model (solid) (b) A, B : two center frequencies of original spectrum. C, D : center frequencies estimated by AR estimator. E : center frequency estimated by AC estimator.

The signal-plus-noise distribution $G(f)$ is then formed by

$$G(f) = - ln[r(f)] \left[\frac{10^{\frac{SNR}{10}} g(f)}{\sum\limits_{f = -0.5PRF}^{0.5PRF} g(f)} + \frac{1}{N} \right] \qquad (17)$$

where noise power with a uniform distribution is added to a scaled version of the signal power distribution. In the above $r(f)$ is a set of random numbers uniformly distributed between zero and one, and the coefficient $- ln[r(f)]$ gives the power spectrum a negative exponential or chi-squared probability distribution, as expected when the underlying process is random and Gaussian.[4]

The real and imaginary components of the vector signal $z(t)$ are computed, respectively, as

$$I(t) = \sum_{f = -0.5PRF}^{0.5PRF} \sqrt{G(f)} \cos[2\pi ft + \phi(f)] \qquad (18)$$

$$Q(t) = \sum_{f = -0.5PRF}^{0.5PRF} \sqrt{G(f)} \sin[2\pi ft + \phi(f)] \qquad (19)$$

where $\phi(f)$ is a random phase uniformly distributed between $-\pi$ and $+\pi$ which is introduced to guarantee the independence of individual frequency generators. A typical Doppler PSD and the corresponding time sequences synthesized following the above procedure are shown in Fig. 3 (a) and (b), respectively, for the case of clutter-to-Doppler ratio (CDR) of 0 dB.

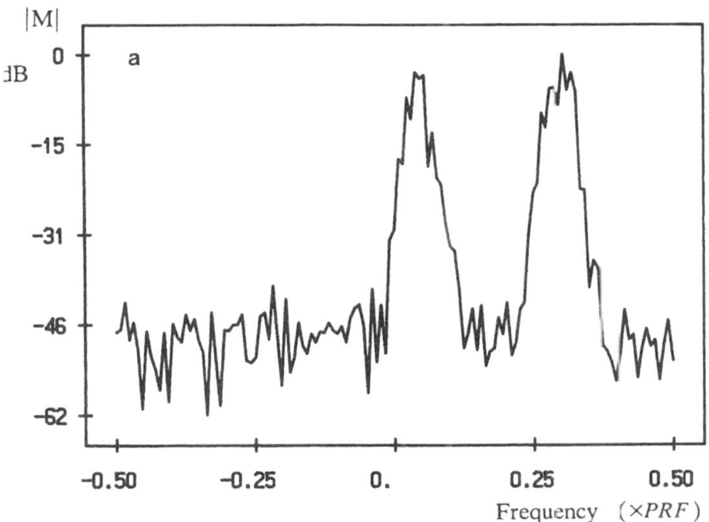

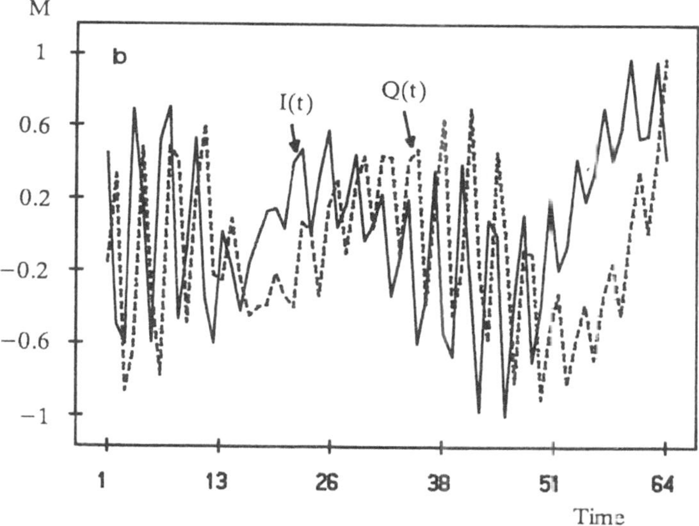

Fig. 3   (a) Power spectrum of Doppler plus clutter (b) Time-domain in-phase signal I(t) and quadrature-phase signal Q(t). CDR = 0 dB, $f_D = 0.3\ PRF$, $f_c = 0.01\ PRF$, $\sigma_D = \sigma_c = 0.015\ PRF$. (Subscripts D and c refers to Doppler and clutter, respectively.)

Estimation of Center Frequency

In the estimation of the center frequency of a Doppler signal we regard the incompletely rejected clutter as a lower Doppler signal to see the effect of clutter on the Doppler frequency estimation by the proposed AR and the conventional AC estimators. The clutter frequency (f_c) was fixed to 0.05 PRF and the Doppler frequency (f_D) was varied from 0.05 PRF to 0.5 PRF. The variances of the clutter (σ_D) and the Doppler signal (σ_D) were also fixed to 0.015 PRF. The simulating results for different numbers N of data and different values of CDR and SNR are plotted from Fig. 4 through Fig. 7, where (a) and (b) are for the AR and AC estimation results, respectively. In Fig. 4, SNR = 30 dB and N = 32; in Fig. 5, SNR = 30 dB and N = 8; in Fig. 6, SNR = 10 dB and N = 32; in Fig. 4, SNR = 10 dB and N = 8. Finally, the case in which both the forward and the reverse Doppler flows exist in the sample volume was investigated and the results are plotted in Fig. 8, where N, SNR, and $\sigma_D = \sigma_c$ are the same as in Fig. 4.

From these figures we make the following observations : 1) In all cases the AR estimator gives better results than the AC estimator, particularly, when the CDR is high, and also when the number of data points is small. 2) The performance of AR estimator is especially good when the Doppler frequency is high (well separated from the clutter), whereas this is not apparent in the AC estimator. (In general, the latter estimate results in greater deviation as the Doppler frequency increases because of the averaging effect.) 3) The AR estimator detect both the forward and the reverse Doppler flows simultaneously, whereas AC estimator gives only the average of the two frequencies (zero value in the worst case). 4) Both estimators are robust to random noise since the results for 30 dB and 10 dB SNRs are almost the same.

Estimation of Variance

The variance of the center frequency of Doppler signal is of another concern. Fig. 9 shows the simulated results by the two estimators under the same conditions as in Fig. 4 (SNR = 30 dB, N = 32, $\sigma_D = \sigma_c$ = 0.015 PRF). We see that the AR estimator (a) gives better results particularly for higher CDR as compared to the AC estimator (b).

Experiments

To conform the performance of the AR estimator, experiments were conducted with the experimental setup shown in Fig. 10. Silicon particles as reflectors with an average diameter 15 μm were mixed with water and circulated at a constant speed around a rubber tube of diameter 15 mm by a DC motor. A burst of 8 pulses focused at 5 cm was generated using a disk type transducer with a center frequency 3.27 MHz. The reflected RF signal was sampled at a rate of 437.5 KHz (thus the axial resolution is 1.7 mm) with a 10 bit A/D converter at 16 points across the tube along the beam axis. The RF data were processed using IBM PC following the algorithms described in this paper. The center frequencies estimated by the AR estimator and the AC estimator using the data of 128 consecutive pulses are shown in Fig. 11(a). A second-order FIR filter was used to reject the clutter before the AC estimator, but not used in the AR estimator. The general appearance of the bell-shaped velocity profiles indicates the adequacy of both estimators. In the detailed level, however, the two profiles differ significantly at the 12th point, where the flow speed is low. Since the clutter filter rejects the Doppler signal as well as the clutter, the center frequency obtained by the AC estimator at this point is zero, whereas a nonzero value is obtained by the AR estimator which does not use the clutter filter. This fact verifies that the AR estimator is capable of estimating lower flow velocity than the AC estimator.

To see the effect of reduced data, we repeated the center frequency estimation with the same experimental data but with 8 sampled signals. The results are shown in Fig. 11(b). Again, the AR estimator gives much better results closer to Fig. 11(a). We conclude that the AR estimator results in better estimation than the AC estimator particularly when the number of available data is small.

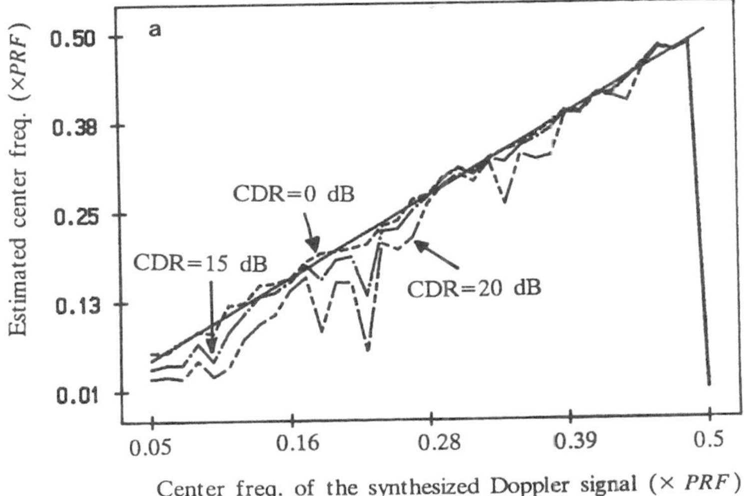

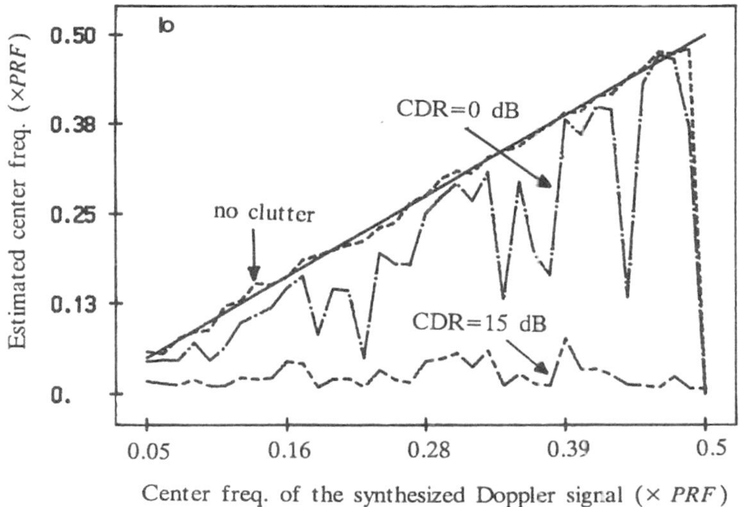

Fig. 4   Estimation of center frequency.   SNR = 30 dB, N = 32, f_c, and $\sigma_D = \sigma_c$ are the same as in Fig. 3.   (a) AR estimator. (b) AC estimator.

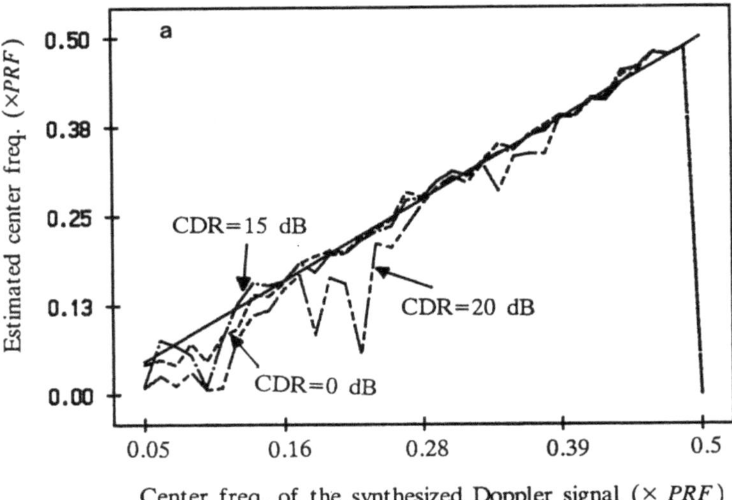

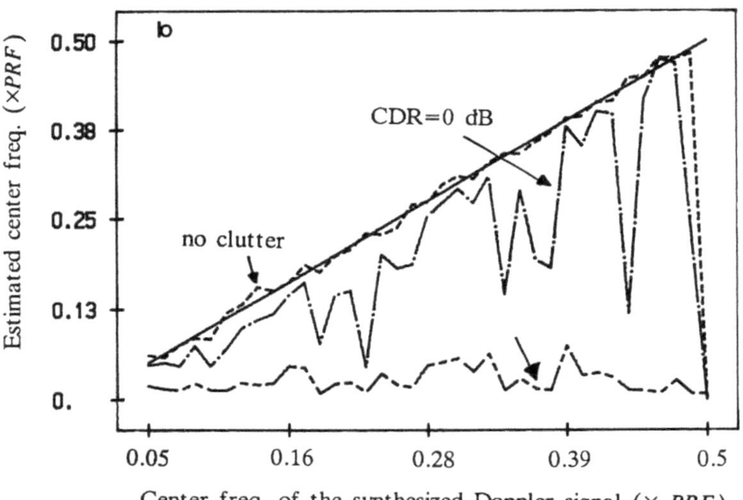

Fig. 5   Estimation of center frequency. SNR $= 10$ dB, $N = 32$. f_c, and $\sigma_D = \sigma_c$ are the same as in Fig. 3.   (a) AR estimator. (b) AC estimator.

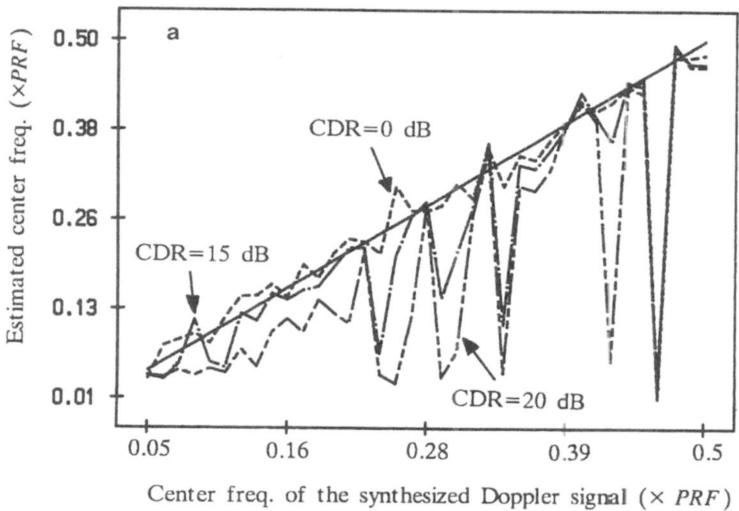

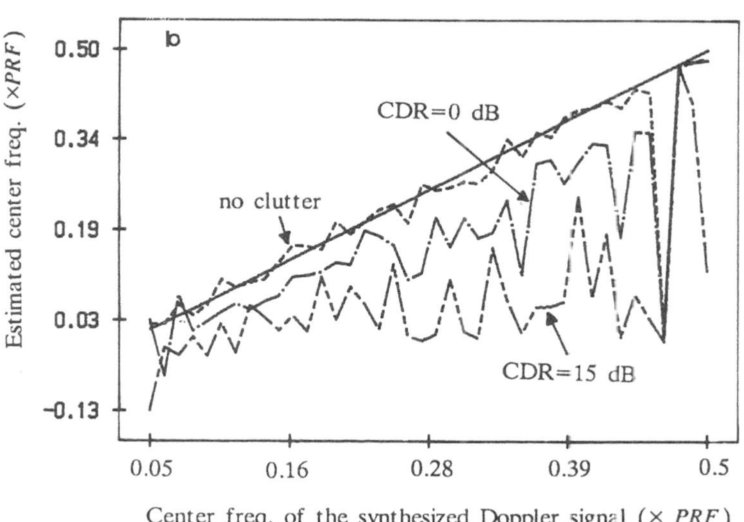

Fig. 6   Estimation of center frequency. SNR $= 30$ dB, $N = 8$. f_c, and $\sigma_D = \sigma_c$ are the same as in Fig. 3. (a) AR estimator. (b) AC estimator.

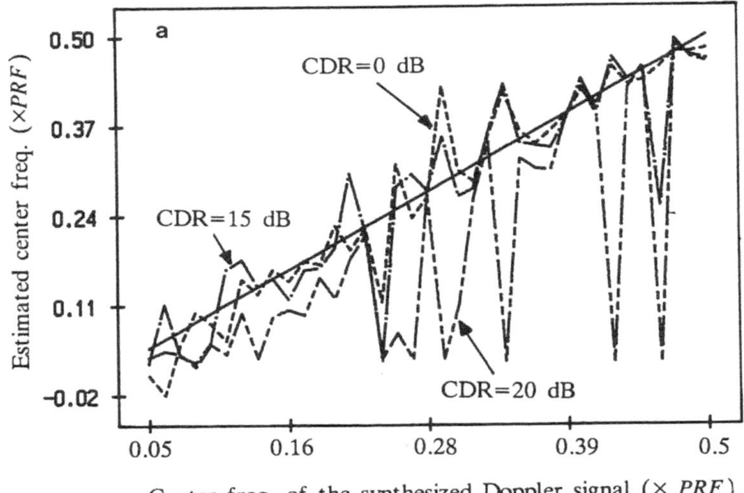

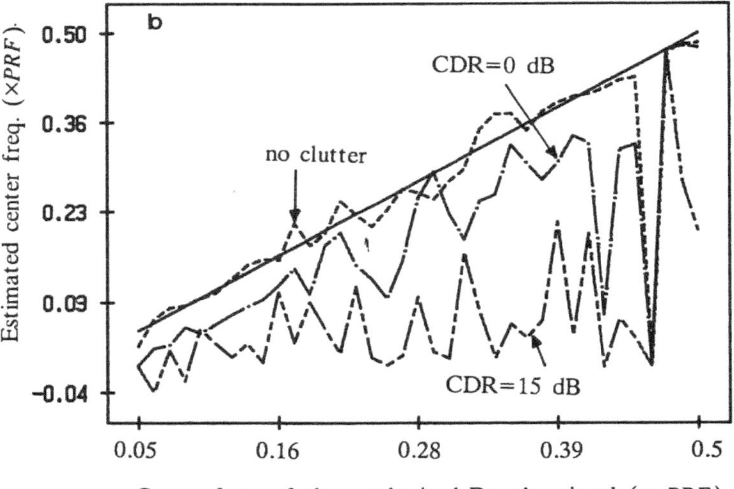

Fig. 7 Estimation of center frequency. SNR = 10 dB, N = 8. f_c, and $\sigma_D = \sigma_c$ are the same as in Fig. 3. (a) AR estimator. (b) AC estimator.

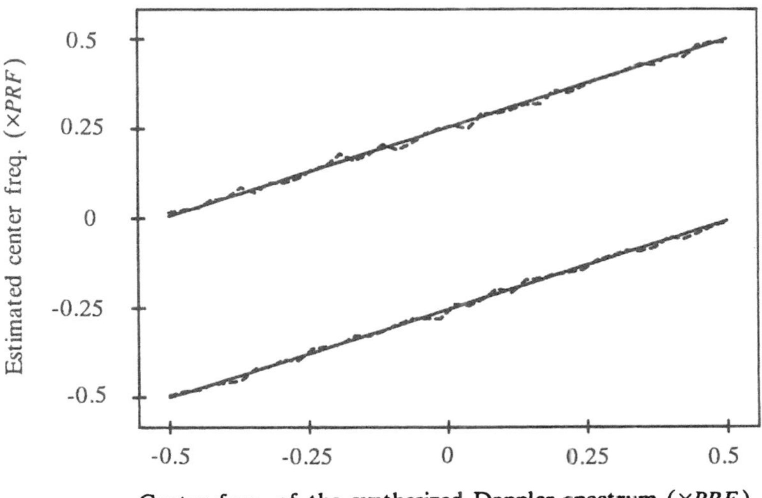

Center freq. of the synthesized Doppler spectrum ($\times PRF$)

Fig. 8   Center frequency estimations of both of the forward and the reverse flows.   Solid line (center frequencies of the synthesized Doppler signal in both directions) and dotted line (center frequencies estimated by AR estimator when $N=32$). $\sigma_D = 0.015$ PRF.

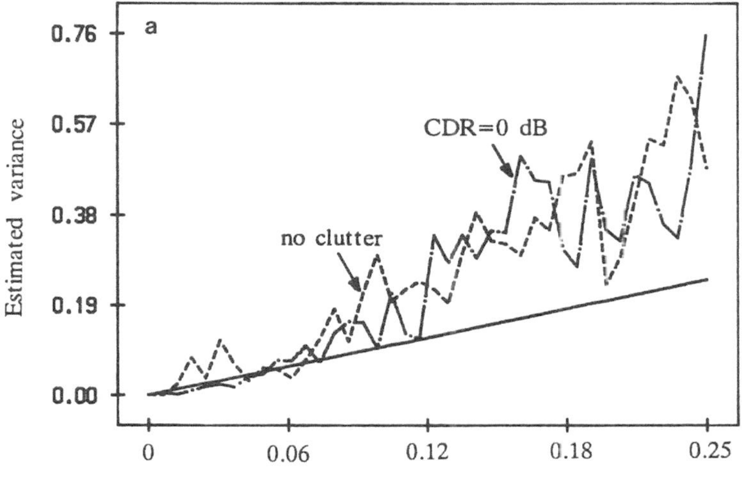

Variance of synthesized Doppler signal ($\times PRF$)

Fig. 9

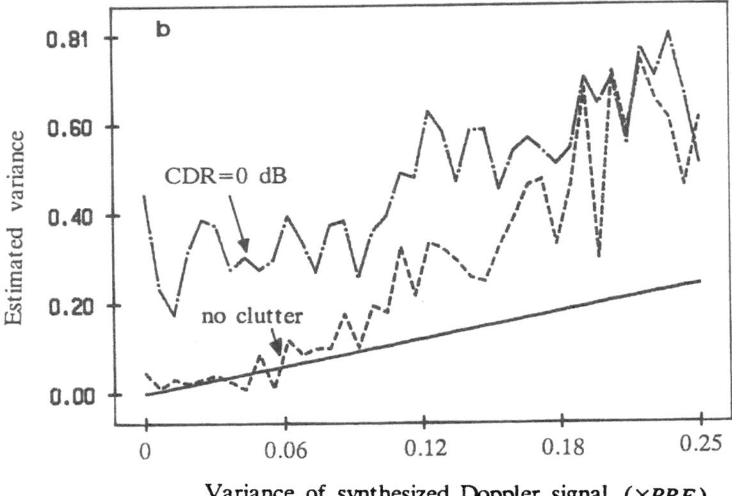

Variance of synthesized Doppler signal ($\times PRF$)

Fig. 9 Variances estimated by (a) the AR estimator and (b) the AC estimator for two different clutter levels.

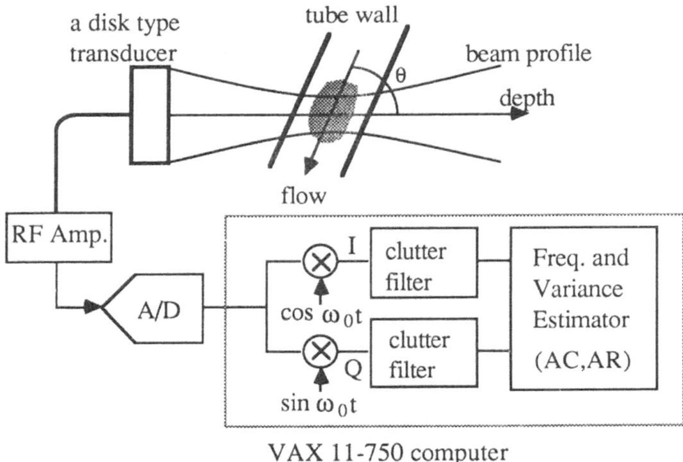

VAX 11-750 computer

Fig. 10 Experimental setup used to get the Doppler signal

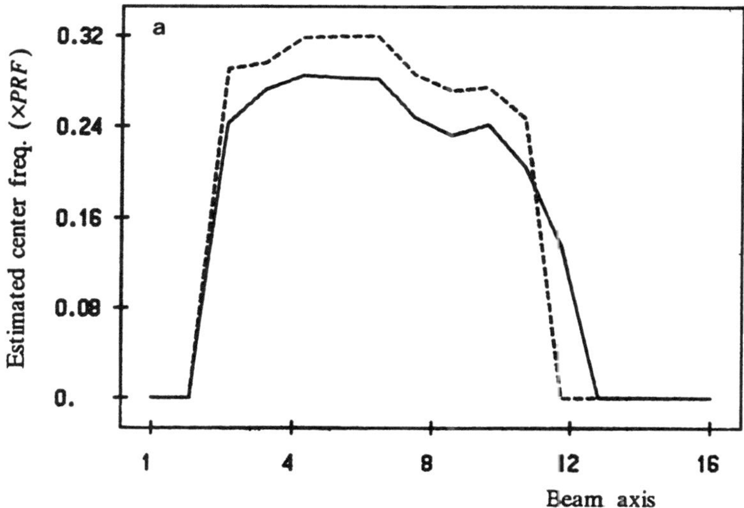

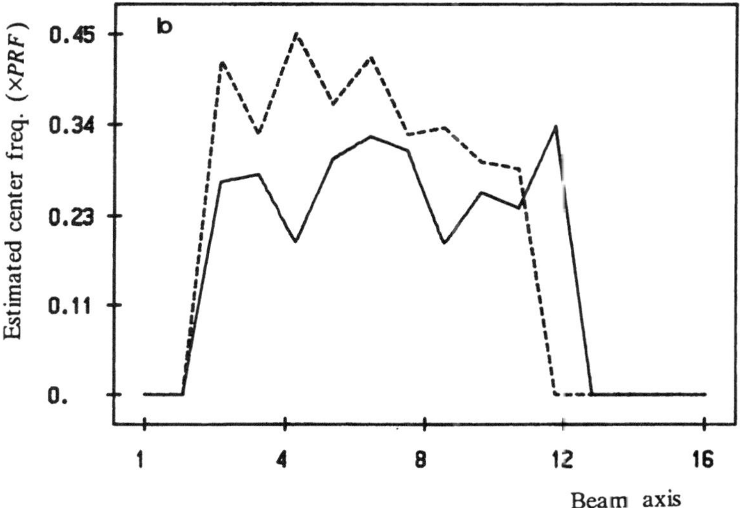

Fig. 11 Experimental results for the center frequency estimation using (a) 128 and (b) 8 sampled data. Solid line for AR estimator and dotted line for AC estimator.

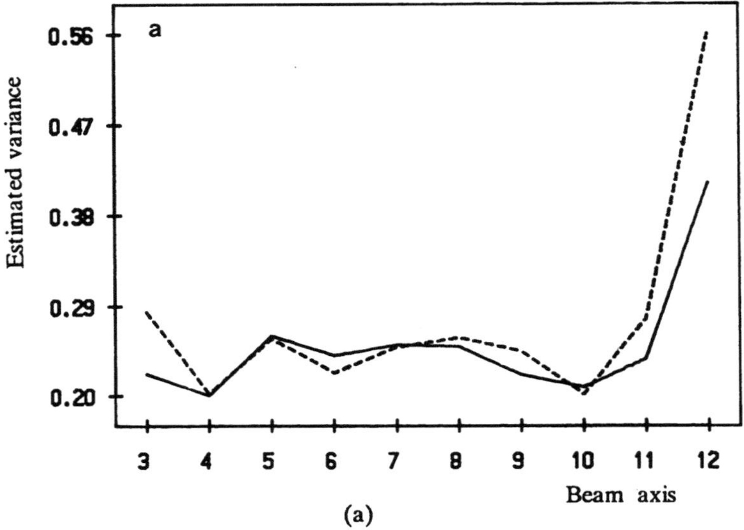

(a)

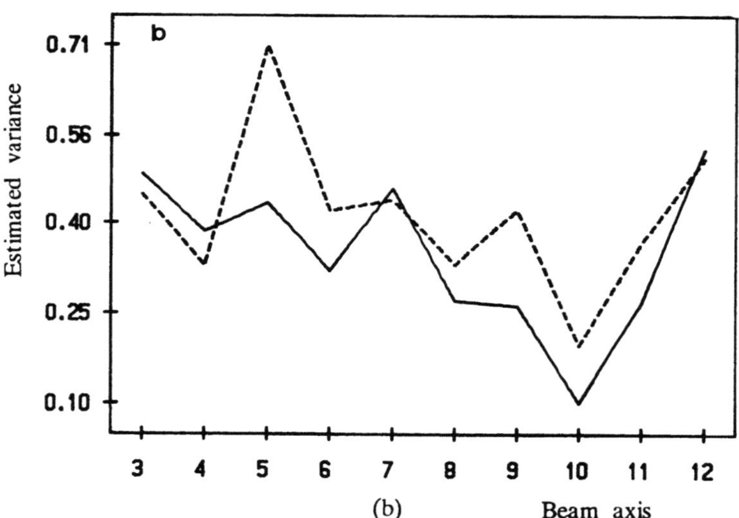

(b)

Fig. 12 Experimental results for the variance estimation using (a) 128 and (b) 8 sampled data. Solid line for AR estimator and dotted line for AC estimator.

The variances obtained by the two approaches using the same experimental data are shown in Fig. 12. Comparisons of the curves lead to a conclusion similar to that for the center frequency estimation.

Practical Consideration

In the realistic situations, we may have the system noise (often random), clutter and forward and/or reverse Doppler signals. Since the second order AR model provides two poles, there may be some confusion as to what to be correspond to the two poles. A good strategy in this case may be to preset a certain threshold for the noise level and another for the clutter frequency, and consider only the pole(poles) located in the unslashed region in Fig. 13. [The pole located in the Region I or II (III or IV) corresponds to the forward (reverse) flow.]

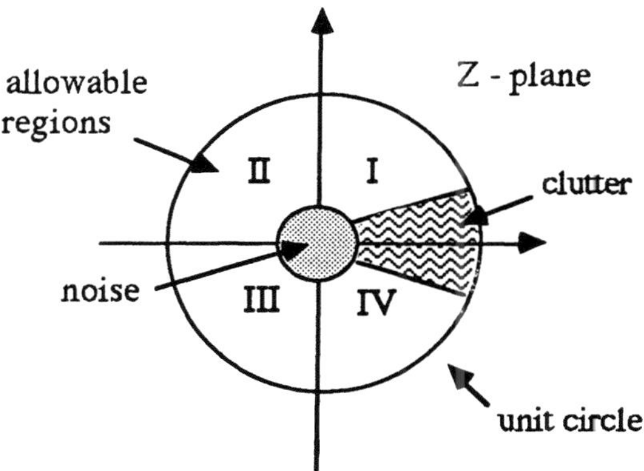

Regions I, II   : pole locations for forward flows

Regions III, IV : pole locations for reverse flows

Fig. 13   Allowable regions for the Doppler pole(s) in the z-plane to exclude clutter and system noise.

CONCLUSIONS

An AR estimator was proposed which estimates the center frequency and the variance of only the Doppler signal from the phase and magnitude, respectively, of the relevant pole of the second-order AR model, even though the clutter is included in the Doppler signal. The AR estimator shows better performance for low flow velocities and requires a smaller number of sampled data than the autocorrelation estimator. Also, the proposed method can estimate the center frequencies and variances of both the forward and the reverse blood flows in multi-gate and 2-D Doppler systems.

The performance of the AR estimator can be degraded by the presence of clutter very large compared with the Doppler signal. Further investigations must be done on this fact and also in other algorithms for the calculation of $a(1)$ and $a(2)$ in Eq. (12).

APPENDIX

The parameters a(1) and a(2) in the second-order AR model using the Burg algorithm[3] are calculated as follows ($z(n)$: input data , N : number of data):

$$k_1 = \frac{2 \sum_{n=2}^{N} z(n) \, z^*(n-1)}{\sum_{n=2}^{N} \{|z(n)|^2 + |z(n-1)|^2\}}$$

$$e^f(n) = z(n) + k_1 \, z(n-1)$$

$$e^b(n) = z(n-1) + k_1^* \, z(n)$$

$$k_2 = \frac{2 \sum_{n=3}^{N} e^f(n) \, e^{b*}(n-1)}{\sum_{n=3}^{N} \{|e^f(n)|^2 + |e^b(n-1)|^2\}}$$

$$a(1) = k_2$$

$$a(2) = k_1 + k_2 k_1^*$$

REFERENCE

1. L. I. Halberg and K. E. Thiele, "Extraction of Blood Flow Information Using Doppler-Shifted Ultrasound," Hewlett-Packard Journal, Vol. 37, No. 6, June 1986.

2. C. Kasai, et al., "Real Time Two-Dimensional Flow Imaging Using an Autocorrelation Technique," IEEE Trans. on Sonics and Ultrasonics, SU-32, No. 3, May 1985.

3. S. Lawrence, Marple. Jr., pp. 189 - 260, in : "Digital Spectral Analysis with Applications," Prentice-Hall. Inc., Englewood Cliffs. New Jersey (1987).

4. G. M. Genkins and D. G. Watts, "Spectral Analysis and Its Applications," Holden-Day, San Francisco (1968)

AN OPTIMIZATION METHOD FOR ACOUSTIC IMPEDANCE ESTIMATION

OF LAYERED STRUCTURES USING PRIOR KNOWLEDGE

Cedric A. Zala* and Kenneth I. McRae†

*Barrodale Computing Services Ltd.
Suite 200, 1677 Poplar Ave.
Victoria, B. C., Canada V8P 4K5

†Defence Research Establishment Pacific
FMO, Victoria, B. C., Canada V0S 1B0

ABSTRACT

We present an optimization method for estimating acoustic impedance profiles of
layered composite materials from ultrasonic pulse-echo data when some prior knowledge
is available. The method assumes that: (1) the defect-free material consists of a small
number of layers with approximately known thicknesses, impedances, and frequency-independent attenuations, and (2) the defects are thin and consist of either disbonds at an
interface or delaminations within a layer. Using the prior knowledge of the impedances
as an initial estimate, the impedances may be optimized for a particular set of layer
thicknesses. The structure of the layers may then be adjusted, in ways consistent with
the nature of expected defects, to improve the fit to the trace. These include altering
the thicknesses of the layers and allowing additional layers in the region of an interface
and within an existing layer. Furthermore, the impedance values within each layer may
be constrained to lie within specified bounds, thus ensuring a solution consistent with
the known physical structure of the material. The method may be applied to noisy and
band-limited data, and may also be formulated so as to allow recursive estimation of
impedance profiles in the absence of prior knowledge. Illustrations of its performance
are given in examples using synthetic and real data.

INTRODUCTION

Ultrasonic techniques are increasingly used in the nondestructive evaluation of layered materials to identify defects within layers or at interfaces between layers. For this
application, normal incidence pulse-echo techniques are often used to collect ultrasonic
traces as the position of the transducer is varied; a series of these traces may be stacked
to form a "B-scan" image of the interior of the material. To aid in the interpretation of
these ultrasonic images and the detection of defects, it is desirable to invert the traces in
order to estimate the acoustic impedance profile of the material and the corresponding
reflection coefficients.

The problem of inversion of normal incidence acoustic data has been extensively studied. Recursive inversion methods have been developed for the ideal case of noise-free data and impulsive wavelets for both the seismic (free surface) and the ultrasonic (non-free surface) cases [1-4]. Special purpose methods have been proposed for cases where the reflections are weak and multiples can be neglected [5]. General inversion methods have also been developed for application to noisy and band-limited data: these have involved recursive estimation [6-9], optimization techniques [10-12] or linear programming [13].

Certain special considerations arise in the inversion of data obtained during ultrasonic inspection of composite materials. Generally only a few layers exist, consisting of metals and composite materials such as graphite-epoxy laminates. The impedance profiles are thus strongly discontinuous, giving rise to substantial multiple reflections, and a low level of noise is present on the traces. The transducer wavelet is band-limited and may be accurately measured. Measurements on composite specimens have also shown substantial attenuation of the wavelet within the composite layers. Finally in many cases in practice, prior information is available about both the defect-free acoustic impedance profile and the types of defect likely to occur.

The special properties of this particular inversion problem (in particular, the incorporation of attenuation and prior knowledge) were not easily accommodated using existing inversion procedures. We therefore developed special purpose inversion procedures, based on optimization, for acoustic impedance estimation and detection of defects in composite materials.

STATEMENT AND SOLUTION OF THE INVERSION PROBLEM

In posing the problem to be solved, the generally used discretization with respect to travel-time rather than distance ([1],[4]) is adopted here. A series of M discrete layers is assumed to provide an underlying model for the data, with the mth layer characterized by an acoustic impedance z_m for $m = 1, \ldots, M$, and by an attenuation coefficient α_m and a two-way travel-time h_m for $m = 1, \ldots, M-1$. An N-element model trace $t(M, z, h, \alpha, w)$ is taken to be the convolution of a transducer wavelet w with an impulse response $r(M, z, h, \alpha)$; i.e., $t(M, z, h, \alpha, w) = w * r(M, z, h, \alpha)$. An observed trace is represented by the sum of the model trace and band-limited or broadband noise.

The estimation of the relative impedances of the layers is equivalent to the computation of reflection coefficients $c_m, m = 1, \ldots, M - 1$ for the $M - 1$ interfaces between the layers through the relations

$$c_m = \frac{z_{m+1} - z_m}{z_{m+1} + z_m} \qquad \text{and} \qquad z_{m+1} = z_m \frac{1 + c_m}{1 - c_m}. \tag{1}$$

Inversion in the Presence of Prior Knowledge

The overall problem in the presence of prior knowledge is: *given a measured trace d, a known wavelet w, initial estimates for M, h and z, and fixed α, obtain final estimates of M, h and z which are consistent with the approximately known structure and types of defects.* An optimization procedure was adopted which minimizes the objective function

$$f(M, z, h) = \sum_{n=1}^{N} [d_n - t_n(M, z, h)]^2. \tag{2}$$

There are two stages involved in the optimization: specification of the layers and computation of the impedances within the specified layers. For the optimization of impedances for fixed layer positions, the computation of the objective function and derivatives was formulated in such a way as to allow the use of a recent algorithm of Powell [14]. This algorithm minimizes a differentiable function of several variables subject to bounds and linear constraints, using an iterative line search method. Thus lower and upper bounds on the impedances may be specified in accordance with prior knowledge. The derivatives required by the algorithm were computed analytically.

Attenuation was modelled separately for each layer, using a simple frequency-independent model. No attempt to jointly estimate impedance and attenuation was made since initial studies suggested that this problem can be ill-conditioned.

The problem of the estimation of the layers (i.e., the number of layers and the thickness of each) is non-linear and non-differentiable, and was approached in the context of the prior knowledge of the structure of the material. The strategy was first to optimize the impedances using the prior knowledge provided as an initial estimate. If the fit was not acceptable, three types of layer manipulations were then allowed. These were:

1. adjustment of the thicknesses of the layers, one unit at a time, until a minimum sum of squares was achieved;

2. addition of a new layer at a boundary;

3. addition of a new layer entirely within an existing layer.

The first manipulation was defined to deal with the fact that the prior knowledge of the interface position is approximate, while the second two provided for two known types of defect: disbonds at an interface and delaminations within a layer.

Recursive Inversion in the Absence of Prior Knowledge

To widen the applicability of the impedance optimization procedure to the general case where no prior knowledge was available, a recursive estimation procedure was developed using the same optimization algorithm [14] to compute the impedances at each stage. The procedure involved the following stages, starting at the first position in the trace.

1. Holding positions and impedance values of previous layers fixed, optimize the impedance for a layer brought in at the current position.

2. If the decrease in sum of squares exceeds a specified threshold, and there is no position within a specified window surrounding the current position which gives a greater decrease, bring in a new layer, optimizing the impedance values for all previous layers.

3. Continue 1 and 2 until a satisfactory fit has been achieved, or until a specified depth has been reached.

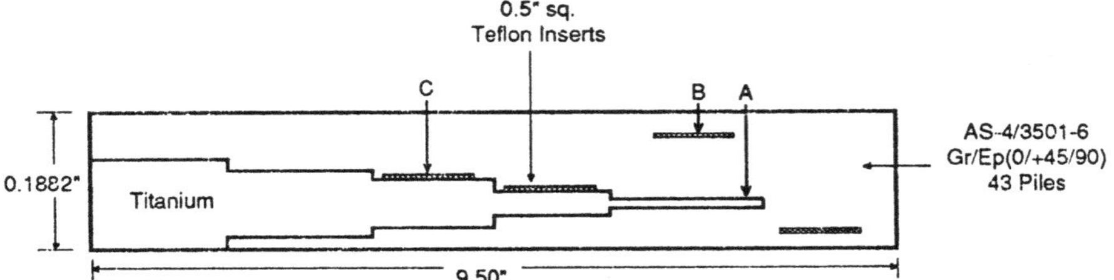

Figure 1. Structure of steplap joint specimen used for acquiring ultrasonic data. Arrows and letters identify the positions where traces were measured.

It was found useful in practice to specify the window to coincide with the first positive maximum of the autocorrelation function of the wavelet. The efficiency of the procedure could also be improved by first performing a local least squares fit of the wavelet to the residual trace for each position (i.e., neglecting multiples for a trial fit), and performing an optimization only if the decrease in sum of squares exceeded a certain threshold.

RESULTS AND DISCUSSION

Inversion of Real Data using Prior Knowledge

Ultrasonic traces were obtained for different regions of the titanium-to-graphite/epoxy steplap joint specimen shown in Fig. 1. This specimen contained two adhesively bonded interfaces as well as four square 0.5×0.5-in folded Teflon inserts placed at various positions within the specimen. These inserts trapped a thin layer of air and were used to simulate regions of disbond or delamination. Locations at which traces were obtained are shown by arrows. Acquisition of the data was done at normal incidence using a 5 MHz transducer; data was digitized at 25×10^6 samples/sec, with 256 points per trace

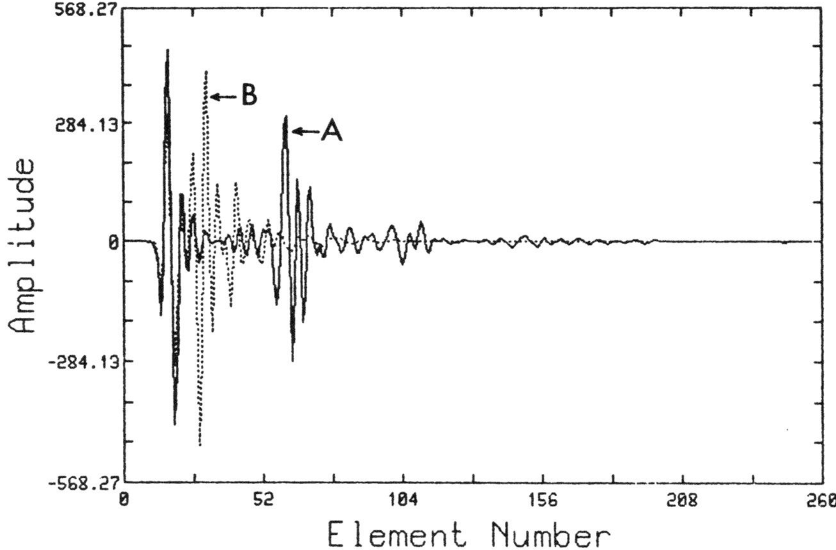

Figure 2. Comparison of traces observed for defect-free region of specimen (position A of Fig. 1: solid line) and simulated delaminated region (position B of Fig. 1: dotted line).

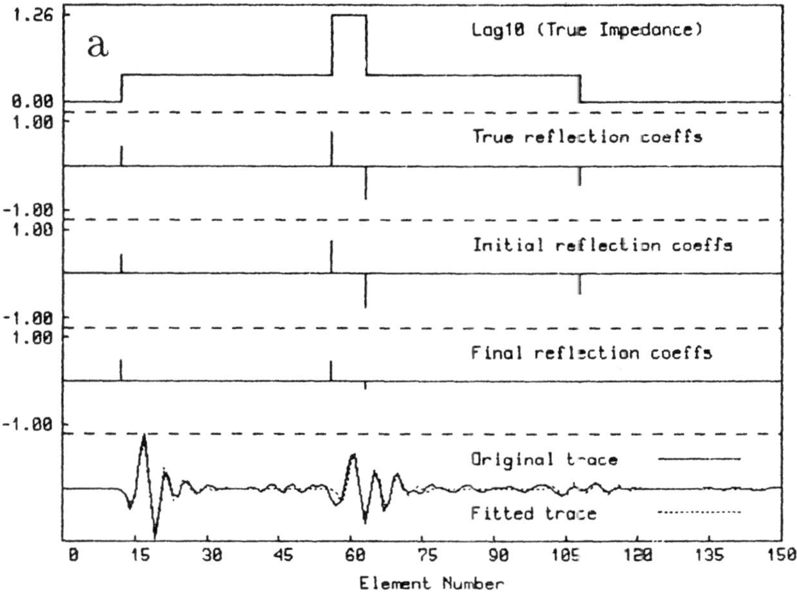

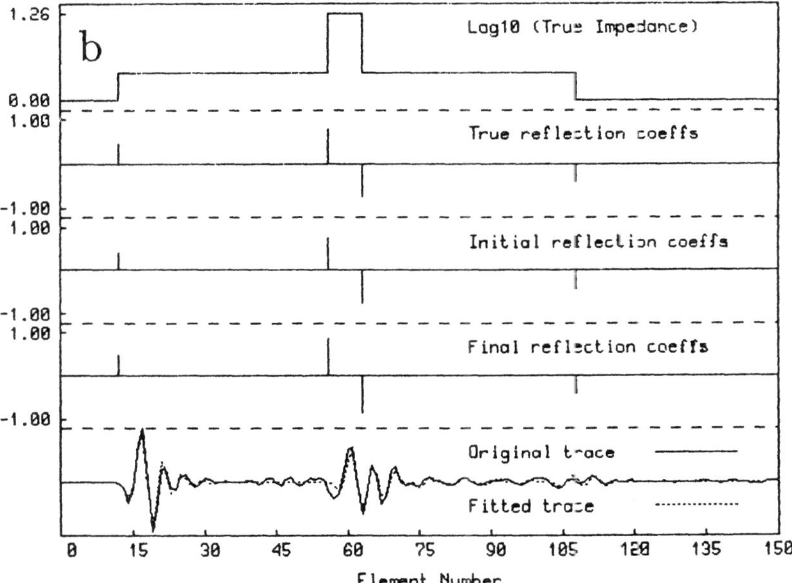

Figure 3. Inversion of ultrasonic traces from defect-free region (position A of Fig. 1). (a) No attenuation in any model layers; (b) With attenuation coefficient of 1.9 cm^{-1} in graphite/epoxy layers.

being collected, of which the first 150 were usually used for the inversion studies. The transducer wavelet was separately obtained from the front surface reflection from an aluminum block. Impedances were expressed relative to water being unity, in which case the relative impedances of graphite/epoxy and titanium are about 2.51 and 18.23, respectively. In the figures which follow, plots of the inversion results are shown with impedance plotted on a logarithmic scale, and reflection coefficients plotted on a linear scale.

The effect of a simulated delamination on the ultrasonic trace is shown in Fig. 2, which compares the data observed at positions A and B of Fig. 1. The solid line corresponds to the defect-free trace (position A), in which the water:graphite/epoxy and graphite/epoxy:titanium reflections are clearly visible. In the trace for the delaminated region (dotted line - position B), an early reflection of reversed phase followed by its multiple are apparent. These correspond to reverberations within the graphite/epoxy layer between the water and Teflon-air interfaces.

The necessity of including attenuation in the modelling is shown in Fig. 3. The trace here was obtained from position A of Fig. 1, and inverted in the absence of bounds on the impedance and using the correct positions for the interfaces. Fig. 3a shows the reflection coefficients computed from the optimized impedances when the attenuation was assumed to be zero; in this case they were severely underestimated. When an attenuation coefficient of 1.9 cm^{-1} was specified for the graphite/epoxy layer, good estimates for the reflection coefficients were then obtained (Fig. 3b).

Fig. 4 illustrates the effects of manipulating the layer thicknesses when the initial estimate for these was in error. In this example, the true thicknesses, in terms of

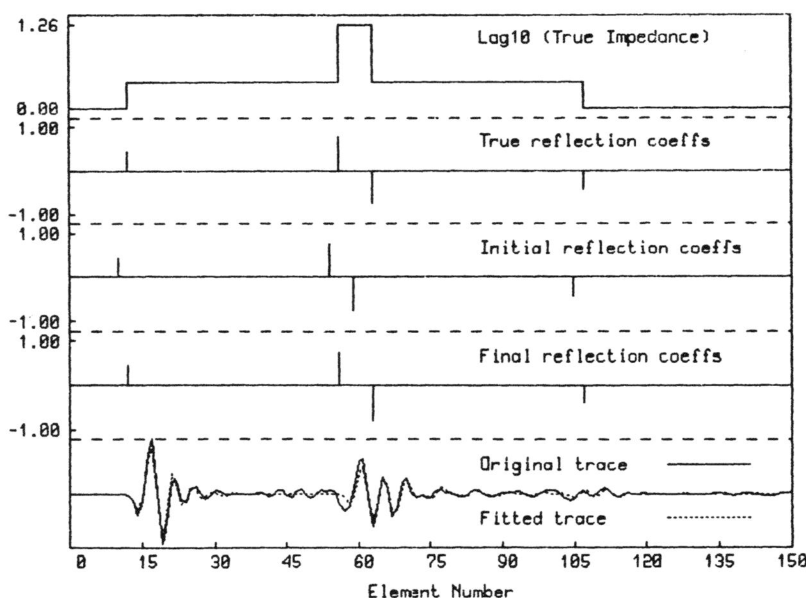

Figure 4. Recovery of actual impedance profile by layer thickness manipulation when the initial estimate provided was in error.

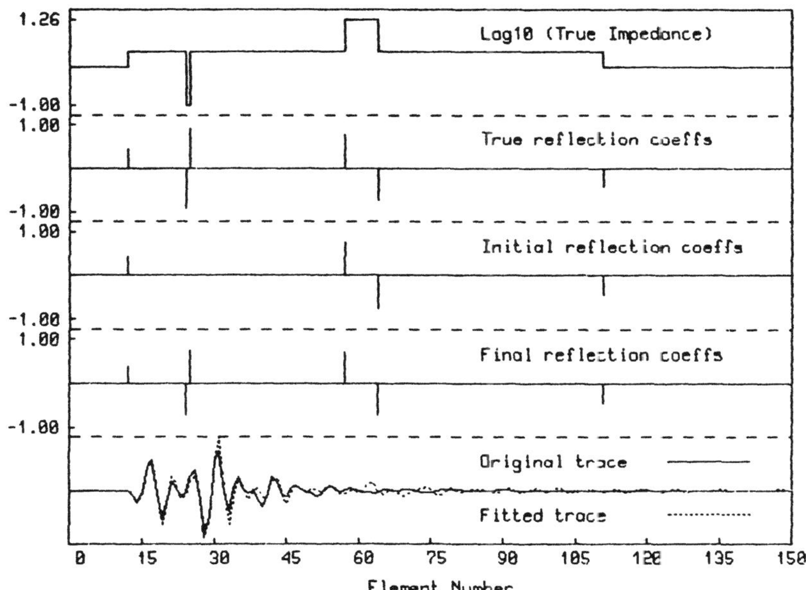

Figure 5. Inversion of a trace containing a simulated delamination (position B of Fig. 1), using the layer-adding procedure for layer manipulation, when the initial estimate was the defect-free profile.

numbers of samples, were 12, 44, 7, and 44, while those for the estimate were 10, 44, 5, and 46. Optimizing the impedance values using the initial estimate and restricting the impedances to within 50% to 200% of their true values gave a relative sum of squares of 1.654, compared to the sum of squares of the trace. (Note that because of the bounds, the sum of squares of the residuals of the fit may actually be greater than that for the original trace.) Following manipulation of the layer boundaries, this value decreased to 0.0821 and the correct positions of the interfaces were in fact identified.

Inversion of a trace with a simulated delamination is shown in Fig. 5. This trace was obtained from position B of Fig. 1, and the prior estimate provided to the algorithm was the defect-free impedance profile. In this example relative bounds for the impedance values of the known layers of 90% to 111% were specified. The relative sum of squares using the original impedance profile provided was 0.946, but after layer manipulation allowing an extra layer to be brought in at the optimal position, this was reduced to 0.162, and a good estimate of the impedance structure in the presence of the defect was obtained.

An example of inversion of a trace from a region containing a simulated disbond at the graphite/epoxy:titanium interface (position C of Fig. 1) is shown in Fig. 6. In this example the bounds on the impedances of the known layers were again restricted to the region 90% to 111% of their correct values, and the initial estimate was the defect-free impedance profile. For this estimate, the relative sum of squares was 0.992, but after a new layer was brought in at the optimal position, this was reduced to 0.0923, and the resulting profile provided a very good estimate of the actual impedance structure.

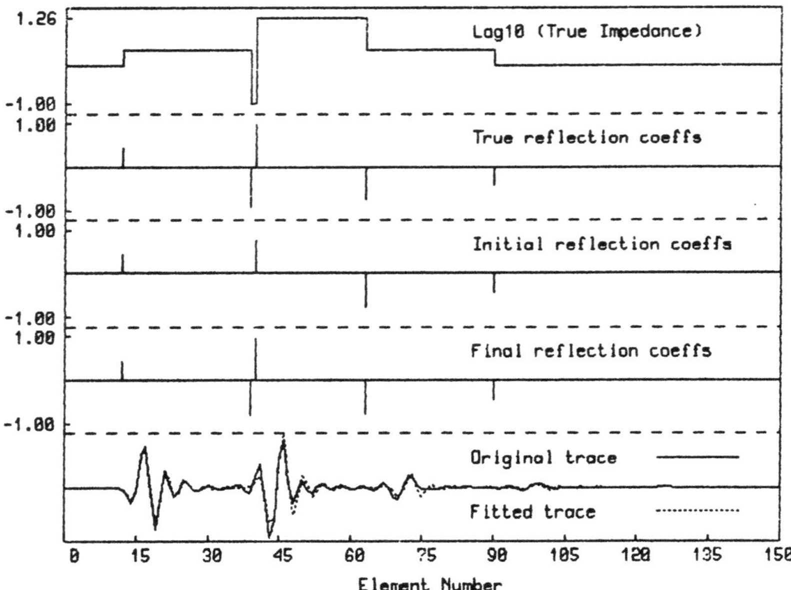

Figure 6. Inversion of a trace containing a simulated disbond (position C of Fig. 1), using the boundary-adding procedure for layer manipulation, when the initial estimate was the defect-free profile.

Inversion of Synthetic Data without Prior Knowledge

The performance of the optimization procedure formulated as described above to allow recursive estimation of the impedance profile is shown in Fig. 7. In this example, the ten-layer system described in the legend to Fig. 7 was defined, with reflection coefficients of at most 0.25 in absolute value. No attenuation was modelled in any of the layers. A zero-phase band-limited 25-element wavelet was generated using the formula $w(j) = \cos(2\pi 0.1 j) \exp(-0.025|j|^2), j = -12, \ldots, 12$. This wavelet was convolved with the impulse response corresponding to the impedance profile, and the resulting noise-free trace and wavelet were used as input to the algorithm. The resulting estimate of the reflection coefficients is a close approximation of the true set, even though one interface corresponding to a weak reflection has been missed. This example illustrates that the present recursive estimation procedure can give reasonable estimates of deeper layers and reflection coefficients even when the structure of earlier stages is somewhat in error.

CONCLUSIONS

The optimization-based inversion procedure described in this paper can provide accurate estimation of impedance profiles for non-destructive evaluation applications. Here prior knowledge of the defect-free structure is incorporated in the form of constraints on the impedance values, and information on the types of defect likely to be present is accommodated by the types of manipulations allowed for the layers. Good estimates for the actual impedance profiles of a composite specimen have been obtained for real data from defect-free and defect-containing regions of a composite material specimen. The optimization method also has the ability to be implemented in a recursive mode, allowing impedance estimation in the absence of prior knowledge.

370

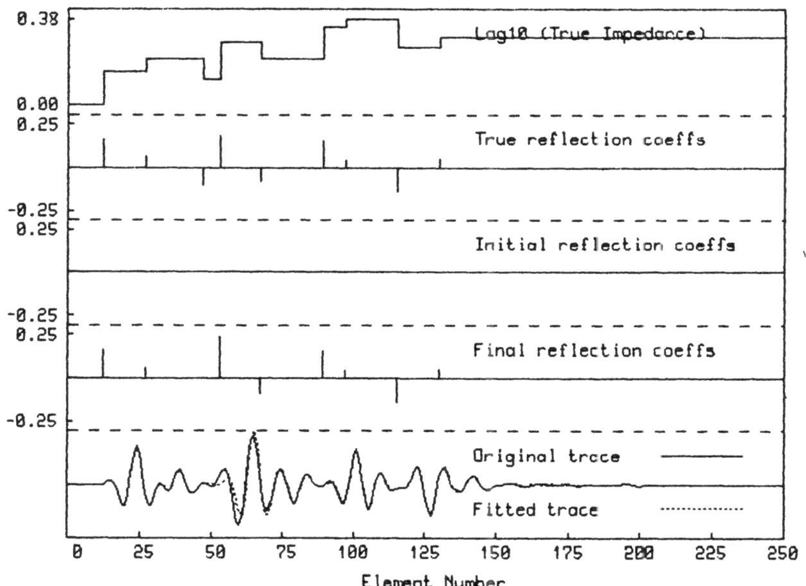

Figure 7. Recursive estimation of layers using optimization method, with no prior knowledge. The 10-layer impedance profile $\{z_m, h_m\}$ used to generate the trace was $[(1.0,12); (1.4,15); (1.6,20); (1.3,6); (1.9,14); (1.6,22); (2.2,8); (2.4;18); (1.8,15); (2.0,-)]$.

ACKNOWLEDGEMENTS

We are indebted to Prof. M. J. D. Powell of the University of Cambridge for his assistance in formulating the inversion problem to allow application of optimization techniques. This work was supported through a contract with the Defence Research Establishment Pacific, Canada, for which funding was provided from the Unsolicited Proposals Program.

REFERENCES

1. P. L. Goupillaud, An approach to inverse filtering of near-surface layer effects from seismic records, Geophysics 26: 754-760 (1961).

2. R. G. Newton, Inversion of reflection data for layered media: A review of exact methods, Geophys. J. R. astr. Soc. 65: 191-215 (1981)

3. E. A. Robinson, Spectral approach to geophysical inversion by Lorentz, Fourier and Radon transforms, Proc. IEEE 70: 455-470 (1982).

4. K. P. Bube and R. Burridge, The one-dimensional inverse problem of reflection seismology, SIAM Review 25: 497-559 (1983).

5. J. P. Jones, Impediography: A new ultrasonic technique for non-destructive testing and medical diagnosis, in "Ultrasonics International 1973 Conference Proceedings," 214-218 (1973).

6. J. M. Mendel and F. Habibi-Ashrafi, A survey of approaches to solving inverse problems for lossless layered media systems, <u>IEEE Tr. Geosci. Rem. Sens.</u> GE-18: 320-330 (1980).

7. F. Habibi-Ashrafi and J. M. Mendel, Estimation of parameters in lossless layered media systems, IEEE Tr. Auto. Contr. AC-27: 31-48 (1982).

8. J. M. Mendel and J. Goutsias, One-dimensional normal-incidence inversion: A solution procedure for band-limited and noisy data, <u>Proc. IEEE</u> 74: 401-414 (1986).

9. I. Koltracht and P Lancaster, Threshold algorithms for the prediction of reflection coefficients in a layered medium, <u>Geophysics</u> 53: 908-919 (1988).

10. A. Bamberger, G. Chaventi, Ch. Hemon and P. Lailly, Inversion of normal incidence seismograms, <u>Geophysics</u> 47: 757-770 (1982).

11. D. Lesselier, Optimization techniques and inverse problems: Probing of acoustic impedance profiles in time domain, <u>J. Acoust. Soc. Am.</u> 72: 1276-1284 (1982).

12. D. A. Cooke and W. A. Schneider, Generalized linear inversion of reflection seismic data, <u>Geophysics</u> 48: 665-676 (1983).

13. D. W. Oldenburg, S. Levy and K. J. Stinson, Inversion of band-limited reflection seismograms: Theory and practice, <u>Proc. IEEE</u> 74: 487-497 (1986).

14. M. J. D. Powell, TOLMIN: A Fortran package for linearly constrained optimization calculations, DAMTP Report 1989/NA2, University of Cambridge (1989). [This subroutine will be made available through the IMSL and Harwell software libraries.]

THE FEASIBILITY OF REAL-TIME 3D-ACOUSTICAL IMAGING

C.J.M.van Ruiten, G.Boersma and A.J.Berkhout[*]

TNO-Institute of Applied Physics, Delft, The Netherlands
[*]Delft University of Technology Fact.of Appl.Physics

ABSTRACT

The concept of signal processing for 3D acoustical imaging should be based on holographic techniques or simplifications of these techniques. The holographic approach is attractive both from the point of flexibility and implementation. With appropriate hardware a real-time 3D-imaging system can only be practically realized by parallel processing on orthogonal vectors in the spatial Fourier domain. Techniques, commonly used in seismic exploration, are proposed for manipulating the acoustic measurements in Fourier oriented domains.
Ultimate image quality is given by the maximum number of elements which can be gathered in one 2D-linear transducer array.

1. INTRODUCTION

Three dimensional acoustical imaging provides opportunities for completely automotive operating vehicles in underwater situations with low visibility. Visualization in these and other hazardous or unstructured environments demands for a new generation of sonar systems. ROV's normally make use of optical or radar systems. The various tasks of ROV's in this field of applications will be: tracking and orientation, classification and identification of objects, inspection of offshore structures, etc.
An acoustical sensor should have comparable properties, with respect to real time operation and field of view, as optic systems and should be flexible and robust in operation. However, attempts to develop a marketable 3D-acoustical imaging system has not been successful thusfar. This challenge to innovation means a concentration of research forces on the field of transducer design, miniaturization/integration of microelectronics, signal processing and system configuration. A feasibility study has been carried out to demonstrate 3D-image forming capabilities and to formulate the system design considerations.

2. BACKGROUND OF 3D-ACOUSTICAL IMAGING

Basically, acoustical imaging may be formulated as 'delineation of the reflectivity' of the insonified volume.
Reflectivity from scattering, reflection or even refraction depends on geometrical shape, surface properties and material of the targets.

In contrast to optical imaging, mainly specular reflections will be
obtained. This is due to the large wavelength relative to the dimensions
of the surface roughness. An other problem is given by speckle. Refecti-
ons suffer from interference between discrete echoes within a single
resolution cell (voxel). The need for high spatial resolution and high
sidelobe suppression increases the complexity of the system considerably.

Before defining the signal processing algorithms of the proposed imaging system, a summary is given of the state-of-the-art with respect to
available techniques.

Three different methods of imaging are known:
a. a linear array with an acoustic lens for beamforming and focussing;
b. a phased array with beamforming by "delay and sum" or spatial Fourier
   transforms;
c. synthetic aperture method, where the image is reconstructed from
   received data in a digital memory by making use of wave theory.

The first two methods are specific realizations of the more general description of method c). They are based on direct imaging of narrowband
plane waves. Method a) is not practical because of the size of the lens;
it is not flexible in operation. Method c), synthetic aperture adopted
from radar, is not adequate for real-time imaging. Combinations of
processing used in methods b. and c. could also be used for image formation in the far field as well as in the very near field of the acoustic
array.
In a homogeneous medium imaging can be elegantly carried out in the
Fourier domain. For this reason a flexible design of the proposed imaging hardware and software will be based on a FFT-method. A major advantage of a system based on FFT-methods is a free choice of the data
domain in which beamforming, focussing or other imaging tools can be
applied.
The basic system for 3D-imaging consist of a 2D-receiver array with a
single source for insonification of the target volume. In figure 1 the
geometry of the array and the resolution cell of the system are illustrated.

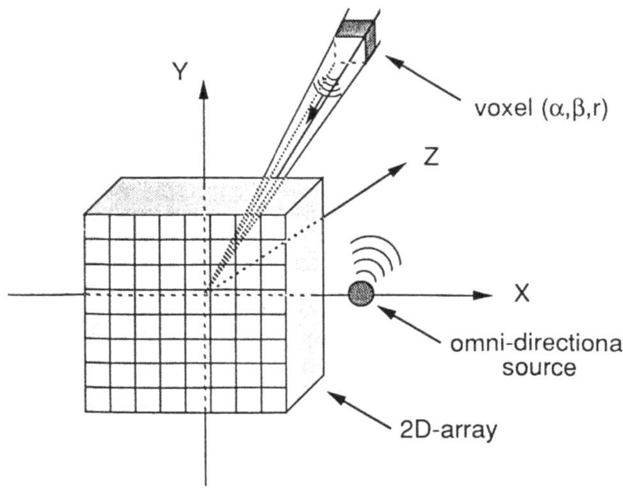

Figure 1. Illustration of geometry of 2D-array with resolution cell

3. ALGORITHMS AND PROCESSING

A generic approach for acoustical imaging is adopted from seismic exploration (see Berkhout (1) and Tatham (2)). The acoustic model for the
experiment which describes the monochromatic response from all depth
levels at the detector array is given by:

$$P(z_0) = \Sigma_i \ D \ W_d(z_0,z_i) \ R(z_i) \ W_s(z_i,z_0) \ S$$

where D = the transducer field pattern matrix
       S = the source vector
W_s,W_d = the propagation matrix respectively for forward and backward propagation
$R(z_i)$ = the reflectivity matrix (generally assumed to be diagonal) at depth level z_i

In the simplified situation of backward propagation in a homogeneous
medium, imaging can be presented in the Fourier domain. Imaging is the
inverse problem which can be expressed as:

$$R(x,y,z_i) = \text{IFT} \ \underset{\omega}{\Sigma} \ \{ \ F^{-1}(k_x,k_y,z_i,\omega) \ . \ P(k_x,k_y,z_0,\omega)\}$$

where IFT = the inverse Fourier transform,
       P   = the monochromatic sound pressure at the 2D-array in the Fourier domain,
     F^{-1} = inverse operator which describes the compensation for forward and backward propagation.

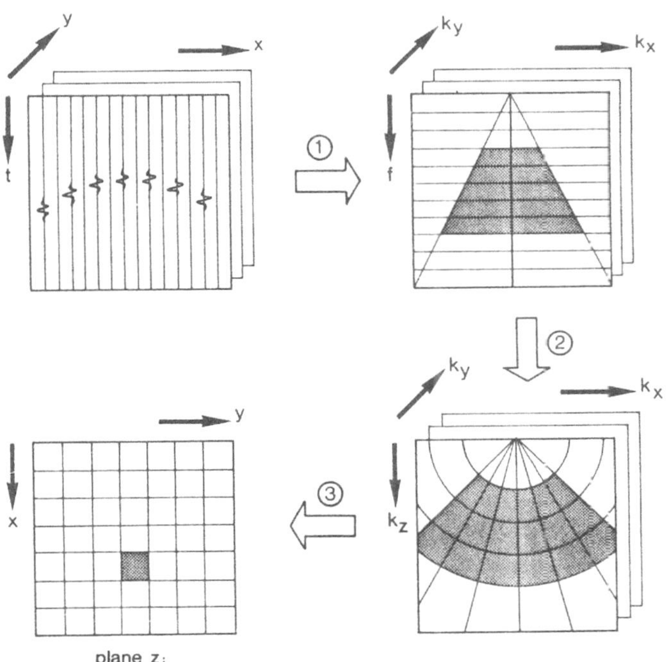

Figure 2.　　Processing steps in imaging of broadband ZO-data:
step 1: 3D-FT of array data p(x,y,t)
step 2: Compensation of propagation by mapping f -> k_z
step 3: 3D-IFT of $P(k_x,k_y,k_z)$ -> R(x,y,z)

The assumptions are made that transducers are omnidirectional and the source vector is unity. In figure 2 schematic presentation of a zero offset broadband experiment illustrates the different processing steps.

This efficient mapping algorithm is only valid for zero-offset acquisition, thus the source and detector have no spatial offset.

Acquisition using a single source position -the recording method for a 3D-camera- will use a more complex wavefield extrapolation method. Backscattering in the far field of the array can be lateral distinguished by plane wave decomposition, which is obtained by 3D-Fourier transforms:

$$P(k_x, k_y, z_o, \omega) = \text{3D-FT}\{ p(x, y, z_o, t) \}$$

The direction of every plane wave component from the insonified volume is known by the wave vector (k_x, k_y). In regions with high axial resolution a technique adopted from seismics, called Tau-Pi mapping, can be used for compensation of the propagation in these directions (see Tatham (2)). The mapping concerns the process of calculation the angle components (p_x, p_y) by interpolation from the k_x, k_y, ω-domain according to:

$$p_x = k_x/\omega = \sin\alpha/c \quad \text{and} \quad p_y = k_y/\omega = \sin\beta/c$$

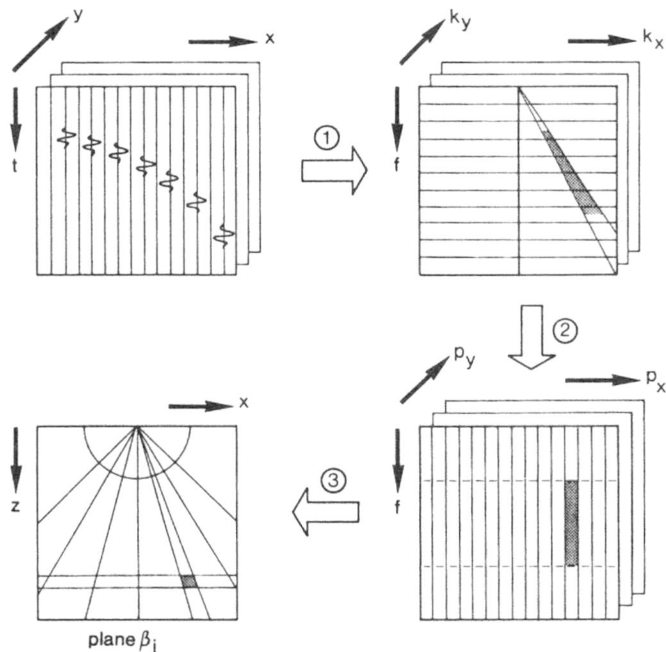

Figure 3.   Processing steps in imaging of broadband plane waves:
step 1: 3D-FT of array data p(x,y,t); in the near field corrections for the curved wavefront must be applied before this step is carried out.
step 2: Compensation of propagation by Tau-Pi mapping.
step 3: Mapping of (p_x, p_y, ω) to R(x,y,z) or $R(x, \beta_i, z)$.

All frequency components for one angle (α, β) are gathered and inverse
FT results in:

$$p(\tau, \alpha, \beta) \quad = \quad IFT\{ \ P(\omega, \alpha, \beta) \ \}$$

where Tau is the intercept time defined by $t = \tau + p_x x + p_y y$.
After geometrical mapping the reflectivity of the volume is imaged in
the spatial domain. In figure 3 these processing steps are illustrated.

The next possibility is extension of this method to the near field of
the array. The Fraunhofer condition can be fullfilled by travel time
correction in the time traces along the array elements (see figure 4).
This alignment step must be done before Tau-Pi mapping. The focal zone
is limited in aperture angle, so the aligned data are only a portion of
the total data. To focus all the data within the near field, different
focussing directions must be used. In the very near field this situation
becomes worse. For imaging very close to the array the holographic
technique mentioned above should be used.

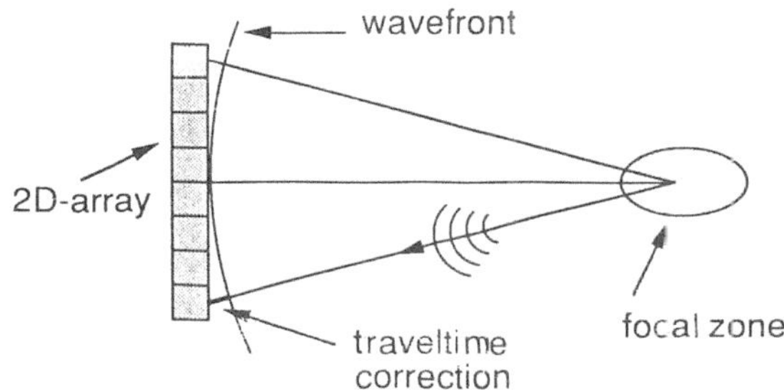

Figure 4. Focussing in the near field by travel time correction.

Table 1 summarizes the processing algorithms and the relevant parameters.

Table 1. Processing algorithms for beamforming.

range	ax.resolution/ (m)		algorithms (bandwidth)	applications
> 10	>0.2m	(1%)	Complex demodulation & plane wave decom	tracking & orientation
3-10	0,05-0,2m	(3-10%)	Tau-Pi transform	object-classifica-tion
1-3	0,02-0,05m	(<40%)	Traveltime correction & Tau-Pi transform	object-identifica-tion
< 1	focal zone	(>40%)	Wavefield extrapolation	inspections

Finally, in a region where axial resolution can be low, the frequency
bandwidth can be small. A single frequency component results directly
in the plane wave vectors after a spatial 2D-FT. By using complex
demodulation of a narrowband signal, the frequency component at every
array element is determined as a function of time. In figure 5 the
processing steps are presented schematically.

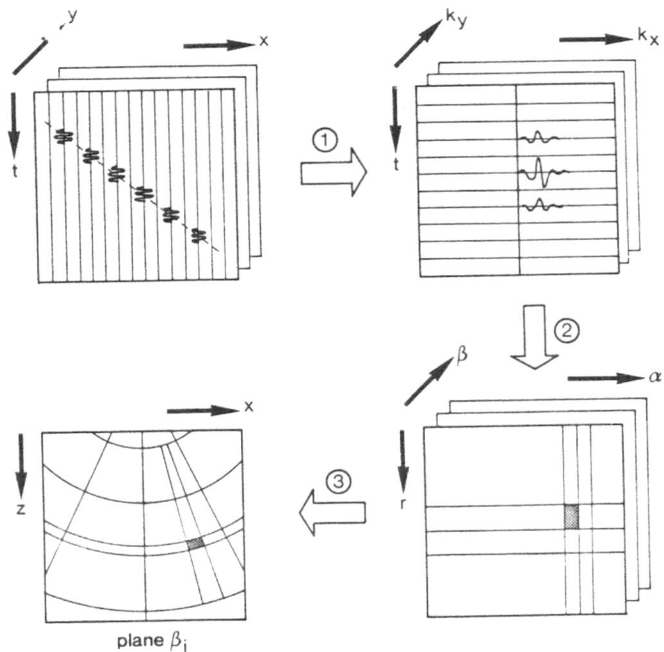

Figure 5. Processing steps in imaging of narrowband far field-data
  step 1: Complex demodulation and 2D-FT of array data $p(x,y,t)$
  step 2: Mapping of $P(k_x,k_y,t)$ to $R(p_x,p_y,r)$
  step 3: Mapping of $R(\alpha,\beta,r)$ to $R(x,y,z)$ or $R(x,\beta_i,z)$.

4. SYSTEM DESIGN CONSIDERATIONS

System specifications follow from user specifications in most applications. Specifications are primarily set by the field of view, the desired measurement range (sonar equation), the real-time requirements and
backscattering function of targets.
The final result of the image formation process will be a 5-dimensional
vector space: $R(x,y,z,\omega,t)$
The first two lateral dimensions are given by beamwidth and focal
distance. The axial resolution in the depth field is depending on the
frequency bandwidth of the signal. The last two dimensions are additional, not necessary for the volume imaging. The frequency dimension
stands for multi-frequency operation, where the image of the target is
assumed to be different at different operating frequencies. The last
dimension is the real-time sequence of images, which can be used for
further processing like multi view imaging.

REAL-TIME REQUIREMENTS

High demands are imposed by the requirement of real-time imaging. If
the frame repetition rate is fixed, there is a maximum range where
real-time definition is violated. If a frame is built up by various
scans, these frames have to be combined in one shot record. The processing time per scan is also limited by the same time of flight. Only
with dedicated hardware and state-of-the-art parallel processing, these
requirements can be realised.

HIGH RESOLUTION

In the field of underwater acoustics, we do not have an equivalent for
a CCD-array, which is a very efficient sensor for local light intensity
distribution. Up to now we still need a complete circuity for each
single element to detect the local phase of the acoustical wave field.
Innovative research will be needed on the integration of the transducer
array with front-end electronics.
If we use a completely filled array to meet the given specifications a
2D-transducer array with at least 10.000 elements is required. The
number of pixels in the frontal image basically equals the number of
elements.

5. EVALUATION

Going from a linear array to a 2D-array with comparable resolution
means a multiplication of the number of components, the scale of integration and a number of processing operations which goes beyond the
feasibility of conventional beamforming techniques.
By using processing algorithms which can operate in parallel processes
and by using the flexibility of software to modify the system for
various applications it is feasible to reach the desired specifications.

Imaging algorithms from seismics are an elegant solution for a real-
time, modular high-resolution 3D-acoustical imaging system.
Flexibility in the system is given by the bandwidth of the signal.
Available bandwidth can be used for axial resolution and multi-frequency operation with respect to reflection properties.

REFERENCES

[1] A.J.Berkhout, "Seismic migration, imaging of acoustic energy by
     wave field extrapolation A: Theoretical aspects", Elsevier Scien-
     tific Publishing Company, Amsterdam, 1982.

[2] R.H.Tatham, "Multidimentional filtering of seismic data", Procee-
     dings of the IEEE, October, 1984.

A COMPARISON OF BROADBAND HOLOGRAPHIC

AND TOMOGRAPHIC IMAGING CONCEPTS

Günter Prokoph* and Helmut Ermert[+]

* Universität Erlangen-Nürnberg
Institut für Hochfrequenztechnik
Cauerstraße 9
D-8520 Erlangen, Federal Republic of Germany

[+] Ruhr-Universität Bochum
Institut für Hoch- und Höchstfrequenztechnik
P.O. Box 10 21 48 / IC 6
D-4630 Bochum 1, Federal Republic of Germany

INTRODUCTION

Originally, holography is a monofrequency technique which also was applied to acoustical waves after its introduction into the optical range[1]. While acoustical holography was not very successful in medical imaging because of the inhomogenity of biological tissue it became a powerful tool for non-destructive testing[7]. Unfortunately, its excellent lateral resolution obtained by the synthetic aperture concept of scanning holographic techniques is accompanied by a poor axial resolution due to its zero bandwidth and large acoustical wavelengths. Axial resolution is considerably better in the case of conventional pulse-echo imaging, here axial positions of scatterers are related to the echo travelling time while axial resolution corresponds to the pulse length. Large signal bandwidth, for example representated by short pulses, leads to good axial resolution. Lateral resolution of conventional pulse-echo imaging is determined by the transducer beam shape and is generally 2 to 3 times lower than axial resolution. Consequently, high resolution with respect to both the axial and the lateral direction can be obtained by a combination of synthetic aperture and bandwidth. The paper deals with an investigation of various holographic imaging concepts including diffraction tomography. Broadband modifications will be presented and significant differences will be pointed out.

SCATTERING

Scalar wave fields shall be considered defined by their time-domain and their frequency domain representations corresponding to

$$\phi(\vec{r}, t) = \mathrm{Re}\{\phi(\vec{r}, \omega)e^{-j\omega t}\} \tag{1}$$

with a spatial vector $\vec{r}$ describing spatial variations of ϕ. Scattering objects in a source-free area are assumed to be surounded by a homogeneous medium where the wave equation

$$(\nabla^2 + k^2)\phi(\vec{r}, \omega) = 0 \tag{2}$$

is valid with $k = \omega/c_o$ the wave-number and c_o the velocity of sound. A wave ϕ_i incident into a scattering region causes a scattered field

$$\phi_s(\vec{r},\omega) = \iiint\limits_{V} q_s(\vec{r}',\omega) \cdot G(\vec{r}'|\vec{r},\omega)dv' \qquad (3)$$

q_s is a distribution of sources induced by the incident field at the location of the scatterers inside a volume V'. G is Green's function and $\phi = \phi_i + \phi_s$ the total field. In the case of objects represented by inhomogenities of sound velocity $c(\vec{r})$ induced sources are determined by

$$q_s(\vec{r},\omega) = H_s(\omega) \cdot O(\vec{r}) \cdot \phi_i(\vec{r},\omega) \qquad (4)$$

with the influence of the scattered field ϕ_s being neglected (BORN-approximation). For BORN-approximation the object function and a spectral function[12] are defined by

$$O_b(\vec{r}) = \frac{1}{c^2(\vec{r})} - \frac{1}{c_o^2} \, , \qquad H_{sb}(\omega) = \omega^2 \qquad (5,6)$$

For scattering discontinuities having totally reflecting surfaces a Physical-Optics approximation assumes $\phi = 0$ on the surface of these scatterers which also are considered to have specular reflectivity. In this case the object function and spectral function are[12]

$$O_p(\vec{r}) = \frac{1}{c} \cdot \Gamma(\vec{r}) \, , \qquad H_{sp}(\omega) = 2j\omega \qquad (7,8)$$

with $\Gamma(\vec{r})$ the distribution of scattering surface elements which are "visible" from the aperture and contribute significantly to the scattered field. Multiple scattering has been neglected and the orientation of scattering elements is approximated to be normal with respect to the direction of wave propagation. It has been shown, that Eq. (3) is not only applicable to longitudinal waves but also to shear-waves approximately[12]. In the following chapters imaging considerations will be based on a 2 dimensional geometry shown in Fig. 1. x_A and x_B are the aperture position of the

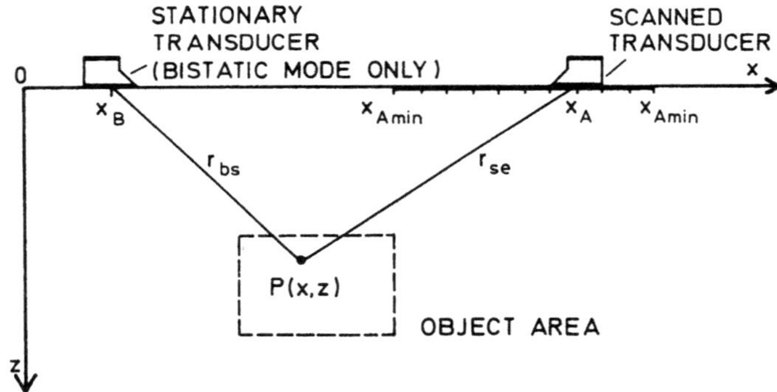

<u>Fig. 1.</u> Imaging geometry

scanned receiving transducer and the stationary transmitting transducer (bistatic concept). In monostatic imaging only one transducer is acting as transmitter and receiver ($x_B = x_A$). Distances between transducer positions and an object point located at x,z are

$$r_{bs} = \sqrt{(x - x_B)^2 + z^2} \qquad \text{and} \qquad r_{se} = \sqrt{(x - x_A)^2 + z^2} \qquad (9a,b)$$

and the corresponding time delays

$$\tau_l = \tau_{bs} + \tau_{se} = (r_{bs} + r_{se})/c \qquad (10)$$

In 2D-geometry scatterers are line scatterers, and Greens-Function is

$$G(\vec{r}\,'|\vec{r},\omega) = \frac{j}{4} \cdot H_O^{(1)}(kr) \approx \sqrt{\frac{2}{\pi k r}} \cdot \frac{e^{j(kr + \pi/4)}}{4} \tag{11}$$

with $r = |\vec{r} - \vec{r}\,'|$ and the approximation being valid for $kr \gg 1$.

MULTIFREQUENCY HOLOGRAPHY

Multifrequency holography is similar to an electromagnetic remote-sensing concept which is called "synthetic-aperture-radar" (SAR)[2]. It was developed for microwave imaging[6,8] before being transferred to the acoustical region[9]. The algorithm is based on a spatial matched filter-concept. Signals F_E measured at aperture positions x_A are compared to a test function F_F which represents the response of the system to a single line-scatterer at that point x,z of the image area which has to be reconstructed numerically. Mathematically, this comparison is a two dimensional correlation of F_E and F_F for all frequencies ω and all aperture positions x_A:

$$B_{MH}(x,z) = \left| \int_{x_{Amin}}^{x_{Amax}} \int_{\omega_{min}}^{\omega_{max}} F_E(x_A,\omega) \cdot F_F^*(x_A,x,z,\omega)\,d\omega\,dx_A \right| \tag{12}$$

B_{MH} is the image function of multifrequency holography which leads to certain values representing image brightness of the corresponding pixel. Assuming a line scatterer at the position x, z and applying the approximation in Eq. (11)

$$F_F(x_A,x,z,\omega) = G_{MH}^*(\omega) \cdot \frac{e^{jk(r_{bs} + r_{se}) + j\frac{\pi}{2}}}{2\pi\sqrt{r_{bs} \cdot r_{se}}} \tag{13}$$

can be calculated which is related to Fig. 1. $G_{MH}(\omega)$ is defined by

$$G_{MH}^*(\omega) = \frac{1}{4k} \cdot F_B(\omega) \cdot H_B(\omega) \cdot H_E(\omega) \cdot H_s(\omega). \tag{14}$$

F_B is the spectrum of the electrical transmitter signal, H_B the transfer function of the transmitting transducer with respect to the electrical input and to the acoustical output. H_E is a corresponding transfer function of the receiving transducer, and H_s is defined in Eq. (4). Using

$$F_E'(x_A,\omega) = F_E(x_A,\omega) \cdot G_{MH}(\omega) \tag{15}$$

a reconstruction algorithm for the time-domain can be derived:

$$B_{MH}(x,z) = \left| \int_{x_{Amin}}^{x_{Amax}} \frac{f_E'(x_A,\tau_l)}{\sqrt{r_{bs} \cdot r_{se}}}\,dx_A \right| \tag{16}$$

As analytical signals have spectral components at positive frequencies only, measured time signals representing the real part have to be supplemented by their imaginary part via a HILBERT-transform. In order to get $f_E'(x_A,t)$ the measured function $f_E(x_A,t)$ has to be filtered (convolved) according to Eq. (17):

$$f_E'(x_A,t) = g_{MH}(t) * [f_E(x_A,t) + j \cdot HI\{f_E(x_A,t)\}] \tag{17}$$

$g_{MH}(t)$ is the time-domain representations of $G_{MH}(\omega)$. In an experimental imaging system this filtering can be obtained by generating prefiltered transmitter signals in a waveform generator. In conclusion, the time domain reconstruction algorithm of multifrequency holography (Eq. (16)) is an integration of analytical signals with respect to their travelling time between the antenna positions and each object point which has to be reconstructed. The algorithm is applicable to the bistatic case (fixed transmitting transducer and scanned receiving transducer) as well as to the monostatic mode with only one transmitting and receiving antenna ($r = r_{bs} = r_{se}$).

RAYLEIGH-SOMMERFELD HOLOGRAPHY

Rayleigh-Sommerfeld holography is based on diffraction theory[1,3,4]. The basic idea is to reconstruct a scattered field caused by scatterers in a homogeneous medium. Knowing the scattered field on the aperture the field intensities in the object area can be computed. The field intensity distribution in the object area is closely related to the shape of a scattering object and to the acoustical discontinuity of its surface. This approach does not allow monostatic operation mode, because in this case the scattered field is changing with the changing transmitter position. Therefore in this chapter monostatic imaging will not be considered. As before only two dimensional fields and objects shall be taken into account, they are considered to be constant with respect to the y-direction. One way to solve the wave equation as to apply a spatial Fourier-transform with respect to the x-direction according to

$$\phi(x,z,\omega) \leftrightarrow \hat{\phi}(K_x,z,\omega) \tag{18}$$

with K_x the space-domain variable. Fourier-transform representation is equivalent to an expansion of scattered field in terms of plane waves. Taking into account that scattered waves are travelling into the negative z-direction only an expression relating $\hat{\phi}_s$ in an arbitrary distance $z > 0$ to the field in the area $z = 0$ can be derived:

$$\hat{\phi}_s(K_x, z > 0, \omega) = \hat{\phi}_s(K_x, z = 0, \omega) \cdot e^{-jz\sqrt{k^2 - K_x^2}} \tag{19}$$

Eq. (19) is the Fourier-domain representation of Rayleigh-Sommerfeld-Holography, also called "backpropagation", with the exponential function in Eq. (19) as the "propagator". A more efficient way to calculate the backpropagated field in the entire object area is to apply two dimensional Fourier-transforms[3] with respect to x and z:

$$\phi(x,z,\omega) \leftrightarrow \tilde{\phi}(K_x,K_y,\omega) \tag{20}$$

For monofrequency wave-fields in homogeneous media a relation between the magnitudes of wave-number k and wave-vectors K_x and K_z is $K_x^2 + K_z^2 = k^2$. Consequently, the two dimensional Fourier-transform of a monofrequency scattered field is located upon a circle with a radius k and a center in the origin of the K-space with variables K_x and K_z ("Ewald-circle"). A relation between 1-D Fourier-transform $\hat{\phi}_s$ and 2-D Fourier-transform $\tilde{\phi}_s$ can be derived according to[3]

$$\tilde{\phi}_s(K_x, K_z = -\sqrt{k^2 - K_x^2}, \omega) = \frac{2\pi}{k} \cdot \sqrt{k^2 - K_x^2} \cdot \hat{\phi}_s(K_x, z = 0, \omega) \tag{21}$$

Following a procedure presented in[3] the reconstruction algorithm of Rayleigh-Sommerfeld holography can be derived for the space-domain:

$$\phi_s(x,z,\omega) = -2 \cdot \int_{-\infty}^{\infty} \phi_s(x_A,\omega) \cdot \frac{\partial}{\partial z}\left[-\frac{j}{4} \cdot H_0^{(2)}(k \cdot r_{se}(x_A,x,z))\right] dx_A \tag{22}$$

with a Hankel-function which represents wave propagation with respect to positive radial direction. Using approximations valid for $k \cdot r \gg 1$ and the relation

$$G_{RS}(\omega) = \frac{(1-j) \cdot \sqrt{\omega} \cdot H_E(\omega)}{2\sqrt{\pi c}} \tag{23}$$

eq. (22) can be transformed into the time domain[13]:

$$\phi_s(x,z,t) \simeq \int_{-\infty}^{\infty} g_{RS}(t) * f_E(x_A, t + \tau_{se}) \cdot \frac{z}{r_{se}^{3/2}} dx_A \tag{24}$$

Eq. (24) allows a calculation of the scattered wave field in the object area. $\phi_s(x,z,t)$ is a time dependent function which is not suitable for representation in an image. This problem can be solved by a calculation of the envelope of the complex signal and taking values of this envelope at certain times. It shall be assumed that the scattered field has its maximal values on the surface of the scatterers. Therefore, in Eq. (24) a time delay τ_{bs} has to be taken into account for a calculation of the image amplitude in the pixel located at x,z. Consequently, a suitable reconstruction algorithm is:

$$B_{RS}(x,z) = \left| \int_{-\infty}^{\infty} f_E'(x_A, \tau_l) \cdot \frac{z \cdot \sqrt{r_{bs}}}{r_{se}^{3/2}} dx_A \right| \tag{25}$$

For practical applications knowledge of the scattered field amplitudes calculated via backpropagation into the object area is of a less interest as apposed to the properties of the scatterers themselves. Usually, the object scene is not insonified by a plane wave, and therefore the incident wave field is not of constant amplitude over the entire object area. Assuming cylindrical wave insonification the amplitude of the incident wave field at the location of the scatterers depends on the distance between these scatterers and the position of the transmitter. In order to reconstruct the scattering properties of the objects an additional geometrical weighting factor $\sqrt{r_{bs}}$ has been inserted into Eq. (25), which is very simular to the corresponding reconstruction algorithm of multifrequency holography (Eq. (16)). f_E' has to be calculated according to Eq. (17) with $g_{MH}(t)$ replaced by $g_{RS}(t)$.

DIFFRACTION TOMOGRAPHY

Finally a spatial-domain broadband reconstruction algorithm based on the concept of diffraction tomography will be derived. The aim of diffraction tomography algorithms is a direct inversion of the integral equation which describes the problem of forward scattering. This inversion is possible in the spatial Fourier-domain as well as in the spatial-domain. Reconstruction in the Fourier-domain is more efficient than in spatial-domain as long as the number of aperture points and the number of pixels in the image which has to be reconstructed are very large. This is due to the availability of efficient FFT-algorithms. In many practical cases with less data computational times are comparable and spatial-domain techniques become more advantageous because they are easier to handle. This chapter will focus on spatial-domain reconstruction.

Bistatic imaging

Foundations of diffraction tomography with a fixed object illumination have been presented in detail in[3,5]. The aim is to reconstruct in a 2D-geometry "induced" line-sources q_s which lead to a scattered field

$$\phi_s(x,z,\omega) = \int_{-\infty}^{\infty} \int_{-\infty}^{\infty} q_s(x',z',\omega) \cdot \frac{j}{4} \cdot H_0^{(1)}(k\sqrt{(x-x')^2 + (z-z')^2}\,)dx'dz' \tag{26}$$

Inversion of Eq. (26) can be obtained in the Fourier-domain by

$$\tilde{q}_s(K_x, K_z = -\sqrt{k^2 - K_x^2}\,, \omega) = \hat{\phi}_s(K_x, z = 0, \omega) \cdot \frac{2}{j} \cdot \sqrt{k^2 - K_x^2} \qquad (27)$$

Consequently, scattered aperture field (ϕ_s) has to be transformed to the spatial Fourier-domain ($\hat{\phi}_s$) and represents after weighting by a k-dependent term image information located on the Ewald circle in the $-K_z$-area. Relation between Eq. (27) and the corresponding reconstruction algorithm of Rayleigh-Sommerfeld holography Eq. (21) is very close, the only difference is a factor $2/j$ instead of $2\pi/k$. As different frequency dependent weightings can be eliminated by suitable prefiltering of transmitted signals the reconstruction problem of bistatic diffraction tomography turns out to be identical to that of Rayleigh-Sommerfeld Holography.

Monostatic imaging

Reconstruction of fields or induced sources does not make sense if the illuminating source is moving and the fields or the induced sources are changing during data collection. Therefore, in the case of monostatic imaging (one transmitting and receiving antenna) object properties have to be chosen for reconstruction which do not change. Reflectively O(x,z) as defined in Eqs. (5,7) approximately causes an aperture signal

$$F_E(x, z = 0, \omega) \simeq F_B(\omega) \cdot H_G(\omega) \cdot \frac{1+j}{\sqrt{8\pi k}} \int\limits_{-\infty}^{\infty} \int\limits_{-\infty}^{\infty} O(x', z') \cdot \frac{1}{\sqrt{r}} \cdot \frac{j}{4} H_0^{(1)}(2kr) dx' dz'$$

$$(28)$$

Because of the influence of the distance $r = \sqrt{(x - x'^2) + z'^2}$ on wave propagation between transducer and object back and forth 2kr appears as an argument of the Hankel-function after approximating the square of the Hankel-function in the exact relation[12]. F_B is the spectrum of the transmitted signal and H_G is the transfer function which takes into account the frequency response of transducer and scattering according to Eqs. (6,8). Eq. (28) is valid for distances r equal or larger than a couple of wavelengths. For further considerations, let us replace the factor $1/\sqrt{r}$ by 1 in Eq. (28) and call the corresponding echo function F_{EM}. With this modification Eq. (28) could be used to derive a spatial-domain reconstruction algorithm as in the case of bistatic diffraction tomography. Based on the assumption, that the filtered transmitter signal $F_B'(\omega) = F_B(\omega) \cdot H_G(\omega) \cdot j/(8\pi k)$ represents a short time domain signal $f_B'(t)$ a detailed analysis[12] leads to a simple relation between F_E and F_{EM} in the time domain:

$$f_{EM}(x, z = 0, t) \simeq \sqrt{\frac{c \cdot t}{2}} \cdot f_E(x, z = 0, t) \qquad (29)$$

which has been verified by numerical simulations for $r > c/\Delta f$ with Δf the pulse bandwidth. The agreement of the approximation to the exact formula improves with growing distance between transducer and scatterer and can be applied for practical situations. Having shown the validity of Eq. (29) a 2D-Fourier transform of the object function can be obtained in analogy to Eq. (27):

$$\tilde{O}(K_x, K_z = -\sqrt{4k^2 - k_x^2}\,, \omega) = \hat{F}_{EM}(K_x, z = 0, \omega) \cdot H_{GM}(\omega) \cdot \sqrt{4k^2 - K_x^2} \qquad (30)$$

with $H_{GM}(\omega) = -2\sqrt{2\pi k} \cdot (1 + j) \cdot F_B^{-1}(\omega) \cdot H_G^{-1}(\omega)$. Before reconstruction, the received signals have to be modified corresponding to Eq. (29). Fig. 2 shows the location of the $\tilde{O}$- values in the k-space. In the monofrequency case they are located on an arc of the Ewald-circle which is determined by the location of the aperture length and the scatterer positions. This arc represents a certain lateral resolution

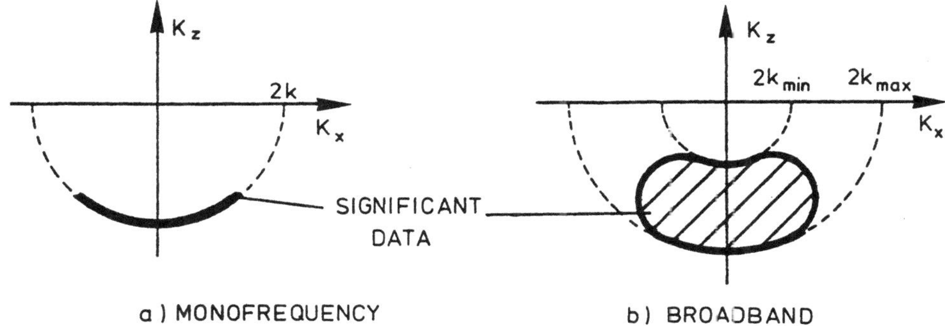

Fig. 2. Ewald circles for diffraction tomography

but a poor axial resolution. Suitable superposition of different frequencies lead to a broadband version with axial and lateral resolutions of the same order of magnitude. Additionally it shall be noted, that Ewald-circles have a radius of 2k in monostatic cases indicating

twice the resolution of bistatic concepts. Starting from Eq. (30) a broadband spatial-domain relation can be derived:

$$O(x,z) = \frac{1}{2\pi} \int\limits_{\omega_{min}}^{\omega_{max}} G_{MDT}(\omega) \int\limits_{-\infty}^{\infty} F_{EM}(x_A,\omega) \cdot \frac{z \cdot e^{-j \cdot 2kr}}{r^{3/2}} dx_A \, d\omega \qquad (31)$$

with $G_{MDT}(\omega) = -2k^2 \cdot F_B^{-1}(\omega) \cdot H_G^{-1}(\omega)$ and $r_{bs} = r_{se} = r$ leading to a final reconstruction algorithm

$$B_{MDT}(x,z) = \left| O(x,z) \right| = \left| \int\limits_{-\infty}^{+\infty} f_E'(x_A,\tau_l) \frac{z}{r} dx_A \right| \qquad (32)$$

with $f_E'(t)$ as defined in Eq. (17) and $g_{MH}(t)$ replaced by $g_{MDT}(t)$.

COMPARISON

Broadband time-domain reconstruction algorithms derived in the preceeding chapters have been summarized in Table 1: All algorithms represent signal summation with respect to their travelling time between antennas and scatterers. Taking into account different temporal filter formulas for f_E all reconstruction algorithms contain the same signal f_E'. Remaining differences are represented by geometrical weighting functions M. A numerical simulation has been carried out in order to visualize significant differences of the algorithms

$$B(x,z) = \left| \sum_{n=1}^{N} M(x_A,x_B,x,z) \cdot f_E'(x_A,\tau_l) \right| \qquad (33)$$

was applied with N the number of equidistant aperture positions along a finite aperture (Fig. 3). Fig. 3a shows the geometry of simulated imaging procedures with $x_B = 0$ in bistatic cases. The object is an arrangement of 5 line scatterers. Signals have been assumed to have a center frequency of 2.5 MHz and a bandwidth of 2 MHz. Velocity of sound was 5.9 mm · MHz (steel). All values of the image functions B(x,z) have been normalized with respect their maximal value and displayed using a grey scale between black (=0) and white (=1). Monostatic concepts lead to a better lateral resolution. Scatterers with larger distance to the aperture appear with reduced

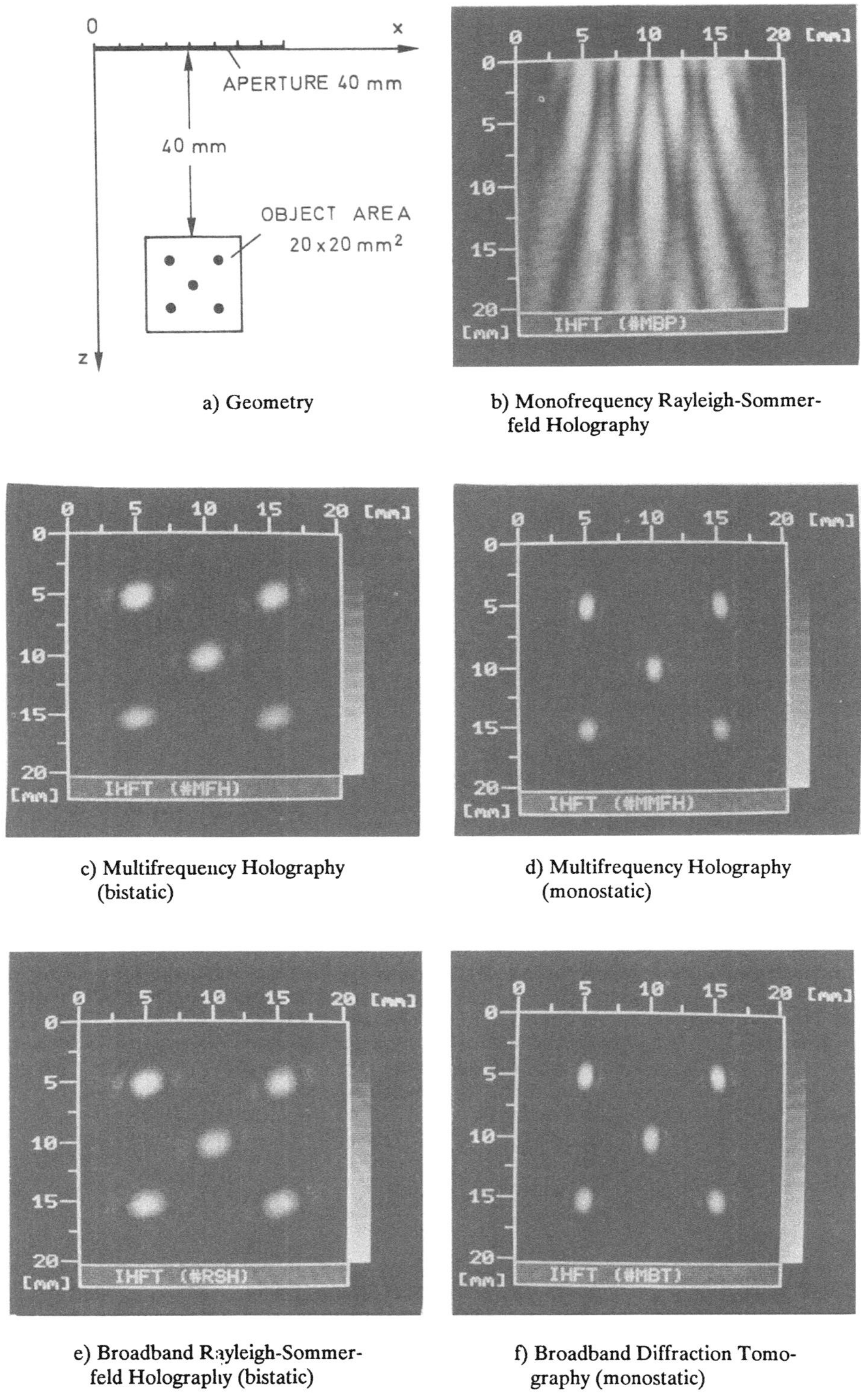

a) Geometry

b) Monofrequency Rayleigh-Sommerfeld Holography

c) Multifrequency Holography
(bistatic)

d) Multifrequency Holography
(monostatic)

e) Broadband Rayleigh-Sommerfeld Holography (bistatic)

f) Broadband Diffraction Tomography (monostatic)

Fig. 3. Comparative reconstruction simulations

Table 1. Comparison of reconstruction algorithms

Method :	Multifrequency Holography	Rayleigh - Sommerfeld Holography	Diffraction - Tomography
Procedure :	spatial matched filter for line scatterers	reconstruction of scattered field in homogeneous medium	inversion of linearized forward problem (Born - approximation)
Reconstruction Algorithm (bistatic) :	$B_{\mathrm{MH}}(x,z) = \left\| \displaystyle\int_{x_{A\min}}^{x_{A\max}} f'_{\mathrm{E}}(x_A,\tau_l)\cdot M_{\mathrm{MH}}\cdot dx_A \right\|$	$B_{\mathrm{RS}}(x,z) = \left\| \displaystyle\int_{-\infty}^{\infty} f'_{\mathrm{E}}(x_A,\tau_l)\cdot M_{\mathrm{RS}}\cdot dx_A \right\|$	$B_{\mathrm{DT}}(x,z) = \left\| \displaystyle\int_{-\infty}^{\infty} f'_{\mathrm{E}}(x_A,\tau_l)\cdot M_{\mathrm{DT}}\cdot dx_A \right\|$
Geometrical Weighting Function :	$M_{\mathrm{MH}} = \dfrac{1}{\sqrt{r_{bs}\cdot r_{se}}}$	$M_{\mathrm{RS}} = \dfrac{z\sqrt{r_{bs}}}{r_{se}^{3/2}}$	$M_{\mathrm{DT}} = \dfrac{z\sqrt{r_{bs}}}{r_{se}^{3/2}}$
Reconstruction Algorithm (monostatic) :	$B_{\mathrm{MMH}}(x,z) = \left\| \displaystyle\int_{x_{A\min}}^{x_{A\max}} f'_{\mathrm{E}}(x_A,\tau_l)\cdot M_{\mathrm{MMH}}\cdot dx_A \right\|$	reconstruction not possible, scattered field changes with changing insonification	$E_{\mathrm{MDT}}(x,z) = \left\| \displaystyle\int_{-\infty}^{\infty} f'_{\mathrm{E}}(x_A,\tau_l)\cdot M_{\mathrm{MDT}}\cdot dx_A \right\|$
Geometrical Weighting Function :	$M_{\mathrm{MMH}} = \dfrac{1}{r}$	—	$M_{\mathrm{MDT}} = \dfrac{z}{r}$

brightness and reduced lateral resolution. Geometrical weighting functions M do not vary significantly for small object areas. Therefore, assuming M=1, a very simple algorithm can be obtained which leads to a good result as shown in Fig. 4. For further comparison, a reconstructed image obtained from RF-signals without demodulation is presented in Fig. 5. An analysis of resolution capability as well as of the influence of noise, variations of phase velocity, medium inhomogeniety, and multiple scattering on the imaging result has been presented in [12] as well as experimental imaging results.

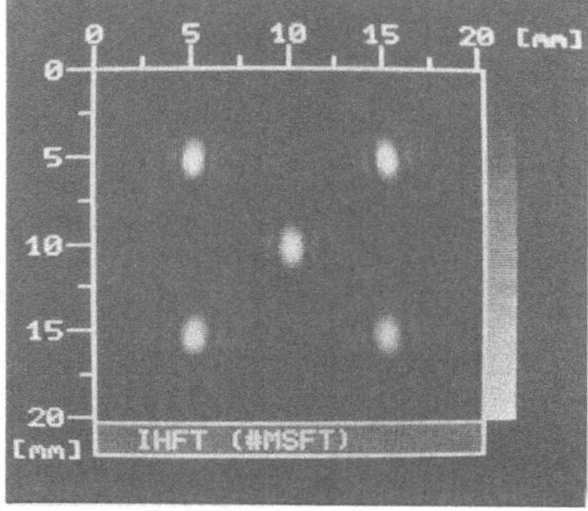

Fig. 4. Reconstruction with Eq. (33) and M=1 (monostatic)

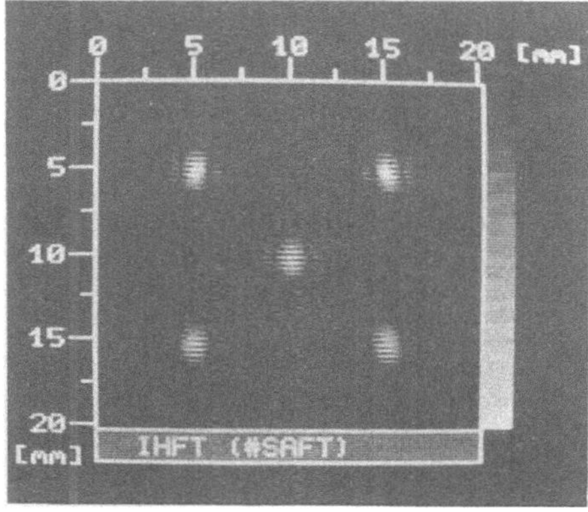

Fig. 5. Reconstruction using RF-signals

CONCLUSIONS

A detailed analysis of different imaging concepts and their extension to broadband modifications lead to reconstruction algorithms representing echo signal summation with respect to their travelling time between antennas and scatterers. Multiple scattering has been neglected. Linearized inversion of scattering formulas has been used taking into account continous inhomogeneities (BORN-approximation) as well as discontinous scatterers (Physical-Optics-approximation). Differences of the imaging concepts are represented by different geometrical weighting functions. All broadband algorithms lead to very similar imaging results.

REFERENCES

1. Goodman, J.W., "Introduction to Fourier Optics," McGraw-Hill, New York (1968).

2. Skolnik, M., "Radar Handbook," McGraw-Hill, New York (1970).

3. Langenberg, K.J., Applied Inverse Problems for Acoustic, Electromagnetic and Elastic Scattering, in: "Basic methods of Tomography and Inverse Problems", P.C. Sabatier, ed. Adam Hilger, Bristol and Philadelphia (1987), pp. 125-467.

4. Devaney, A.J., Diffraction Tomography, in: "Inverse Methods in Electromagnetic Imaging," W.M. Boerner et al., edts., Part 2, pp. 1107-1135, D. Reidel Publ. Co., Dordrecht (1985).

5. Kak, A.C., and Slaney, M., "Principles of Computerized Tomographic Imaging," IEEE Press, New York (1988).

6. Leith, E.N., Quasi Holographic Techniques in the Microwave Region, Proc. IEEE, Vol. 59, pp. 1305-1318 (1971).

7. Kutzner, J., and Wüstenberg, H., Akustische Linienholographie, ein Hilfsmittel zur Fehleranzeigeninterpretation in der Ultraschallprüfung, Materialprüfung, No. 6, pp. 189-194 (1976).

8. Karg, R., Multifrequency Microwave Holography, AEÜ, Vol. 31, pp. 150-156 (1977).

9. Ermert, H., and Karg, R., Multifrequency Acoustical Holography, IEEE Trans. on Sonics and Ultrasonics, Vol. SU-26, pp. 279-286 (1979).

10. Doctor, S.R., Hall, T.E., and Reid, L.D., SAFT - The Evalution of a Signal Processing Technology for Ultrasonic Testing, NDT International, Vol. 19, No. 3, pp. 163-167 (1986).

11. Prokoph, G., Ermert, H., and Kröning, M., A Broadband- Holography Imaging System for Nondestructive Evaluation, in: "Acoustical Imaging," Vol. 15, H.W. Jones, ed., Plenum, New York (1987), pp. 547-557.

12. Prokoph, G., "Breitbandige, holographische Abbildungsverfahren und ihre Anwendung in der zerstörungsfreien Werkstoffprüfung mit Ultraschall," Ph.D. thesis, University of Erlangen-Nürnberg (1988).

AN EXPERIMENTAL STUDY OF DIFFRACTION TOMOGRAPHY UNDER THE BORN APPROXIMATION

Brent S. Robinson* and James F. Greenleaf

Mayo Clinic/Foundation, Rochester, Minnesota
* Advanced Technology Laboratories, Bothell, Washington

ABSTRACT

Diffraction tomography under the Born approximation was studied experimentally to help assess its potential utility for clinical imaging, particularly in the detection and classification of breast disease. The Born approximation is convenient to use since, being linear, it leads to simple and efficient inversion algorithms. However, its appropriateness for clinical imaging is questionable. In these experiments, particular care was taken to minimise errors of the data acquisition and image reconstruction processes in an effort to isolate the effects of the choice of the Born scattering model on the overall performance of diffraction tomography. Diffraction tomograms of a tissue mimicking breast phantom are compared with both X-ray and ultrasound computed tomograms. The results indicate that, even under conditions when the assumptions of the Born approximation are violated, useful images can be obtained. Such images, though not quantitatively accurate maps of the complex refractive index, allow identification of the major internal structures.

INTRODUCTION

We have been studying diffraction tomography because of our interest in using ultrasound for the detection and classification of breast disease. In particular, the hope is that it could be useful in screening for breast cancer. However, in spite of its initial promise [1-7], diffraction tomography has not yet found use in routine clinical imaging. Furthermore, it is currently unclear what its ultimate role, if any, will be.

In diffraction tomography a wave equation serves as the model of the propagation and scattering processes. The theory and procedures for image reconstruction are reviewed in a number of articles [1, 8-11]. Most treatments begin with the inhomogeneous Helmholtz wave equation,

$$\nabla^2 \psi + k_0^2 \psi = k_0^2 (1 - n^2) \psi \ ,$$

(1)

(where ψ is some descriptor of the wave motion such as excess pressure, k_0 is the free space wave number, and n is the (complex) refractive index) since this is commonly assumed to describe propagation and scattering of ultrasound in biological soft tissues with useful accuracy.

Equation (1) inherently accounts for scattering due to diffraction, refraction, and absorption. In contrast, other acoustical imaging methods such as ultrasound computed

Acoustical Imaging, Vol. 18, Edited by H. Lee and G. Wade
Plenum Press, New York, 1991

tomography and pulse-echo (ie: B-Scan) imaging are based on simpler models. Both assume straight-ray propagation and usually neglect diffractive and refractive effects. The validity of any type of ray model, let alone straight-ray models, is questionable when the inhomogeneities within the object have sizes comparable to the incident wavelengths.

It seems reasonable to assume that diffraction tomography, being based on a more sophisticated model, should be able to produce images superior to those obtained by computed tomographic or pulse echo techniques. Other potential advantages of diffraction tomography (as well as computed tomography) over pulse echo imaging arise from the characteristics of the transmitting and receiving apertures. Diffraction tomography typically employs multiple large apertures. The apertures are "enclosing" in the sense that incident waves may arrive from, and scattered waves may be observed from, all angles. Such apertures furnish a great quantity of independent data concerning the scattering properties of the object. Furthermore, tomography based on forward scattering measurements provides "DC" and associated low spatial frequency information about the object and is therefore inherently quantitative [5, 6]. Conversely, pulse echo imaging tends to utilise single "peep-hole" apertures. Less total data about the object is acquired and, since the data is related to backscatter, quantitation, if attempted at all, must be by indirect methods.

The principal difficulty using Equation 1 is that ψ depends on n (yet both ψ and n are unknown in the interior of the object). Equation (1) is therefore non-linear and hence difficult to invert (ie: reconstruct the refractive index distribution based on measurements of ψ exterior to the object). If ψ is partitioned into the sum of incident and scattered waves, denoted by ψ_0 and ψ_S respectively, then the Born approximation allows Equation 1 to be transformed into the form [4, 6]

$$\nabla^2 \psi_S + k_0^2 \psi_S = k_0^2 (1 - n^2) \psi_0 \quad , \tag{2}$$

The incident wave, ψ_0, is the wave that would exist with the object removed. Since ψ_0 is therefore independent of n, Equation 2 is linear and readily inverted [1, 9, 11]. However, the Born approximation requires that $\psi_S \ll \psi_0$, and hence it (and, consequently Equation (2)) is only valid under conditions of very weak scattering. For the types of scatterers encountered in clinical imaging, the validity of the Born approximation is highly questionable.

Most research into diffraction tomography has been concerned with the accuracy and efficiency of the method of image reconstruction, formulating higher order solutions, and alternative data acquisition/reconstruction strategies. The comparatively limited amount of experimental work to date [2, 4-8, 10, 12-14] has been restricted to simple objects in the sense of weak scattering and/or simple two-dimensional geometries. It is the purpose of this paper to report on the results of diffraction tomography under the Born approximation of a clinically relevant and realistic object (specifically, an anthropomorphic breast phantom). This work both documents the image quality that can currently be achieved in practice using diffraction tomography and suggests some potentially fruitful avenues for further research.

METHODS

Factors that may degrade the quality of images obtained in these and other diffraction tomography experiments include :

(a)   Data acquisition errors, eg : data can be contaminated by noise, bias, non-linearities, drift, mechanical mis-registration, imperfections in the transmitting and receiving transducers, etc.

(b)   Data inversion errors. The image reconstruction procedure involves mapping the acquired data from a measured data space into an image space. This typically introduces round-off and interpolation errors.

(c)   Inappropriate model. An exact wave equation for propagation through biological tissue is not known but in any case would possibly be too complicated to be directly useful [3, 15]. Therefore diffraction tomography is necessarily based on approximate wave equations. Even in Equation (1), phenomena such as anisotropy, non-linearity, and mode-conversion between compressional and shear waves, etc, are neglected [10]. Equation (2), is a further approximation to the exact wave equation. A second problem is that a two-dimensional geometry is assumed when in fact the object and insonifying waves both have three-dimensional structure.

The approach used in this study was to minimise, or at least characterise the errors introduced by the first two factors in order to isolate as much as possible the effects of the last factors on the overall performance of diffraction tomography under the Born approximation.

<u>Data Acquisition</u>

A translate/rotate scanning mechanism reminiscent of "first generation" X-ray computed tomography was used to acquire the data. The experimental apparatus, depicted in Fig. 1, is described in more detail elsewhere [16, 17]. The object, in this case the "second breast phantom" constructed by Madsen and Zagzebski [18], was immersed in degassed water at room temperature. Notice that the phantom had been damaged allowing some solvent leakage. Therefore, its acoustical properties were probably different from those in the original description [18]. Also, an external wire was added to the phantom to provide an easily recognisable "point" target.

A cylindrical foam-backed PVDF transducer was used to insonify the object. This transmitted substantially plane wavefronts (at least over the horizontal extent of the transducer). The phantom was located within the vertical focal region of the transducer. The height of the wavefronts in the focal region, and hence the slice thickness of the insonification, was roughly half a centimetre. The receiving hydrophone was horizontally translated to obtain data for each view. The phantom was rotated to obtain data for subsequent views.

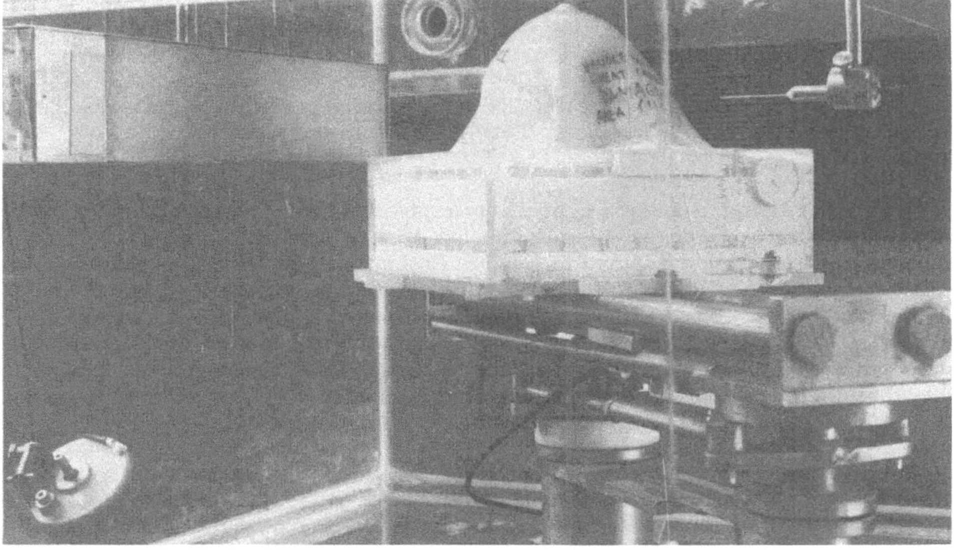

Fig. 1. *The experimental apparatus.*

A standard phase quadrature receiver was used to detect the complex amplitude of the wave field at each sample point [16, 17]. The receiver bandwidth of 10 kHz provided high signal to noise ratio. The data was normalised by a view acquired through water only (ie: with the object removed). The ability of this apparatus to accurately measure forward scattering was investigated by experiments with cylinders in which close agreement between theory and experiment was found [16, 17]. It should be noted however, that those experiments were performed for two-dimensional geometries rather than for objects with three-dimensional geometries such as the breast phantom.

A large amount of data was acquired in an effort to completely sample the scattered fields. For each tomogram,

$$240 \text{ views} * 512 \text{ complex samples/view} * 12 \text{ bits/sample} = 360 \text{ KBytes}$$

of data were acquired. The sample spacing was 0.508 mm. Since the transmit frequency was 1 MHz (λ = 1.48 mm in water), the field was sampled approximately 3 times per wavelength.

Considerable care was taken to obtain the cleanest possible set of data. Scanning was performed slowly to avoid disturbing the temperature stratification in the tank. Total time to acquire the data was a little under one hour (although this could of course be considerably reduced by use of transducer arrays). Compensation was made for backlash in the scanning mechanism. Offsets and gain mismatches of the two quadrature channels were calibrated and corrected. To avoid reverberations from the hydrophone mount (which would interfere with the accuracy of the measurements), the duration of the transmitted waveform was limited to 50 µSec. Possible ramifications of this are discussed in the final section.

<u>Image Reconstruction Algorithm</u>.

Reconstruction algorithms have been the subject of considerable research [9, 11, 19, 20]. In these experiments, the image was reconstructed onto a 256 * 256 complex rectangular grid whose spacing was 1.016 mm (ie: double that of the scanning step size) and a Fourier domain [9, 19, 20] reconstruction algorithm was utilised (since the reconstruction was much faster than for image domain methods yet the errors were acceptably small).

The data from each view was first normalised and fast-Fourier transformed. Each discrete Fourier component, interpreted as lying on the resultant "Ewald semi-circle" [11], was multiplied by an appropriate complex weighting factor [1, 7, 8] and mapped to the nearest point on a rectangular grid. Repeating this procedure for all 240 views completed coverage of the rectangular grid. Finally, an image was obtained by an inverse two-dimensional Fourier transform. Since the weighting functions and the interpolation co-ordinates were precomputed, images could be reconstructed in just under one minute using a (rather archaic by today's standards) FPS AP-120B array processor ported to a 32 bit mini-computer.

In spite of using nearest neighbour interpolation, the reconstruction algorithm performed well for two reasons. Firstly, for low spatial frequencies, the data were over-determined, easing the interpolation task. Secondly, high spatial frequencies were handled in an ad hoc fashion. Because the interpolation was implemented as a 1:1 nearest neighbour mapping, not all locations of the rectangular grid were assigned non-zero values. Specifically, locations removed from the origin corresponding to high spatial frequencies were sparsely sampled. In practice we found it better to let these samples remain zero rather than fill them with erroneously interpolated samples. As well as improving speed, this method apparently achieves a beneficial low-pass filtering effect.

The performance of the reconstruction algorithm was tested by providing as input the Fourier transforms of objects consisting of collections of points and circles. Although not reproduced here, the reconstructions showed that the artefacts introduced by the algorithm were only just discernible. Hence the reconstruction algorithm is discounted as a significant source of error in the final diffraction tomograms.

RESULTS

The breast phantom was imaged at the level of the "anterior frontal plane" [18]. The particular breast phantom used had an abnormally low concentration of backscatterers [18] and, consequently, compound B-scans exhibited low contrast. However, it was felt that the phantom had realistic forward scattering properties. Fig. 2 reveals the characteristics of the forward scattered data acquired from the breast phantom. Patterns arising from scattering from internal structures are readily identifiable in both the magnitude and phase sinograms. In the phase sinogram, $-\pi$ radians is depicted as black and $+\pi$ radians is white.

The black to white transitions evident in Fig. 2 indicate that the phantom must shift the phase of the incident wave by more than π radians. Such large phase shifts imply that the scattered field is at least as large as the incident field at some locations. Therefore, the Born approximation is grossly invalid for this set of data.

Fig. 3 shows a diffraction tomographic reconstruction of the breast phantom obtained under the Born approximation. In this figure, as in subsequent figures, each image is scaled so that the minimum value appears black and the maximum value white. Only the real and imaginary parts are displayed. In principle, the speed and absorption of the object can be obtained from the real and imaginary parts of the reconstruction. However, since the Born approximation is violated, "cross-talk" between the real and imaginary parts results in mutual contamination [17] and therefore no quantitative information about the speed and absorption is available from the reconstruction. In spite of this, Fig. 3 depicts the internal structure of the phantom reasonably faithfully.

Fig. 3 should be compared with Figs. 4 and 5 as well as the images in Reference [18] (although these are of the undamaged phantom and are rotated 180 degrees relative to the images presented here). Allowance must be made for geometric distortions caused by the differing aspect ratios of the various video screens from which these figures were reproduced. Fig. 4 is an x-ray computed tomogram [9] at the same level and orientation as for Fig. 3 and reveals with exquisite fidelity the boundary between the tissue mimicking fat and glandular parenchyma, the two simulated tumours and 9.5 mm cyst, and the obliquely orientated Cooper's ligaments. Fig. 5 shows acoustic speed and attenuation images obtained by state-of-the-art ultrasound computed tomography [8, 10]. These have higher artefact levels and are not as well resolved as the diffraction tomogram. In particular, the simulated Cooper's ligaments and wire target are only just visible. Whereas diffraction tomography has estimated the sizes and borders of the tumours quite accurately, ultrasound computed tomography in general under or over estimates sizes depending on the acoustic speed [8, 9].

Comparison of Figs. 3 and 4 reveals that the main artefacts in the Born diffraction tomogram are the streaks emanating from the Cooper's ligaments into the parenchyma, and the filling-in of the concavity beneath the wire target.

There is disagreement in the literature concerning the sensitivity of diffraction tomographic reconstructions to errors in the nominal geometry of the measurement apparatus [2, 7]. We investigated this subject by altering *a posteriori* the geometrical parameters used by the reconstruction algorithm and examining the associated effects on the reconstructions. For example, it was found that an error of 3 mm in the distance of the perpendicular from the center of rotation to the scan line (or, equivalently, an error of 4 % in the assumed speed of propagation) had a negligible effect. The error could be as large as 15 mm (about 10 λ or 19 %) before degradation of the image was noticeable. Sensitivity to lateral displacements of the nominal center of the scan line was found to be somewhat greater. While an offset of 1.5 mm was inconsequential, 3 mm was noticeable, and the reconstructions were seriously degraded by a 7.5 mm offset. However, these tolerances are easily managed at typical diagnostic ultrasound wavelengths. Therefore, it appears that diffraction tomography, at least in the form described herein, is quite a robust imaging technique.

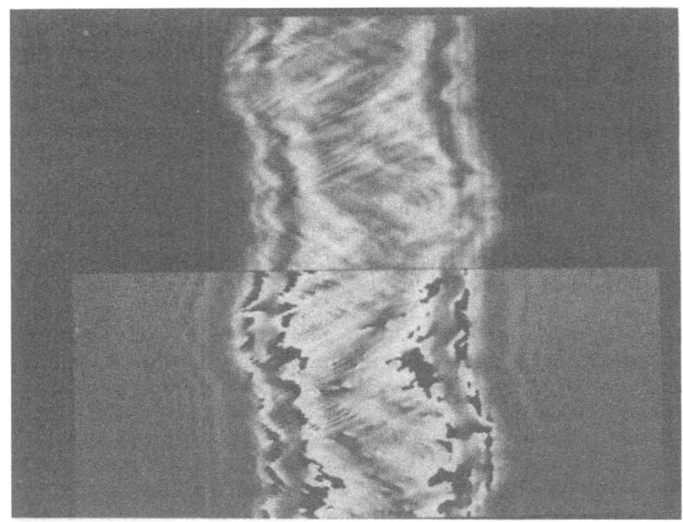

Fig. 2. *Data set acquired from the breast phantom.   Magnitude of the scattered field (upper) and phase of the total field (lower) are displayed as a function of sample number (horizontal) and view number (vertical).*

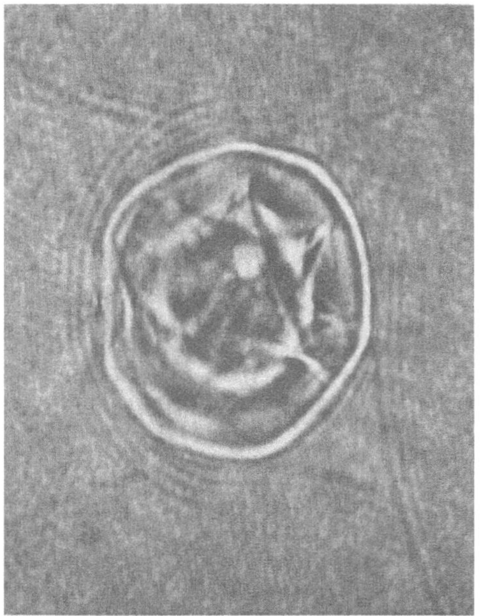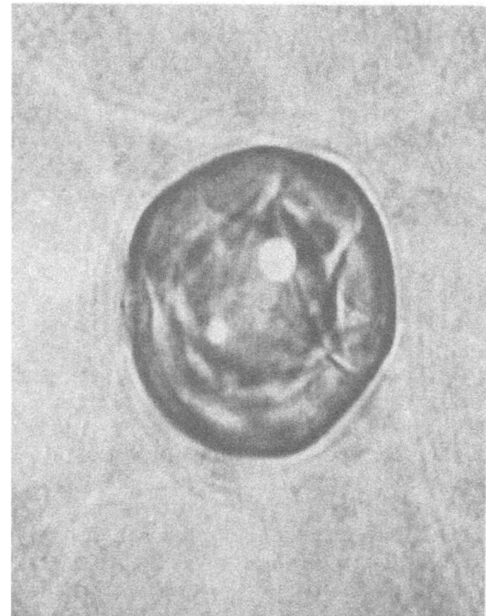

Fig. 3. *Diffraction tomogram of the breast phantom using the Born approximation.   Real part (left) and imaginary part (right).*

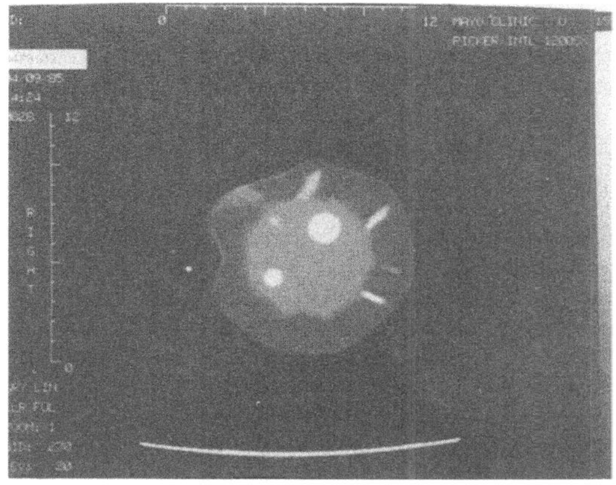

Fig. 4.   *X-Ray computed tomogram of the breast phantom.*

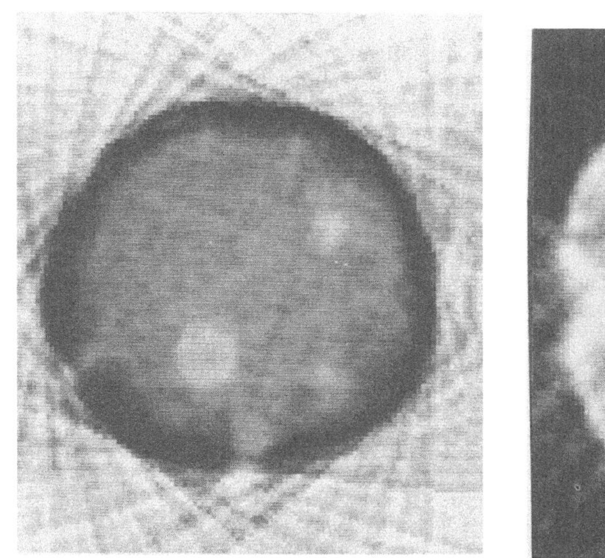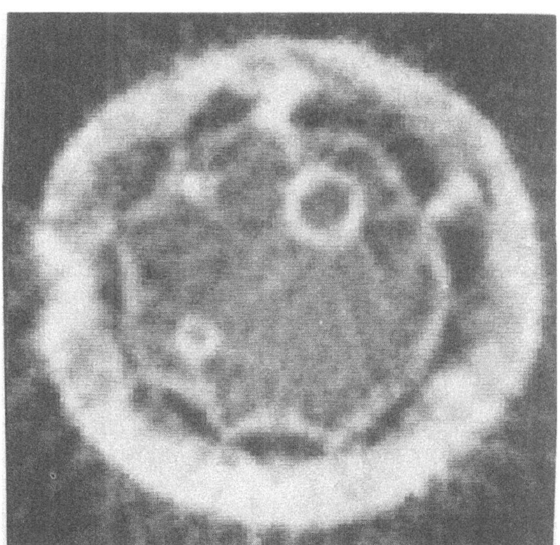

Fig. 5.   *Ultrasound computed tomograms of acoustic speed (left) and attenuation (right) for the breast phantom.*

SUMMARY AND CONCLUSIONS

In this study, care was taken to minimise errors in the data acquisition and image reconstruction processes. We have previously demonstrated that the inhomogeneous Helmholtz wave equation accurately describes scattering from simple test objects and have suggested that it is probably a valid basis for diffraction tomography of biological soft tissues [16, 17]. The same studies also confirmed that our experimental apparatus accurately measures forward scattering from simple, two-dimensional, test objects. The errors introduced by the reconstruction algorithm are demonstrably small. Therefore, the major remaining causes of image artefact in these experiments are the assumption of a 2-D geometry (when in fact the object and insonifying waves have three-dimensional structure), and the use of the Born approximation to the inhomogeneous Helmholtz wave equation.

The relatively good quality of the diffraction tomographic reconstructions obtained using the Born approximation is somewhat surprising. It is clear from the phase data presented in Fig. 2 that the Born approximation is invalid for these experiments. Yet the reconstructions shown in Fig. 3 usefully depict details of the internal structure. The relative positions and sizes of the imbedded tumours and connective tissues are accurate and their boundaries are well resolved. However, as earlier studies [17] predict, the real and imaginary parts of the reconstructions are mutually contaminated so that quantitative maps of acoustic speed and absorption cannot be obtained.

Also puzzling is the robustness of the Born reconstructions with respect to errors in the geometrical parameters as well as the lack of speckle in the images. Both these properties seem particularly noteworthy when it is recalled that a large aperture, fully coherent, narrow bandwidth imaging technique is being used.

We offer the following conjecture as to why the Born reconstructions were relatively successful. Firstly, the limited height of the plane wave insonification may be advantageous in counteracting three-dimensional scattering effects [21].

Secondly, we have observed by time-of-flight measurements that the propagation times through the central regions (though not the edges) of the breast phantom are close to those of water. In fact, the phase shifts through the central regions are somewhat less than $\pi / 2$ radians. In terms of the spatial domain interpretation of diffraction tomography [11, 20], this suggests that, for at least some views and for some internal structures, the Born reconstruction procedure is able to usefully "focus" during the back-propagation process.

Thirdly and finally, in these experiments, the durations of the transmit bursts were limited to 50 µSec (nominally to avoid reverberations inherent in the measurement apparatus). The complex samples were taken at the times that the quadrature receiver outputs peaked for the water path signals (ie: at approximately the trailing edges of the received bursts). Diffraction tomography however assumes that CW fields are measured. This is not strictly the case in these gated measurements. One effect of this discrepancy would be to reduce or eliminate energy arising from multiple scattering paths (such as from internal reverberations between the water/fat/parenchyma interfaces) from the measured data set. We speculate that this effect should be beneficial to the Born reconstruction procedure (since it neglects multiple scattering).

These experimental results are preliminary but encouraging. Much further work is needed to establish the ultimate utility of diffraction tomography for clinical imaging. It is common to dismiss the Born approximation as being unsuitable for clinical diffraction tomography on theoretical grounds. However, it appears that a re-examination of this view point [22] together with a better understanding of the results presented here could help further the progress of diffraction tomography towards the ultimate goal of being a useful diagnostic imaging modality.

ACKNOWLEDGEMENTS

This work was conducted at the Mayo Clinic and was supported in part by grants CA-24085 and RR-02540 from the National Institutes of Health.

REFERENCES

[1]   R. K. Mueller, M. Kaveh, and G Wade, Reconstructive tomography and applications to ultrasonics, Proc. IEEE, 67:567 (1979).

[2]   M. Kaveh, R. K. Mueller, R. Rylander, T. R. Coulter, and M. Soumekh, Experimental results in ultrasonic diffraction tomography, in: "Acoustical Imaging", K. Y. Wang, ed., Plenum, New York, 9:433 (1980).

[3]   R. K. Mueller, Diffraction tomography I : The wave equation, Ultrason. Imaging, 2:213 (1980).

[4]   M. Kaveh, M. Soumekh, and R. K. Mueller, A comparison of Born and Rytov approximations in acoustic tomography, in: "Acoustical Imaging", J. P. Powers, ed., Plenum, New York, 11:325 (1982).

[5]   A. P. Anderson and M. F. Adams, Synthetic aperture tomographic imaging for ultrasonic diagnostics, in: "Acoustical Imaging", E. A. Ash and C. R. Hill, eds., Plenum, New York, 12:565 (1982).

[6]   M. Kaveh, M. Soumekh, Z. Q, Lu, R. K. Mueller, and J. F. Greenleaf, Further results on diffraction tomography using Rytov's approximation, in: "Acoustical Imaging", E. A. Ash and C. R. Hill, eds., Plenum, New York, 12:599 (1982).

[7]   S. K. Kenue and J. F. Greenleaf, Limited angle multifrequency diffraction tomography, IEEE Trans., SU-29:213 (1982).

[8]   J. F. Greenleaf, Computerized tomography with ultrasound, PROC. IEEE, 71:331 (1983).

[9]   A. C. Kak, Tomographic imaging with diffracting and non-diffracting sources, in: "Array Signal Processing", S. Haykin, ed., Prentice Hall, New Jersey, 351, (1984)

[10] B. S. Robinson and J. F. Greenleaf, Computerized ultrasound tomography, in: "Three-Dimensional Biomedical Imaging", R. A. Robb, ed., CRC Press, Boca Raton, 2:76 (1985).

[11] A. J. Devaney, Diffraction tomography, in: "Inverse Methods in Electromagnetic Imaging II", W. M. Boerner, ed., Reidel, Dordrecht, 2: 1107 (1985).

[12] J. F. Greenleaf and A. Chu, Multifrequency diffraction tomography, in: "Acoustical Imaging", M. Kaveh, R. K. Mueller, and J. F. Greenleaf, eds., Plenum, New York, 13:43 (1984).

[13] B. Duchene, D. Lesselier, and W. Tabbara, Experimental investigation of a diffraction tomography technique in fluid ultrasonics, IEEE Trans., UFFC-35:437 (1988)

[14] M Kaveh, M. Soumekh, and J. F. Greenleaf, Signal processing for diffraction tomography, IEEE Trans., SU-32:230 (1984).

[15] S. A. Johnson, F. Stenger, C. Wilcox, J. Ball, and M. J. Berggren, Wave equations and inverse solutions for soft tissue, in: "Acoustical Imaging", J. P. Powers, ed., Plenum, New York, 11:409 (1982).

[16] B. S. Robinson and J. F. Greenleaf, Measurement and simulation of the scattering of ultrasound by penetrable cylinders, in: "Acoustical Imaging", M. Kaveh, R. K. Mueller, and J. F. Greenleaf, eds., Plenum, New York, 13:163 (1984).

[17] B. S. Robinson and J. F. Greenleaf, The scattering of ultrasound by cylinders : Implications for diffraction tomography, J. Acoust. Soc. Am., 80:40 (1986).

[18] E. L. Madsen, J. A. Zagzebski, G. R. Frank, J. F. Greenleaf, and P. L. Carson, Anthropomorphic breast phantoms for assessing ultrasonic imaging system performance and for training ultrasonographers : Part II, J. Clin. Ultrasound, 10:91 (1982).

[19] M. Soumekh, M. Kaveh, and R. K. Mueller, Fourier domain reconstruction methods with application to diffraction tomography, in: "Acoustical Imaging", M. Kaveh, R. K. Mueller, and J. F. Greenleaf, eds., Plenum, New York, 13:17 (1984).

[20] S. X. Pan, and A. C. Kak, A computational study of reconstruction algorithms for diffraction tomography : Interpolation versus filtered-backpropagation, IEEE Trans., ASSP-31:1262 (1983).

[21] Z. Q. Lu, M. Kaveh, and R. K. Mueller, Diffraction tomography using beam waves : Z-average reconstruction, Ultrason. Imaging, 6:95 (1984).

[22] R. H. T. Bates, Renaissance inversion, in: "Inverse Problems of Acoustic and Elastic Waves", F. Santosa, ed., SIAM, Philadelphia, 350 (1984).

INITIAL TESTING OF A CLINICAL ULTRASOUND MAMMOGRAPH

Nils Sponheim and Ingvild Johansen

Norwave Development A.S.
Forskningsveinen 1, N-0371
 Oslo, Norway

and

Anthony J. Devaney†
Department of Electrical and Computer Engineering
Northeastern University
Boston, MA USA 02115

ABSTRACT

This paper presents initial results from experimental tests of a clinical ultrasound mammograph that is designed to perform tests on the female breast for the purpose of early detection of cancerous tumors. The results reported in the paper are limited to non-clinical evaluation of the mammograph on a simple test object which has been specifically selected to evaluate the performance of the system to identify small, low contrast anomalies representative of cancerous tumors in the early stages of development. The paper includes a brief description of the mammograph and the associated diffraction tomographic reconstruction algorithm employed in the evaluation.

INTRODUCTION

The use of a tomographic ultrasound scanner for early detection of cancerous tumors was first suggested by J.F. Greenleaf and co-workers in the mid 1970's [1,2]. This pioneering work had to contend both with experimental difficulties associated with implementing the tomographic procedure in an ultrasound environment and with mathematical difficulties associated with the reconstruction algorithms employed to process the ultrasound data. In this early work the standard reconstruction procedures of X-ray computed tomography (CT) [3-5] were employed and it was only later discovered [4-11] that these algorithms were not appropriate for ultrasound tomography due to the inherent diffraction and coherent scattering that occurs when an ultrasound beam passes through an obstacle. The difficulties introduced by these effects in the reconstruction process have resulted in the name "diffraction tomogra-

---

†Also with A.J. Devaney Associates, 355 Boylston St., Boston, MA 02116)

phy" being applied to any tomographic application that employs coherent wavefields such as ultrasound tomography.

Reconstruction algorithms for diffraction tomographic applications were developed extensively during the last decade [4-15]. These algorithms are capable of dealing with the wave effects associated with ultrasound tomography with the result that viable ultrasound tomographic scan systems are now a real possibility. This possibility has lead to a joint venture between Norwave Development A.S. and A.J. Devaney Associates to develop a clinical mammograph in the form of an ultrasonic tomographic scanner. The hardware and experimental work in this project is performed in Oslo, Norway by Norwave Development, while the algorithm development and software implementation is done jointly by the two companies. The scanner discussed herein is a second generation scanner that has evolved over the past two years and that is now scheduled for clinical tests in a hospital in Oslo. A description of the first generation system is presented in reference 16.

EXPERIMENTAL SETUP

The ultrasonic mammograph consists of a water tank which is 400 m.m. by 400 m.m. in area and 250 m.m. deep as shown in Fig. 1. Two spindles with the same rotation center pass through the bottom of the tank and are connected to a stepping motor. One spindle rotates the hydrophone holder and the other spindle rotates the transducer holder. The holders are made so that both the transducer and the hydrophone can be placed at an arbitrary depth inside the tank. In this way both the hydrophone and the transducer can be rotated in circles around the immersed object at a selected depth.

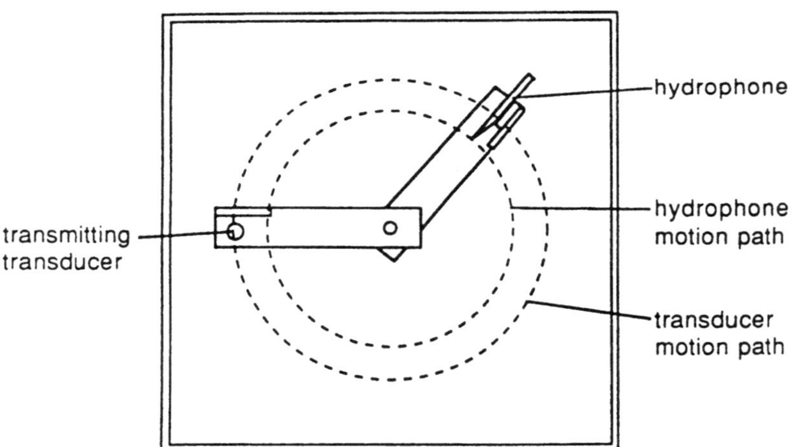

Fig. 1 Top view of the water tank. The two spindles rotate the hydrophone and the transmitting transducer independently in circular arcs having radii of 100 m.m. and 123 m.m., respectively.

The stepping motors are controlled from a PC which automatically carries out the tomographic measurements. To record a data set for one view the transmitting transducer is moved into position and the hydrophone then scans over a specified sector opposite to the transducer and records the amplitude and phase of the transmitted ultrasound wavefield. The transducer is then moved to a new position and the procedure is repeated for each view until the full tomographic recording is completed.

The transmitting transducer is made of a cylindrical PZT tube of diameter 6.3 mm and length 12.6 mm. The PZT tube vibrates in the radial mode with a center frequency of 3.6 MHz and has an electrical input impedance of 30 ohms at resonance. The radiation pattern from this transducer is cylindrically symmetric around the tube axis and the opening angle in the orthogonal direction is about two degrees. To a good approximation the transducer can thus be regarded as a cylindrical source as regards the central plane orthogonal to the transducer axis.

In the so-called "classical configuration" of tomography the background wave is taken to be a plane wave. This gives certain advantages in the mathematical treatment of the inversion problem and the resulting inversion algorithms [4-15]. However, it is not possible to generate a perfect plane wave because of diffraction effects from the edges of the transducer. A cylindrically symmetric transducer has no edges and will give a rotationally symmetric wave field in the image plane that is a close approximation to a cylindrical wave. For the direction orthogonal to the image plane there will be some diffraction effects due to the finite length of the transducer. However, as long as the cylinder is symmetric on each side of the image plane these diffraction effects will not greatly influence the wave field in the image plane. For these reasons the second generation scanner described in this paper employed a cylindrical wave source rather than the plane wave source used in the first generation scanner described in reference 16.

The hydrophone is a microprobe made of PZT whose active area is a disc of diameter 0.5 m.m. At the center frequency of 3.6 MHz this gives the hydrophone an opening angle of about 50 degrees. The reason for choosing a PZT hydrophone rather than a PVDF hydrophone is that we prefer high efficiency rather than large bandwidth.

The signal to the transducer is generated from a stable sine wave signal generator. A switch controlled by the PC turns the signal into a CW-burst and a power amplifier feeds the transducer as shown in Fig. 2. For each view angle and for each hydrophone position the PC turns the signal on and off. The hydrophone signal is amplified before the in-phase (real) and quadrature (imaginary) parts are detected in a quadrature component detector. The real and imaginary components of the wavefield are then digitized and stored in the PC. Thereafter the hydrophone is moved to a new position which is less than half a wavelength from the previous position so as to fulfill the Nyquist sampling criterion before the signal is switched on again. After moving the hydrophone in a sector opposite to the fixed transducer the measured data for one view is recorded. To record a new view the transducer is moved to a new position and the procedure is repeated.

The quadrature component detector is shown in Fig. 3. The input signal from the hydrophone is amplified and then divided into two channels. In one channel the signal is multiplied with a reference signal from the signal generator and in the other channel the signal is multiplied with the reference signal delayed by 90 degrees. This yields the sine and cosine projections of the complex wave field. Finally, the quadrature components are lowpass filtered and amplified to obtain a noise bandwidth of only 20 KHz, which makes the measuring system rather immune to broadband noise.

EXPERIMENTAL EVALUATION

In order to evaluate the ultrasound scanner it was decided to use a test object

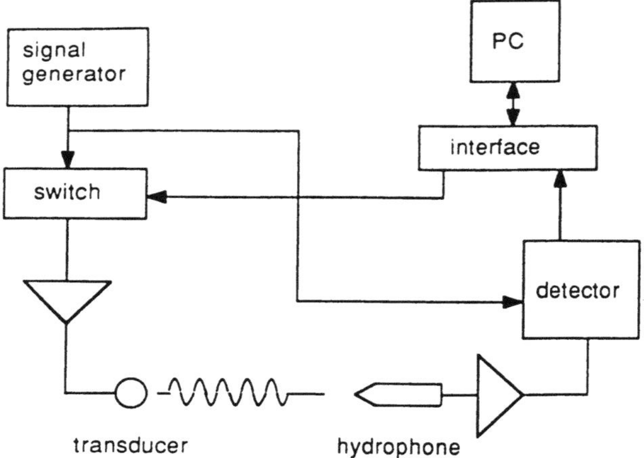

Fig. 2 A stable sine wave generator, a switch controlled by the PC
and a power amplifier supply the transmitting transducer with a
continuous wave burst. The signal from the hydrophone is amplified and the components of the complex wave field are detected
in the quadrature component detector before they are digitized
and stored in the PC.

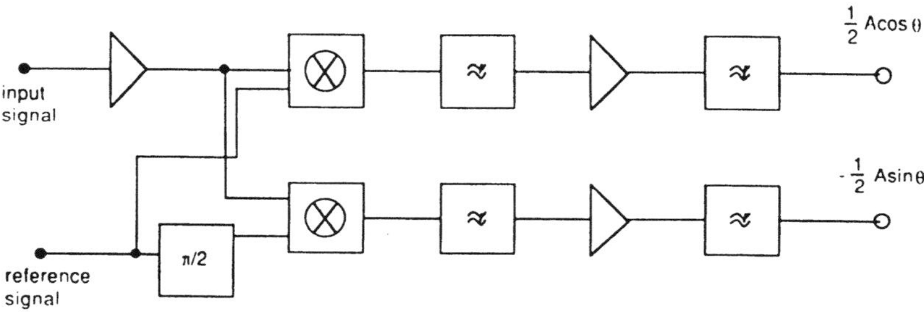

Fig. 3 The detector multiplies the signal from the hydrophone
with two reference signals which are separated by a delay of 90
degrees to detect the sine and cosine projections of the signal.
The signals are also filtered and amplified.

having small size and contrast with the background fluid. This would then presumably
be somewhat representative of early forms of cancerous tumors and, hence, give an
indication of what to expect in later clinical tests of the device. The use of such a
test object has a second advantage in that the image generated in the reconstruction
is an approximation to the impulse response function of the system; i.e, the image
of an extended object is given by the two-dimensional convolution of the test image
with the object function characterizing the test object.

The test object selected was a cylindrical tube of Agar having a diameter of approximately 5.5 m.m. and a velocity of 1530 m./sec. The object was immersed in the water bath, which has a nominal velocity of 1480 m./sec, and was oriented so that its axis was parallel to the axis of the transmitting transducer and was centered as well as possible in the scanner. Both the background wave (insonifying wavefield without the object present) and the total transmitted wave were measured at a total of twenty five view angles.

The total transmitted wavefield at the center frequency of the transducer is represented in terms of its complex phase, vis a vis,

$$U(\xi) = e^{ik(W_0(\xi) + \delta W(\xi))} \tag{1}$$

where $U(\xi)$ is the transmitted wavefield at point ξ along the hydrophone motion path (see Fig. 1), $k = \frac{2\pi}{\lambda}$ is the wavenumber in the fluid, and W_0 and δW are, respectively, the complex phase of the insonifying wave and the perturbation in the (complex) phase of the insonifying wave that is introduced by the presence of the test object. The real and imaginary parts of the complex phase perturbation are computed via the formula

$$\delta W(\xi) = \frac{1}{ik} \log\{\frac{U(\xi)}{U_0(\xi)}\} \tag{2}$$

where $U_0(\xi) = e^{ikW_0(\xi)}$ is the *measured* insonifying wave at the point ξ. It is important to note that the formula (2) gives only the principle value of the real part of the phase $(-\pi \leq k\Re\delta W \leq \pi)$ so that a *phase unwrapping algorithm* [17] generally has to be used to determine the actual phase from the wrapped phase. Fortunately, in the test described herein the unwrapping was easily performed.

Shown in Fig. 4 are the real and imaginary parts of the phase computed using Eq.(2) at a typical view angle. Note that although the object is assumed to be non-attenuating that the imaginary part of the phase (associated with attenuation) is quite large. This is of course due to the wave nature of the ultrasound wave and is a consequence of the fact that propagation tends to inter-mingle the real and imaginary parts of the phase. Also shown in these figures is the theoretical phase as predicted by the Rytov approximation [4-15]. It is seen that there is very poor agreement between the theoretical phase and the experimental phase: a result that caused a great deal of consternation since the reconstruction algorithms of diffraction tomography are based on this approximation [4-15].

To investigate the cause of the apparent breakdown of the Rytov approximation the measured ultrasound field was backpropagated [18] to a line passing through the center of the test object. The process of backpropagation basically undoes the forward propagation process and, hence, generates an approximation of the ultrasound wavefield that would be measured directly across the object. The backpropagation was performed using a modified form of the Rayleigh Sommerfeld diffraction integral and yielded the phase shown in Fig. 5. Also, shown in this figure is the theoretical phase along this line as predicted by the Rytov approximation. Clearly, the experimental and theoretical results are very close in this case. Note also that, as expected, the phase perturbation is very nearly pure real in agreement with the assumption that the test object is non-attenuating.

The fact that the Rytov approximation to the phase and the backpropagated phase were in excellent agreement across the center of the test object lead to the supposition that the breakdown in the Rytov approximation was caused by the large propagation distance (100 m.m.) from the test object to the hydrophone motion path.

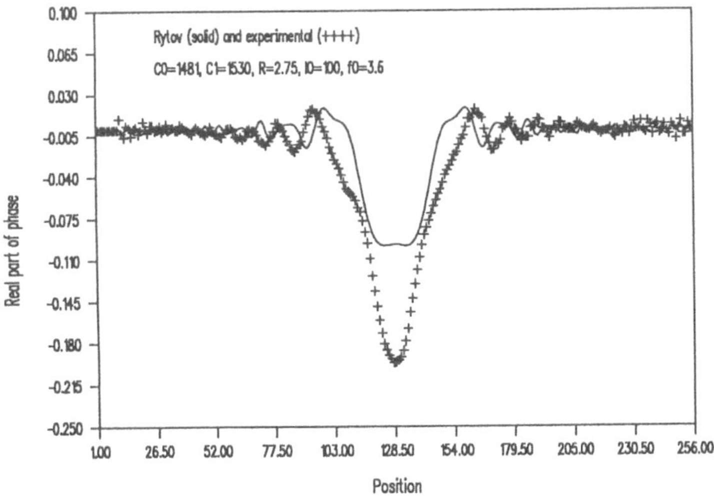

Fig. 4(a) Real parts of the measured (+) and theoretical (solid) phase perturbations $\delta W(\xi)$ across the hydrophone motion path.

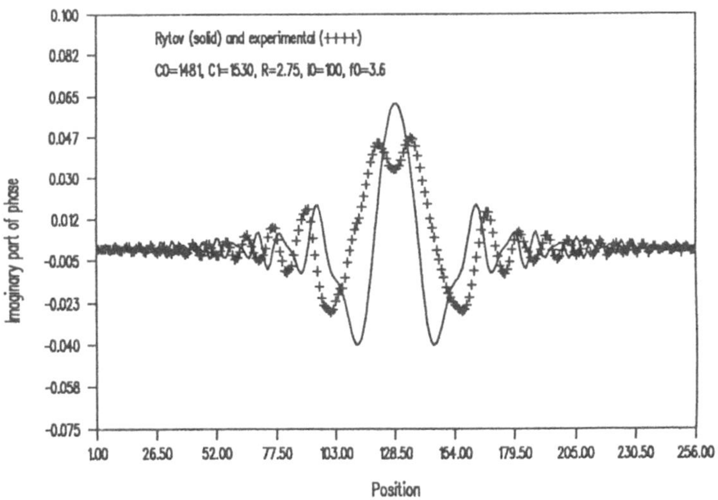

Fig. 4(b) Imaginary parts of the measured (+) and theoretical (solid) phase perturbations $\delta W(\xi)$ across the hydrophone motion path.

Indeed, upon computing the *field* over the center of the object via Eq.(1) with the phase set equal to the Rytov phase shown in Fig. 5 and forward propagating this field to the hydrophone motion path and then, finally, computing the phase of this forward

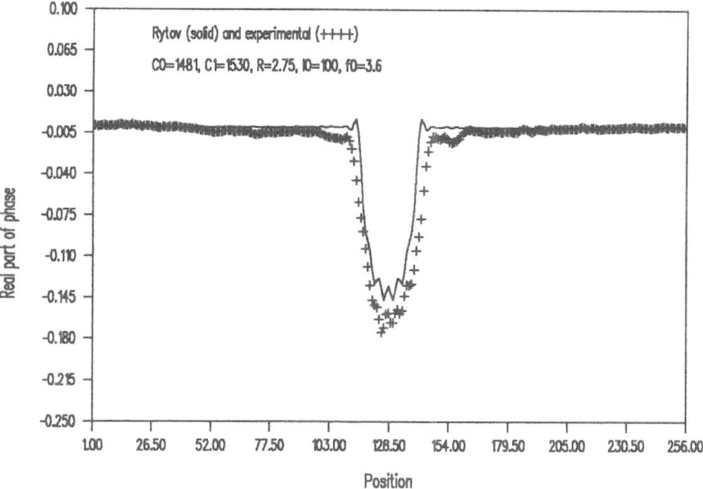

Fig. 5(a) Real parts of the measured ($+$) and theoretical (solid) phase perturbations $\delta W(\xi)$ across a line passing through the center of the test object. The "measured" phase was computed from the backpropagated measured field as described in the paper.

propagated phase via Eq.(2) we obtained a result which was in very close agreement with the measured phase shown in Fig. 4. The conclusion to be drawn from this is that the Rytov approximation is valid *so long as it is employed in the near field of the objects being scanned.* The first step in the processing of the ultrasound scan data then must consist of a backpropagation of the measured field data to a reference line in the vicinity of the object followed by computation of the phase perturbation over this reference line using Eq.(2).

The excellent agreement between the phase of the backpropagated field and the Rytov phase indicated that a diffraction tomographic reconstruction algorithm should perform quite well. Since the object was cylindrically symmetric a special form of the filtered backpropagation algorithm was employed that makes use of known symmetry of the object. This algorithm, which is readily derived from the results presented in reference 11, generates a reconstruction of the deviation of the (complex) index of refraction of the object δn from that of the background fluid at any radial point r according to the equation

$$\delta n(r) = \frac{1}{2\pi} \int_0^k dK \, K \widetilde{\delta W}(K) e^{i(k-\gamma)l_0} J_0(\sqrt{2k}\sqrt{k-\gamma}r). \tag{3}$$

Here, J_0 is the zero order Bessel function of the first kind, $\gamma = \sqrt{k^2 - K^2}$ and

$$\widetilde{\delta W}(K) = \int_{-\infty}^{\infty} d\xi \, \delta W(\xi) e^{-iK\xi} \tag{4}$$

is the Fourier transform of the phase perturbation over the measurement line assumed located at a distance of l_0 from the object's center. Eq.(3) is easy to implement on

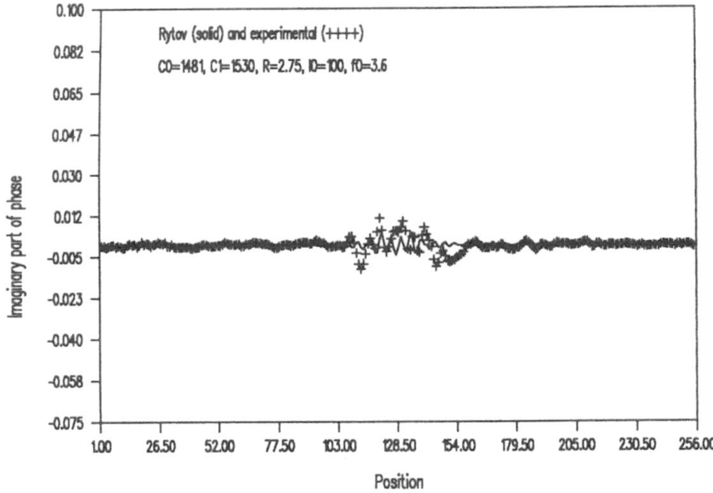

Fig. 5(b) Imaginary parts of the measured $(+)$ and theoretical (solid) phase perturbations $\delta W(\xi)$ across a line passing through the center of the test object. As in Fig. 5(a) the "measured" phase was computed from the backpropagated measured field as described in the paper.

a computer and executes much faster than the standard filtered backpropagation algorithm [11].

The algorithm defined in Eq.(3) was employed on the backpropagated phase shown in Fig. 5 and on the Rytov approximation to this phase shown in this same figure. Thus, the "measurement line" for this case consists of the line passing through the center of the object which then corresponds to a value of $l_0 = 0$ in the reconstruction algorithm (3). The algorithm generated the reconstructions shown in Fig. 6. The deviation of the index of refraction of the test object from the background fluid is assumed to be real and given by $\delta = n - 1 \approx -.0325$ which is seen to be very close to the value generated by the reconstruction algorithm. The size and shape of the object (tube of radius 2.75 m.m.) are also seen to be accurately reconstructed. It is important to note that the imaginary part of the reconstruction is quite small in agreement with the assumption that the object is non-attenuating so that $\Im \delta n = 0$.

SUMMARY AND CONCLUSIONS

We have described an ultrasound mammograph that is designed to perform clinical tests on the female breast for the purpose of early detection of cancerous tumors. The system was evaluated on a simple cylindrically symmetric test object having a small velocity contrast with the background fluid and negligible attenuation. The complex phase of the ultrasound field as measured at a radial distance of 10 c.m. from the test object was found to differ markedly from the complex phase as predicted by the Rytov approximation. This apparent breakdown of the Rytov approximation was determined to be caused by the large measurement distance (10 c.m.) and was circumvented by first backpropagating the measured field to a reference line in the

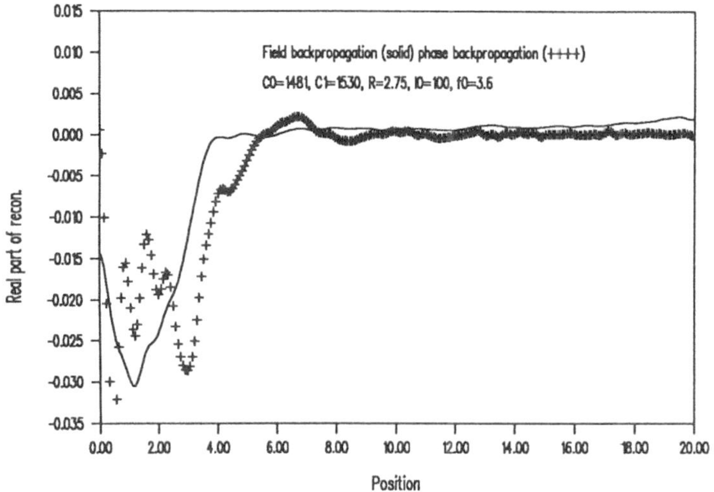

Fig. 6(a) Real parts of the reconstructions generated from the experimental data (+) and synthetic (Rytov) data (solid).

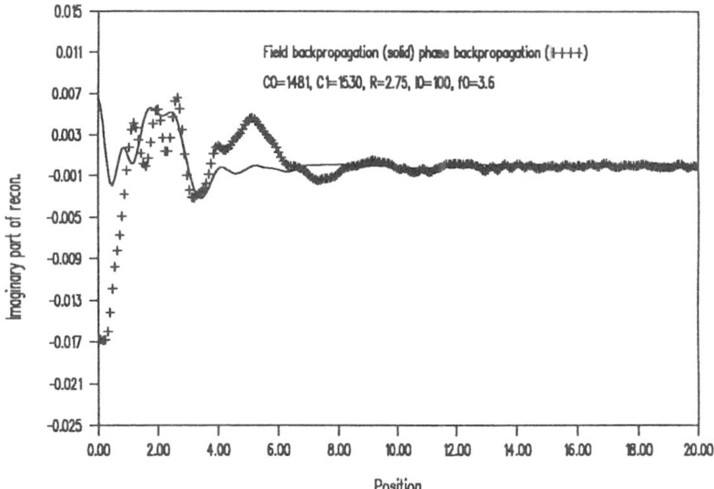

Fig. 6(b) Imaginary parts of the reconstructions generated from the experimental data (+) and synthetic (Rytov) data (solid).

vicinity of the object before computing the complex phase. The "measured" phase so-determined was then found to be in excellent agreement with the Rytov phase and an excellent reconstruction of the index perturbation of the test object was obtained using a modified version of the filtered backpropagation algorithm [11].

We feel that the initial test described in this paper indicates that the mammograph has great potential. The system is currently being evaluated on a number of more complex test objects after which it will be used in a series of clinical evaluations. It is expected to report the results of these tests in the near future.

ACKNOWLEDGEMENT

Norwave Development A.S. received funding for the research reported here from the Royal Norwegian Council for Scientific and Industrial Research (Grant # IT 662.23249)

A.J. Devaney acknowledges support from a Phase II SBIR research grant from the Division of Applied Mathematics of the National Science Foundation. He would also like to acknowledge the support provided by a NATO Fellowship [Grant # SA. 5-2-05 (R.G. 0577/88)] that allowed him to spend the month of August, 1989 at Norwave Development A.S.

REFERENCES

1. J.F. Greenleaf, S.A. Johnson, S.L. Lee, G.T. Herman, and E.H. Wood, "Algebraic reconstruction of spatial distributions of acoustic absorption in tissues from their two-dimensional acoustic projections," in *Acoustical Holography* 5, P.S. Green, ed. (Plenum, New York, 1974), 591-603.
2. J.F. Greenleaf, S.A. Johnson, W.F. Samayoa, and F.A. Duck, "Algebraic reconstruction of spatial distributions of acoustic velocities in with tissues from their time-of-flight profiles," in *Acoustical Holography* 6, N. Booth, ed. (Plenum, New York, 1975), 71-90.
3. G.T. Herman, *Image Reconstruction from Projections. The Fundamentals of Computerized Tomography* (New York: Academic Press) 1980.
4. J.F. Greenleaf, "Computerized transmission tomography," in *Methods of Experimental Physics* 19 (Academic Press, New York, 1981).
5. A.C. Kak, "Computerized tomography with x-ray emission and ultrasound sources," *Proc. IEEE*, Vol. 67, pp. 1245-1272, 1979.
6. J.F. Greenleaf and R.C. Bahn, "Clinical imaging with transmissive ultrasonic computerized tomography," *IEEE Trans. Biomed. Eng.* **BME-28**, 177-185, 1981.
7. J.F. Greenleaf, "Computerized tomography with ultrasound," *Proc. IEEE* **71**, 330-337, 1983.
8. R.K. Mueller, M. Kaveh, and G. Wade, " Reconstructive tomography and applications to ultrasonics", *Proc. IEEE*, **67**, 567-87, 1979
9. R.K. Mueller, M. Kaveh, and R.D. Inverson, "A new approach to acoustic tomography using diffraction techniques," *Proc. Acoustical Imaging* 8, 615-628, 1980.
10. S.X. Pan and A.C. Kak, "A computational study of reconstruction algorithms for diffraction tomography," *IEEE Trans. Acoustics, Speech and Signal Processing*, Vol. ASSP-31, pp. 1262-1275, 1982.
11. A.J. Devaney, "A filtered backpropagation algorithm for diffraction tomography," *Ultrasonic Imaging*, Vol. 4, pp. 336-350, 1982.
12. A. Witten, J. Tuggle, and R.C. Waag, "Ultrasonic imaging with a fixed instrument configuration," *Appl. Phys. Letts.* **xx**, 1988.
13. B. Duchene, D. Lesselier, and W. Tabbara, "Experimental investigation of a

diffraction tomography technique in fluid ultrasonics," *IEEE Trans. Biomed. Eng.* **BME-35**, 437-444, 1988.

14. A.J. Devaney, "A computer simulation study of diffraction tomography," *IEEE Trans. Biom. Eng.* **BME-30**, 377-86, 1982.

15. K.J. Langenberg, "Applied inverse problems for acoustic, electromagnetic, and elastic wave scattering," in *Basic Methods of Tomography and Inverse Problems*, ed. P.C. Sabatier (Adam Hilger, Philadelphia, 1987).

16. N. Sponheim and I. Johansen, "Experimental results in ultrasonic tomography using a filtered backpropagation algorithm," submitted to *Ultrasonic Imaging*.

17. M.Kaveh, M.Soumekn and J.F. Greenleaf, "Signal processing for diffraction tomography," *IEEE Trans. Sonics and Ultra.* **SU-31**, 230-238, 1984.

18. J.R. Shewell and E. Wolf, "Inverse diffraction and a new reciprocity theorem," *J. Opt. Soc. Amer.* **58**, 1596-1603, 1968.

MODIFIED FRESNEL APPROXIMATION
AND ACOUSTICAL HOLOGRAPHY

J.S.Meng[†] and H.W.Jones[‡]

[†] Cancer Treatment & Research Foundation of Nova Scotia, 5820
University Avenue, Halifax N.S., Canada B3H 1V7
[‡] Mechanical Engineering Department, University of Calgary
Calgary, Canada T2N 1N4

ABSTRACT

The backward-propagation algorithm, based on the angular spectrum formulation of diffraction theory, plays a central role in acoustical holography systems. This procedure requires twice two-dimensional FFT's to propagate the wavefront back to its origin. In contrast, the Fourier/Fresnel transformation algorithm needs only one two-dimensional FFT. This means great saving on computation time. The Fourier/Fresnel transformation algorithm, however, is seriously limited in some applications, since it does not apply in near-field region. Near field imaging is of great interest in some applications of practical reasons and offers potential improvement of the spatial resolving power of systems. Two modified approximations are discussed in this paper. They have much less error than the traditional Fresnel approximation in the near field. Simulation results show that the performance of the algorithms based on these two modified approximations may improve the performance of an acoustical holography system in the near field, even if the distance between acoustical array and object is much less than the array dimension.

INTRODUCTION

The backpropagation algorithm plays a central role in acoustical holography. It applies for any distance between the object and the lens, i.e. in both far field and near field. The usual backpropagation algorithm needs twice 2-D FFT [1]. A modified algorithm based on Fresnel approximation needs only one 2-D FFT may be used and this means great saving on computation time. Unfortunately, however, Fresnel approximation does not apply to an object in the near field.

The near field image is of great interest in acoustical holography for practical reasons and there is always an interest in method improvement of spatial resolving power. Two modified approximations are discussed in this paper. Reconstruction algorithms based on these approximations may improve the performance of the imaging system in the near field, even if the distance between the object and the receiving array is much less than the array dimensions.

1. FRESNEL APPROXIMATION AND NEAR FIELD DISTORTION

We investigated the case which is shows in Fig.1. Plane waves of monofrequency in negative Z direction were used to illuminate an object. The reflected waves are received by a planar array which is located in the receiving plane. The reflected wave front in the object plane is that which is to be imaged and is referred to as the object in this paper.

The acoustic field in the receiving plane is

$$U(x_1,y_1)= \iint U(x_0,y_0)\, \exp(jkr)/r\; dx_0 dy_0 \qquad \text{Eq.1}$$

where $U(x_0,y_0)$ is the object to be imaged, and k is the wave number. The distance from a point (x_0,y_0) on the object plane to a receiving element position (x_1,y_1) on the receiving plane is given exactly by

$$r=[Z_1^2+(x_1-x_0)^2+(y_1-y_0)^2]^{1/2}, \qquad \text{Eq.2}$$

where Z_1 is the distance between the object plane and the receiving plane. Under far field condition , i.e. Z_1^2 is much greater than $(x_1-x_0)^2+(y_1-y_0)^2$, Eq.2 may be approximated by

$$r'=Z_1+[(x_1-x_0)^2+(y_1-y_0)^2]/2Z_1 \qquad \text{Eq.3}$$

which is known as the Fresnel approximation [2]. The Fresnel approximation (Eq.3) allows Eq.1 to be rewritten as

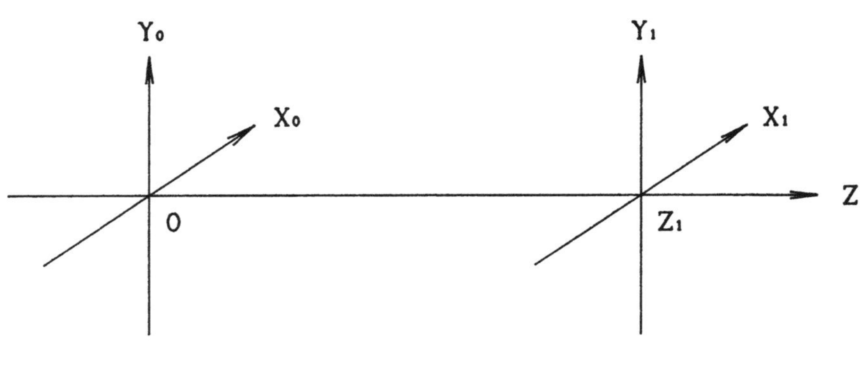

Fig. 1 The geometry under investigation

$$U(x_1,y_1)=\exp\{jk[Z_1+(x_1^2+y_1^2)/2Z_1]\}/Z_1 \; \iint \xi(x_0,y_0)\exp[-j2\pi(x_0u+y_0v)]dx_0dy_0$$

where

$$\xi(x_0,y_0)=U(x_0,y_0)\exp[jk(x_0^2+y_0^2)/2Z_1],$$

$$u=x_1/\lambda Z_1,$$

and

$$v=y_1/\lambda Z_1.$$

Eq.4

where r^{-1} in the integrand of Eq.1 has been approximated by Z^{-1} and moved out of the integration. Eq.4 indicates that the acoustic field on the array plane is approximately the Fourier transform of the two dimensional function $\xi(x_0,y_0)$ except for a complex coefficient. Therefore, an inverse transform can restore the function $\xi(x_0,y_0)$. After simple phase compensation of $\exp[-jk(x_0^2+y_0^2)/2/Z_1]$, the two-dimensional amplitude and phase images of the acoustical field in the object plane are thus available to be displayed or for further processing.

Therefore Fresnel approximation constitutes the basis of a reconstruction algorithm for holographic images. In the near field, however, the small value of Z_1 may cause serious image degradation if the Fresnel approximation is used. In order to get no-aliasing imagings, the maximum dimension D_{max}, of the observed object is limited by the element interval d, wavelength λ and the distance Z_1 between the object plane and the array plane by [3]

$$D_{max}= \lambda Z_1/d .$$

This means that the maximum permissive object dimension is proportional to Z_1 for an imaging system. Thus there should be no problems caused by the object's dimensions in the near field, if the object subtends some finite angle at the receiving array; this angle may, of course, be determined by the illuminating system.

2. MODIFIED FRESNEL APPROXIMATIONS

Two modified approximations to be discussed are as follows

$$r_1=(Z_1^2+x_1^2+y_1^2)^{1/2}+(x_0^2+y_0^2)/2Z_1 - (x_1x_0+y_1y_0)/Z_1$$

Eq.5

and

$$r_2=(Z_1^2+x_1^2+y_1^2)^{1/2}+(x_0^2+y_0^2)/2Z_1 - (x_1x_0+y_1y_0)/(Z_1^2+x_1^2+y_1^2)^{1/2}$$

Eq.6

both arising from the binomial theorem. These two modified approximations are more exact than the original Fresnel approximation. Fig.2 shows the maximum relative error of the distance for the Fresnel and the two modified approximations for the case that the object dimension is 40% of the distance Z_1. The horizontal axis is the ratio of the distance between the object and array planes to the array dimension. The vertical axis is the relative error of the distance compared to the array dimension. It is obvious according these three curves that the modified approximations have much smaller errors than that given by the Fresnel approximation when the distance between array and object is less than the array dimension. Especially, the approximation of Eq.6 has an error which goes to zero monotonically when the distance between the receiving array and the object goes to zero. In contrast, the error of the Fresnel approximation goes to infinity.

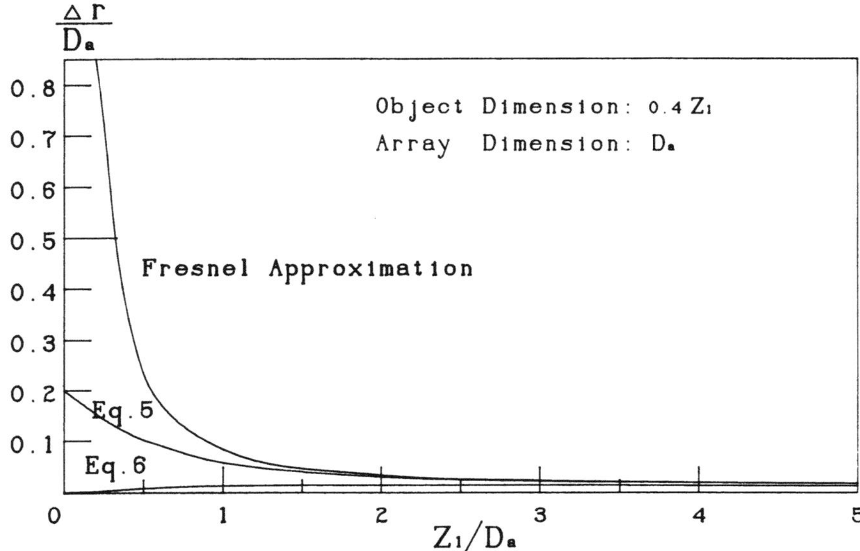

Fig.2 Relative distance errors

3. NEAR-FIELD IMAGING BY MODIFIED APPROXIMATIONS

Either of the approximations of Eq.5 and 6 can be used to reconstruct images. The acoustical signal received at (x_1,y_1), approximated by Eq.5, is

$$U(x_1,y_1)=[Z_1^2+(x_1^2+y_1^2)]^{-1/2}\exp\{jk[Z_1^2+(x_1^2+y_1^2)]^{1/2}\}$$

$$\iint \xi(x_0,y_0)\exp[-j2\pi(x_0u+y_0v)]dx_0dy_0$$

Eq.7

According to Eq.7, a specific amplitude modification, $[Z_1^2+(x_1^2+y_1^2)]^{1/2}$, as well as a phase shift, should be given to the signal at each receiving element before an image can be obtained by inverse Fourier transform.

The other approximation, Eq.6, is more precise when objects are close to the receiving array. In this case, the received signal is

$$U(x_1,y_1)=[Z_1^2+(x_1^2+y_1^2)]^{-1/2}\exp\{jk[Z_1^2+(x_1^2+y_1^2)]^{1/2}\}$$

$$\iint \xi(x_0,y_0)\exp[-j2\pi(x_0u+y_0v)]dx_0dy_0$$

$$Eq.8$$

where

$$u=x_1/\lambda[Z_1^2+(x_1^2+y_1^2)]^{1/2}$$

$$v=y_1/\lambda[Z_1^2+(x_1^2+y_1^2)]^{1/2}$$

$$Eq.9$$

In order to reconstruct the image by an inverse Fourier transform, a non-linear transform of x_1 and y_1 of Eq.9 is needed.

4. SIMULATED RESULTS

A simulation was conducted using an IBM-PC-XT. Coherent illumination of monochromatic sound of 1 mm wavelength was simulated. The receiving array was a planar rectangular array of 16X16 point receiving elements with a 2 mm interval between adjacent elements.

Fig.3 to 5 show some of the simulation results. In these figures, Figs.(a) are the far-field images of the object. When the array is far from the object, say several times the array dimension, the reconstruction algorithm according to any of the three approximations of Eq.3, 5, and 6 can produce images of high fidelity. When the array moves towards the object, the images by different approximations are different. Figs.(b), (c), and (d) are the amplitude images by the approximations from Eqs. 3, 5 and 6, respectively. The distance between the object and the array are 14 mm in Figs.3 and 4 and 18 mm in Fig.5. At the distance of 14 or 18 mm, which is less than the array dimension (30mm), the Fresnel approximation cannot give correct images. In comparison, the modified algorithms by Eq.5 and 6 give better images.

Fig.3 shows the images of a point object which locates at the center of the observed region. Either of the modified approximations gives an image similar to the far-field image. When the object to be imaged consists of several points, as in Figs.4 and 5, the modified approximation Eq.5 gives a much better image by comparison with the Fresnel approximation, but the object dimension (the distance between points) is reduced a little.

5. SUMMARY

Two modified Fresnel approximations are discussed in this paper for near-field imaging. They are less susceptible to near field errors than is the usual Fresnel approximation. The modified approximations offer the possibility of reconstructing

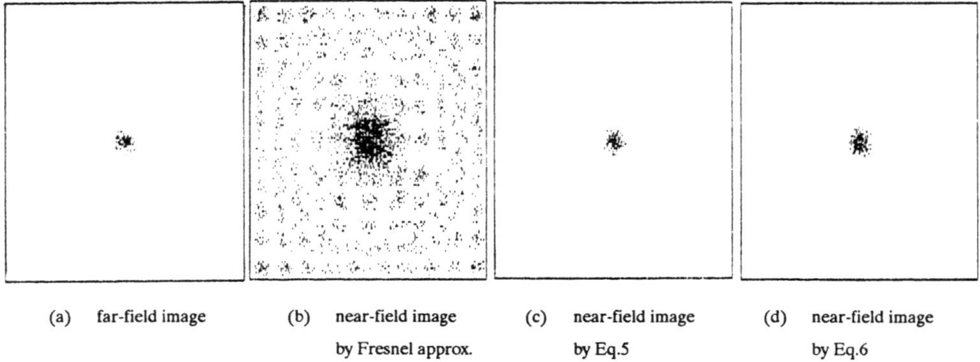

(a) far-field image (b) near-field image by Fresnel approx. (c) near-field image by Eq.5 (d) near-field image by Eq.6

Fig.3 Simulation Results: Z=14mm

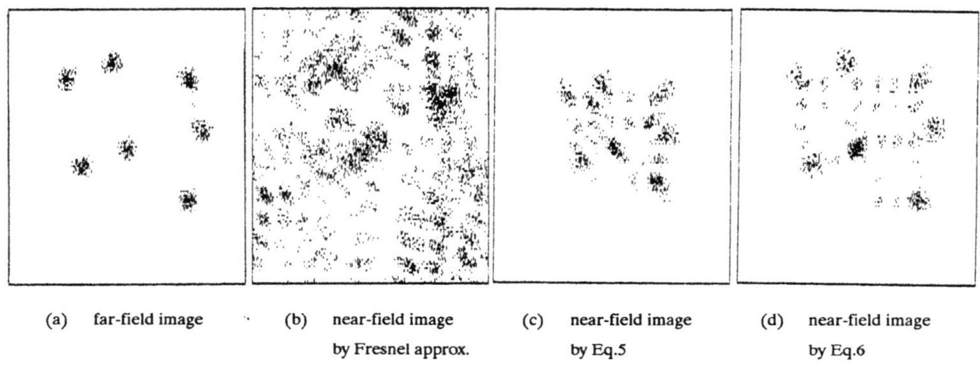

(a) far-field image (b) near-field image by Fresnel approx. (c) near-field image by Eq.5 (d) near-field image by Eq.6

Fig.4 Simulation Results: Z=14mm

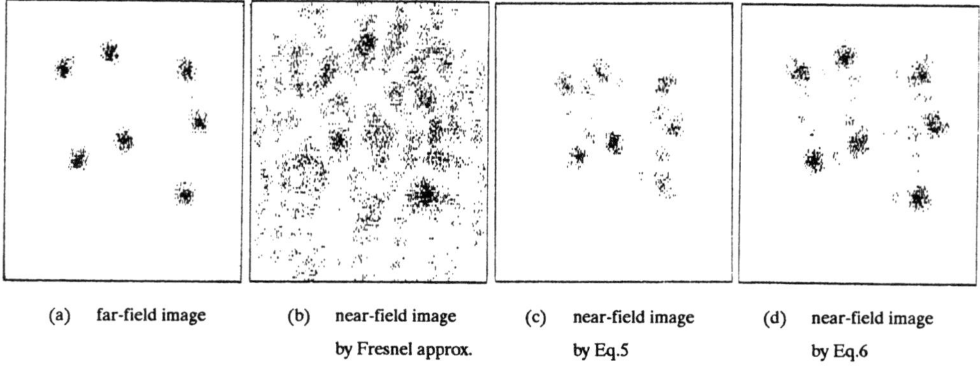

(a) far-field image (b) near-field image by Fresnel approx. (c) near-field image by Eq.5 (d) near-field image by Eq.6

Fig.5 Simulation Results: Z=18mm

holographic images by only one 2-D FFT, even if the distance between the objects and the receiving array is much less than the array dimension.

By comparison to conventional Fourier/Fresnel transformation, the reconstruction algorithm based on the modified approximation Eq.5 does not change much, but the quality of near-field images will be improved quite remarkably. Second modified approximation discussed in this paper, Eq.6, may provide even better near-field images, but the use of this approximation requires more computation time due to the non-linear transform of Eq.9. This approximation indicates that equispaced elements of the receiving array means non-equispaced sampling of the spatial spectrum plane of object which locates in the near field.

REFERENCES

[1]   C.F.Schueler,   H.Lee   and   G.Wade,   "Fundamentals   of   digital   ultrasonic imaging",   IEEE Trans. on Sonics and Ultrasonics,   Vol.SU-31,   No.4,   pp.195-217, July 1984.

[2]   J.W.Goodman, "Introduction to Fourier Optics", McGRAW HILL, 1968.

[3]   J.S.Meng and H.W.Jones, "Effects of receiving element size on acoustic imaging",   accepted by 1989 Meeting of the Canadian Acoustical Association, Halifax, Canada, October 1989.

IMPROVEMENTS OF THE MULTILAYER HOLO-ACOUSTIC TOMOGRAPHY

BY ITERATIVE AND ALGEBRAIC TECHNIQUES

C.B. Ahn and Z.H. Cho

Department of Radiological Sciences
University of California, Irvine
Irvine, CA 92717

ABSTRACT

The scanning tomographic acoustic microscopy (STAM) using the holographic principle has been recently proposed and the algorithm known as "back-and-forth propagation"(BFP) has been devised for the reconstruction of multi-planar objects. The BFP algorithm is a holographic reconstruction coupled with a tomographic processing of multiple data set obtained by the variation of insonification frequency or transducer angle. The tomographic processing is a superposition process applied to the holographically reconstruction images based on the principle that the structures on the reconstruction plane is insonification-frequency or angle invariant, while the diffraction noise (blurred image) coming from out-of-reconstruction planes is frequency or angle dependent. In practice, often the cancellation is incomplete and reconstructed image is blurred by the diffraction from other image planes. In this paper, two methods of restorations, namely iterative and algebraic restorations, have been proposed with which diffraction noise can be largely removed. Simulation results of the iterative method show substantial improvements in image quality by only a few iteration steps. The algebraic method which solves the simultaneous equations of the coefficients related with the propagation transfer function has better computational efficiency than the iterative algorithm. The algebraic technique provides accurate restoration if a noise free environment is available. On the other hand, the algebraic method appears to be more susceptible to random noise. Study of these two methods by computer simulations is performed and results are shown to demonstrate the efficacy of the two algorithms when they are applied to the BFP algorithm.

INTRODUCTION

The concept of the computerized tomography (CT) has been revolutionized the entire field of imaging sciences in general and the medical diagnostic imaging in particular. Currently various types of computerized tomography are available, e.g., x-ray CT, positron emission tomography (PET), single photon emission tomography (SPET), nuclear magnetic resonance (NMR) imaging, etc.[1] The concept of tomography has also been applied to

---

* Also at Dept. of Electrical Science, Korea Advanced Institute of Science, P.O. Box 150, Cheongyangni, Seoul, Korea.

acoustic imaging,[2-7] where diffraction phenomena should be considered due to the relative long wavelengths of ultrasound. Recently several reconstruction algorithms have been developed for the diffraction tomography. Among these algorithms, the reconstruction method known as "back-and-forth propagation" (BFP), is an efficient algorithm for the reconstruction of well-defined planar objects.[5-7] The BFP algorithm is basically a holographic reconstruction similar to the focal plane reconstruction. The tomographic processing is a least-square estimation of multiple data set obtained by the variation of either frequency or transducer angle. If the holographic reconstruction is applied for a particular plane of the object, the structure, more specifically the object transmittance function, on that plane appears clear and invariant to the insonification frequency or incident angle, while the diffraction noise coming from other planes appears blurred and variant depending on the frequency or angle. Since the reconstruction process is holographic, therefore phase is still remained, diffraction noise can be cancelled out if the holographically reconstructed images with different insonification frequencies or angles are added together. If the undesirable diffraction pattern in each holographically reconstructed image is independent and is identically distributed, the least-square estimation becomes simply a superposition process. Although the holographic reconstruction is a matched filtering process for each measured field, the distribution of insonification frequency or transducer angle is not considered in the BFP algorithm. Thus, the degree of cancellation is dependent upon the interplane correlation of the diffraction noise.

The algorithm described in this paper is a deterministic model to remove diffraction noise by an either iterative or algebraic method. Similar concept has been applied for the restoration of tomosynthetic x-ray images, e.g., constrained iteration method.[8-9] Although the restoration concepts are similar, detailed approach of the algorithm as well as the extent of the application are very specific to each image formation principles. For example, image blurring is mainly related with the aperture geometry in the x-ray tomosynthesis, while the main source of blurring is diffraction in the acoustic imaging. Following this introduction, back-and-forth propagation algorithm will be briefly reviewed. The two approaches, namely iterative and algebraic restorations, will be formed, and the efficacy of the proposed algorithms will be tested by computer simulations.

REVIEW OF THE BACK-AND-FORTH PROPAGATION METHOD

In this section, the theory of back-and-forth propagation method with a plane-wave insonification is reviewed shortly. Since the algorithm can be considered as a combination of holographic reconstruction and tomographic processing, these two processings will be reviewed one by one.

Holographic Reconstruction

The holographic reconstruction is similar to the focal plane reconstruction. For a plane wave insonification, the wavefield can be written in spherical coordinates as

$$u(x, y, z) = u_o \exp[\, i 2\pi(f_x^i x + f_y^i y + f_z^i z)\,] \tag{1}$$

where u_0 is the amplitude of the plane wave, and (f_x^i, f_y^i, f_z^i) is the incident wave vector given by

$$f_x^i = \sin\,\theta\,\cos\,\phi\,/\,\lambda$$

$$f^i_y = \sin\theta \sin\phi / \lambda \qquad\qquad\qquad (2)$$

$$f^i_z = \cos\theta / \lambda$$

and λ is the acoustic wavelength. At the object plane $z=z_k$, the transmitted wavefield is given in Fourier domain as

$$V(f_x, f_y; z_k) = V_o O(f_x - f^i_x, f_y - f^i_y; z_k) \qquad\qquad (3)$$

where V_o is a constant. Note the modulation caused by the slant insonification. The wavefield detected at $z=z_D$ is further modulated by the propagation transfer function given by[10]

$$H(f_x, f_y; z_D - z_k) = \begin{cases} \exp[i\ 2\pi\dfrac{(z_D - z_k)}{\lambda}\sqrt{1 - (\lambda f_x)^2 - (\lambda f_y)^2}\ ] \\[2mm] \qquad\qquad\qquad\qquad , f_x^2 + f_y^2 < \dfrac{1}{\lambda^2} \\[3mm] 0 \qquad\qquad\qquad\qquad , \text{otherwise} \end{cases} \qquad (4)$$

Thus the holographic reconstruction is an inverse filtering of the propagation transfer function (Eq.(4)) together with a demodulation (Eq.(3)) given by

$$O(f_x, f_y; z_k)_{Holo} = H(f_x + f^i_x, f_y + f^i_y; z_k - z_D) V(f_x + f^i_x, f_y + f^i_y; z_k) \qquad (5)$$

Note that the signal out of the focal plane $z \neq z_k$ appears blurred with phase dependent on the acoustic wavelength insonified.

<u>Tomographic processing</u>

Although there is a depth resolving capability in the holographic reconstruction by focusing and blurring, it may not be very useful if the object to be scanned is complex and adjacent planes are located closely. In order to improve depth resolution, a tomographic processing has been proposed. Suppose M data set (simply called "projection" by the analogy with x-ray CT) was acquired with the variation of insonification frequency or transducer angle, then the least-square error estimate of $o(x,y,z_k)$ is given by

$$\hat{o}(x, y, z_k)_{LSE} = \frac{1}{M}\sum_{m=1}^{M}\hat{o}_m(x, y, z_k)_{Holo} \qquad\qquad (6)$$

where $\hat{o}_m(x, y, z_k)_{Holo}$ is the holographic reconstruction with m-th projection as shown in Eq.(5). In deriving Eq.(6), diffraction noise in different projections is assumed to be mutually independent. Thus the tomographic reconstruction is nothing but a superposition process of the holographically reconstructed images for various projections. The plane definition will be improved due to the fact that the signal in the focused plane is linearly increasing with the superposition, while the standard deviation of the noise or blurred diffraction would be increasing in proportion to the square root of number of projections.

For a computer generated planar object composed of four layers, the holographic and tomographic reconstructions are tried as shown in Fig.1. The object size was about 2mm and the separation between layers was 75 μm corresponding to 5λ at 100MHz with velocity 1500m/s (see Fig.4 for the whole

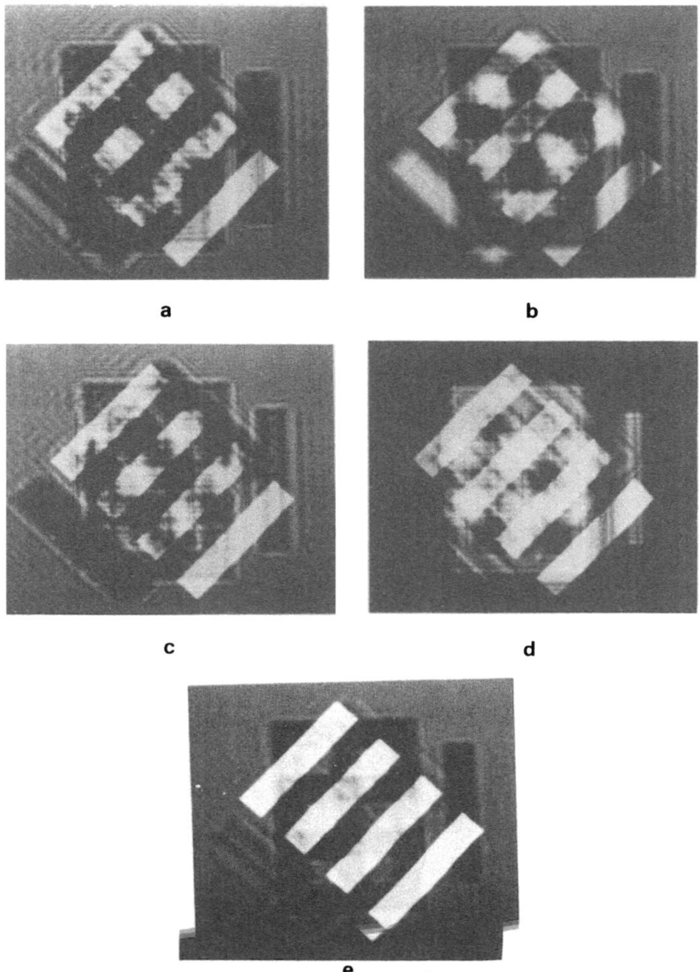

Fig.1   A set of holographically reconstructed images are obtained for z_3 plane by varying insonification frequency: (a) 95 MHz, (b) 100 MHz, (c) 105 MHz, and (d) 110MHz.   A summation of these four images makes (e).

object).   Although the reduction of   diffraction noise by the tomographic processing is significant,   still there exists   noticeable diffraction   noise having shapes similar to the   structures in out-of-focus planes.   A further reduction of this remained diffraction noise is the subject of the next section.

ITERATIVE AND ALGEBRAIC RESTORATION OF THE BACK-AND-FORTH PROPAGATION

Although the inner plane definition is largely improved with the tomographic processing as shown in the previous section, often the cancellation of diffraction noise by the superposition alone is   not   an efficient method,   especially when the number of projections or the bandwidth of insonification frequency is limited.   The remained background diffraction noise degrades image contrast as well as resolution. In this section two algorithms, namely iterative and algebraic restorations, have been proposed.

Iterative Restoration

The algorithm is composed of iterative estimation and cancellation of the diffraction noise from the diffraction-resident-reconstructed images. The diffraction subtracted images will be used for the next-stage estimation of the diffraction. For initial values, the reconstructed images by the BFP method can be used. Using the propagation transfer function in Eq.(4), the diffraction noise at $z=z_k$ can be estimated at n-th iteration as

$$\hat{D}^{(n)}(f_x, f_y; z_k) = \sum_{j \neq k} \sum_{m} \begin{cases} \hat{O}^{(n-1)}(f_x, f_y; z_j) \exp[i2\pi \dfrac{(z_k - z_j)}{\lambda_m} \sqrt{1 - (\lambda_m f'_x)^2 - (\lambda_m f'_y)^2}] & , f'^2_x + f'^2_y < \dfrac{1}{\lambda^2_m} \\ \\ 0 & , \text{otherwise} \end{cases} \qquad (7)$$

where $f'_x = f_x + f^i_x$ and $f'_y = f_y + f^i_y$, and $\hat{O}^{(n-1)}$ is the estimated object function at (n-1)-th iteration. The inner summation with the index m denotes the tomographic processing for all the applied insonification frequencies, and the outer summation j is the calculation of diffraction coming from all the planes except $z=z_k$. Note that f^i_x and f^i_y are also functions of λ_m as previously defined in Eq.(2). The estimated object function at n-th iteration is given by

$$\hat{O}^{(n)}(f_x, f_y; z_k) = \hat{O}^{(0)}(f_x, f_y; z_k) - \hat{D}^{(n)}(f_x, f_y; z_k) \qquad (8)$$

Note that the estimated diffraction pattern is subtracted from the initial reconstruction, rather than the previous restored image, since the diffraction effect was already subtracted in the previous restored image.

As an example, the proposed iterative algorithm is applied to the reconstructed images by the BFP algorithm, and representative results for the reconstruction plane $z=z_3$ are shown in Fig.2(a)-(d) according to the iteration number of from 0 to 3. As observed in these results, the diffraction noise is decreased substantially as the iteration number increases. For example, the normalized mean square error (NMSE) of the initially reconstructed image shown in (a) (same as Fig.1(e)) was 0.209, while that of the restored image (iteration=3) was 7.55×10^{-3} with a reduction ratio more than 20. Similar amounts of noise reductions were also observed for other planes. The improvement of the depth resolution by the iterative restoration is further investigated in Fig.3 by varying the distance between the object layers. The NMSE in Fig.3 is generally increased due to the reduced distance between layers, however, the iterative algorithm provides much improved plane definitions. There seems a limitation in decreasing NMSE by the iterative method if the distance between layers approaches about λ (theoretical limitation). In such a case, there appears rectangular background pattern as the iteration number increases.

Algebraic Restoration

The algebraic restoration is to solve the simultaneous equations in the spatial frequency domain whose matrix coefficients are related with the

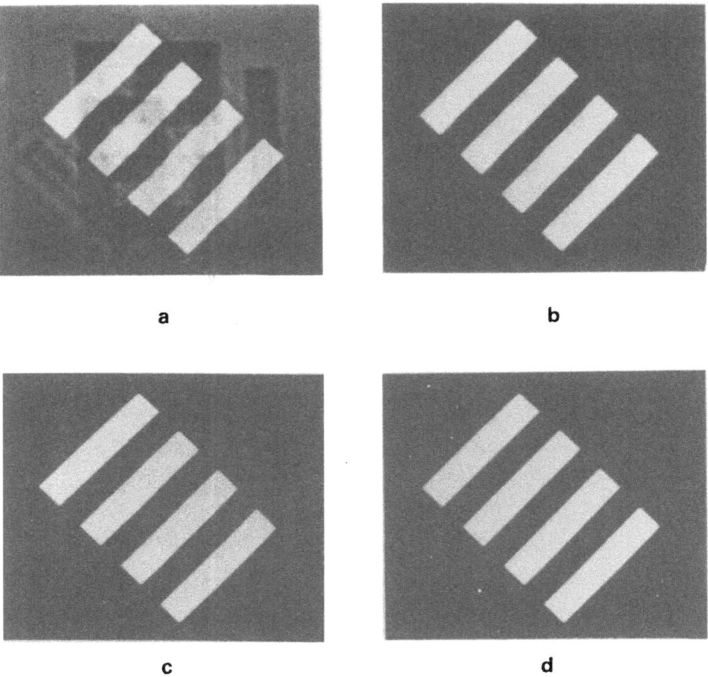

Fig.2 (a)Reconstruction by BFP(NMSE=0.209), (b) 1st iteration(NMSE = 3.53×10^{-2}),
(c) 2nd iteration (NMSE = 1.30×10^{-2}), and (d) 3rd iteration (NMSE = 7.55×10^{-3}).

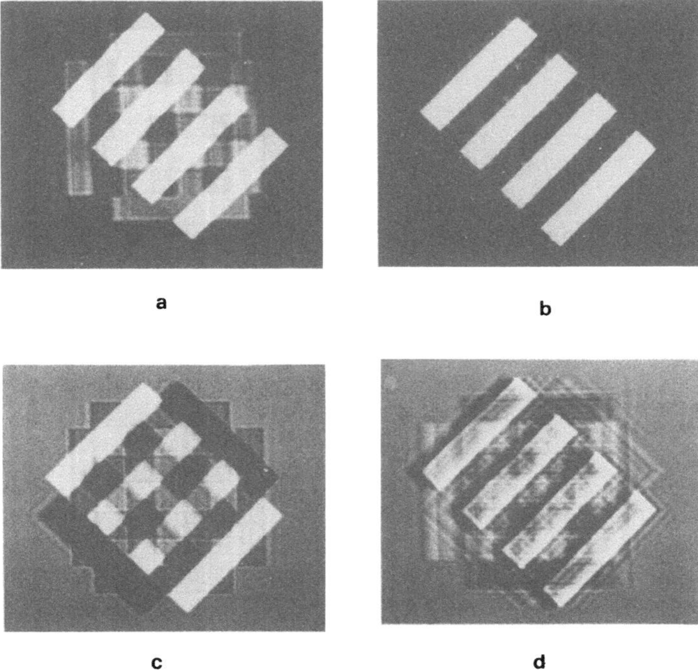

Fig.3   (a) Reconstruction(NMSE=0.295) and (b) restored image with 3 iterations
$(NMSE=2.92 \times 10^{-2})$ for the separation of 3λ , (c) reconstruction(NMSE= 0.950)
and   (d) restored image with 1 iteration (NMSE= 0.661) for the separation of 1λ.

propagation transfer function. For a set of received wavefield with various insonification frequencies, Eqs.(3) and (4) can be written in a matrix form as

$$
\begin{bmatrix}
V_1(f'_x, f'_y) / \Omega_1 \\
V_2(f'_x, f'_y) / \Omega_2 \\
\cdot \\
\cdot \\
\cdot \\
V_M(f'_x, f'_y) / \Omega_M
\end{bmatrix}
=
\begin{bmatrix}
1 & \Omega_1 & \Omega_1^2 & \ldots & \Omega_1^{M-1} \\
1 & \Omega_2 & \Omega_2^2 & \ldots & \Omega_2^{M-1} \\
\cdot & \cdot & \cdot & & \cdot \\
\cdot & \cdot & \cdot & & \cdot \\
\cdot & \cdot & \cdot & & \cdot \\
1 & \Omega_M & \Omega_M^2 & \ldots & \Omega_M^{M-1}
\end{bmatrix}
\begin{bmatrix}
O(f_x, f_y; z_1) \\
O(f_x, f_y; z_2) \\
\cdot \\
\cdot \\
\cdot \\
O(f_x, f_y; z_M)
\end{bmatrix}
\tag{9}
$$

where $\Omega_m = \exp\{i2\pi\dfrac{\delta}{\lambda_m}\sqrt{1 - (\lambda_m f'_x)^2 - (\lambda_m f'_y)^2}\}$ with the layer separation δ,

$f'_x = f_x + f^i_x$, $f'_y = f_y + f^i_y$, and the spatial frequency domain is confined to the region where the propagation transfer function is nonzero for all the applied insonification frequencies. The matrix Ω in Eq.(9) is known as Vandermonde matrix[11] whose complexity for the matrix inversion is M^2, rather than M^3 for a regular matrix inversion, or $M\log(M)$ for a Fourier matrix inversion by a fast algorithm. In deriving Eq.(9), object planes are assumed to be equally spaced, and the number of projections is identical to the number of object planes. Once $O(f_x, f_y; z_k)$ is calculated, restored image can be obtained simply by the inverse Fourier transformation and then by taking the real part. Figure 4 depicts four slices of the restored images obtained by the algebraic method.

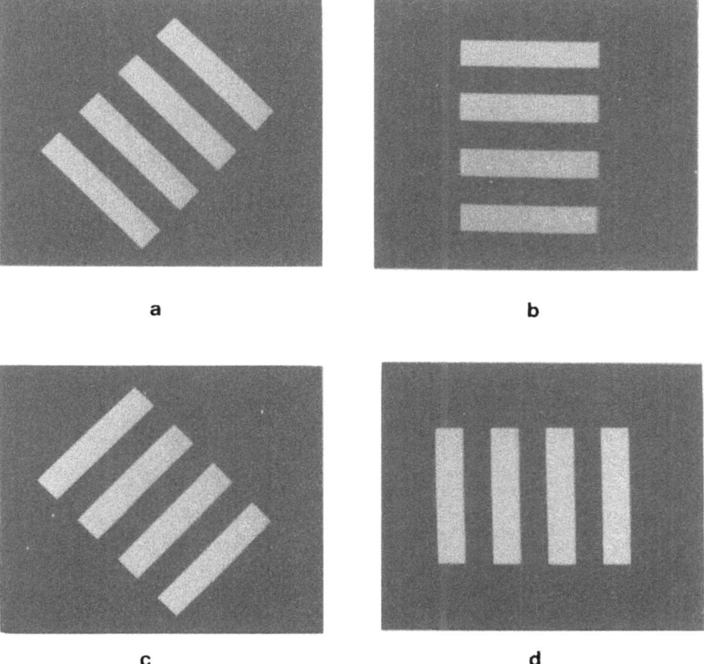

Fig.4　Four slices of images obtained by the algebraic restoration (NMSE<7.30x10^{-3}). The separation between layers was 3λ.

Although the algebraic method provides better restoration (NMSE < 7.30x10^{-3}) in a noise-free environment, often the Vandermonde matrix is ill-conditioned under a noisy environment. In this context, noise denotes statistical random noise independent to the object since the diffraction terms were already considered in the matrix formula in Eq.(9). If we assume a unit noise variance in the measured wavefield for all the insonification frequencies, then the noise variance in the k-th slice after solving the simultaneous equations is given in spatial frequency domain by

$$\sigma_k^2 (f_x, f_y) = \sum_m \left| \Omega_{km}^{-1}(f_x, f_y) \right|^2 \tag{10}$$

where $\Omega_{km}^{-1}(f_x, f_y)$ is the element in the k-th row and m-th column in the inverse matrix of Ω. Equation (10) is evaluated point by point for the layer distance of 3λ and the results are shown in Fig.5 with gray scales proportional to the variance. In Fig.5, an unsymmetrical bell-shaped noise distribution centered at d.c. is observable. Also there are intensive peaks around the boundary of the circle. In the latter case, the matrix was almost singular. For a white gaussian noise with a standard deviation of 0.3% of the d.c. component, two restoration algorithms are tested and their results are shown in Fig.6(a) and (b) for the algebraic method and in (c) for the iteration method, respectively. Due to the nonuniform noise variation in the algebraic

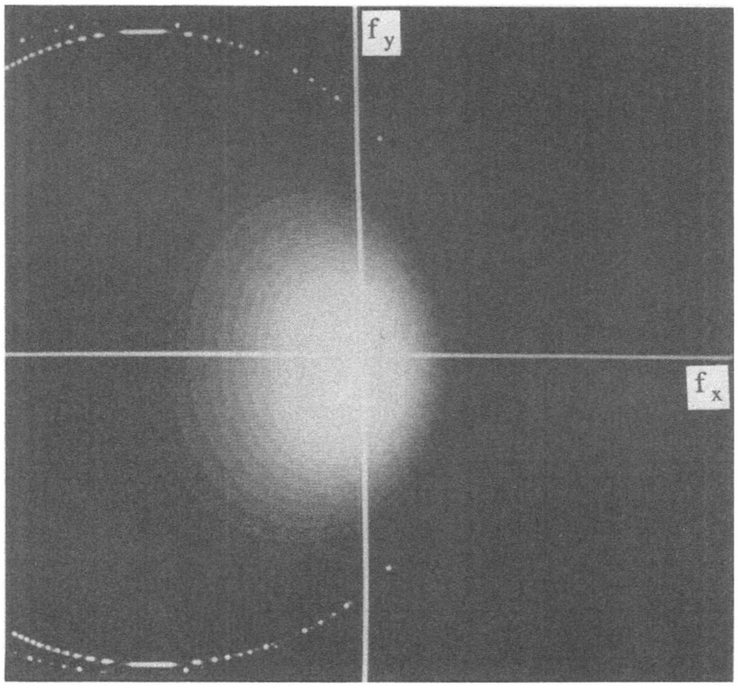

Fig.5 The noise variance in the algebraic restoration method is shown for a unit noise variance in the measured wavefield. Twenty gray scales are allocated in proportion to the variance from 0 (black) to 60 (white).

428

method as shown in Fig.5, there appears distinct noise pattern with particular spatial frequency components in Fig.6(a) with the NMSE as high as 1.53. The restored image after filtering out those intensive frequency peaks surrounding the circular boundary is shown in (b), where relatively low frequency noise is, then, dominant due to the bell-shaped noise distribution. Since only a few frequency components are removed in (b), virtually there is no resolution loss in the filtered image, while a substantial SNR improvement is obtained (NMSE=0.130). The corresponding restored image by the iterative method is shown in (c), where no particular noise pattern is observable (NMSE=3.59×10^{-2}). Comparing these results with previous ones in Fig.4, the NMSE by the algebraic method is largely amplified from 7.30×10^{-3} in a noise-free environment to 0.130 in a noisy condition, while the increase of the NMSE by the iterative method is relatively small (from 2.92×10^{-2} to 3.59×10^{-2}). Thus, in high SNR, algebraic method provides better restoration with less computational time, while in poor SNR, iterative method is more recommendable due to the noise immunity.

CONCLUSION

The back-and-forth propagation (BFP) method has been known as an efficient algorithm for the reconstruction of well-shaped planar objects, and

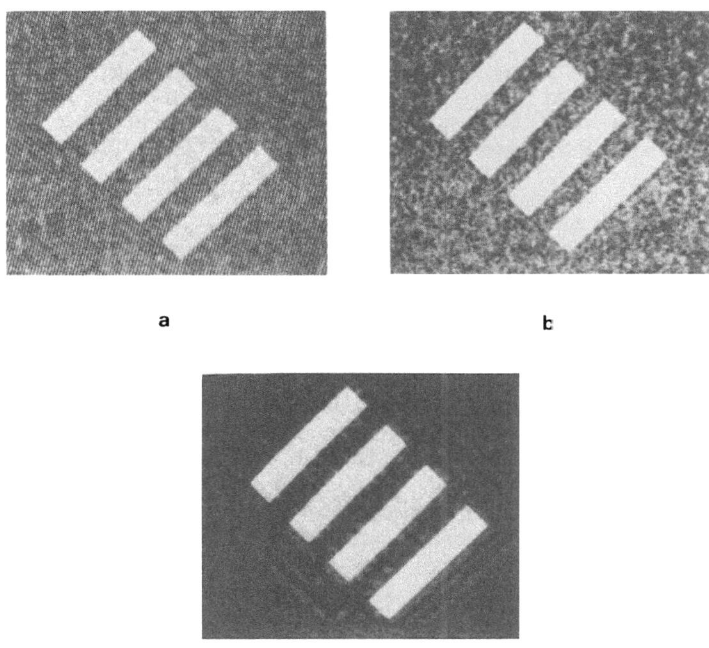

Fig.6 The proposed iterative and algebraic restoration algorithms are examined under a noisy environment. (a) Restored image by the algebraic method (NMSE=1.53). (b) Restored image after suppression of intensive peaks around the circular boundary (NMSE=0.130). (c) Restored image by the iterative method (NMSE=3.59×10^{-2}). In this simulation, the separation between object layers was 3λ.

has gained successful results in the area of scanning tomographic acoustic microscopy (STAM). In this paper, a short coming of the BFP algorithm, i.e., diffraction noise remained after statistical cancellation by a superposition (tomographic processing), is investigated and two approaches to remove diffraction noise are proposed, namely the iterative and algebraic restorations. The former is the algorithm of the iterative estimation and cancellation of diffraction noise from the reconstructed images which are also updated by the removal of the estimated diffraction noise at each iteration step. Simulation results show a fast convergence of the algorithm, and substantial improvements are usually obtained within a few iteration steps. The algebraic restoration is to solve the simultaneous equations whose matrix coefficients are related with the propagation transfer function. For equally spaced object layers, the matrix becomes in the form of the Vandermonde matrix whose complexity for the inversion is M^2 for MxM matrix, rather than M^3 as in a regular matrix inversion. In a noise free environment, the algebraic approach provides more accurate restoration with less computational time, while in a noisy environment, there appears a distinct noise pattern due to the nonuniform noise variation caused by the matrix inversion in the spatial frequency domain. In contrast to the algebraic method, the iterative algorithm appears to be relatively robust to the external noise. By computer simulations, the efficacy of the proposed algorithms are demonstrated when they are applied to the BFP algorithm.

REFERENCES

1. Z.H. Cho, Computerized tomography, in " Encyclopedia of physical science and technology," vol.3, pp. 507-544, Academic press, New York (1987).
2. J.F. Greenleaf, S.A. Johnson, S.L. Lee, G.T. Herman, and E.H. Wood, Algebraic reconstruction of spatial distributions of acoustic absorption with tissues from their two-dimensional acoustic projections, in "Acoustical Holography," P.S. Green, ed., vol.5. pp. 591-603, Plenum, New York (1974).
3. R.K. Mueller, M. Kaveh, and G. Wade, Reconstructive tomography and applications to ultrasonics, Proc. IEEE, 67:567 (1979).
4. A.J. Devaney, A Filtered backpropagation algorithm for diffraction tomography, Ultrason. Imag., 4:336 (1982).
5. H. Lee, C.F. Schueler, G. Flesher, and G. Wade, Ultrasound planar scanned tomography, in "Acoustical Imaging," John Powers, ed., vol.11, pp.309-323, Plenum, New York (1982).
6. Z.C. Lin, H. Lee, and G. Wade, Back-and-forth propagation for diffraction tomography, IEEE Trans. Sonics Ultrason. SU-31:626 (1984).
7. H. Lee and R.Y. Chiao, Phase error estimation and correction in acoustic microscopy, to be published in IEEE Trans. ASSP, (1989).
8. D.G. Grant, Tomosynthesis: A three-dimensional radiographic imaging technique, IEEE Trans. Biomed. Eng. BME-19:20 (1972).
9. R.W. Schafer, R.M. Mersereau, and M.A. Richards, Constrained iterative restoration algorithms, IEEE Proc. 69:432 (1981).
10. J.W. Goodman, "Introduction to Fourier optics," p.54, McGraw Hill, San Francisco (1968).
11. W.H. Press, B.P. Flannery, S.A. Teukolsky, W.T. Vetterling, Numerical recipes in C, pp.52-54, Cambridge University Press, Cambridge, (1988).

DEVELOPMENT OF AN UNDERWATER FRONTAL IMAGING SONAR,

CONCEPT OF 3-D IMAGING SYSTEM

Pierre Alais, Pascal Challande and Lilia Eljaafari

URA-CNRS 868, Laboratoire de Mécanique Physique
Université Pierre et Marie Curie
78210 Saint-Cyr, France

FRONTAL IMAGING PRINCIPLE

As it is shown in the Fig. 1, a frontal imaging device obtains its
information from the reverberation of an acoustical beam furnished by the
submarine bottom at low site angles in the same way than side scan
sonars. The acoustical beam is wide open in the vertical plane and
focused in the horizontal direction at the intercept of the ground floor.
Information is achieved through the modifications of the reverberation or
through the shadows created by obstacles laid on the floor. These shadows
reveal the relief in an exaggerated way as low angled light does. As for
side scan sonars, the distance R of the obstacles is related to the time
of flight of the acoustical information and instead of the parallel

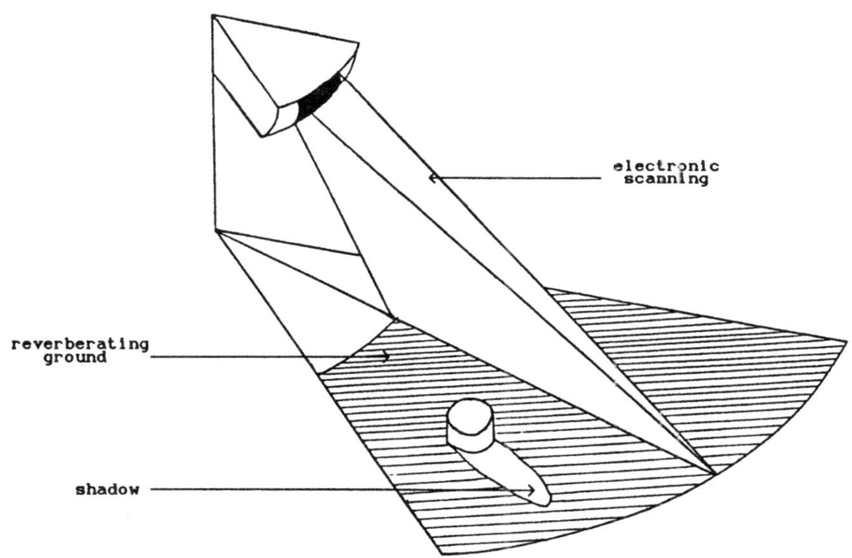

Fig. 1, Frontal imaging principle : objects are detected by their
projected shadow.

scanning of the beam obtained by the motion of the side scan sonar
vehicle, a rotation of the beam is operated electronically with an array
of transducers in the θ-azimutal direction to obtain a sectorial R-θ
image of the ground floor. For theoretical and practical reasons, we have
chosen to operate the azimutal rotation of the beam with the electronic
scanning of a focused aperture along a circular array of transducers.
This technique permits to use relatively low delays to focus the beam and
reduces the electronic rotation of the beam to a simple commutation of
the transducers retained for the focused aperture. The focusing operation
depends only on the R variable and may be controlled in a better way.

The geometry of the experimental array is defined by 64 transducers
set at a .5 degree pitch on a radius of 800 mm. The focused aperture is
obtained from a set of 32 adjacent transducers which means a width of
nearly 220 mm. The range of frequencies of 600-900 kHz which may be
operated by our transducers is convenient for imaging distances from a
few meters to nearly 100 meters. The Fig. 2 shows the directivity curve
obtained at 600 kHz from this aperture focused at infinity with different
weighting laws. The Fig. 3 and 4 show the evolution versus frequency of
the 3 dB angular resolution and of the main sidelobe level for the same
ponderation laws. The angular pitch of .5 degree is equivalent to a
linear pitch of 7 mm for the transducers which means that grating

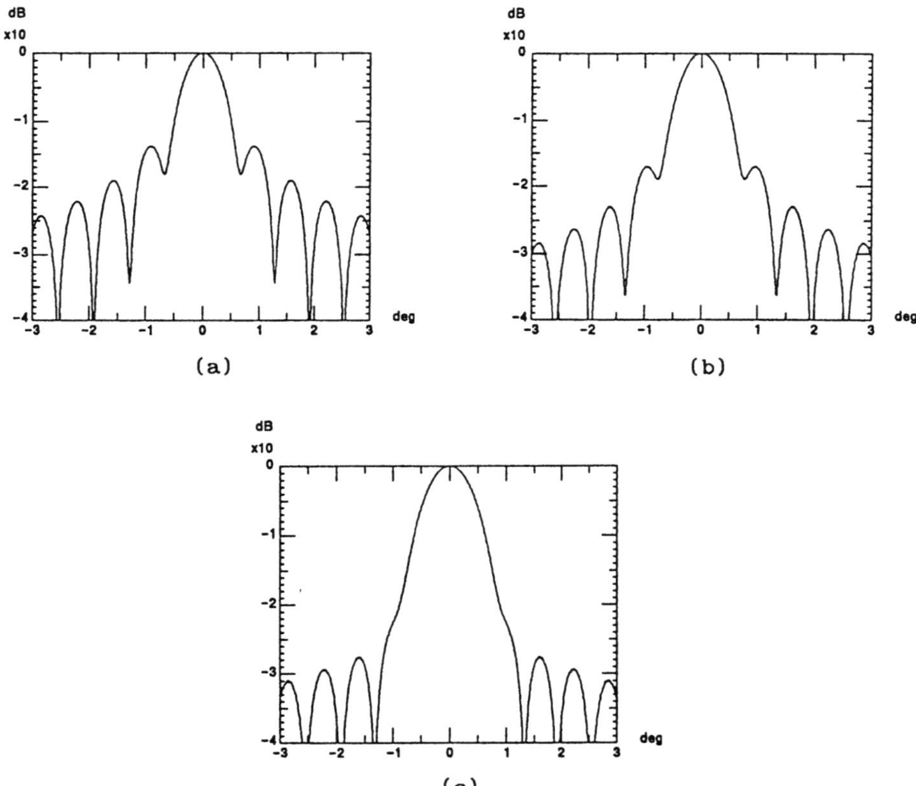

(a)   (b)

(c)

Fig. 2, Shading influence : theoretical directivity patterns of a
circular antenna made of 32 tranducers working at 600 kHz
with a pitch of 7 mm, the focus is 24 m. (a) no shading;
(b) parabolic aperture; (c) raised cosine aperture.

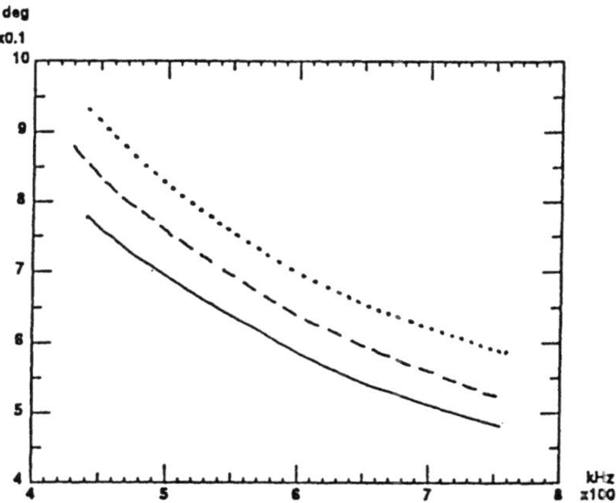

Fig. 3, Evolution of the theoretical 3 dB lateral resolution versus frequency when focusing at 24 m. (—) no shading; (--) parabolic aperture; (⋯) raised cosine aperture.

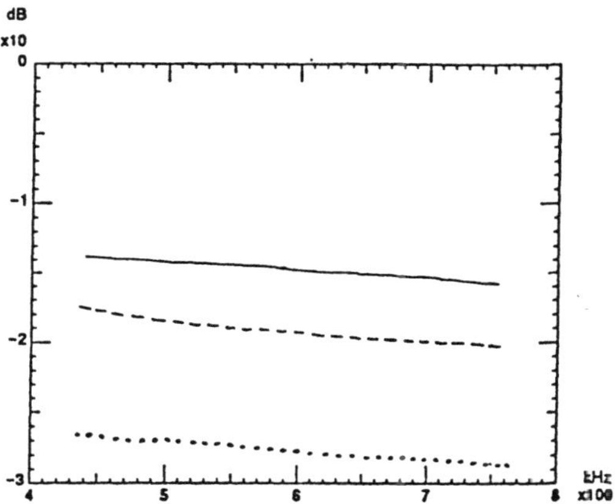

Fig. 4, Evolution of the theoretical first sidelobe level versus frequency when focusing at 24 m. (—) no shading; (--) parabolic aperture; (⋯) raised cosine aperture.

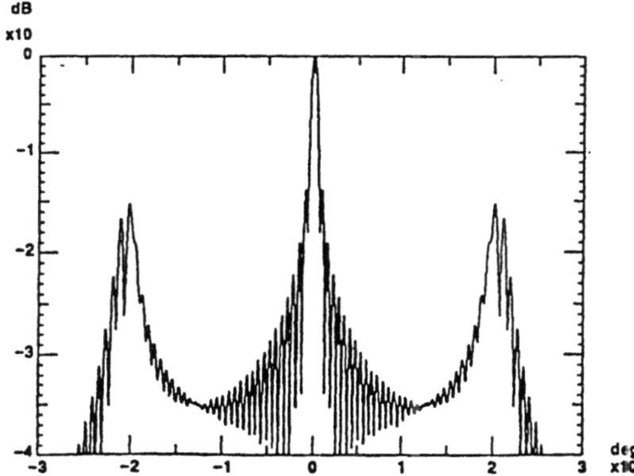

Fig. 5, Presence of grating lobes : theoretical directivity pattern of the antenna defined in Fig. 2. focusing at 24 m with no shading.

sidelobes exist for our range of wavelengths (1.7 mm - 2.5 mm) as shown in the Fig. 5, but we shall show later how to avoid their artefacts.

THE BEAM FOCUSING AND SCANNING TECHNOLOGY

As well for the antenna as for the electronic focusing and scanning technique, our technology has been largely inspired by our previous research in the medical echography field. The antenna has been developped like a medical array i.e. with subdivised elements[1] except that for obtaining a wide aperture for the angle of site (40°), each transducer has a cylindrical geometry and their association on a circular line confers a toroidal shape to the antenna as shown by the Fig. 6. This technology permits to obtain a wide band of working frequencies of 600-900 kHz, with a directionality of each elementary transducer in the azimutal direction not far from the *sinc* law which has been assumed for the results exposed in the Fig. 2, 3, 4 and 5.

The electronic commutation ensuring the rotation of the beam is operated as in the medical echographic devices using linear arrays, and permits a random access of the azimutal position of the beam among 64 positions identical to the 64 positions of the elementary transducers. The focusing aperture remains constant and relies on 32 transducers except for the extremal values of θ for which the aperture decreases linearly from 32 to 16 elements, which means a corresponding deterioration of the focusing capabilities for the lateral parts of the image.

A dynamic control of the focusing operation versus distance or time of flight is operated as in medical echographic devices with an original technique developed by our laboratory for medical applications[2]. A simple multiaccess delay line is electronically controlled to furnish the parabolic set of delays required in the Fresnel approximation frame, with an accuracy of nearly 10 ns for ranges of imaging from 2.5 m to infinity. The delay line relies on inductance capacitance cells where the capacitance is electronically controlled by a D-C polarisation. This technique is powerful to obtain the best resolution at any distance. The

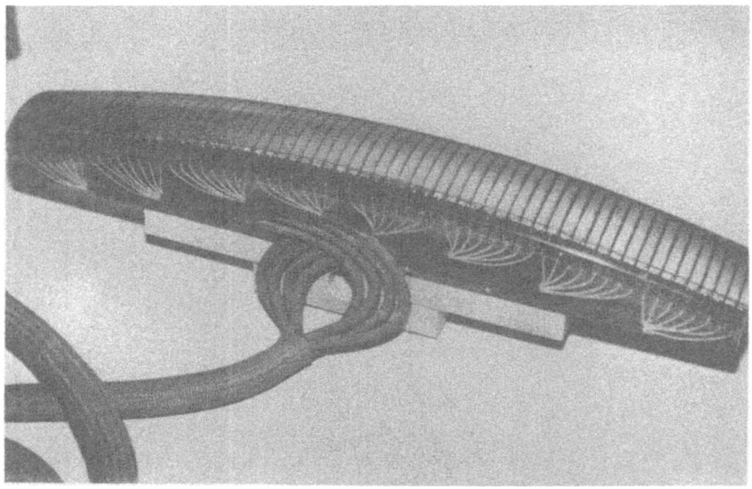

Fig. 6, View of the antenna

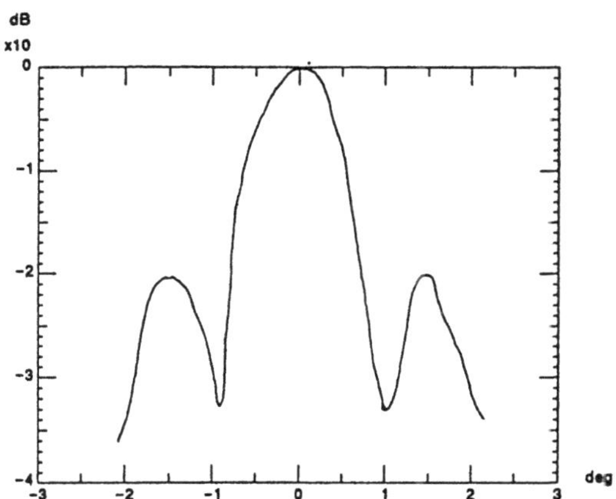

Fig. 7, Experimental directivity pattern of the antenna obtained
with a small spherical target located at 2.5 m.

Fig. 7 shows the experimental results obtained at 600 kHz from a target
set at 2.5 m i.e. the lower range of the imaging technique. The
experimental shading law is something between the parabolic and the
raised cosine law and shows a relatively good accordance with the
theoretical predictions of the Fig. 3 and 4.

IMAGE FORMATION

In relation with its focusing capabilities, the imaging device has
been conceived to image several ranges of distances i.e. 6, 12, 24, 48
and 96 meters. For each range, the sectorial image is built from 64 rays
with an angular pitch of .5 degree and the echographic signal resulting
from the focused aperture i.e. the delay line, is fed through an adequate
time gain controlled circuit to compensate the attenuation and
diffraction effects. It is then numerized in 8 bits, detected and
compressed numerically to give 512 pixels of 4 bits for each ray at a
chosen range. The 64x512 (θxR) information is then written in a X-Y
memory of 256x512 pixels in a way ensuring a good interpolation specially
according to the θ-direction to give a better comfort of visualization on
a video monitor at a classical video frame rate whatever rate the
acoustical information is obtained.

A specific problem of submarine imaging is the time formation of the
image which may become much too long due to the range and the time of
flight. A pure sequential operation like in medical devices is inadequate
and we have retained two ways for increasing the acoustical frame rate :
    - the first solution is to operate in parallel 4 adjacent focused
apertures controlled by 4 identical delay lines adequatly commuted. This
technique permits to obtain a frame rate multiplied by 4, each image
being obtained by 16 shots instead of 64 which means about 2
frames/second at a range of 24 m.
    - A second way consists in rotating the focusing aperture(s) on
several positions during the time of reception of the echographic
information i.e. the time of emission pulse. We have experimentated this
technique for a number of 4 different positions in conjonction with the
first solution so as to get the image with only 4 beams, i.e.
8 frames/second at a range of 24 m. It is obvious that this last solution
at the difference of the first one   deteriorates the conditions of

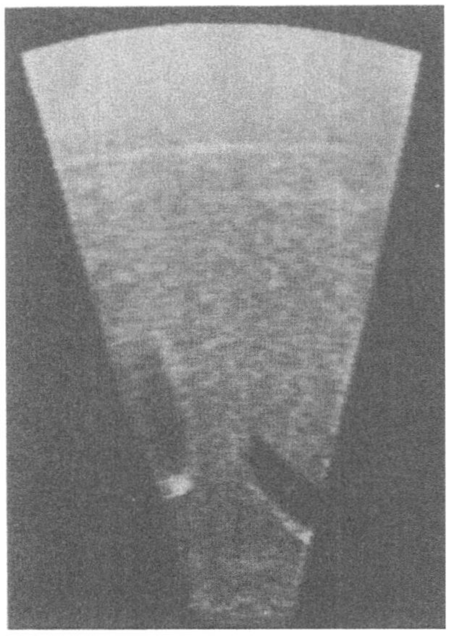
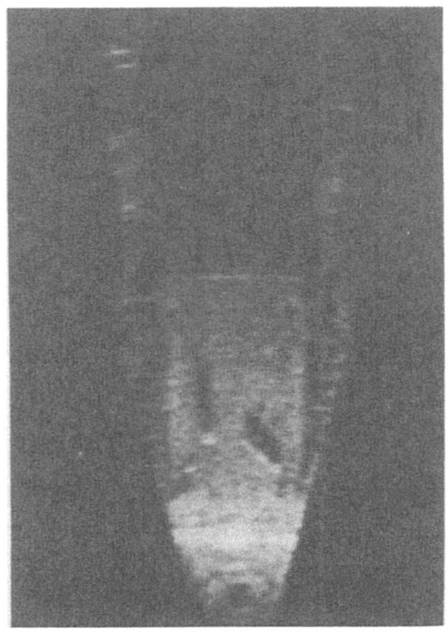

(a)   (b)

Fig. 8, Images obtained with the frontal imaging system : the targets, a concrete cylinder (ø 0.55 m, L 3 m) and a sphere (ø 1 m) led on sediments under 6.5 m of water. (a) targets located at 10 m, range : 24 m; (b) targets located at 25 m, range : 48 m. We can see the steps of stairs at each edge of the sediment tank.

reception and gives images of lower quality but may be useful when the submarine vehicle is moving rapidly.

In any case the emitted beam is delivered with poor directivity to cover the 2 degrees associated with the 4 parallel focused apertures or the 8 degrees corresponding to the azimutal excursion of these apertures on 4 adjacent positions but its directivity is well controlled to eliminate the artefacts of the grating sidelobes associated to the focusing aperture and shown on the Fig. 5.

The Fig. 8 shows images obtained in a test station of the French Navy in Brest. They show two concrete spherical and cylindrical obstacles set on a sediment tank immersed under a few meters of water. It is possible to recognize easily the shape of the obstacles through the shape of the corresponding shadows which are well given by the obtained radial and lateral resolutions. The homogeneity of reverberation gives also an idea of the quality of the image formation.

3-D IMAGING PRINCIPLE

Obtaining a view of the submarine environment with an acoustical beam as we do with a light projector and an optical camera is an old dream which has been fullfilled only partially up to now. A first idea, the oldest one, presented by Sokolov in 1929, is to use a projector, a lens and an ultrasonic retina which may be of the Sokolov kind[3] or a

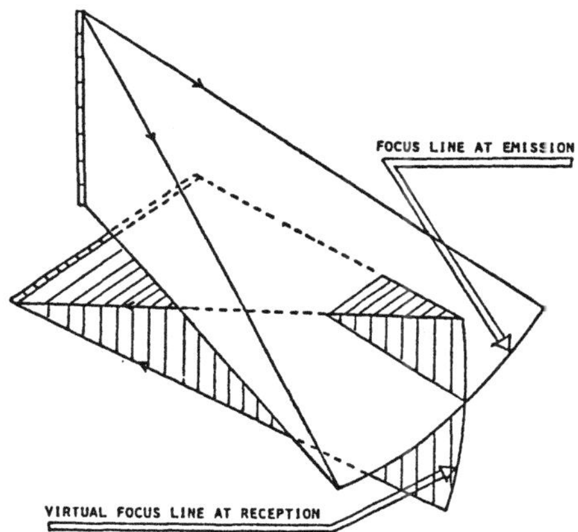

Fig. 9, Principle of the 3-D imaging : a first linear array focuses in a plane at emission, a second one focuses in the orthogonal plane at reception.

matricial one[4]. It is also possible to control a focusing matricial aperture in a θx-θy direction but the involved technology is complicated in comparison with a linear focused aperture because much more transducers and delays must be controlled. A simpler technique consists in using crossed linear arrays as shown in Fig. 9. One of them, the vertical one for example is used to emit a focused conical beam centered on a θy site angle. The other one used as a receiving linear focused aperture then permits to select the echographic information from the azimutal direction θx so that a θx-θy selection of the echographic information may be operated.

We have developped two crossed array systems conceived both for achieving acoustical focusing and scanning in an angular θx-θy aperture of 30x30 degrees. The first one (Fig. 10) is an association of an emitting linear array of 64 cylindrical transducers and a receiving circular array of 64 transducers covered by a toroidal lens ensuring them

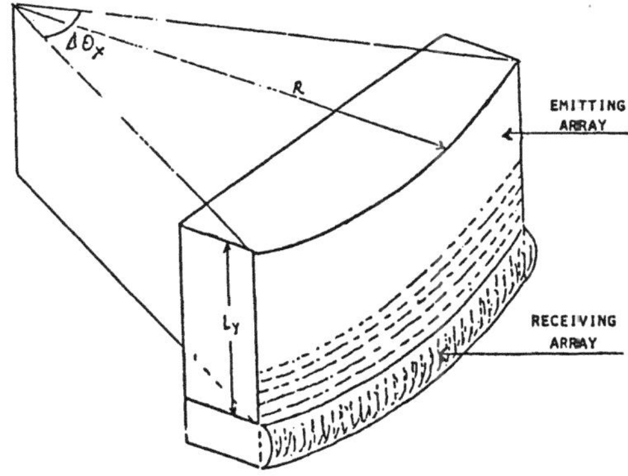

Fig. 10, First crossed array system : the emitter is made of 64 cylindrical transducers, the reception is ensured by a circular array of 64 transducers covered by a toroidal lens.

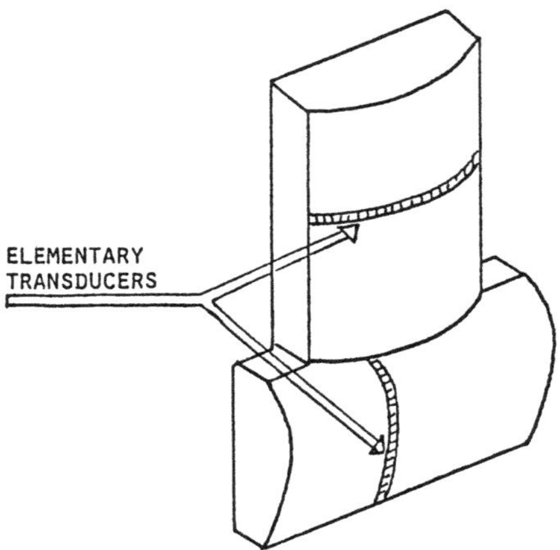

Fig. 11, Second crossed array system : the emission and reception
arrays are identical, made of 64 cylindrical transducers.

to receive correctly in a 30 degrees angle of site. This device has
presented some difficulties for its achievement and no solution was found
for obtaining a fast scanning of the space. It is obvious that the
fastest operation available with this technology consists in obtaining a
sectorial tomographic θ_x-R information for each emitted beam focused at
differents θ_y angles. But such an operation cannot be performed simply
specially for the few meters or even the few decimeters range which
concern the 3-D imaging.

We have built the second system described in Fig. 11 associating two
identical linear arrays of 64 cylindrical transducers to try to achieve a
3-D imaging of the space (limited to 30x30° angular aperture) in a time
$N\tau$ independant of the range, relying only on the number N of pixels
obtained in the three dimensions θ_x-θ_y-Z of the space i.e. N = $n_x.n_y.n_z$
and a time τ representing the minimum time for obtaining a correct signal
representative of the echographic information delivered by the concerned
θ_x-θ_y-Z pixel. Both arrays have 64 transducers with a pitch of 1 mm and
work at 1.6 MHz. To obtain a relatively fast exploration of the 3-D
space, we have reduced the number of pixels to n_x = n_y = 32 and n_z = 128
i.e. 4 k pixels for a sectorial tomographic acquisition for each value of
the site angle θ_y. The time τ = 8 μs of reception has been chosen to
enclose enough oscillations of the acoustical frequencies. In these
conditions, the minimum range corresponds to a time of flight of 128 τ
i.e. nearly Z_0 = 0.8 m. for which 32 shots are used sequentially to
obtain a 32x128 sectorial tomographic information in 4 K τ $\cong$ 32 ms and a
3-D information in about 1 s. For higher ranges Z_n = 2^n Z_0, the
acquisition time remains the same because an azimutal exploration of the
receiving focusing aperture permits to reduce the number of shots to
$64:2^n$ for each sectorial plane while the time of reception for each shot
is multiplied by 2^n. The ultimate range is Z_5 = 32 Z_0 $\cong$ 25 m and in this
case the tomographic information is obtained with only 1 shot.

3-D IMAGING TECHNOLOGY

The transducers have been obtained by segmenting PZT-4 ceramic plate
of 3 MHz nominal frequency[5,6]. The segmentation reduces the working
frequency to 1.6 MHz and the adaptation front plate is made consequently.

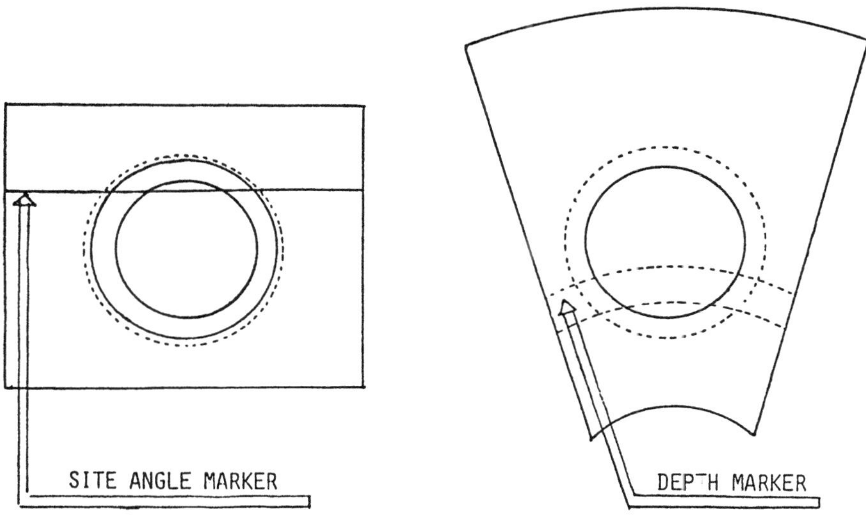

Fig. 12, Principle of the image formation

At this time, it remains difficult to obtain a good homogeneity of the directivity in the curvature plane of the cylindrical elementary transducer.

The electronic circuitry is hybrid. The reception is done by heterodyning the received signals with a reference wave the phase of which is chosen numerically among a choice of 16 with a circuitry of EPROM memories which allows to select the (n_x, n_y, n_z) pixel in terms of n_y at the emission and of n_x and n_z at the reception.

The visualization of a 3-D information is difficult to achieve. We have conceived a technique which permits to appreciate simultaneously on a video monitor the real time evolution of the sectorial tomographic θ_x-R image and a X-Y elevation image by integrating or detecting the information in a selected depth of field and refreshed continuously in 1 s (Fig. 12). For the moment being experimental results have been

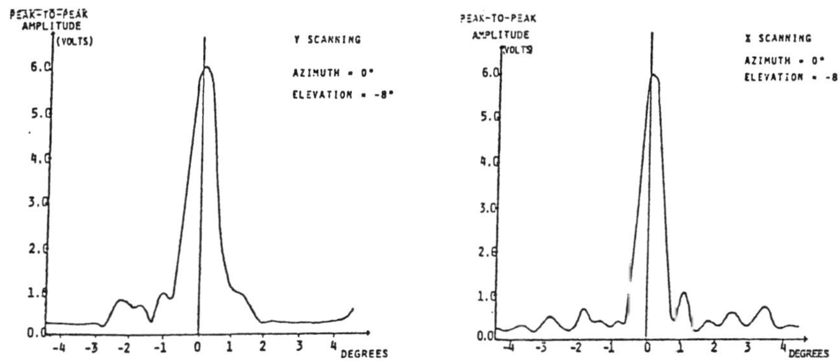

Fig. 13, Experimental directivity pattern : focusing at 40 cm, the 3 dB lateral resolution is about 1° in both planes.

obtained only in the appreciation of the acoustical focusing from a single small target set at 40 cm i.e. at the sequential minimum range Zo. The Fig. 13 shows the results for a central location of the target in the angular aperture of the camera. The high level of sidelobes is probably due to the inhomogeneity of the cylindrical transducers in their curvature plane and this certainly is the major problem to be solved in the future. The sectorial image obtained in these conditions is too poor to be presented and no elevation picture may be done correctly.

CONCLUSION

We have presented two different imaging projects. The frontal imaging device is giving good pictures in correlation with the quality of the achieved antenna. The 3-D imaging project remains very limited at the experimental level certainly due to the inhomogeneities of the arrays which have been built, and work has to be done in the future to correct these defects.

We thank the D.R.E.T and the I.F.R.E.M.E.R. for their financial support. We thank also the G.E.S.M.A. for having allowed us to use the installations of the French Navy when testing the frontal imaging sonar.

We thank very much R. LALIMAN and C. SASSIER for their technical cooperation in this research.

REFERENCES

1. M. Th. Larmande, P. Alais, "A theoretical study of the transient behaviour of ultrasonic transducers in linear arrays," in : "Acoustical Imaging," 12, Plenum Press, London (1982).
2. French Patent n°80-22-880 (October 1980).
3. S. Sokoloff, "Means for indicating flaws in materials," U.S.S.R. Patent n°49 (1936), U.S. Patent n°2164125 (1939).
4. P. Alais, "Real time acoustical imaging with a 256x256 matrix of electrostatic transducers," in : "Acoustical Holography," 4, Plenum Press, Santa Barbara (1972).
5. P. Alais, P. Challande, C. Kammoun, B. Nouailhas, F. Pons, "A new technique for realizing annular arrays or complexed shaped transducers," in : "Acoustical Imaging," 13, Plenum Press, Minneapolis (1984).
6   P. CHALLANDE, "Etude de réseaux de transducteurs annulaires et modélisation de cavités piezoélectriques par la méthode des éléments finis," Thèse de l'Université Pierre et Marie Curie, Paris (1987).

Signal Processing in the 1988 Monterey Bay Acoustic Tomography Experiment

Robert C. Dees, James H. Miller, Kevin P. Schaaff, Sönke Paulsen,
Ching-Sang Chiu, Laura Ehret

Naval Postgraduate School, Code 62Mr
Monterey, CA 93943

and

James F. Lynch

Woods Hole Oceanographic Institution
Woods Hole, MA 02543

I. Introduction

The main obstacle to understanding ocean circulation is our inability to observe it. The ocean is opaque to most of the electromagnetic spectrum. In situ measurements are time-consuming and expensive. However, the ocean is transparent to low frequency acoustic waves. The navies of the world have utilized this fact to detect submarines with sound since 1918.[1] Recently, oceanographers and engineers have looked to acoustics for assistance in observing the circulation of the ocean.

In 1979, Munk and Wunsch proposed a scheme for large scale monitoring of the deep ocean and named it *ocean acoustic tomography*.[2] Ocean acoustic tomography can observe the mesoscale dynamics of the ocean (eddies, fronts, and currents) by measuring changes in the travel time of narrow-band acoustic signals transmitted over various paths. The ocean's fluctuations can then be estimated from these travel time perturbations using mathematical inverse methods.[3]

The work described in this paper is part of a effort by the Naval Postgraduate School (NPS) and Woods Hole Oceanographic Institution (WHOI) to use tomography in coastal regions. In particular, an experiment was held in Monterey Bay, California in December, 1988. Among the experiment goals were to:

1.   *Determine experimentally the relationship between surface waves and fast fluctuations in acoustic arrival times of tomography signals;*

2.   *Determine the effect of internal waves on tomography signals in a coastal environment;*

3.   *Determine the effect of the complex bathymetry of the Monterey Submarine Canyon on acoustic propagation, and;*

Acoustical Imaging, Vol. 18, Edited by H. Lee and G. Wade
Plenum Press, New York, 1991

4.   *Test a real-time shore-based acoustic tomography data acquisition system.*

The experiment had a single acoustic tomography transmitter (with a source level of 177 dB re 1µPa) located 36 km west of Point Sur and ten receivers located around the rim of Monterey Canyon between Point Piños and Santa Cruz as illustrated in Figure 1. Between the transmitter and most of the receiver positions is the Monterey Submarine Canyon, the largest canyon on the California continental shelf. The extreme bathymetry of the canyon can give difficult three-dimensional propagation paths for the sound traveling through the canyon. The receivers were placed on the continental shelf with the transducer resting on the bottom (to reduce motion) in about 100 meters of water. The principal investigators were James H. Miller and Ching-Sang Chiu of NPS and James F. Lynch of WHOI. Also involved in the experiment were the National Data Buoy Center, Sparton Corporation, and Mitre Corporation.

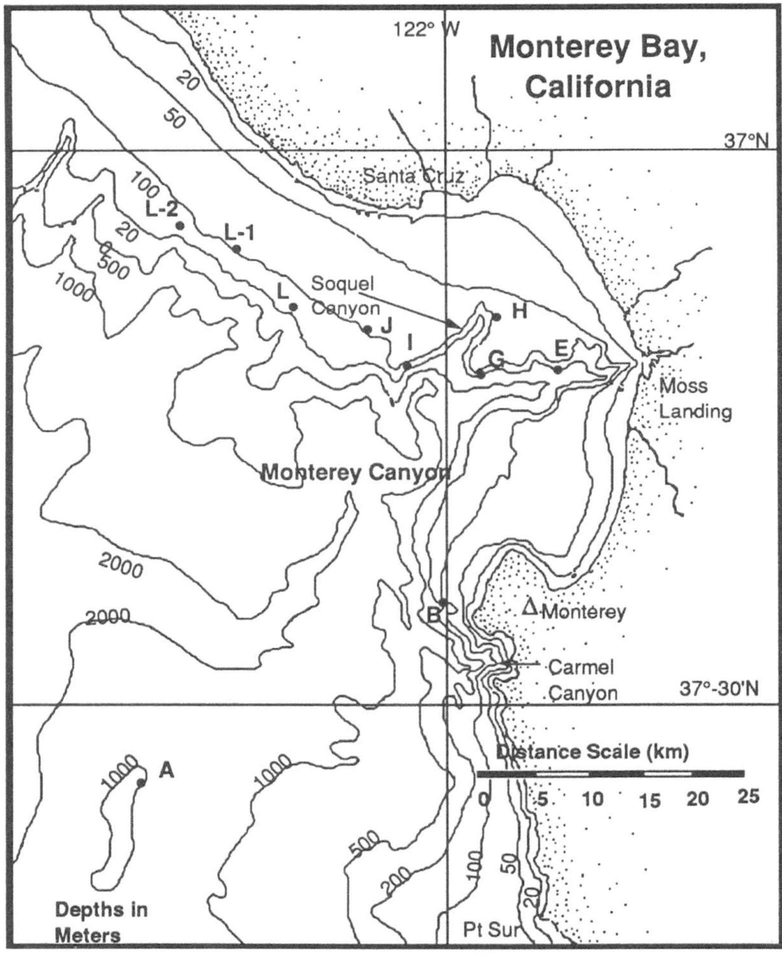

Figure 1. Monterey Bay showing the positions (marked with •) of the acoustic tomography source (A) and receivers (B-L2). The shore station position is marked by △.

This paper concentrates on signal design using maximal-length sequences, data recording, and a fast algorithm for a data-synchronous digital correlator receiver in this experiment. The new tomographic data recording system has demonstrated its effectiveness and the fast algorithm using the Fast Hadamard Transform performed the matched-filtering for the pulse compression very quickly (300 times faster than a program using Discrete Fourier Transforms). Examples of the results of the data analysis are given, including power spectra for the arrival time perturbation series in the 0.01 to 0.26 Hz (surface wave) frequency band. These spectra correlate well with surface wave spectra obtained from a wave-measuring buoy.

II. The Ocean Acoustic Tomography Problem

Treating the ocean medium as a large, time-varying distortionless filter, the impulse response of the source-receiver channel is just the sum of the impulse responses of the individual paths

$$h(t) = \sum_{i=1}^{P} a_i \, \delta(t-\tau_i) \; , \tag{1}$$

where P is the number of paths, a_i is the amplitude, and τ_i is the total travel time along the path. If the transmitted signal is an impulse then the received signal will be the impulse response. The separate paths that connect source and receivers (eigenrays) can be predicted from ray theory. The restrictions placed on the sound speed structure and acoustic environment for ray theory to be reasonable are well-known and can be found in Ref. 4. If the conditions for the validity of ray theory cannot be met, other methods must be used, and "full wave" or modal solutions can be attempted.[4,5] Ray solutions of the acoustic paths are appropriate to this experiment, although the solutions may not be confined to a single plane.

The travel time for a ray path can be found by integrating the sound slowness (inverse speed) over the specific ray path denoted by P_i (the ith ray path)

$$\tau_i = \int_{P_i} \frac{ds}{c(x,y,z,t)} \; . \tag{2}$$

If the sound speed is represented as a perturbation on a constant base speed depending on depth along the path, such that $c(x,y,z,t) = c_o(z) + \delta c(x,y,z,t)$. For $\delta c \ll c_o$, the perturbation to the travel time from Eq. (2) can be written as

$$\delta\tau_i \cong -\int_{P_i} \frac{\delta c\,(x,y,z,t)}{\left[c_o(z)\right]^2} \, ds \; . \tag{3}$$

So the tomography problem is this: given the travel time perturbations $\delta\tau_i$ of rays between a number of sources and receivers, and the background sound speed field $c_o(z)$, estimate the sound speed field perturbation $\delta c(x,y,z,t)$ using a linear inverse method such as singular value decomposition.[6] Equation 3 ignores the effect of current. Variations in the travel time due to current are generally an order of magnitude less than the variations due to changes in the sound speed and so can be ignored in a first order treatment.[7] Also note that a change in the path length (as occurs with surface waves) can have a significant effect on the arrival times.[8]

III. Tomography Signal Design

In an ideal world, we would transmit an impulse of sound of very high amplitude and very short duration. However, practical sound sources are inherently resonant devices with a finite bandwidth and peak-power limitations. One solution to this problem is to transmit a coded narrowband signal and use pulse compression to get the desired SNR. The transmitted, narrowband signal that has been used in ocean acoustic tomography experiments is given by

$$s(t) = A \sin [2\pi f_o t + \theta(t)] \qquad (4)$$

where A is the amplitude, f_O is the center frequency and $\theta(t)$ is the phase modulation and is given by

$$\theta(t) = \theta_o m(t) \qquad (5)$$

and m(t) is a maximal length sequence of +1's and -1's. The width of one of the bits of the sequence is d while the total length or period of the sequence is

$$L = Nd \qquad (6)$$

where N is the number of bits in the m-sequence. See Appendix A for a more complete discussion of m-sequences. Spindel[9] provides a well-written discussion on tomography signal design and processing.

For this experiment, f_O = 224 Hz and the bandwidth of the signal (determined by the bandwidth of the source) is 1/d = 16 Hz. The number of bits in the m-sequence is N = 31 so that the total length of the transmitted sequence is L = 1.9375 seconds, allowing sampling above the Nyquist frequency of dynamic ocean processes with frequencies below 0.258 Hz (periods greater than 3.875 sec), which includes the longer period surface gravity waves.

IV.   Signal Reception and Processing

At each of the receiver locations shown in Figure 1, a modified sonobuoy was moored, most with hydrophones directly on the bottom. These sonobuoys transmitted all acoustic signals received in the band between 20 Hz and 20 kHz frequency modulated on RF carriers between 160 and 170 MHz. The RF signals were received at a shore station overlooking Monterey Bay. At the shore station, the acoustic signals were taped for later processing.

After the experiment, each channel of acoustic data was played back, quadrature demodulated to baseband and then sampled at twice the Nyquist frequency before processing with the Fast Hadamard Transform as illustrated in Figure 2. See Appendix A for more discussion of the Fast Hadamard Transform. With the transmitted signal bandwidth of 16 Hz, the sampling rate was 64 Hz. The points of the received signal are separated by the sample period of 15.625 milliseconds. The points can be interpolated to a smaller separation by using curve fitting. A cubic spline curve fitting routine adapted from Press, et al., generates points separated by 0.976 milliseconds.[10]

In summary, the entire signal processing system estimates the arrival time perturbations from the data recordings through the following procedure:

1. The signal passes through a band-pass filter to remove any out-of-band noise.

2. The signal is quadrature-demodulated to baseband and low-pass filtered to remove the high frequency components.

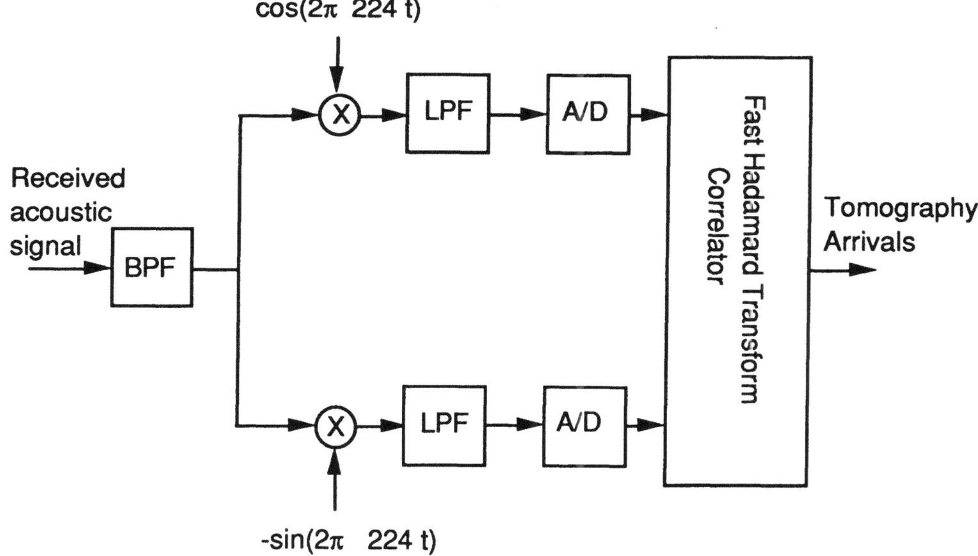

Figure 2. Block diagram of quadrature demodulator and Fast Hadamard Transform Correlator.

3. The signal in-phase and quadrature components are sampled at 64 Hz and digitized.

4. The Fast Hadamard Transform is used to matched-filter for the maximal-length sequence code and the result is stored.

5. Given a certain window around an eigenray arrival, the arrival time of the ray is estimated with respect to an arbitrary code starting position, and the signal-to-noise ratio is calculated.

6. The geophysical time (clock time) of the data point, time of arrival, peak magnitude, and signal-to-noise ratio are stored.

This stored data contains the fluctuations due to path length and sound speed perturbations and will be the input data for the tomographic inversion to estimate the ocean conditions.

V. Experimental Results

After matched-filtering the ray arrivals are very apparent in the data. An example one hour of the arrivals is shown in Figure 3. The arrival time estimates can be thought of as tracking the motion of the peak from one code period to another to get the arrival time fluctuations. Figure 3 shows four ray arrivals: the first at about 1.00 sec, the second, high amplitude arrival at 1.32 sec, a third at 1.45 sec, and a fourth, unstable arrival at 1.62 sec. All of the arrivals are resolved and arrivals 1-3 are stable for this hour. It was predicted that the power spectral density of the arrival time fluctuations caused by the surface waves would reflect the power spectral density measured by the NDBC surface wave measurement buoy. The power spectral density of the arrival time perturbations of arrival 2 from Figure 3 (except at slightly different time) and the surface waves were estimated using a segmented Fast Fourier Transform with results as shown in Figure 4. The travel time spectrum resolution bandwidth is 0.00806 Hz and the sea surface spectrum resolution is 0.01 Hz. A comparison of the arrival time and surface wave power spectra immediately shows agreement in the general shape and frequency distribution with the largest concentration of power in the long period swell frequency region of 0.07 to 0.09 Hz.

The arrival time spectrum also shows a smaller but still significant peak at about 0.03 Hz, longer than is normally observed for sea swell in the Pacific. However, 0.03 Hz is higher than can be attributed to internal waves and must be due to a path length change. A possible explanation is "beating" between two systems of long period swell propagating in slightly different directions producing a long modulation on the sea surface. A source of this surf beat could be swell that has been reflected or refracted off the shallow water or shoreline along the north side of the Bay.

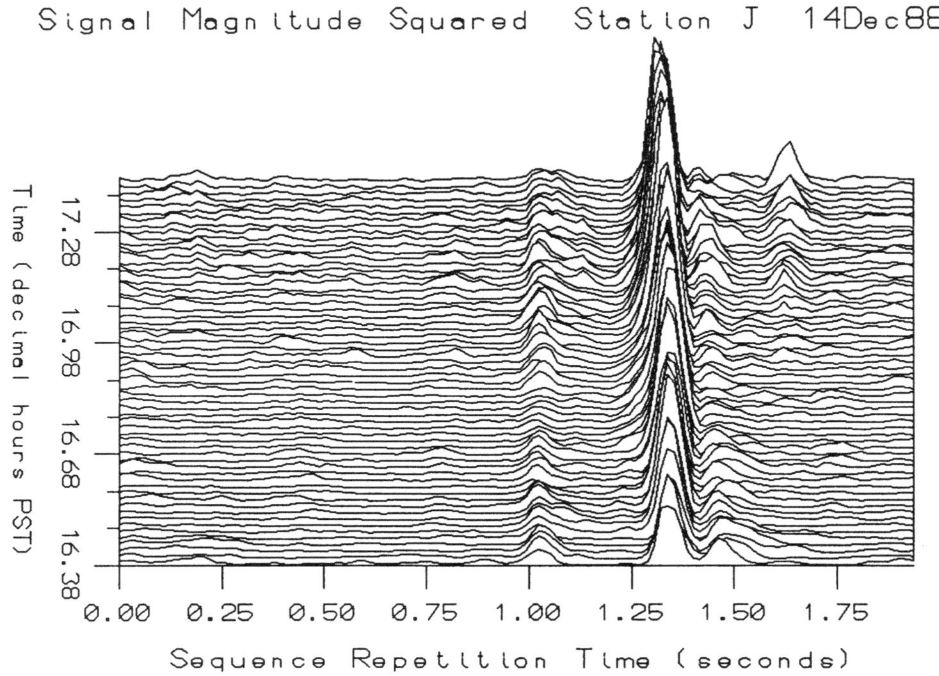

Figure 3. Acoustic arrivals generated by Fast Hadamard Transform Correlator.

The arrival time spectrum in Fig. 4 shows a nearly white noise floor. This is due in part to the random uncertainty in the estimation of the arrival time. The spectrum for the surface wave buoy data does not show this kind of noise. The algorithm for estimating the surface wave spectrum calculates a noise correction factor from the two lowest frequency data points, 0.01 and 0.02 Hz, and applies this to the rest of the data. Since the accelerometers onboard the NDBC buoy are most sensitive to noise and least sensitive to motion at the lower frequencies this is convenient. Unfortunately, the energy seen by the tomography signal may indicate that "zeroing" the low frequencies may not always be correct.

VI. Conclusions

The system of signal processing using data recorded with a time synchronization signal meets its requirement for accurate timing with an estimated accuracy of under ± 1 millisecond loss over 6 hours, and sufficient signal-to-noise after processing with 7 - 12 dB SNR for most channels. This allows adequate precision in arrival time estimation for tomographic analysis. In this experiment, the uncertainty in the estimate was usually between 2 and 4 milliseconds. The FORTRAN programs developed for processing the received acoustic signal were fast enough enough to digitize and matched-filter two channels in real-time because of the efficiency gained through using the Fast Hadamard Transform in place of Discrete Fourier Transforms. The system used a Zenith Z-200 PC (6MHz, 80286 machine) to store the data on 20 Mbyte IOMEGA Bernoulli Box cartridges, each of which will hold about 6 channel-hours of data. The travel time estimation program reliably interpolated to find the peak amplitude and recorded the arrival estimate as well as the signal-to-noise ratio, which was used to calculate the uncertainty in each measurement.

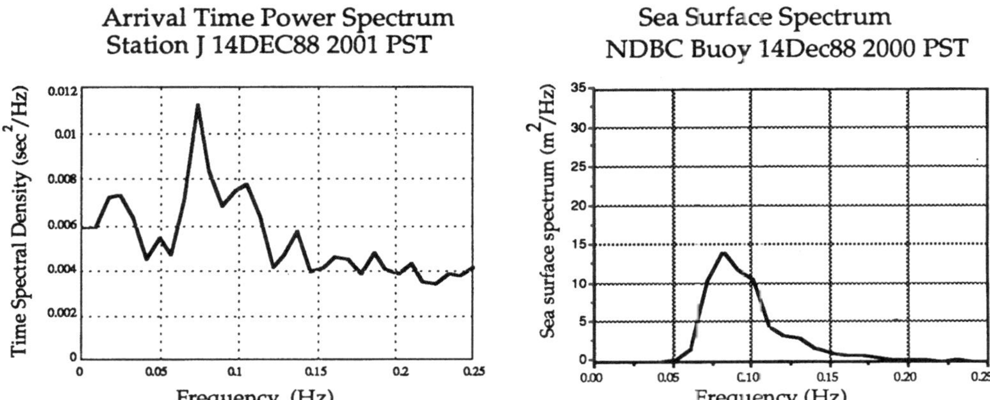

Figure 4. Comparison of arrival time power spectrum and the sea surface power spectrum.

In summary, this system of telemetered acoustic data, recording with a synchronizing signal, data-synchronous quadrature demodulation, and digital matched-filtering based on the Fast Hadamard transform, is efficient and effective for acquiring ocean acoustic tomography data. While the data analysis is still incomplete, the preliminary results have already demonstrated that continuous transmission of a relatively short period maximal-length sequence can measure the surface wave spectrum. This is the first experiment where the link between the arrival time power spectrum and the sea surface spectrum has been demonstrated. Of particular interest is the energy appearing around 0.02 and 0.03 Hz which is not measured by the surface wave buoys. The data also shows internal wave frequency arrival time perturbations and with extended synchronization between tapes, could show tidal fluctuations and mesoscale eddy features. This system could be the basis for a real-time tomography system in which the signal is transmitted to shore for immediate processing.

Appendix A. Maximal-Length Sequences and the Fast Hadamard Transform

The maximal-length sequence (m-sequence)[11] consists of a number of digits determined by the order of the sequence and generated from the successive contents of a binary shift register, as shown in Figure A1. The sequence is transmitted continuously, phase modulating a carrier for the period of the sequence. If the code is transmitted at the maximum rate allowed by the bandwidth of the transmitter, the length will be determined as a compromise between two characteristics:

1. The shorter the code length, the greater the repetition frequency and the higher the sampling frequency for ocean data. This determines the highest frequency which may be observed.

2. The longer the code, the greater the increase in signal-to-noise ratio of the signal and the more accurately the arrival time of the signal can be estimated.

The driving consideration in this experiment is the period of the surface waves to be investigated - fully developed seas of greater than 5 seconds period. To sample the fluctuations due to the surface waves at the Nyquist frequency, the period of the signal must be less than 2.5 seconds. A maximal-length sequence 31 digits long transmitted at a digit frequency of 16 Hz has a period of 1.9375 seconds. This length was chosen for the Monterey Bay experiment.

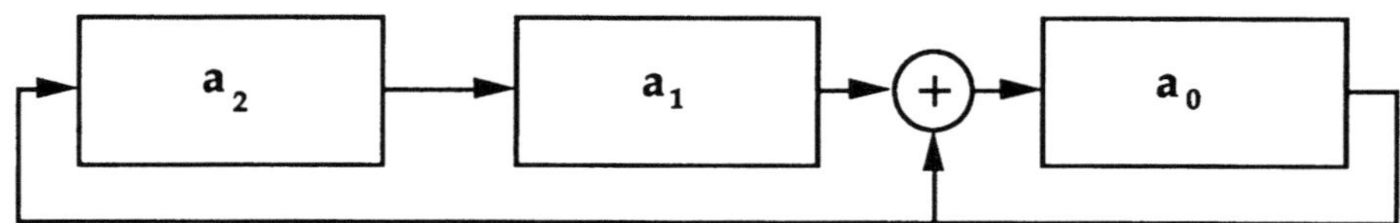

Figure A1. Binary shift register for generating a third order maximal-length sequence.

A third-order m-sequence will be discussed as an example for the signal processing. Loading the initial state is arbitrary since the register will cycle through all possible combinations before repeating. For an initial state $a_2=1$, $a_1=0$, $a_0=0$ and the shift register in Figure A1, one period of the sequence as shown in Table A1.

The characteristics of the autocorrelation are unaffected by whether the m-sequence is read in one direction or the reverse, but the method for formulating the Hadamard demodulation does change. The codes are

$$\text{forward code } \quad S = \quad 1001110,$$

$$\text{and} \quad \text{reverse code } \quad S = \quad 0111001. \tag{A1}$$

In use, the m-sequence digits are transformed by replacing 1 with -1 and 0 with 1. When dealing with the structure and mathematics it is easier to use 0 and 1 because of the simplicity of modulo-two mathematics.

Table A1. Shift register contents when generating m-sequence. Note that the eighth cycle is only included to show that the register begins to repeat.

Cycle	a_2	a_1	a_0
1	1	0	0
2	0	1	0
3	0	0	1
4	1	0	1
5	1	1	1
6	1	1	0
7	0	1	1
8	1	0	0

The received signal has an unknown time delay and so must be correlated with all possible shifts of the code. Let the seven shifted sequences form the matrix **M**:

$$\mathbf{M} = \begin{bmatrix} 1 & 0 & 0 & 1 & 1 & 1 & 0 \\ 0 & 1 & 0 & 0 & 1 & 1 & 1 \\ 1 & 0 & 1 & 0 & 0 & 1 & 1 \\ 1 & 1 & 0 & 1 & 0 & 0 & 1 \\ 1 & 1 & 1 & 0 & 1 & 0 & 0 \\ 0 & 1 & 1 & 1 & 0 & 1 & 0 \\ 0 & 0 & 1 & 1 & 1 & 0 & 1 \end{bmatrix} \tag{A2}$$

When this matrix and the code are transformed to + and -1's , multiplying the signal by the matrix will result in the correlation,

$$\mathbf{R_{sm}} = \mathbf{MS} \ . \tag{A3}$$

This is the entire goal of the initial signal processing, all that remains is to develop a fast, efficient algorithm to accomplish this multiplication.

To describe the fast algorithm,[12,13,14] it is necessary to introduce the Hadamard matrix. The Sylvester-type Hadamard Matrix has a recursive form for higher orders given by

$$\mathbf{H_1} = [1], \qquad \mathbf{H_{2i}} = \begin{bmatrix} \mathbf{H_i} & \mathbf{H_i} \\ \mathbf{H_i} & -\mathbf{H_i} \end{bmatrix} . \tag{A4}$$

The third degree matrix **H** is

$$\mathbf{H} = \begin{bmatrix} 1 & 1 & 1 & 1 & 1 & 1 & 1 & 1 \\ 1 & -1 & 1 & -1 & 1 & -1 & 1 & -1 \\ 1 & 1 & -1 & -1 & 1 & 1 & -1 & -1 \\ 1 & -1 & -1 & 1 & 1 & -1 & -1 & 1 \\ 1 & 1 & 1 & 1 & -1 & -1 & -1 & -1 \\ 1 & -1 & 1 & -1 & -1 & 1 & -1 & 1 \\ 1 & 1 & -1 & -1 & -1 & -1 & 1 & 1 \\ 1 & -1 & -1 & 1 & -1 & 1 & 1 & -1 \end{bmatrix} \tag{A5}$$

or, represented by ones and zeros,

$$
\mathbf{H} \;=\; \begin{bmatrix} 0 & 0 & 0 & 0 & 0 & 0 & 0 & 0 \\ 0 & 1 & 0 & 1 & 0 & 1 & 0 & 1 \\ 0 & 0 & 1 & 1 & 0 & 0 & 1 & 1 \\ 0 & 1 & 1 & 0 & 0 & 1 & 1 & 0 \\ 0 & 0 & 0 & 0 & 1 & 1 & 1 & 1 \\ 0 & 1 & 0 & 1 & 1 & 0 & 1 & 0 \\ 0 & 0 & 1 & 1 & 1 & 1 & 0 & 0 \\ 0 & 1 & 1 & 0 & 1 & 0 & 0 & 1 \end{bmatrix} . \tag{A6}
$$

One way to form the matrix is by multiplying matrices formed of the binary 'counting' matrix from 0 to 7,

$$
\mathbf{H} = \mathbf{A}\,\mathbf{A}^{T} = \begin{bmatrix} 0 & 0 & 0 \\ 0 & 0 & 1 \\ 0 & 1 & 0 \\ 0 & 1 & 1 \\ 1 & 0 & 0 \\ 1 & 0 & 1 \\ 1 & 1 & 0 \\ 1 & 1 & 1 \end{bmatrix} \begin{bmatrix} 0 & 0 & 0 & 0 & 1 & 1 & 1 & 1 \\ 0 & 0 & 1 & 1 & 0 & 0 & 1 & 1 \\ 0 & 1 & 0 & 1 & 0 & 1 & 0 & 1 \end{bmatrix} . \tag{A7}
$$

The matrix $\mathbf{M}$ can be factored in the same fashion, but not as simply, and involves the shift register used to generate the code. Form the first matrix $\mathbf{B}$ from the successive contents of the shift register, but bit reversed (from right to left) and in reverse order (from bottom to top). The original order is then preserved by shifting the rows of the matrix to bring the 3x3 identity matrix to the top,

$$
\mathbf{B} \;=\; \begin{bmatrix} 1 & 0 & 0 \\ 0 & 1 & 0 \\ 0 & 0 & 1 \\ 1 & 1 & 0 \\ 0 & 1 & 1 \\ 1 & 1 & 1 \\ 1 & 0 & 1 \end{bmatrix} . \tag{A8}
$$

Form the second matrix $\mathbf{C}$ from three shifted versions of the m-sequence

$$
\mathbf{C} \;=\; \begin{bmatrix} 1 & 0 & 0 & 1 & 1 & 1 & 0 \\ 0 & 1 & 0 & 0 & 1 & 1 & 1 \\ 1 & 0 & 1 & 0 & 0 & 1 & 1 \end{bmatrix} . \tag{A9}
$$

It is easy to verify that

$$
\mathbf{BC} = \mathbf{M}. \tag{A10}
$$

Note that $\mathbf{M}, \mathbf{B},$ and $\mathbf{C}$ matrices must be expanded by a leading row and/or column of zeros to be of the proper size. The new matrices will be denoted with a prime. If mapping matrices can be found such that $\mathbf{QA} = \mathbf{B'}$ and $\mathbf{A^{t}P} = \mathbf{C'}$ then the same matrices will map the Hadamard matrix to the m-sequence matrix as

$$\mathbf{M'} = \mathbf{B'C'} = \mathbf{QAA^t P} = \mathbf{QHP} . \tag{A11}$$

Recall that the correlation for the signal with the output code is given by multiplication, as in Eq. (A3), which now becomes

$$\mathbf{R'_{sm}} = \mathbf{M'S'} . \tag{A12}$$

(S' because the leading zeros must be added.) Combining equations (A11) and (A12) results in

$$\mathbf{R'_{sm}} = \mathbf{QHPS'} . \tag{A13}$$

This gives the signal correlation that is required. The initial entry is removed to change $\mathbf{R'_{sm}}$ to $\mathbf{R_{sm}}$, and the correlation is complete.

The mapping of one matrix to another is easily accomplished. Notice that if each three digit binary row or column is considered as a number, then that number occurs only once. Since these operate on vectors (1×2^n dimensional) the mapping matrix just re-orders (or permutes) the input vector. The binary number formed by the rows or columns of $\mathbf{A}$ and $\mathbf{A^t}$ are in numerical order while those in $\mathbf{B'}$ and $\mathbf{C'}$ are not. The binary number can be treated as an index. With the indices, the permutation matrices do not have to be constructed. The 'multiplication' by the permutation matrices is accomplished by shuffling the order of the signal vector, rather than direct multiplication. For a given code the permutations can be evaluated once and the result stored as an index array to be applied to each vector.

There exists an efficient method of performing the multiplication by the Hadamard matrix. If a vector is multiplied by the Hadamard matrix (the normal Hadamard matrix of $\{+1,-1\}$). The result is a vector of sums of all the components of the vector with various + and - weighting.

Define a vector $\mathbf{V}$ such that

$$\mathbf{V} = \begin{bmatrix} a \\ b \\ c \\ d \\ e \\ f \\ g \\ h \end{bmatrix} . \tag{A14}$$

After multiplying this by the Hadamard matrix the vector becomes

$$\mathbf{HV} = \begin{bmatrix} 1 & 1 & 1 & 1 & 1 & 1 & 1 & 1 \\ 1 & -1 & 1 & -1 & 1 & -1 & 1 & -1 \\ 1 & 1 & -1 & -1 & 1 & 1 & -1 & -1 \\ 1 & -1 & -1 & 1 & 1 & -1 & -1 & 1 \\ 1 & 1 & 1 & 1 & -1 & -1 & -1 & -1 \\ 1 & -1 & 1 & -1 & -1 & 1 & -1 & 1 \\ 1 & 1 & -1 & -1 & -1 & -1 & 1 & 1 \\ 1 & -1 & -1 & 1 & -1 & 1 & 1 & -1 \end{bmatrix} \begin{bmatrix} a \\ b \\ c \\ d \\ e \\ f \\ g \\ h \end{bmatrix}$$

$$
= \begin{bmatrix}
a + b + c + d + e + f + g + h \\
a - b + c - d + e - f + g - h \\
a + b - c - d + e + f - g - h \\
a - b - c + d + e - f - g + h \\
a + b + c + d - e - f - g - h \\
a - b + c - d - e + f - g + h \\
a + b - c - d - e - f + g + h \\
a - b - c + d - e + f + g - h
\end{bmatrix} \cdot \qquad (A15)
$$

When calculating correlations, let a=0 so that no new information is added. The zeroth position result is the sum of all the elements of the code and is therefore equal to the DC pedestal. This pedestal can be removed by subtracting this sum of all elements, or not, depending on the application.

Compare this result to the result using a flow diagram identical to the procedure used with the Fast Fourier Transform, except that all the 'twiddle' factors are equal to one, as shown in Figure A2. The result of the Fast Hadamard Transform is the same as for multiplication. The algorithm used for the Fast Fourier Transform is trivialized in this case - there is no bit reversal or multiplication by a phase factor. Because the method requires only additions, the exact computational speed increase is difficult to calculate. The speed improvement for FFT over DFT is usually calculated by comparing the number of multiplications required. The 'multiplication' by P and Q has been replaced by reordering, so that there is no multiplication required. The speed of execution now depends on other statements in the program as well as the correlation because loop increments and tests for completion may take as long as the additions.

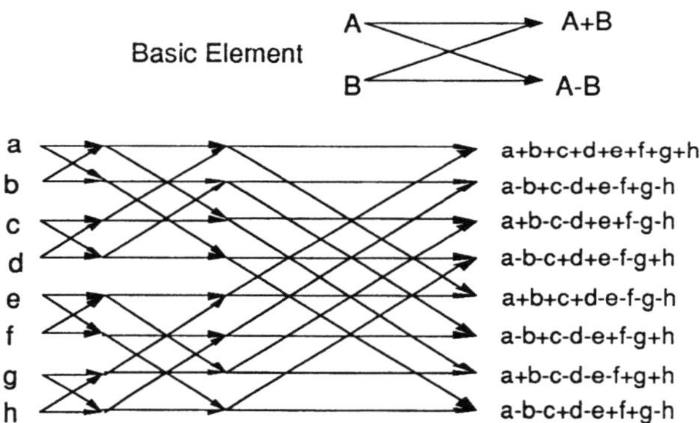

Figure A2. Basic Fast Hadamard Transform element for cascading additions and the full diagram for an eight point FHT.

Acknowledgements

The authors wish to thank Kurt Metzger from the University of Michigan for his sound advice on fast Hadamard transform algorithms. Also, thanks go to Paul Boutin, Steve Liberatore, Arthur Newhall, John Kemp, and John Bouthilette of WHOI, Tim Stanton, Cal Dunlap, and Floyd York of NPS, Tom Walton of Mitre Corporation, Jim Widenhofer and Dave Sparks of Sparton Corporation, Ken Steele and Larry Clayton of NDBC, Mike Lee, Khosrow Lashkari, and Leslie Rosenfield of the Monterey Bay Aquarium Research Institute (MBARI). This work was supported by the Naval Postgraduate School Research Council, MBARI, and ONR.

References

1.   Clay, C. S. and H. Medwin, *Acoustical Oceanography*, p. 2, John Wiley & Sons, 1977.

2.   Munk, W. and C. Wunsch, "Ocean acoustic tomography: a scheme for large scale monitoring," Deep-Sea Res., vol. 26A, pp. 123-161, 1979.

3.   Chiu, C. S. and J. F. Lynch, "Tomographic resolution of mesoscale eddies in the Marginal Ice Zone: a preliminary study," J. Geophys. Res., vol. 92(C7), pp. 6886-6902, 1987.

4.   Kinsler, L.E., A. R. Frey, A. B. Coppens, and J. V. Sanders, *Fundamentals of Acoustics*, 3rd ed., pp. 117 - 120, John Wiley & Sons, 1982.

5.   Flatté, S. M., ed., *Sound Transmission through a Fluctuating Ocean*, pp. 3 - 61, Cambridge University Press, 1979.

6.   Menke, W., *Geophysical Data Analysis: Discrete Inverse Theory*, Academic Press, 1984.

7.   Spindel, R.C., "Ocean Acoustic Tomography: A Review," *Current Practices and New Technology in Ocean Engineering*, v. 11, pp. 7 - 13, 1986.

8.   Lynch, J. F., J. H. Miller, and C. S. Chiu, "Phase and travel-time variability of adiabatic acoustic normal modes due to scattering from a rough sea surface, with applications to propagation in shallow-water and high-latitude regions," J. Acoust. Soc. Am., vol. 85(1), pp. 83-89, 1989.

9.   Spindel, R. C., "Signal Processing in Ocean Tomography," *Adaptive Methods in Underwater Acoustics*, ed. H.G. Urban, pp. 687-710, D. Reidel Publishing Company, 1985.

10.   Press, W. H., B. P. Flannery, S. A. Teukolsky, and W. T. Vetterling, *Numerical Recipes, The Art of Scientific Computing*, pp. 86-89, Cambridge University Press, 1986.

11.   Ziemer, R. E. and R. L. Peterson, *Digital Communications and Spread Spectrum Systems*, pp. 385-404, Macmillan Publishing Company, 1985.

12.   Cohn, M. and A. Lempel, "On the fast m-sequence transform," IEEE Trans. Info. Theory, pp. 135-137, January, 1977.

13.   Borish, J. and J. B. Angell, "An efficient algorithm for measuring the impulse response using psuedorandom noise," J. Audio Eng. Soc., vol. 31(7), pp. 478-487, 1983.

14.   Metzger, K., personal communication with R. C. Dees and J. H. Miller, November, 1988.

RESULTS FROM AN EXPERIMENTAL SYNTHETIC APERTURE SONAR

Michael P. Hayes and Peter T. Gough

Department of Electrical and Electronic Engineering
University of Canterbury
Private Bag, Christchurch, New Zealand

INTRODUCTION

Imaging sonars can be subdivided into two main categories: real
aperture and synthetic aperture. Real aperture sonars obtain a high azimuth
resolution by radiating very narrow beams and therefore require large
apertures and/or high acoustic frequencies.

Synthetic apertures radiate a wider beamwidth and subsequently
synthesise a narrow beam by coherently combining returns from along the
aperture. Unlike traditional imaging techniques, synthetic aperture systems
achieve high along-track resolution using small real apertures. Synthetic
aperture radar (SAR) methods are widely recognised. Whereas SAR is a mature
technique with sophisticated methods of processing, the unclassified
Synthetic Aperture Sonar (SAS) literature is much less abundant and often
less advanced (de Herring, 1984). The only practical examples of imaging
synthetic aperture sonars are an experimental rail-based system (Loggins
et al., 1982) and a short range low frequency system designed primarily for
sub-bottom imaging (Dutkiewicz and Denbigh, 1987).

There are a number of differences between a synthetic aperture radar
and its sonar equivalent. The major difference is the large disparity in
the speed of microwave propagation through the atmosphere (nominally
3×10^8 m s^{-1}) compared to the speed of sound through seawater (nominally
1500 m s^{-1}). To achieve a comparable range resolution, radars must therefore transmit a signal bandwidth many orders of magnitude greater than sonar
signals. Thus the basic signal processing and storage requirements are many
times greater than sonar. On the other hand, a slow speed of acoustic
propagation results in a very slow mapping rate for sonars. This is because
the propagation delays are much longer for sonar, even when compared with
spaceborne radar. Consequently, many of the approximations (and underlying
assumptions) made in the analysis of synthetic aperture radars are not
applicable to synthetic aperture sonars.

The direct application of SAR techniques to a practical synthetic
aperture sonar is, in addition, complicated by the following factors:

(a)   Slow mapping rates due to the relatively slow speed of sound in
      seawater.
(b)   Coherence problems resulting from the difficulty of maintaining a
      straight trajectory over the length of the synthetic aperture.
(c)   Coherence problems related to an inhomogeneous and turbulent medium.

In general, both these coherence problems are compounded by a slow
mapping rate. The slower the speed that the aperture is traversed, the more
likely it is that the survey vessel will deviate from a straight track.

It has long been thought that the underwater medium could not support
a coherent acoustic synthetic aperture. However, experiments have shown
that the medium is not as bad as has been thought and is in fact a lot
better (Williams, 1976)(Gough and Hayes, 1989). Nevertheless, it is the
medium coherence that limits the performance of any synthetic aperture
sonar.

Whereas the size constraints of a satellite or aircraft impose a
maximum physical aperture length, with sonars it is possible to tow long
hydrophone arrays (eels) astern of the survey vessel. Therefore, some of
the advantages of synthetic aperture processing are lost on sonar. In
addition, a towed array can be configured to provide multiple horizontal
beams thus alleviating the mapping rate problems. However, all-range
focusing with towed arrays involves a considerable hardware complexity.
A combination of towed array and synthetic aperture techniques may thus be
useful for reducing this hardware complexity or for improving system
performance. One idea that comes to mind is to tow a 'thinned' array and to
use synthetic aperture processing to remove the azimuth ambiguities inherent
with this array.

BROADBAND METHODS OF SYNTHETIC APERTURE SONAR

To achieve a resolution in range comparable to the resolution in along-
track it is necessary to transmit broadband signals. These are usually
produced using either amplitude modulation (e.g. impulse-like waves), phase
modulation (e.g. M-sequence waves), or frequency modulation (e.g. linear
FM). In general, a higher signal to noise ratio is obtained by modulating
the phase or frequency rather than the amplitude, since more energy may be
transmitted. With impulse-like signals, for example, the number of cycles
can be so few that detection of the phase and amplitude of the echo signals
becomes difficult (Sato and Ikeda, 1977 , p341).

Farhat (1977, p1020) shows that coherent imaging with broadband
illumination, whether produced by impulsive, swept frequency, or white
noise source, obey similar principles. As well as improving range resolution, broadband signals reduce the effects of coherent artefacts (grating
lobes), reduce resonance effects in the image and provide increased object
information (assuming that the object is non-dispersive and non-resonant).
In addition, a single frequency pulse sonar tends to support the fine lobe
structure of a target diffraction pattern, while broadband signals tend to
give more of an average value, thereby improving detection at a random
aspect (Winder, 1975, p294).

The disadvantages of broadband signals include increased signal
processing requirements, more difficult transducer design, and a variable
beam pattern. With a slowly changing linear frequency sweep the signal/
beam pattern interaction is predictable, but the behaviour of acoustic
arrays is markedly different for pulsed signals. Rather than relying on
steady state theory, the transient theory must be considered (Fink, 1980).

The processing complexity of broadband systems is often simplified by
filtering the echoes into a number of narrowband frequency components that
are processed independently. This technique is termed *spectral decomposition*
because many narrowband signals are derived from a single wideband signal,
and has been suggested for both impulsive signals (Nagai, 1984) and linear
FM sweeps (Robinson, 1982, §4.3).

Since sonar/target motion has a greater effect on longer duration
signals, it is also desirable, if possible, to split long duration signals
into shorter components for processing. The processing of these shorter
length components is often less complicated than processing the entire
signal, and then afterwards they can be recombined to produce the desired
resolution.

Smearing the Grating Lobes

There are many different methods of determining the maximum towed speed
(Rihaczek, 1969) for SAR. Most assume narrowband operation. If this limit
is exceeded, then artefacts due to grating lobes appear in the synthesised
image. However, for broadband operation, the grating lobes tend to be
smeared since their positions are frequency dependent. The broader the
bandwidth, the better the overall beam pattern.

When the second grating lobe of the synthetic beam pattern at the
highest transmitted frequency is at or beyond the first grating lobe of the
lowest transmitted frequency, the grating lobes are completely smeared
(defocused) (de Heering, 1984). This condition requires the transmitted
signal to have a bandwidth of at least an octave, i.e. a quality factor
$Q \geq \sqrt{2}$, assuming that the echo strength is constant over the transmitted
bandwidth. Note that the grating lobes are smeared, but not cancelled.
Although they are reduced in relationship to the desired main lobe, the
smeared grating lobes contribute to the background reverberation and therefore limit the dynamic range of the resultant image. However, the aperture
may now be sampled at a faster rate than that required for a narrowband
SAS. Thus with a broadband SAS there is a trade-off between mapping rate
and image dynamic range. In addition, traversing the aperture more rapidly
may improve the overall image quality in conditions where the temporal
coherence is affected by turbulence.

CTFM Sonars

One type of broadband signal is that based on continuous transmission
frequency modulation (CTFM). It has some remarkably advantageous properties, one
being that it is continuously transmitting acoustic energy. A second
advantage is that by passing the reflected echos through various bandpass
filters (or their digital equivalent) a set of narrowband chirped sonars
can be simulated where each sonar operates over a different band of
frequencies.

The current sonar sweeps over a complete frequency band of 30 to 15 kHz
and sweeps down through this band in 0.8 sec. When set up to simulate
narrowband sonars, the equivalent bandpass filters were arranged to cover
a 750 Hz bandwidth. At the nominated sweep rate, this narrowband was
traversed in 40 msec. With this arrangement we could configure the system
to look like twenty different chirped pulse sonars or a single 30 to 15 kHz
CTFM sonar.

THE EXPERIMENT

To investigate the imaging capability of our sonar, a large air filled
buoy was tethered to the sea floor as shown in Figure 1. The sonar was
constrained to pass by the buoy with its nearest approach at some 66 m.
Various sonar velocities were used and the results shown here were obtained
at tow speeds of 0.51 and 0.7 m s^{-1}. The returned echoes were preprocessed
(demodulated to constant frequency baseband) and stored for later processing.

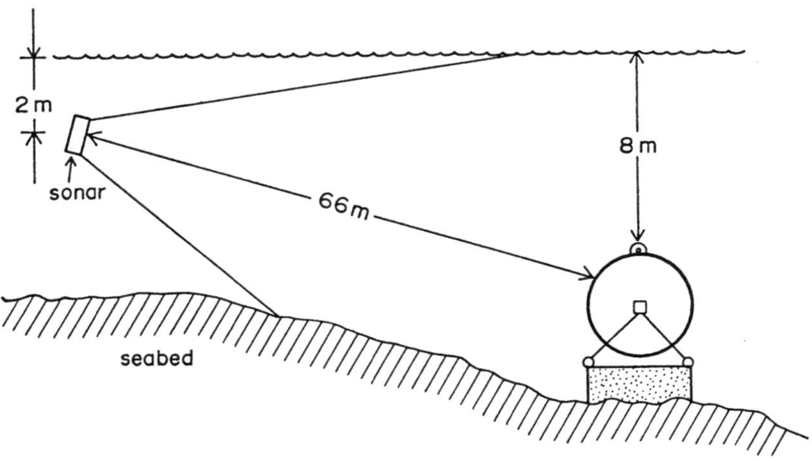

Fig. 1. The experimental geometry

THE ALGORITHM

The image reconstruction algorithm is a matched filtering operation where the matched filter is a function of cross-track and frequency but not along-track. A single wideband matched filter may use all the data in one step or there may be an ensemble of different frequency narrow-band matched filters where the final image is computed only after the matched filtering operations. This enables us to use either coherent or non-coherent addition of the intermediate narrowband images to produce a final image.

RESULTS OF IMAGE RECONSTRUCTIONS

The intensity distribution of the measured data is shown in Figure 2. Although the along-track resolution is predictably poor, at least two targets at different ranges are apparent. The closer target (66.5 m) is the desired test-target, but the identity of the other target is unknown. This could be a reflection from the floating marker buoy (and coil of unused rope) used to mark the presence of the test target. This mysterious target appears to lie almost normal to the end of the cableway, and is therefore only detected at the low end of the frequency sweep when the beamwidths are wider, and when the sonar is near the end of the cableway.

The combined effect of undersampling the aperture and using narrowband reconstruction algorithms is clearly shown in Figure 3. This confusing image is the result of moving the towfish along the aperture about three times faster than the maximum speed required to obey the sampling criterion. The raw data for this image distribution is taken from the demodulated echoes of a sequence of chirped pulses, 40 ms long, separated in time by 0.8 s, with the frequency of the chirped pulse centred at 29 KHz and sweeping down over 750 Hz during the transmitted pulse. At the towfish speed of 0.51 m s^{-1}, the samples are separated along the aperture by approximately 0.4 m. The image reconstruction algorithm is based on a 0.5 m cross-track by 0.1 m along-track pixel grid, and the resultant intensity distribution covers a sea floor area of 16 m cross-track by 32 m along-track, which includes the test target. In all the figures (which show only a small portion of the available swath), the sonar track is off the lower left-hand edge of the picture, and the bottom line of the along-track pixels are at a distance of 65 m from the track.

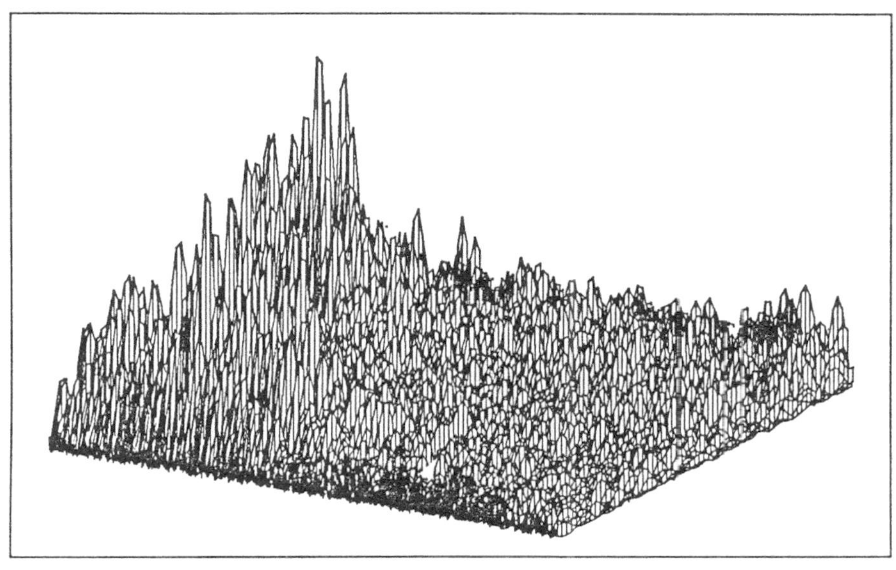

(a)

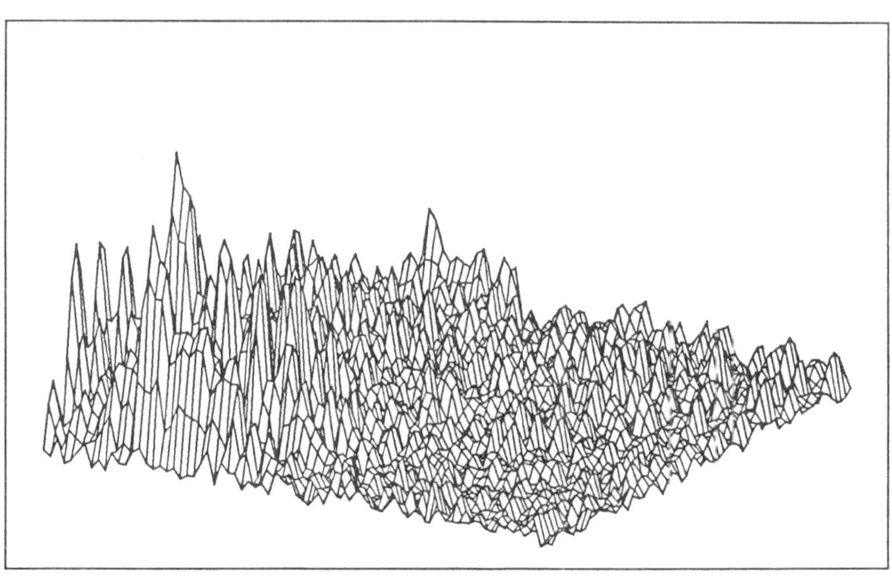

(b)

Fig. 2. The intensity distribution of the range-compressed data, with
the along-track direction going across the page to the left.
Note that this image distribution only covers a small
portion of the swath, with the near range at 65 m and the far
range at 71 m. (a) All twenty frames, side by side (b) Frame
8 (29 kHz). Note that the presence of the far range target
is not so obvious in this image distribution.

Although an image of the test target is evident in the displayed
intensity distribution (Figure 3), the undersampled aperture has produced
three artefacts, and what is worse, one of the artefacts is stronger than
the desired image. In addition, the whole image shows the effects of
coherent speckle. This image distribution is obviously unacceptable.

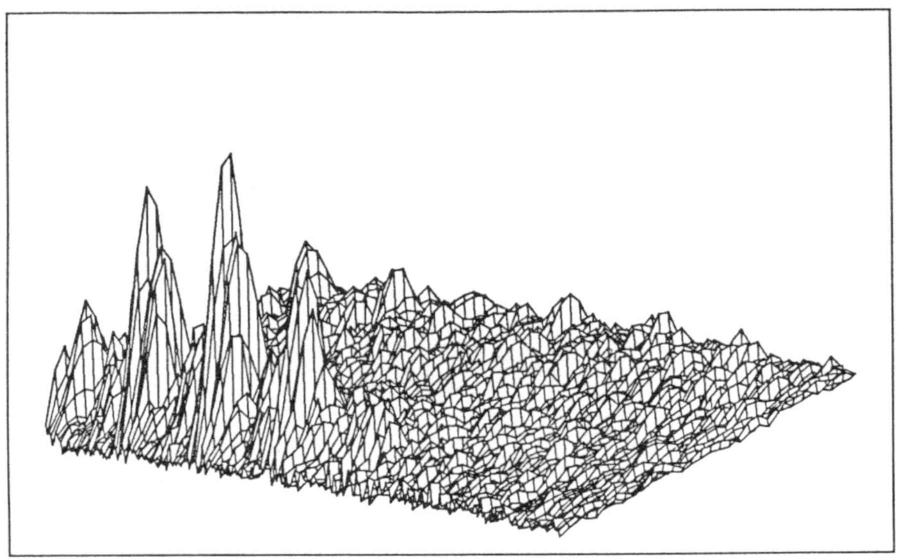

Fig. 3. The intensity distribution of a narrowband image of the test
        target calculated at 29.0 kHz. The image distribution is
        calculated on a pixel grid of 0.5 m cross-track by 0.1 m
        along-track, and covers a total area of 16 m by 32 m.

The image distribution shown in Figure 4 was generated by adding all
twenty narrowband image distributions non-coherently. That is, the intensity of each narrowband image distribution was added. The effect of this
addition is to smear out the aretfacts while adding the intensities of the
desired image. Although the combined desired image is in the correct
position, its resolution is poor and some aretfacts are still apparent.
Simulations indicate that the artefacts could be further reduced if the
echo strength of the target is held constant over the entire transmitted
bandwidth. Despite these artefacts, addition on an intensity basis is
robust and may prove to be the algorithm-of-choice for fast or real-time
imaging.

Figure 5 shows the effect of coherently combining the twenty narrowband
images. Only the centre portion of the reconstructed image distribution
is displayed (since the higher resolution requires a magnified display to
see the details). This image shows the buoy and the concrete blocks where
the along-track display now covers only 6 m and the cross-track display
from 65 m to 68 m.

The image details of the main target (the buoy) are demonstrably better
than 0.5 m by 0.5 m, which is not surprising since the theoretical resolution
of the sonar is 0.05 m in cross-track by 0.125 m in along-track, for a
transmitted bandwidth of 15 kHz and a transducer length of 0.25 m. Unfortunately, time and money restrictions prevented the verification of whether
this image quality could be maintained at longer operating ranges.

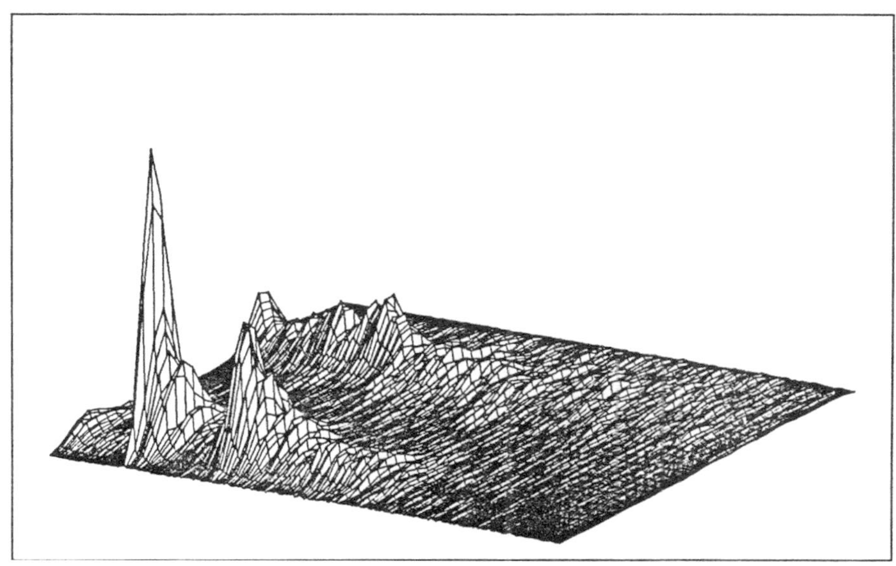

Fig. 4. The resultant intensity distribution when all 20 narrowband
images are added non-coherently (0.5 m by 0.1 m pixel grid).

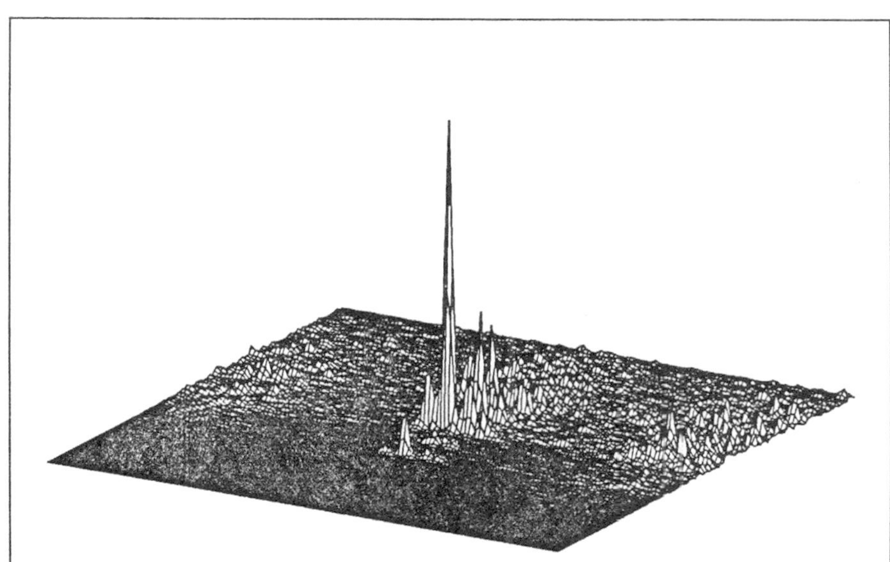

Fig. 5. The resulting intensity distribution when all 20 narrowband
images are added coherently to create an image on a 0.05 m
cross-track by 0.1 m along-track pixel grid covering an area
of 6 m (cross-track) by 12 m (along-track), where the target
is now in the centre of the displayed area.

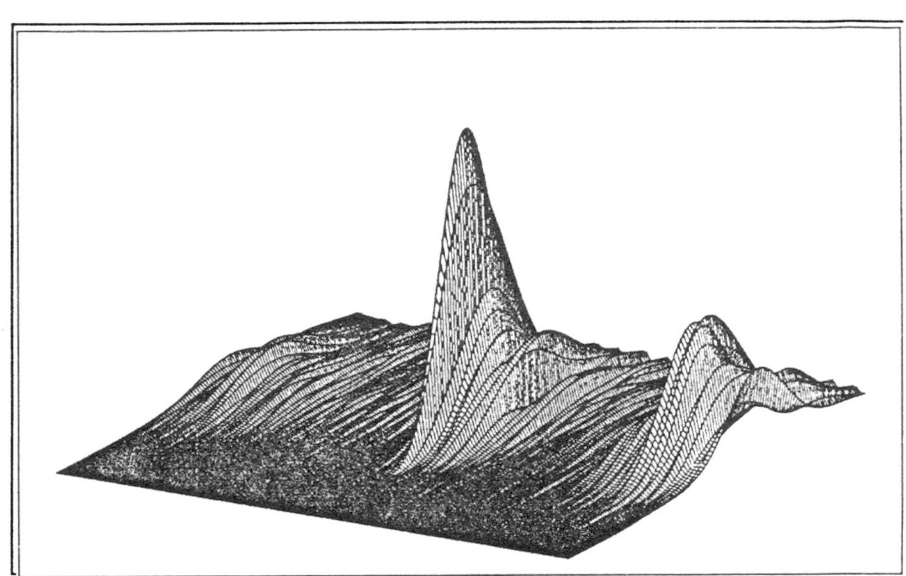

Fig. 6. The centre portion of the non-coherent image shown in Figure
7.4, but now displayed to the same degree of magnification as
Figure 7.5.

Figure 6 shows the central portion of the image created by the non-coherent addition algorithm displayed to the same magnification as that used in Figure 5. Note that the cross-track resolution is noticeably worse, but the along-track resolution is maintained. Comparing Figure 5 and Figure 6, it is apparent that the fully coherent image has high resolution, smaller artefacts, but a more speckled (noisy) appearance.

The Effects of Speed of Sound and Tow Speed Errors

The images presented in this section were generated from data obtained during the second run, in which the towfish was pulled along the aperture at a slightly faster speed of $\nu = 0.70$ m s^{-1}. All the displayed image distributions were generated by coherently adding all twenty narrowband images, and are similar to Figure 5. Furthermore, each image distribution covers an area of 3.2 m (cross-track) by 6.4 m (along-track), and they are all centred about the expected target position, for comparison.

Figures 7 and 8 illustrate the effect of incorrectly assessing the along-track tow speed. In both these figures, the top image distribution is a reference that was reconstructed using the tow speed estimated from stopwatch measurements. Consequently, this image distribution has the sharpest response. However, there are artefacts in cross-track, because of incomplete cancellation of the narrowband image sidelobes. This is a direct result of a mistake that occurred while on location in Scotland, in which each of the demodulated echo frames were irrevocably Hamming windowed before they were range compressed.

Incorrectly estimating the along-track tow speed has two effects. Firstly, the image distributions are incorrectly scaled in the along-track direction, as revealed by the shift in position of the peak response. Secondly, and more importantly, the along-track response becomes defocused. Note how an error, as small as 0.1 m s^{-1}, in the along-track speed can produce a significant defocusing effect (see Figure 8(b)); the along-track sidelobe level has increased, the width of the response is broader, and the peak response has decreased. Overestimating the tow speed even further has caused the image distribution in Figure 8(c) to break up.

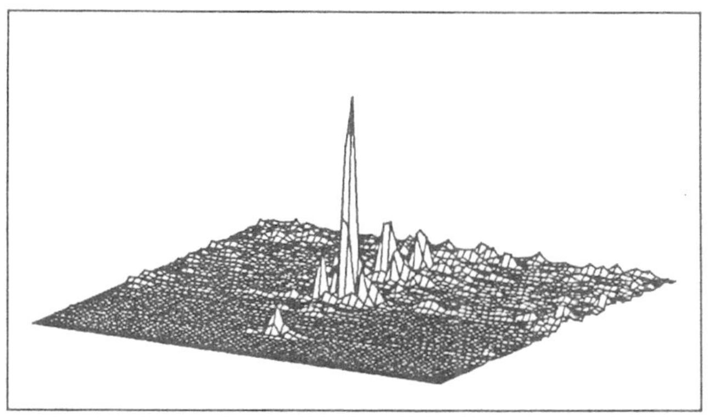

(a)

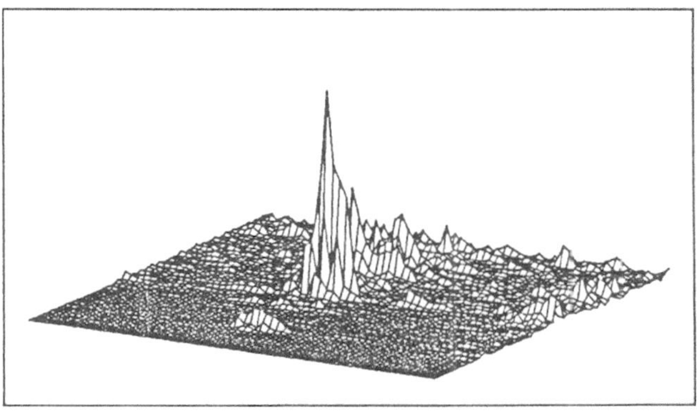

(b)

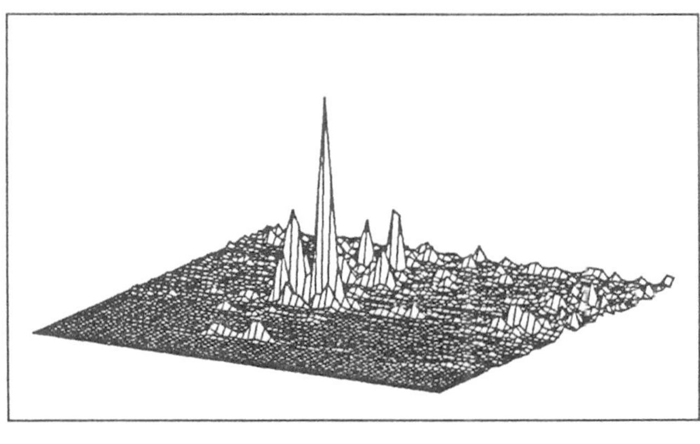

(c)

Fig. 7. The effect of varying the tow speed in the reconstruction of the test sphere below the true speed. $c = 1500$ ms^{-1}, $\Delta x = 0.05$ m, $\Delta y = 0.1$ m. (a) $v = 0.70$ ms^{-1}, (b) $v = 0.69$ ms^{-1}, (c) $v = 0.66$ ms^{-1}.

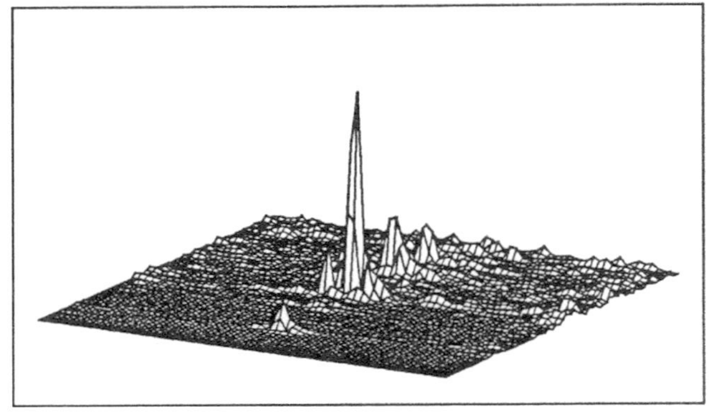

(a)

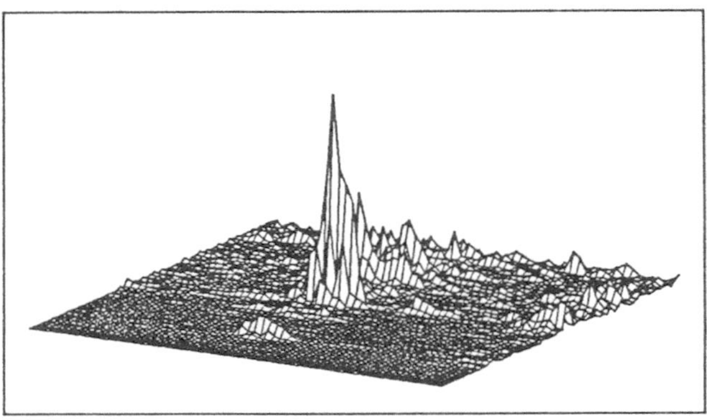

(b)

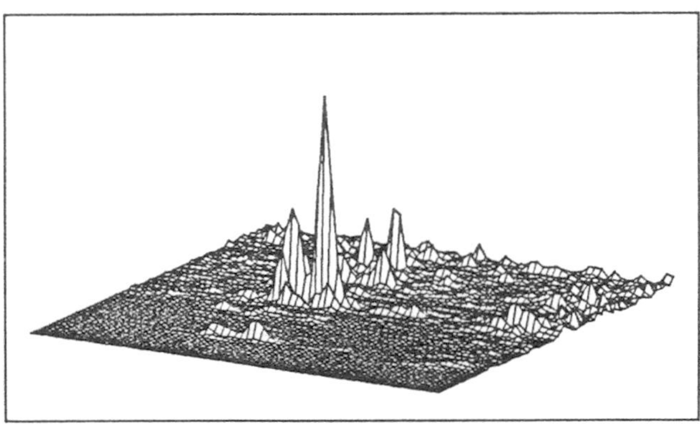

(c)

Fig. 8. The effect of varying the tow speed in the reconstruction of the test sphere above the true speed. $c = 1500$ ms^{-1}, $\Delta x = 0.05$ m, $\Delta y = 0.1$ m. (a) $\nu = 0.70$ ms^{-1}, (b) $\nu = 0.71$ ms^{-1}, (c) $\nu = 0.73$ ms^{-1}.

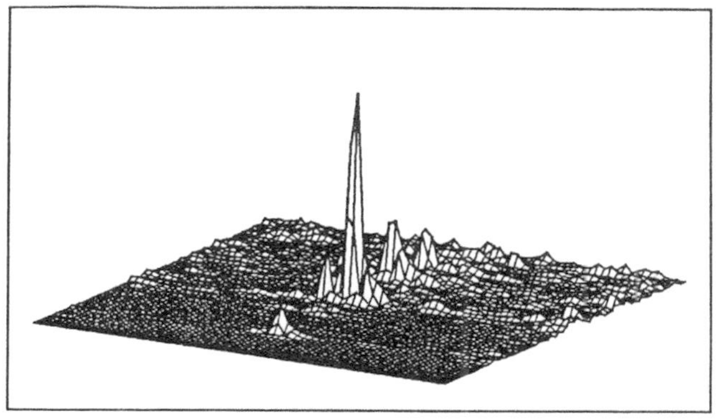

(a)

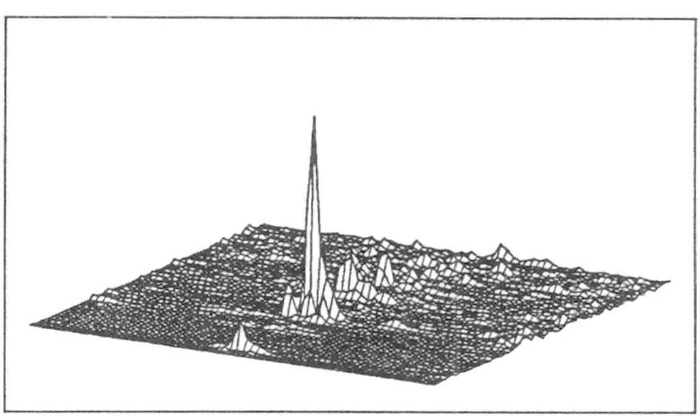

(b)

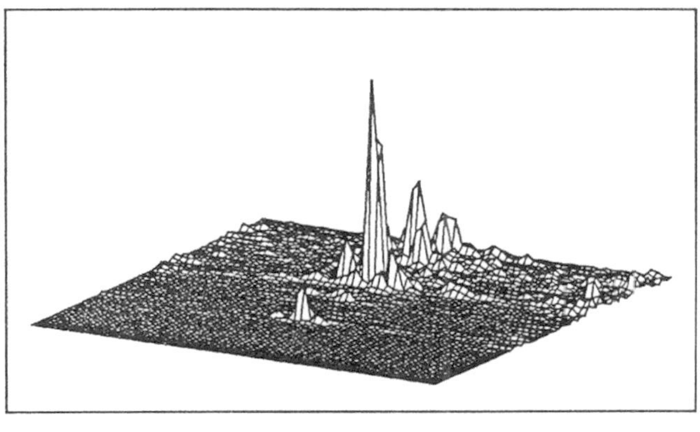

(c)

Fig. 9. The effect of varying the speed of sound in the reconstruction of the test sphere. $\nu = 0.70$ ms^{-1}, $\Delta x = 0.05$ m, $\Delta y = 0.1$ m. (a) $c = 1500$ ms^{-1}, (b) $c = 1490$ ms^{-1}, (c) $c = 1510$ ms^{-1}.

Incorrectly estimating the speed of sound is not so severe as incorrectly
estimating the along-track tow speed, since in practice the speed of sound
can be estimated sufficiently accurately from measurements of the water
temperature, depth, and salinity (cf. Urick, 1975, §5.4).

The effect of incorrectly assessing the speed of sound is much the same
as incorrectly assessing the tow speed, except that the image is now
incorrectly scaled in the cross-track direction. Figure 7.9 illustrates
this effect for a speed of sound error of ± 10 m s^{-1} (0.7%). Note that
while the peak response has shifted several pixels in cross-track, the
defocusing has produced only a slight degradation in the along-track
resolution. Furthermore, it appears that the true speed of sound is closer
to 1490 m s^{-1} than 1500 m s^{-1} as estimated.

CONCLUSION

On comparing Figures 2 and 5, it is apparent that the synthetic aperture
processing has significantly improved the along-track image resolution.
The theoretical along-track resolution of the prototype sonar is $\Delta y = 0.125$ m,
which compares favourably to the value of 0.2 m estimated from Figure 5.

Although the aperture was significantly undersampled, the suppression
of image artefacts using broadband CTFM has been demonstrated. While
artefacts resulting from the aperture undersampling are apparent in each
of the narrowband images, they occur at different positions and hence are
smeared in the resultant image.

Noncoherent addition of the narrowband images produces a smoother, less
speckly image, but with a reduced cross-track resolution and an increased
background due to self-clutter. However, noncoherent addition is robust
with regard to frequency dependent phase errors introduced by the sonar.

Finally, for accurate high resolution synthetic aperture imagery, the
along-track tow speed must be accurately estimated, however, the speed of
sound is not so critical.

REFERENCES

Dutkiewicz, M.K. & Denbigh, P.N., 1987, Synthetic aperture sonar for sub-
    bottom imaging, Acoust. Im., Plenum, NY: 585.
Farhat, N.H, 1977, Principles of broad-band coherent imaging, J. Opt. Soc.
    Am., 67: 1015.
Fink, M., 1980, Theoretical aspects of the Fresnel focusing technique,
    Acoust. Im., Plenum, NY: 149.
Gough, P.T. & Hayes, M.P., 1989, Measurements of acoustic phase stability in
    Loch Linnhe, Scotland, J. Acoust. Soc. Am., In print.
de Herring, P., 1984, Alternative schemes in synthetic aperture sonar
    processing, IEEE OE-9: 277.
Loggins, C.D., Christoff, J.T. & Pipkin, E.L., 1982, Results from rail
    synthetic aperture experiments, J. Acoust. Soc. Am., Supp. 1: S85.
Nagai, K., 1984, Multifrequency acoustic holography using a narrow pulse,
    IEEE Su-31: 151.
Rihaczek, A.W., 1969, Principles of High-Resolution Radar, McGraw, NY.
Robinson, B.S., (1982), Speckle Processing for Ultrasonic Imaging, PhD, UofC, NZ.
Sato, T. & Ikeda, O., 1977, Super-resolution ultrasonic imaging by combined
    spectral and aperture synthesis, J. Acoust. Soc. Am., 67:341.
Urick, R.J., 1975, Principles of Underwater Sound, McGraw, NY (2nd ed.).
Williams, R.E., 1976, Creating an acoustic synthetic aperture in the ocean,
    J. Acoust. Soc. Am., 60: 60.
Winder, A.A., 1975, Sonar system technology, IEEE SU-22: 291.

HIGH RESOLUTION WIGNER DISTRIBUTION
FOR SONAR SIGNALS

L.A. Ferrari, P.V. Sankar, D. Pang, and H. Masahara

Department of Electrical and Computer Engineering
University of California
Irvine, California

Abstract

In this paper we propose new algorithms for high resolution estimation of the Wigner distribution, based on classical AR and MUSIC spectral estimators. We illustrate with examples that these tech[Aniques do provide high resolution estimates of closely spaced chirps better than WD estimates based on DFT techniques. We also identify potential problems with the use of Wigner distributiopn in general, namely cross term artifacts and large computational complexity of the estimation algorithms.

1 Introduction

Classical signal representations such as correlation and spectral density functions are not generally useful for the time varying signals encountered in areas such as sonar, radar, medical ultrasound and geophysics. These signals are characterized by changes in amplitude and frequency as a function of time and classified as time varying. Changes in frequency with time occur because of doppler effect induced by moving targets and because of changes in rpm of rotating machinery. The ocean is a time varying medium which induces additional nonstationarity in the received signals, eg. temperature fluctuations and multipath propagation.

A simple example is given by the time limited, linearly frequency modulated signal (chirp signal)

$$s(t) = rect(\frac{t - \frac{T}{2}}{T})sin2\pi t(f_0 + at) \tag{1}$$

Spectral analysis cannot easily display the appropriate information contained in time varying signals such as s(t). The example portrays a simple sonar signal model and demonstrates the need for time varying signal analysis tools. The need for high resolution techniques is shown by the example given in Figures 1 and 2 where the discrete Fourier transform is unable to resolve two closely spaced sinusoids. While several time-frequency distributions have been described in the literature for use in signal processing [1, 2, 3, 4, 5], widespread utilization of these techniques has not yet been realized. Reasons for this include the increase in computational complexity and data handling associated with two dimensional representations of one dimensional signals, the loss of frequency resolution induced by "short time"

Acoustical Imaging, Vol. 18, Edited by H. Lee and G. Wade
Plenum Press, New York, 1991

approaches, and the inability of users to interpret time-frequency distributions of complex signals in noise. Other difficulties specifically associated with sonar singnal processing include incomplete knowledge of the signals, sensitivity of the algorithms to noise and clutter and errors in the location of the signal sensors. In this paper we propose new algorithms for high resolution Wigner distribution. In section 2 we propose an autoregressive estimate of the Wd, and in section 3 we propose an eigen space approach for estimating the WD. In section 4 we discuss the merits and demerits of using the WD and suggest possible improvements.

2 Time-Frequency Representations

The short time (time dependent) discrete Fourier transform (STDFT) has been one of the predominant tool for the analysis of time varying signals because of its ease of implementation using the fast Fourier transform. A definition of the short time Fourier transform is given by

$$X_n(e^{j2\pi f}) = \sum_{m=-\infty}^{\infty} w(n-m)x(m)e^{-j2\pi fm} \tag{2}$$

where, $w(n-m)$ is a real window sequence which determines the portion of the input signal used to compute X_n. The underlying assumption is that the signal $x(n)$ is stationary over the extent of the window function $w(n-m)$. In order to increase the time resolution of the representation for nonstationary signals it is necessary to shorten the extent of the window function. Unfortunately, this has the undesired effect of decreasing the frequency resolution. An example of the difficulty associated with this approach is shown in Figures 1 and 2 where the sum of two sinusoids closely spaced in frequency (Figure 1) cannot be resolved using a windowed STDFT algorithm (Figure 2). Autoregressive methods tend to improve the PSD estimate in the case of sinusoidal signals by providing improved spectral resolution. This is demonstrated by the result shown in Figure 3.

The WD offers an alternative to short-time representations and offers improved time resolution with no loss in frequency resolution. The WD of a continuous analytic signal $x(t)$ is defined as

$$WD_x(t,f) \triangleq \int_{-\infty}^{\infty} x(t-\frac{\tau}{2})x(t+\frac{\tau}{2})e^{-j2\pi f\tau}d\tau \tag{3}$$

A discrete version of the WD can be obtained with no difficulty if the signal $x(t)$ is oversampled by a factor of two [5]. Hence, we obtain

$$WD_x(n,f) = 2\sum_{k=-\infty}^{\infty} e^{-k4\pi f}x(n+k)x^*(n-k) \tag{4}$$

An example of the WD of a gaussian modulated signal is shown in Figure 4. The properties of the WD are well documented in the literature $[1, 4, 5]$. We list a few of the more important properties below:

1. $WD(t,f)$ is real for all t and f.

2. Integrating $WD_x(t,f)$ over f yields the signal power at time t.

$$\int_{-\infty}^{\infty} WD_x(t,f)df = |x(t)|^2$$

3. Integrating $WD_x(t,f)$ over t yields the power spectral density of $x(t)$ at frequency f

$$\int_{-\infty}^{\infty} WD_x(t,f)dt = |X(f)|^2$$

4. $WD_x(t,f)$ is both time and frequency invariant.

$$y(t) = x(t-t_0) \implies WD_y(t,f) = WD_x(t-t_0,f)$$

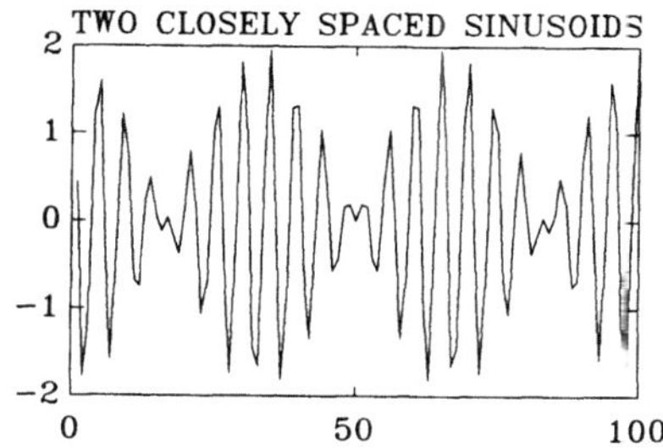

Figure 1. Two Closely Spaced Sinusoids

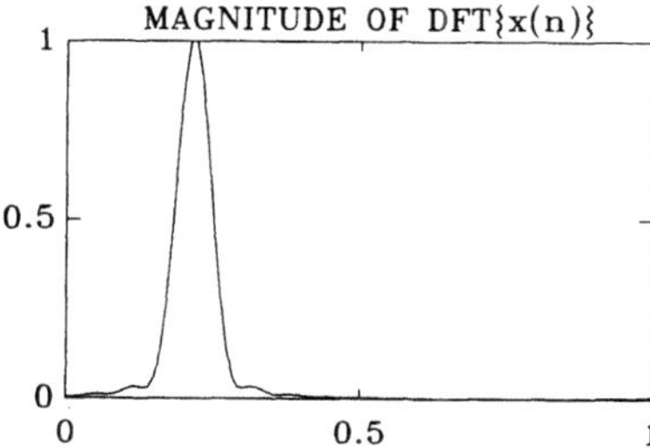

Figure 2. Resolution Using a Windowed STDFT

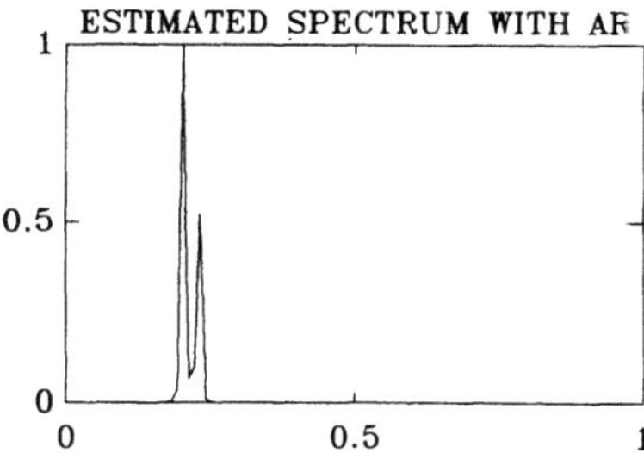

Figure 3. Autoregressive Methods to Improve PSD

5. The first order moments of WD_x provide the signal instantaneous frequency and the signal group delay.

$$\frac{\int_{-\infty}^{\infty} f \cdot WD_x(t,f)df}{\int_{-\infty}^{\infty} WD_x(t,f)df} = f_i(t) \tag{5}$$

$$\frac{\int_{-\infty}^{\infty} t \cdot WD_x(t,f)dt}{\int_{-\infty}^{\infty} WD_x(t,f)dt} = t_g(f) \tag{6}$$

6. All other time-frequency distributions can be obtained from the WD.

When the signal $x(t)$ is time limited the frequency resolution of the WD is inversely related to the size of the time window. The WD of the sum of two time limited closely spaced chirp signals is shown in Figure 5. Because of the short window used to time limit the signals we are unable to resolve the two chirp signals. In the next section we propose several new approaches for computing the discrete WD for time limited, time varying signals which provide better estimates of the WD than the STDFT.

2.1 Autoregressive (AR) Estimates of the WD (ARWD)

It is well known that the AR modeling approach to spectral estimation can yield dramatic improvements in the spectral estimate, especially for short records. One of the difficulties associated with the periodogram approach to power spectral density (PSD) estimation is the assumption that the signal is zero outside of the window function $w(n)$. The multiplication of the signal $x(n)$ by the window function in the time domain causes a convolution of their Fourier transforms in the frequency (spectral) domain. Hence, the resolution of the periodogram is limited to the main-lobe width of $W(f)$, the Fourier transform of $w(n)$.

The AR spectral estimation approach is directly related to linear prediction theory and maximum entropy spectral estimation. These relationships allow more realistic assumptions about the data outside the window interval other than the classical cyclic or zero assumptions. This eliminates the need for window functions and leads to much higher resolution estimates of the PSD. We first provide a brief overview of AR modeling and several techniques used for AR PSD estimation.

The AR model of a signal is given by

$$x(n) = -\sum_{k=1}^{p} a_k x(n-k) + u(n) \tag{7}$$

where, $u(n)$ is an input driving sequence and a_k are the AR model parameters. We can interpret (7) as a linear regression of $x(n)$ on itself with $u(n)$ representing the error. The AR PSD can be written as

$$P_{AR}(f) = \frac{\sigma^2 \Delta t}{|1 + \sum_{k=1}^{p} a_k e^{-j2\pi f k \Delta t}|^2} \tag{8}$$

The AR PSD estimation problem requires estimating the AR parameters $\{a_1, a_2, \ldots, a_p, \sigma^2\}$ from the data sequence $x(n)$. Comprehensive discussions of the techniques used to estimate the the AR model parameters are given in the recent texts by Marple and Kay [6, 7].

For narrowband processes, we propose the modified covariance method (MCM) [7] to generate the AR parameters. It has been observed that in the case of sinusoidal signals in noise this approach provides peaks in the AR PSD estimate at the true frequency locations. Many AR estimators experience shifting of the peaks in the presence of noise. Spectral line splitting in which a single sinusoidal produces two distinct peaks in the PSD has not been observed with use of the modified covariance method, but does occur using other AR estimators.

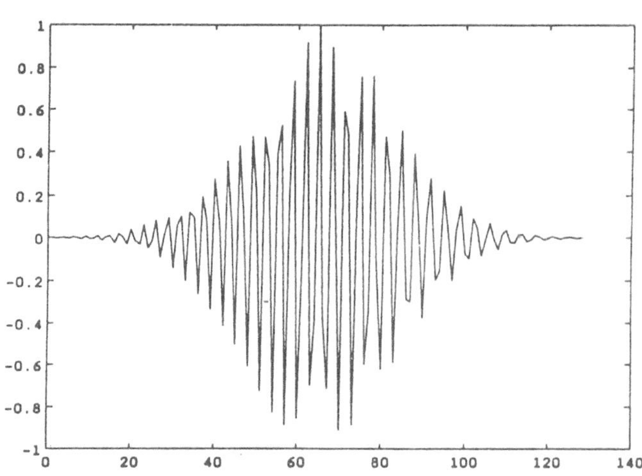

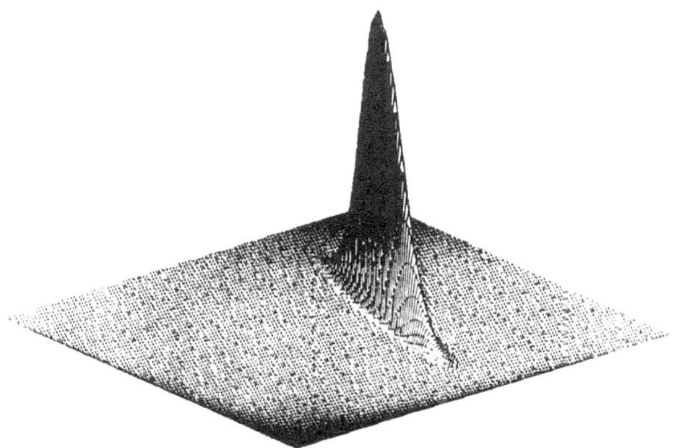

Figure 4. The WD of a Gaussian Modulated Signal

Although this technique appears more accurate for narrowband PSD estimation, the associated matrix equation is not of the Yule-Walker form. That is, the matrix is not Toeplitz, so we are unable to use the computationally efficient Levinson recursion to solve the linear equations leading to the estimate of the AR parameters. A suggested approach [7] uses Cholesky decomposition to solve the matrix equations.

For wideband processes we propose the use of Burg's algorithm [7] which is based on the relationships between AR modeling and linear prediction. The parameters in this case are estimated directly from the data using forward and backward prediction errors. In the Yule-Walker and MCM approaches the data is first used to form an autocorrelation or a covariance matrix which then provides a set of linear equations containing the unknown AR parameter $\{a_1, \ldots, a_p, \sigma^2\}$. Burg's algorithm solves the Yule-Walker equations without actually finding the associated autocorrelation matrix R_{xx}.

The Yule-Walker equations are given by

$$
\begin{bmatrix}
R_{xx}(0) & R_{xx}(-1) & \ldots & R_{xx}(-p) \\
R_{xx}(1) & & & \\
\vdots & & & \vdots \\
R_{xx}(p) & R_{xx}(-1) & \ldots & R_{xx}(-p)
\end{bmatrix}
\begin{bmatrix}
1 \\
a_1 \\
\vdots \\
a_p
\end{bmatrix}
=
\begin{bmatrix}
\sigma^2 \\
0 \\
\vdots \\
0
\end{bmatrix}
\tag{9}
$$

The Levinson recursion is initialized by

$$
a_{11} = -\frac{R_{xx}(1)}{R_{xx}(0)}
\tag{10}
$$

$$
\sigma_1^2 = (i - |a_{11}^2|)R_{xx}(0)
\tag{11}
$$

and is given by

$$
a_{kk} = -[R_{xx}(k) + \sum_{l=1}^{k-1} a_{k-1,l} R_{xx}(k - l)][\sigma_{k-1}]^{-2}
\tag{12}
$$

$$
a_{ki} = a_{k-1,i} + a_{kk} a_{k-1,k-1}^*
\tag{13}
$$

$$
\sigma_k^2 = (1 - |a_{kk}^2|)\sigma_{k-1}^2 \text{ for } k = 2, 3, \ldots, p
\tag{14}
$$

The AR parameters are found from

$$
a_{pi} = a_{p-1,i} + a_{pp} a_{p-1,p-1}^* \quad i = 1, \ldots, p
\tag{15}
$$

The variables a_{kk} for $k = 1, \ldots, p$ are termed the reflection coefficients and are computed using (12) from the estimated autocorrelation sequence $R_{xx}(n)$. Burg's algorithm computes the reflection coefficients from the forward and backward linear prediction errors. The details of the Burg algorithm are contained in [6, 7]. An outline of the procedure is given in Figure 6

We define the autoregressive estimate of the WD using the AR PSD estimate as

$$
|ARWD(t,f)|^2 \triangleq \frac{\Delta t \sigma^2}{|1 + \sum_{k=1}^{p} a(t,k)e^{-j2\pi f k \Delta t}|^2}
\tag{16}
$$

where,

$$
\begin{aligned}
\Delta t \quad & \text{is the sampling interval} \\
p \quad & \text{is the AR model order} \\
a(t,k) \quad & \text{are the time varying AR parameters and} \\
\sigma^2 \quad & \text{is the estimated noise power.}
\end{aligned}
$$

In Figure 7 we show the ARWD of the same two chirp signals used to generate the STDFT estimate of WD contained in Figure 5. We find it encouraging that we are easily able to identify the chirp signals and to extract the characterizing quantitative signal information from the ARWD.

We also notice the presence of a small spurious peak in Figure 7b. This peak is produced by cross terms which are naturally present in the computation of the WD. For example,

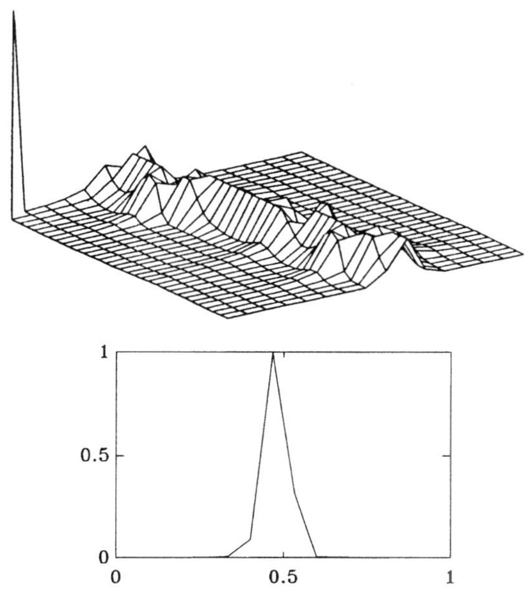

Figure 5. The WD of Two Time Limited Chirp Signals

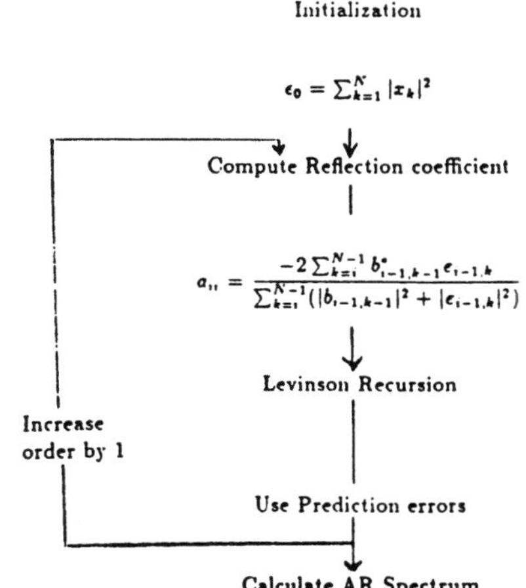

Figure 6. Outline of the Burg Algorithm

$$WD_{x_1+x_2} = WD_{x_1} + WD_{x_2} + 2Re[WD_{x_1x_2}] \tag{17}$$

The number of cross terms increases with the number of signals present in the composite signal and can lead to noisy and difficult to interpret WD displays. The cross terms do not generally interfere with the auto-terms of the WD so we are optimistic that techniques to separate these components can be developed.

In the next section we propose the use of eigenspace spectral estimation for the estimate of the WD.

3   Eigenspace WD Estimation

If the input signal, x(n) is composed of multiple sinusoids then an eigenanalysis of the autocorrelation matrix can be used to provide better estimates of the PSD than the periodogram and AR techniques. We require that the information in the autocorrelation matrix be decomposed into two vector spaces, a signal subspace and a noise subspace. We discuss two noise subspace frequency estimators, the MUSIC algorithm [8] and the Eigenvector algorithm [9] and propose the application of these techniques for improved estimation of the WD for narrowband signals.

We assume the signal x(n) is composed of M sinusoids plus noise. That is

$$x(n) = \sum_{k=1}^{M} A_k e^{j2\pi f_k n \Delta t} + u(n) \tag{18}$$

We further assume that an appropriate technique has been used to determine the autocorrelation matrix R_{xx} for the data sequence $x(n)$.

$$R_{xx} = \begin{bmatrix} r_{xx}[0] & \cdots & r^*_{xx}[p] \\ \vdots & & \vdots \\ r_{xx}[p] & \cdots & r_{xx}[0] \end{bmatrix} \tag{19}$$

R_{xx} can be decomposed into a signal autocorrelation matrix S_{xx} and a noise autocorrelation matrix U_{xx}. Hence,

$$R_{xx} = S_{xx} + U_{xx} \tag{20}$$
$$\tag{21}$$

For the case of M complex sinusoids S_{xx} and U_{xx} can be written as

$$S_{xx} = \sum_{i=1}^{m} A_i s_i s_i^H \quad U_{xx} = \sigma^2 I \tag{22}$$

where,

$$
\begin{aligned}
H &\quad \text{denotes conjugate transpose} \\
I &\quad \text{is a } (p \times 1) \times (p \times 1) \text{ identity matrix} \\
\text{and } s_i &\quad = [1 \; e^{j2\pi f_i \Delta t} \dots e^{j2\pi p f_i \Delta t}]
\end{aligned}
$$

It is clear that S_{xx} has rank M while the noise matrix U_{xx} must have rank $p + 1$. Using singular valued decomposition (SVD), R_{xx} can be written as

$$R_{xx} = \sum_{i=1}^{M} (\lambda_i + \sigma^2) v_i v_i^H + \sum_{i=M+1}^{p+1} \sigma^2 v_i v_i^H \tag{23}$$

where,

$$
\begin{aligned}
v_i &\quad \text{are the computed eigenvectors} \\
\text{and } \lambda_i &\quad \text{are the computed eigenvalues.}
\end{aligned}
$$

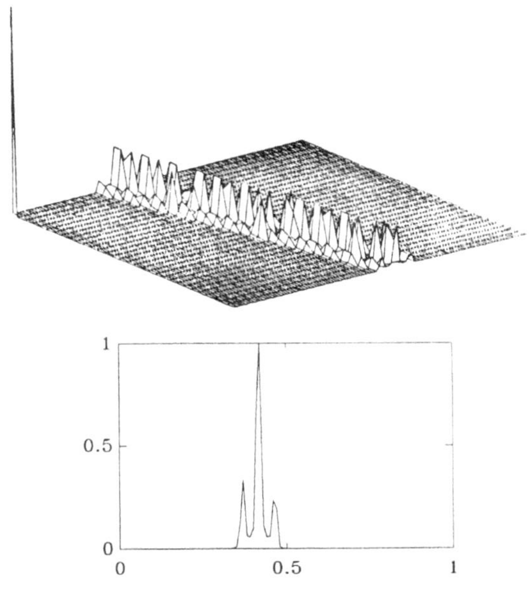

Figure 7. ARWD Estimate of Closely Spaced Chirps

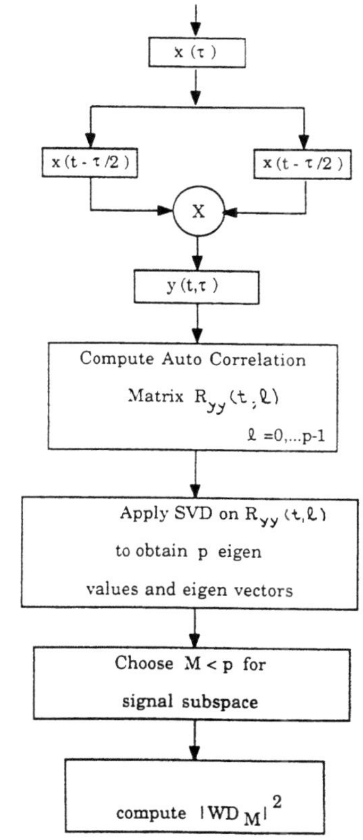

Figure 8. the WD MUSIC estimator

It is clear that s_i (for all $i = 1, \ldots, M$) is orthogonal to v_i (for all $i = M + 1, \ldots, p$). This is the central idea behind the MUSIC and Eigenvector spectral estimators.

The eigenvectors $\{v_i\}_{i=M+1}^{p}$ are used to form the product-combination

$$\sum_{k=M+1}^{p} \alpha_k |e^H(f)v_k|^2 \qquad (24)$$

where,

$$e(f) \quad = [1 \ e^{j2\pi f \Delta t} \ldots e^{j2\pi fp\Delta t}]^T$$

$$\text{and } \alpha_k \quad \text{are arbitrary constants}$$

Selecting $\alpha_k = 1$ for $k = M + 1, \ldots, p$ leads to the MUSIC frequency estimator [8] while selecting $\alpha_k = \lambda_k^{-1}$ for $k = M + 1, \ldots, p$ leads to the Eigenvector frequency estimator [9]. These estimators are given by

$$P_{MUSIC}(f) \triangleq \frac{1}{e^H(f)(\sum_{k=m+1}^{p} v_k v_k^H)e(f)} \qquad (25)$$

$$P_{EV}(f) \triangleq \frac{1}{e^H(f)(\sum_{k=m+1}^{p} \lambda_k^{-1} v_k v_k^H)e(f)} \qquad (26)$$

Because of the orthogonality of $\{e(f_i)\}_{i=1}^{M}$ and $\{v_i\}_{i=M+1}^{p}$, (25) and (26) become extremely large at $f = f_i$. Hence, P_{MUSIC} and P_{EV} have sharp peaks at the signal frequencies $\{f_i\}_{i=1}^{M}$ and provide a high resolution estimate of the PSD of $x(n)$.

Equation (25) can be used to estimate the $|WD|^2$ estimate using

$$|WD_M(n,f)|^2 \triangleq \frac{1}{e^H(f)(\sum_{k=m+1}^{p} v_k(n)v_k^H(n))e(f)} \qquad (27)$$

The complete process for the WD MUSIC estimator is shown in Figure 8. Figure 9 show a MUSIC spectral estimate of a single sinusoid and Figure 10 shows the MUSIC estimate of two closely spaced sinusoids. Figure 11a is the plot of $|WD_M(t,f)|^2$ for a chirp signal. Figure 11b is a plot of $|WD_M(t,f)|^2$ for a particular t. $|WD_M(t,f)|^2$ for two chirp signals is shown in Figure 12a. $|WD_M(t,f)|^2$ for a particular t is shown in Figure 12b. In this case, the number of peaks is equal to the number of signals. The dimension of the signal subspace was set equal to two which corresponds to the original number of chirp signals.

In a similar manner, we propose the WD estimate using the Eigenvector algorithm

$$|WD_{EV}(t,f)|^2 \triangleq \frac{1}{e^H(f)(\sum_{k=m+1}^{p} \lambda_k^{-1} v_k(t)v_k^H(t))e(f)} \qquad (28)$$

where λ_k are the noise eigenvalues.

The EV algorithm tends to produce fewer spurious peaks than the MUSIC algorithm because of the inverse weighting of the eigenvalues [9].

4 Discussion

In this paper we proposed new algorithms for high resolution estimate of the Wigner distribution of a time varying signal. However, the Wigner distribution suffers from an artifact called the cross term artifact. This arises from the fact that the WD of the sum of two signals is not equal to the sum of the WD's of the respective signals. (See 17). Recently [10] proposed a new kernal function for computing the WD of a signal with very minimal cross term artifacts. They defined a modified Wigner distribution as follows:

$$WD_x(t,f) = \frac{1}{4\pi^{\frac{3}{2}}} \int_{-\infty}^{\infty} \int_{-\infty}^{\infty} \sqrt{\frac{\sigma}{\tau^2}} e^{\frac{-\sigma(u-t)^2}{4\tau^2} - j\tau\omega} s^*(u - \frac{1}{2}\tau) \ s(u + \frac{1}{2}\tau)dud\tau \qquad (29)$$

The new kernal function was so chosen to minimize the cross term artifacts. However, computation of the Choi - Williams distribution requires the computation of a double inte-

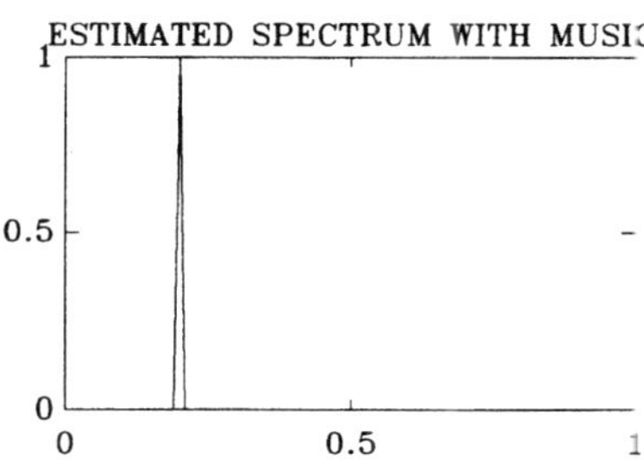

Figure 9. MUSIC Estimation of a Single Sinusoid

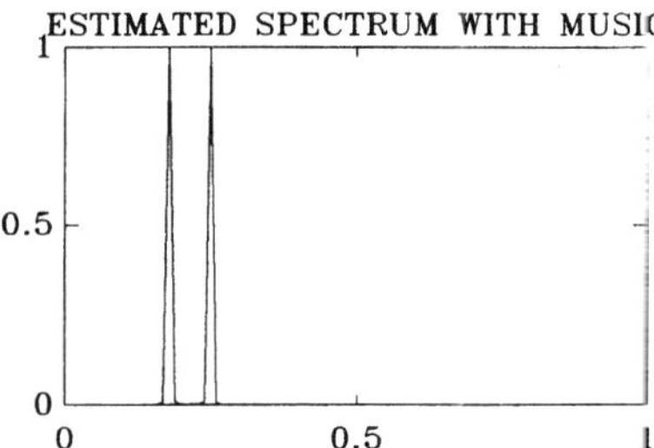

Figure 10. MUSIC Estimation of Two Closely Spaced Sinusoids

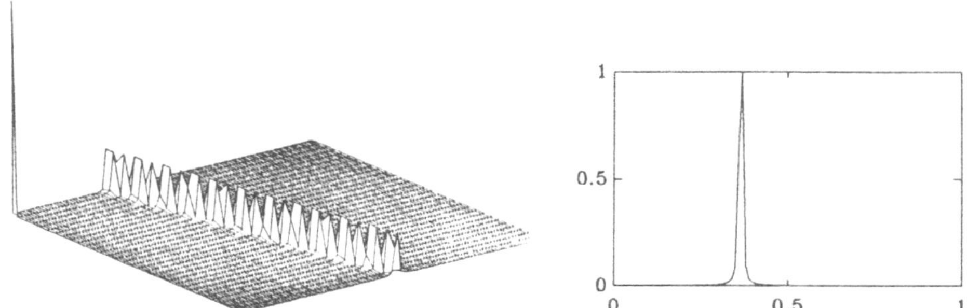

Figure 11. $|WD_M(t,f)|^2$ for a chirp and with a particular t

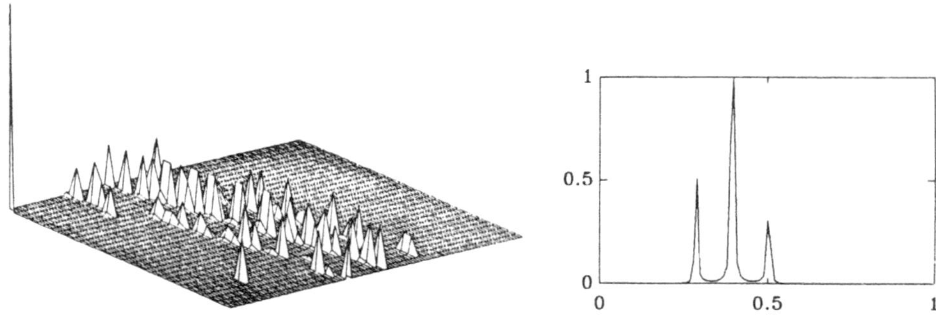

Figure 12. $|WD_M(t,f)|^2$ for two chirps and with a particular t

gral as opposed to a single integral for the traditional Wigner distribution. Also the MUSIC estimate of the WD requires the SVD of the input signal which is a time consuming process. Hence in order to speed up the SVD we require parallel algorithms; one such algorithm in the literature is due to [11] in which the SVD is formulated in such a way that each plane rotation of the matrix can be done in parallel. We are currently investigating a parallel implementation of the Hestenes algorithm and also considering extending it to a MUSIC - Choi - Williams estimate.

References

[1] T.A.C.M. Classen and W.F.G. Mecklenbrauker, "The Wigner Distribution – A Tool for Time–Frequency Signal Analysi", Parts I, II, III, *Philips J. Res*, 35, pp 217–250, 276–300, 372–389, 1980.

[2] A.W. Rihaczek, *Principles of High Resolution Rader*, McGraw–Hill, New Yord, 1969.

[3] R.A. Altes, "Detection, Estimation and Calssification with Spectrograms", *JASA*, V67(4), April 1980.

[4] N.M. Marinovic, "The Wigner Distribution and the Ambiguity Function: Generalizations, Enhancement, Compression and Some Applications"

[5] B. Bouachache, "Représentation Temps-Fréquence: Application à la Mesure de lÁbsorption du Sous–sol", Th'ese LÍnstitut National Polytechnique de Grenoble.

[6] S.L. Marple Jr., *Digital spectral Analysis with Applications*, Prentices–Hall, Englewood Cliffs, N.J., 1987.

[7] S.M. Kay, *Modern Spectral Estimation*, Prentice–Hall, Englewood Cliffs, N.J., 1988.

[8] R.O. Schmidt, "A Signal Subspace Approach to Multiple Emitter Location and Spectral Estimation", Ph.D dissertation, Dept of Electrical Engineering, Stanford University, Stanford University, Stanford, Calif., Novermber 1981.

[9] D.H. Johnson and S.R. DeGraaf, "Improving the Resolution cf Bearing in Passive Sonar Arrays by Eigenvalue Analysis", *IEEE Trans. ASSP*, vol. ASSP-30, pp 638-647, August 1982.

[10] H.C. Choi and W.J. Williams, "Improved Time- Frequency Representation of Multicomponent Signals Using Exponential Kernels"' *IEEE Trans ASSP*, vol. ASSP-37, pp. 862-871, 1989.

[11] M.R.Hastenes, "Inversion of Matrices by Biorthogonalization and Related Results", *Journal SIAM*, Vol. 6, pp. 51-90, 1980.

AN IMAGING OPERATOR FOR A HIGH SPEED HOLOGRAPHIC SONAR

WHICH USES AN INCOMPLETELY ORTHOGONALIZED WAVEFRONT

Yasutaka Tamura, Masanobu Takahashi, and Takao Akatsuka

Department of Information Engineering, Yamagata University,
Yonezawa 992, Japan

ABSTRACT

In this paper, we introduce an imaging operator for a high speed holographic SONAR which uses transmitting pulses modulated with a system of orthonormal functions. This operator is based on a singular value decomposition of the transfer matrix of the digital system. The operator reduces the artifact which emerges when a conventional operator is used. The calculations are simplified using an appropriate model and approximations. The operator obtained can be implemented as multiple filters associated with the directions of view. The point spread function of the 2-D imaging system which uses the pulses modulated by Walsh functions is also demonstrated.

INTRODUCTION

It is important to construct a real-time and artifact-free imaging operator for a holographic imaging system.

The artifact in an echographic imaging system is regarded as a result of nonorthogonal correlations of the wavefield signatures corresponding to a distinct target location [4]. These nonorthogonal correlations are caused by both the receiving and transmitting condition. The operator which reduces the artifact induced by a receiving condition has been proposed in [4 ~ 7]. These operators are based on the discrete form of transfer function of holographic observation, and designed for a 1-D imaging system.

In turn, the operator which reduces the effect of the incompletely orthogonalized transmitting wavefront is required, when the main spatial resolution is obtained with the transmitting array. Such an operator is desirable especially with the high speed holographic SONAR proposed in previous works [1~3] , because the nonorthogonal transmitting wavefront is the main cause of the degradation of the image quality of this system.

This SONAR system transmits pulses modulated with a system of orthonormal functions via multiple transmitters simultaneously. Both the transmitting and receiving directivity is formed indirectly using digitized echo waveforms. The system has high 2-D or 3-D spatial resolving power even when the pulses are modulated with functions of a small time-bandwidth product such as Walsh functions. Therefore, the data-length and the number of receivers required in the system is small, and the system can realize a high speed data acquisition property with a reasonable complexity of hardware. However, the transmitted wavefront is not strictly uncorrelated using such signals. This generates undesired peaks in the point

spread function (P.S.F.) and degrades the quality of the resultant image when the number of
the receivers is small.

In this paper, we introduce a new imaging operator for the 2-D or 3-D imaging system
which uses an incompletely orthogonalized transmitting wavefront. This operator is based on
the singular value decomposition (SVD,[9]) of the transfer matrix of the SONAR.

The size of the matrix tends to be large because the system utilizes wide-band pulses
and the system concerned is a 2-D or 3-D imaging system. This makes the calculation of the
SVD extremely complicated. Furthermore, if the obtained operator has the form of a matrix
operator, the required memory capacity and the calculation cost is impractically large. In
order to simplify both the calculation of SVD and the implementation of the operator, an
appropriate model and approximation is introduced. Using the model and the approximation,
the operator takes the form of multiple filters corresponding to distinct view angles. Hence,
high speed imaging operation is possible with special purpose hardware or FFT algorithm.
The point spread function of the 2-D imaging system which uses the pulses modulated with
a system of Walsh functions is also demonstrated.

PRINCIPLE OF THE IMAGING

Encoded wavefront

The geometry of the array is illustrated in Fig.1. Transmitters are arranged in an equally
spaced linear array, and a receiver is positioned at the center of the array. The driving signal
for the k-th transmitter is $u_k(t)$. Here we assume that $u_k(t)\,(k = 1, 2, ..., N)$ forming a system
of orthonormal functions , and the following relation is approximately satisfied:

$$\varphi_{ij}(\tau) = \int_{-\infty}^{\infty} u_i(t)u_j(t - \tau)^* dt$$
$$= \begin{cases} \varphi_{ii}(\tau) & \text{for } i=j \\ 0 & \text{for } i \neq j \end{cases} .$$

$$(1)$$

where φ_{ij} and φ_{ii} are the cross-correlation and auto-correlation functions of the transmitting
pulses. The orthogonal transmitting pulses construct the wavefront of omni-directional energy
propagation. The echo waveforms returned from distinct points are uncorrelated when the
condition (1) is satisfied [1]. In other words, an encoded wavefront is generated with the pulses
modulated with a system of orthonormal functions, and the imaging operation is regarded as
a decoding process for this wavefront.

The discrete model

Consider that object spatial function and received waveforms are represented in L-
dimensional vector $\mathbf{s}$ and K-dimensional vector $\mathbf{r}$ respectively. The linear relation between $\mathbf{s}$
and $\mathbf{r}$ can be characterized by

$$\mathbf{r} = H\mathbf{s}$$

$$(2)$$

where a $K \times L$ matrix

$$H = [\mathbf{h}_1, \mathbf{h}_2, ... \mathbf{h}_L]$$

$$(3)$$

is the transfer matrix of the holographic SONAR system, and column vector $\mathbf{h}_i\,(i = 1, ... L)$
is the digitized echo corresponding to the point target at position $\mathbf{x}_i$.

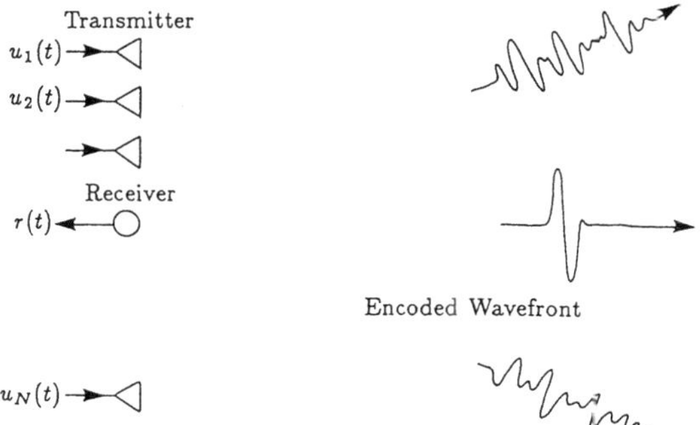

Fig. 1. An encoded wavefront generated with a transmitting array.

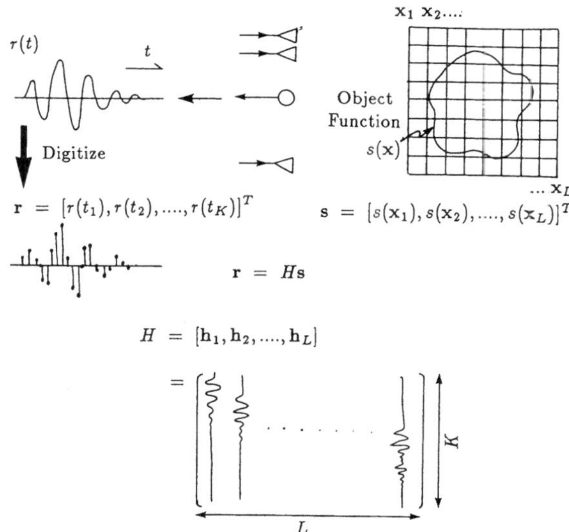

Fig.2. Discrete model of the SONAR.

With a conventional imaging operator, the reconstructed object function $\hat{s}$ is given by

$$\hat{s} = H^* \mathbf{r} \ . \tag{4}$$

where H^* denotes the conjugate and the transpose of H . When the transmitted wavefront is completely orthonormalized, i.e. :

$$\mathbf{h}_i^* \cdot \mathbf{h}_j = \begin{cases} 1 & \text{for } i=j \\ 0 & \text{for } i \neq j \end{cases} , \tag{5}$$

we can obtain a complete image as

$$\hat{s} = H^* H \mathbf{s} = \mathbf{s} \ . \tag{6}$$

However, condition (5) is incompletely satisfied, and undesired peaks emerge at the off-diagonal elements of P.S.F. matrix $H^* H$. A desired operator $G = [\mathbf{g}_1, \mathbf{g}_2, ..., \mathbf{g}_L]^T$ must approximately satisfy

$$\mathbf{g}_i^T \cdot \mathbf{h}_j = \begin{cases} 1 & \text{for } i = j \\ 0 & \text{for } i \neq j \end{cases} \tag{7}$$

for the artifact reduction and image enhancement. The operator satisfies (7) and is known as a pseudo inverse operator; and this can be obtained on the basis of the SVD of H.

Singular Value Decomposition of H

The matrix H is decomposed as

$$H = \sum_{i=1}^{K_0} \lambda^{\frac{1}{2}} \mathbf{u}_i \cdot \mathbf{v}_i^* \ , \tag{8}$$

where $\mathbf{u}_i \, (i = 1, ..., L)$ are L-dimensional eigenvectors of HH^*, $\mathbf{v}_i \, (i = 1, ..., K)$ are K-dimensional eigenvectors of $H^* H$, and K_0 is a number of non-zero singular values of H.

Using the decomposition, the pseudo inverse operator H^+ is given by

$$H^+ = \sum_{i=1}^{K_0} \lambda^{-\frac{1}{2}} \mathbf{v}_i \mathbf{u}_i^* \ . \tag{9}$$

Although the pseudo inverse operator gives a optimal reconstruction for noise-free data, this operator has some operational demerits. The sizes of the matrix H and H^+ are very large in the case of 2-D or 3-D imaging. Therefore, the calculating H^+ is much too complicated to be practical , and the imaging with the operator is impractical because of its large size and high calculation cost. Robustness of the operator to noise is also a problem.

APPROXIMATION AND MODEL

Using an appropriate approximation and a model, the calculation for the SVD can be simplified. This model utilizes a paraxial approximation : the received echoes from points in the same view direction have similar forms , but different delay times in this approximation. In this section , we discuss the 2-D cross-sectional imaging system with a single receiver. The results , however, can be extended to give results for a 3-D system.

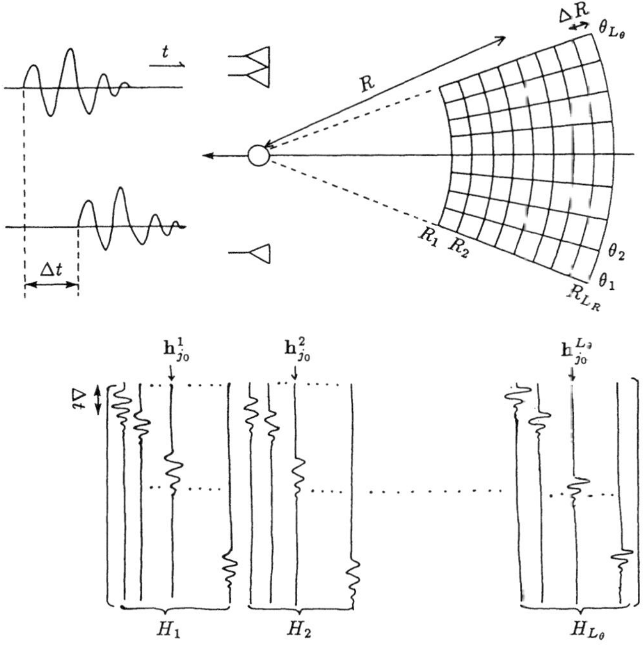

Fig.3. Model and approximation for the design of the operator.

In order to calculate the SVD of H, it is necessary to obtain the eigenvalues and the corresponding eigenvectors of H^*H or HH^*. We calculate the SVD from $K \times K$ matrix HH^* because K is smaller than L.

Fig.3 shows the geometry of the model. In this model, the imaging space is discretized into sector-like $L_\theta \times L_R$ mesh points. The intervals of the mesh are ΔR for range, and the center of the region is separated from the array at a distance of R. Here, we assume that R is sufficiently large and echoes returned from the points along the same view direction have similar but delayed waveforms.

Using this approximation, H can be written as

$$H = [H_1, H_2, ..., H_{L_\theta}]$$

$$(10)$$

$$H_i = [\mathbf{h}_1^i, \mathbf{h}_2^i, ..., \mathbf{h}_{L_R}^i]$$

$$(11)$$

where $\mathbf{h}_j^i$ $(j = 1, ..., L_R)$ are digitized echo waveforms corresponding to the i-th direction angle. If it is supposed that the waveforms are sampled with time-intervals Δt, the following relation is satisfied

$$\Delta R = \frac{C \Delta t}{2} n, (n = 1, 2, 3, ...) ,$$

$$(12)$$

where C is the velocity of acoustical wave. Then the column vectors $\mathbf{h}_j^i$ $(j = 1, 2, ..., L_R)$ are generated from a single vector with a shifting operation. The above situation is illustrated in Fig.3.

Calculation of SVD

Using (10) and (11), the HH^* is reformed as

$$HH^* = \sum_{i=1}^{L_\theta} H_i H_i^*$$

$$= \sum_{i=1}^{L_\theta} \sum_{j=1}^{L_R} \mathbf{h}_j^i \mathbf{h}_j^{i*}$$

$$(13)$$

The above equation shows that HH^* is decomposed into $\mathbf{h}_j^i \mathbf{h}_j^{i*}$. These $\mathbf{h}_j^i$ $(j = 1 \sim L_R)$ can be obtained from an echo vector $\mathbf{h}_{j_0}^i$ returned from the point of range R and azimuth θ. Thus, it is unnecessary to obtain all the elements of H for the calculation of HH^*. It requires only L_θ vectors. A K-dimensional vector and a $K \times K$ matrix for accumulation are required in this numerical procedure.

In the next step, we calculate the eigenvalues $\lambda_1 \sim \lambda_{K_0}$ and the corresponding eigenvectors $\mathbf{u}_1 \sim \mathbf{u}_{K_0}$. The vectors $\mathbf{v}_1 \sim \mathbf{v}_{K_0}$ can be obtained using the relation

$$\mathbf{v}_i = \lambda^{\frac{1}{2}} H^* \mathbf{u}_i .$$

$$(14)$$

However, it is unnecessary to calculate the $\mathbf{v}_i$ when the pseudo inverse operator is the objective operator.

We construct the pseudo inverse operator using the valuables $\mathbf{h}^i_{j_0}$, λ_i , and $\mathbf{u}_i$. The pseudo inverse operator H^+ is given by

$$H^+ = H^*(HH^*)^+$$
$$= H^* \sum_{i=1}^{K_0} \lambda_i^{-1} \mathbf{u}_i \mathbf{u}_i^*$$

$$(15)$$

Here, we rewrite H^+ in the same form as eq.(10) and (11) :

$$H^+ = [H_1^+, H_2^+, ..., H_{L_\theta}^+] \, ,$$

$$(16)$$

$$H_i^+ = [\mathbf{h}_1^{i+}, \mathbf{h}_2^{i+}, ..., \mathbf{h}_{L_R}^{i+}] \, ,$$

$$(17)$$

where $\mathbf{h}_j^{i+}$ $(j = 1 \sim L_R)$ are the operating waveforms corresponding to $\mathbf{h}_j^i$ $(j = 1 \sim L_R)$. Although H^+ is not a circular matrix, we approximate it by a circular one ; $\mathbf{h}_j^{i+}$ $(j = 1 \sim L_R)$ are generated with a shifting operation from $\mathbf{h}_{j_0}^{i+}$ corresponding to $\mathbf{h}_{j_0}^i$. Using this approximation, we obtain the whole operator by calculating only

$$\mathbf{h}_{j_0}^{i+} = \mathbf{h}_{j_0}^{i*}(HH^*)^+.$$

$$(18)$$

Since the operator H_i^+ for the view angle θ_i is approximated to a circular matrix, this operator can be implemented as a filter. The impulse response of this filter is the time-reverse of the waveform $\mathbf{h}_{j_0}^{i+}$.

We summerize the calculation procedure for the SVD as follows:

i) The echo waveform $\mathbf{h}_{j_0}^i$ corresponding to the point at the polar co-ordinate (R, θ_i) is calculated.

ii) Using the $\mathbf{h}_{j_0}^i$, $\sum_{j=1}^{L_R} \mathbf{h}_j^i \mathbf{h}_j^{i*}$ is calculated. The obtained values are accumulated into a $K \times K$ matrix A .

iii) Procedures i) and ii) are carried out for $i = 1 \sim L_\theta$. Finally, we obtain HH^* as the accumulating matrix A .

iv) From HH^*, $\lambda_1 \sim \lambda_{K_0}$ and $\mathbf{u}_1 \sim \mathbf{u}_{K_0}$ are calculated using the eigenvalue analysis, and $(HH^*)^+$ is obtained.

v) Operating waveforms $\mathbf{h}_{j_0}^{1+} \sim \mathbf{h}_{j_0}^{L_\theta+}$ are obtained by operating $(HH^*)^+$ to the echo wave-forms $\mathbf{h}_{j_0}^1 \sim \mathbf{h}_{j_0}^{L_\theta}$. These operating waveforms are time-reversed and the L_θ filters are obtained as the imaging operator.

Fig.4. shows an example of the implementation of the operator. Note that the obtained operator is designed for the distance R. When the range is considerably different from the value used in the design, the performance of the operator is degraded. For wide range imaging, the parameter of the filters must be changed with delay time : the improved operator is implemented as time varying filters.

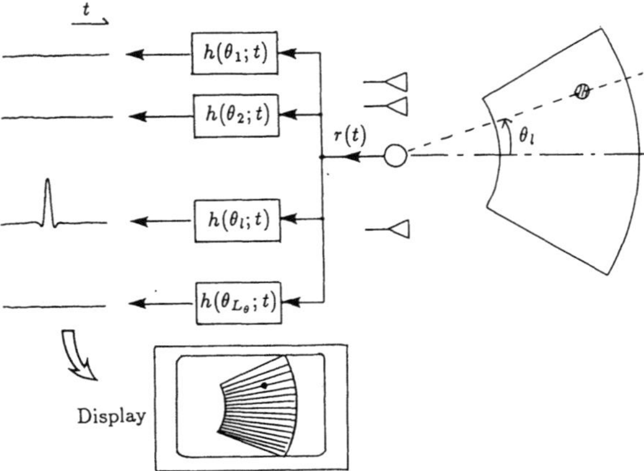

Fig.4. The implementation of the operator using multiple filters.

SIMULATION

In this section, we demonstrate the point spread function of a 2-D cross-sectional imaging system using a computer simulation. We assume the system uses the pulses modulated by a system of Walsh functions. A transmitter array of 16 elements and a single receiver are assumed. The intervals of the elements are selected to be single wavelength of the carrier wave, and the elements are driven by the sinusoidal carrier signals modulated by the system of Walsh functions. The width of the clock for the Walsh functions is equal to two times as long as the period of the carrier signal.

The region for the construction of the operator is the sector of 30° width and 19λ depth. The region is discretized into 32×19 mesh points ,that is $L_\theta = 32$, $L_R = 19$.

The operator is constructed as described in the previous sections, and the P.S.F. is calculated using the obtained 32 filters. Fig.5 shows the P.S.F.s at the center of the sector region. The P.S.F. using the conventional operator H^* is shown for comparison. The ability of the new operator to reduce the artifact is clearly observed.

DISCUSSION

The results of the computer simulation shows the ability of an approach using the SVD of the transfer matrix. It is , however, only a preliminary test for the feasibility of this operator. We must solve several problems to obtain an operator which can be used in a practical system. In this section, we discuss the problems.

It is most important to establish the theoretical criterion for the operator, and to construct an operator which satisfies these criterion. The pseudo inverse operator forces the nonorthogonal wavefront to become an orthonormalized one. In other words, the P.S.F. matrix $H^+ H$ is forced into an identical matrix I. Although it is desirable in noise-free and ideal cases, the pseudo inverse operator causes an unexpected response in the waveforms deviating from ideal ones. This operator tends to amplify undesired noise when unnecessarily fine grids are used. Deviation is caused by the targets which fall off the grid points or the sector region, or by the induced noise. Therefore, criterion which take the robustness, total signal to noise ratio, resolving power, and the artifact level into account, are absolutely necessary. Similarly, it is also important to construct an operator which satisfies these criterion.

In this paper, we demonstrate an imaging system with a single receiver. For ease of implementation, a single receiver system is desirable. However, for the purposes of noise reduction , a system using multiple receivers is preferable if it is available. Of course, it is possible to extend the proposed approach to be used with a multiple receiver system. More

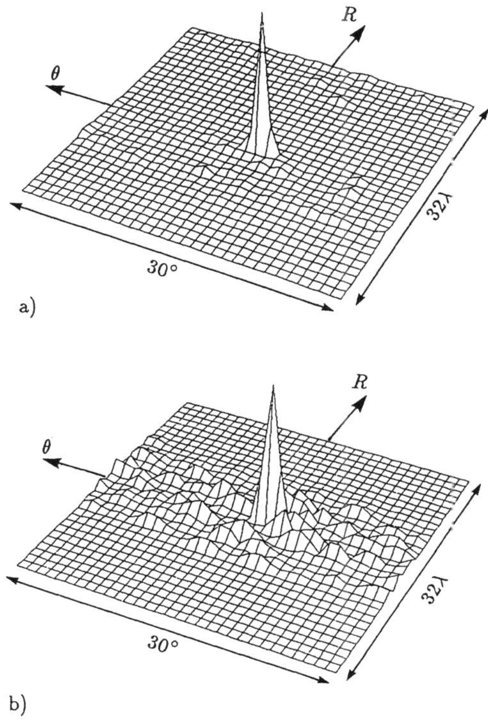

Fig.5. The simulated PSFs
a) : with proposed operator, b) : with H^*

easily, beam-forming with a receiver array can be carried out using a conventional delay-and-sum operation. The detection of the received echo is performed by the same filter designed for the single receiver system. It is also necessary to design an optimal operator for a multiple receiver system.

The approach used for the construction of the operator can be extended for use with a 3-D imaging system. The received waveforms are calculated for the grid points on a spherical surface at a given distance. From the waveforms, the matrix HH^* is obtained and the operator is calculated using the SVD of this matrix. The size of HH^* is determined by the temporal sample length for the echoes. Hence, the calculation cost for the SVD is comparable to that in a 2-D system , when the same duration time of transmitting the pulses is used.

When the above problems are solved, the approach can also be used for an imaging system with an aperture of spatially distinguishable impulse response [10,11].

CONCLUSION

We have introduced an imaging operator for a high speed holographic SONAR which uses an incompletely orthogonalized wavefront. We have proposed a method for the calculation of the operator. The calculation is simplified using an appropriate approximation and model The constructed operator can be represented as multiple filters, and it can be implemented using a digital filter and a digital signal processing technique. We have demonstrated the point spread function of a 2-D imaging system which uses pulses modulated by a system of

Walsh functions. The ability of the operator to reduce the artifact is shown by the simulation. However, the criterion and the design theory to produce a feasible operator are still required.

REFERENCE

1. Y. Tamura and T. Akatsuka, A digital beam-former using uncorrelated ultrasound pulses, Proc. of SPIE. 768, 85/92(1987)
2. Y. Tamura and T. Akatsuka, Ultrasound holographic imaging system using high speed scanning with pulses modulated by Walsh functions, Trans. of SICE. 24-3.213/219(1987)
3. Y. Tamura and T. Akatsuka, Holographic sonar using orthogonal transmitting pulses. Acoustical Imaging,17,753/760(1989)
4. Hua Lee. Resolution enhancement of backward propagated images by wavefield orthogonalization. J. Acoust. Soc. Am.,77-5,1845/1848(1985)
5. Hua Lee, Resolution enhancement by wavefield extrapolation, Trans. of IEEE. SU-31, 6. 642/645 (1984)
6. Hua Lee and Gen Wade, Constructing an imaging operator to enhance resolution, J. Acoust. Soc. Am.. 75-2,499/504(1984)
7. Hua Lee, Formation of the generalized backward projection method for acoustical imaging, Trans. of IEEE. SU-31,3,157/161(1984)
8. Hua Lee and D. I. Sullivan, Fundamental limitation of resolution enhancement by wavefield extrapolation, J. Acoust. Soc. Am., 84-2, 611/615(1988)
9. G. H. Golub and C. Reinsch, "Singular value decomposition and least squares solutions", Numer. Math.,14, (1970)
10. Y. Tamura and T. Akatsuka, An echolocation and imaging using transducers of directionally distinguishable impulse response, Acoustical Imaging, 18
11. K. Koyama, H. Ito, K. Tanaka, and Y. Tamura, Polymer transducer system for imaging using pseudo inverse matrix detection, Acoustical Imaging, 18

THREE-DIMENSIONAL DISPLAY TECHNIQUE FOR

FISH-FINDER WITH FAN-SHAPED MULTIPLE BEAMS

Yoshinao Aoki, Tomoyuki Sato, Pei-kai Zeng and
Kohji Iida*

Department of Information Engineering,
Hokkaido University Sapporo, Japan
* Department of Fisheries, Hokkaido University
Hakodate, Japan

ABSTRACT

A three-dimensional(3-D) echo-sounder is used as a fish-finder to
identify the 3-D location of fish in the sea. The image data are collected by scanning the ocean using an echo-sounder with multiple fan-
shaped beams. The recorded data are processed and a 3-D image is reconstructed. These volume images are stored in a memory system with 2
M byte LSI memories. Algorithms, such as a depth-shading algorithm
for displaying 3-D images in 2-D CRT scope, are discussed. Experiments
using this algorithm were carried out using data obtained from both
metallic balls and real fish.

INTRODUCTION

A fan-shaped multi-beam technique was discussed as a way of improving the resolution in a fish-finder.[1] Using this technique the resolution of the proceeding direction of a vessel is determined by beam
sharpness of the fan-shaped beam, whereas the resolution of the azimuth
direction (the direction perpendicular to the proceeding direction) is
given by the number of the divided beams as illustrated in Fig.1. The
location of targets with respect to the depth direction is determined
by the principle of the ordinary echo-sounder, where the delay time of
the pulse reflected from the targets is measured to obtain depth information.
Usually in such an echo-sounder system, resolution is not high
enough to enable indentification of the 3-D distribution of the school
of fish and special diagnosing techniques for data of low resolution
is necessary. In this paper we propose a technique to display 3-D
images of the fish-finder in a 2-D CRT scope. Algorithms and hardware
of high-speed processing necessary for projecting 3-D image data onto
a 2-D screen are discussed. Data acqisition experiments were carried
out using an experimental fish-finder system with 5 fan-shaped beams.
For our image display we used hardware specially developed for processing volume image data.

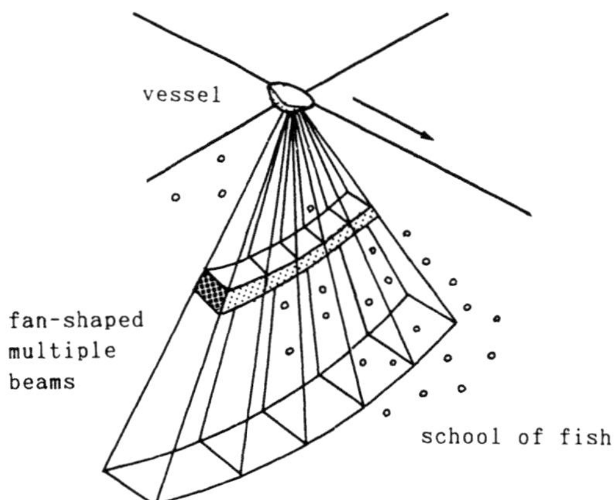

Fig.1   Principle of fish-finder using
fan-shaped multiple beams.

PRINCIPLE OF THE FISH-FINDER USING FAN-SHAPED MULTIPLE BEAMS

In the system discussed in this paper, a fan-shaped beam is divided into several beams in order to obtain higher resolution in the
azimuth direction, and allowing a more accurate diagnosis of the distribution of fish schools.   The principle of the multiple beams is shown
in Fig.1.   Multiple beams are constructed by synthesis of transmitting
and receiving beams.   The transmitting beam has a wide beam-pattern with
respect to the proceeding direction as shown in Fig.2(a), while the
multiple beam-pattern of the receiver is shown in Fig.2(b).

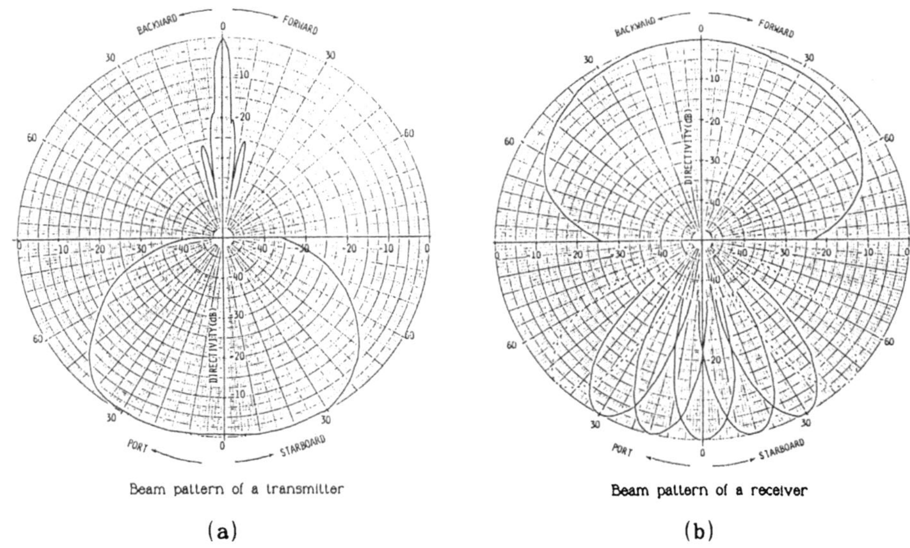

(a)   (b)

Fig.2   Beam patterns of a transmitting transducer (a)
and a receiving transducer (b).

The angles of the targets located in overlapping regions of the beams as shown in Fig.2(b) can be determined by calculating the ratio of the target strengths of the neighboring beams. Target strength Ts is defined by the following equation [2]

$$Ts = 20 \log Vrms + (40 \log R + 2\alpha R) - 20 \log D_T(\theta, \psi) D_R(\theta, \psi) - K \qquad (1)$$

where Vrms is the root mean square voltage of echo from the fish, R is the distance between the receiver and the fish, α is an absorption attenuation coefficient, $D_T(\theta, \psi)$ and $D_R(\theta, \psi)$ are directivity functions of the transmitter and receiver respectively with variables θ, ψ of the angular location of the target and K is the total gain of the system. The synthesized beams with both transmitting and receiving beams of beam-patterns of Fig.2(a) and (b) are fan-shaped multiple beams as illustrated in Fig.1.

CONFIGURATION OF THE FISH-FINDER SYSTEM

The configuration of our system's signal processing circuit is shown in Fig.3. Pulse waves from the transmitting transducer are generated by signals from a tone-burst generator with a frequency of 50kHz and a time duration of 1 msec. The tone-burst generator is synchronized with timer-controller. The receiving transducer is divided into 16 parts and output from the pre-amplifiers are fed into delay lines, forming 5 beams. For distance compensation, output from delay lines is fed into TVG(Time Varying Gain) amplifiers. These amplifiers have the gain characteristics of 40log R (where R is the distance). Output from TVG amplifiers are detected and the detected signals of 5 channels are recorded by a tape recorder as data of the fish-finder.

The experimentally constructed transducers in our fish-finder were T-shaped as shown in the photograph of Fig.4, where the transmitting transducer radiates a narrow fan-shaped beam with respect to the direction of the large aperture side. The receiving transducer is a linear array of 200 array elements. Using both the transmitting and receiving transducers, 5 beams with different angles are synthesized.

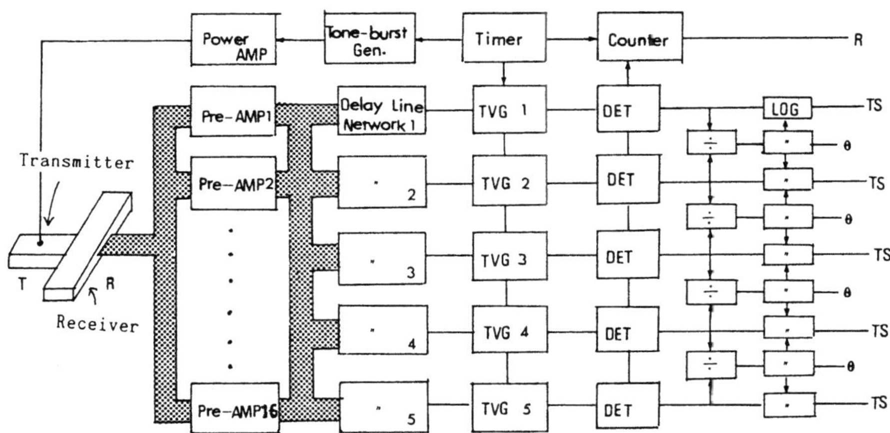

Fig.3 Block diagram of signal processing circuit.

Fig.4　Photograph of the transmitting
and receiving transducers.

PROCEDURE OF IMAGE RECONSTRUCTION

First the region where the targets are located is determined.The
regions are divided by 5 beams as described previously. The directional
angles of each beam are known before the experiment and the directional
angle θ of a target is obtained by comparing levels of signals received
between neighboring channels by the following equation.

$$\theta = a_n \log \frac{V_{n+1}}{V_n} + \Phi_n \tag{2}$$

where Vn+1 amd Vn are the strengths of the signals received by (n+1)-th
and n-th channels. a_n is a coefficient and Φ_n is the given n-th directional angle. Once the angle of the target is obtained. the strength
of the signal received from the target can be calibrated by the characteristics curve of directionality of the beam. Compensation of
attenuation due to the different depths can be carried out using the
TVG amplifiers. A 2-D tomographic image can thus be obtained by this
procedure for any ocean depth. Since the vessel with the fish-finder
system moves along at a fixed speed. a 3-D image can be obtained by recording the tomographic image data at each sampling point along the path
of the vessel. From this image data. we can construct a 3-D image of
the ocean.

DATA AQUISITION EXPERIMENT

The constructed experimental system was set up on a vessel and experiments were conducted using both metallic balls and on real schools
of fish. Six metallic balls of 38 mm diameter were suspended under a
buoy at 10 meter intervals in depth. Data acquisition was carried out
as the vessel passed near the bouy. The depth range to be measured was
100 meters and 200 tomographic images were recorded.
Figure 5 shows one of the echo images of channel 3. where the vertical axis represents water depth and the horizontal axis is the time axis
corresponding to the location of the vessel. Two images showing the
sequences of spots are displayed in Fig.5. This is because the metallic balls are at first observed on the left hand side of the vessel as

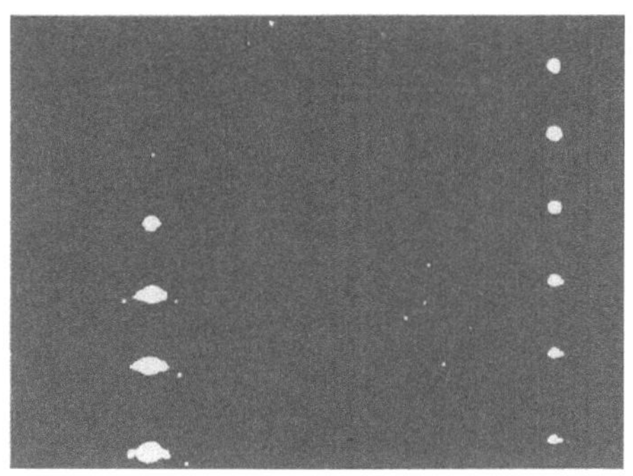

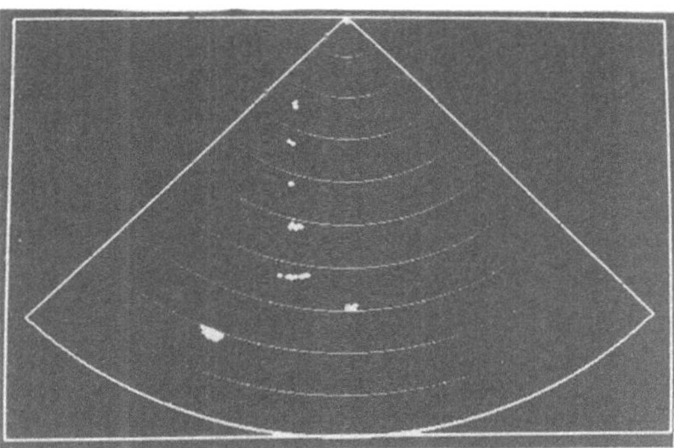

Fig. 5   Echo images of metallic balls by the signals of one channel.

Fig. 6   Superposed image of metallic balls by 5 channels.

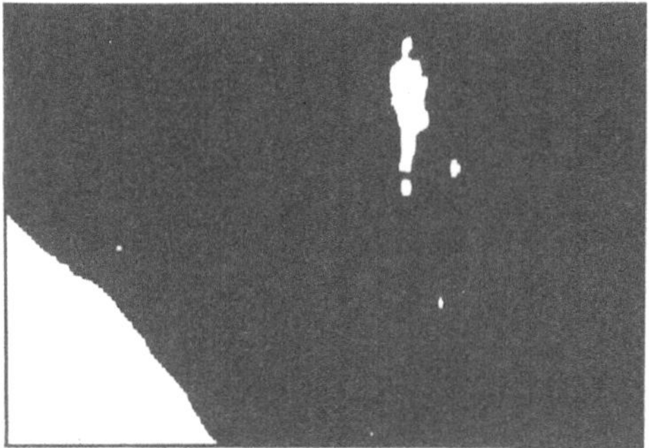

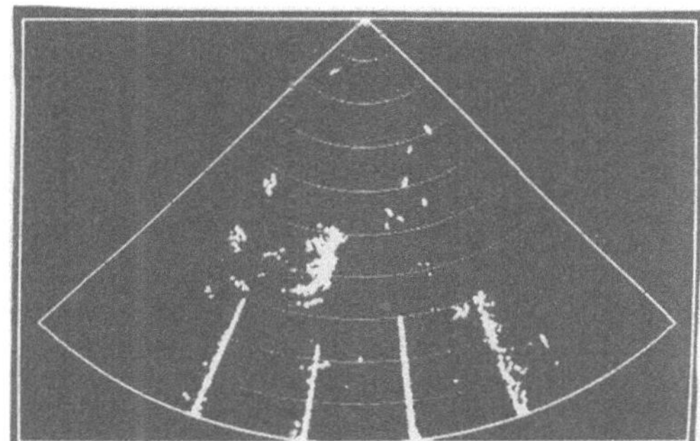

Fig. 7   Echo image of a school of kokanee and the bottom of the lake by one channel.

Fig. 8   Superposed image of a school kokanee and the bottom of the lake by 5 channels.

it passes near the buoy and on the vessel's return leg, they are observed on the vessel's right hand side. By processing the data from all channels, a tomographic image can be obtained at a sampling point along the path of the vessel as shown in Fig.6.

The same experiment of data acquisition was carried out on a school of kokanee in Kuttara lake, Hokkaido island. The depth of the lake ranges from 40 to 80 meters. The experiment was carried out at a depth of 50 meters and 100 sampled tomographic images were taken. Figure 7 shows an image displayed by data from channel 3 and Fig.8 is the tomographic image reconstructed using the processed data from channels 1-5.

PROCESS OF SEARCHING DATA IN IMAGE SPACE AND PROJECTION

We constructed a 3-D image space by storing a sequence of 2-D tomographic images. The images of the fish-finder are then reproduced by searching significant data in the 3-D image space using the algorithms of reproducing images. In other words, this is a technique for projecting 3-D image data onto a 2-D screen. Several algorithms for image projection have previously been reported.[3,4] Here, a direct data projection mehod was used, where image data to be projected is searched along the view line connecting a pixel of the screen and voxels on the image space as shown in Fig.9.

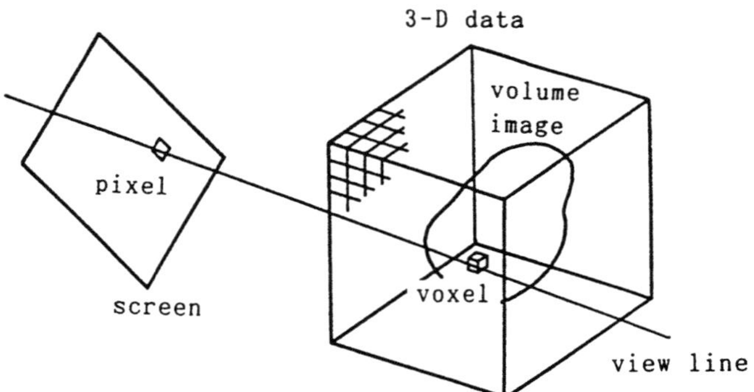

Fig.9   Princiiple of projecting 3-D
image data onto a 2-D screen.

To construct high-speed processing display system, 3-D image data
are processed in the following way.   Image space is assumed to be cubic
as shown in Fig.9.
1)   Since a maximum of only 3 planes of the cubic image space can be
projected onto the screen, we must determine the planes to be projected.
By this process we select the view lines coming through the planes of
the cubic image space.
2)   We determine the effective region by projecting one of planes onto
the screen and we disregard data search which is out of this region.
3)   We search the voxel whose value exceeds the threshold data along
the view line coming through the pixel in the effective region in the
screen.   If the number of voxels searched exceeds the fixed number, we
stop searching along the view line, and then change the view line and
repeat the same searching process .
4)   We repeat the process of 3) until the view line scans all pixels
in the effective region.
5)   We repeat the processes of 1) and 2), changing the plane to be
projected.

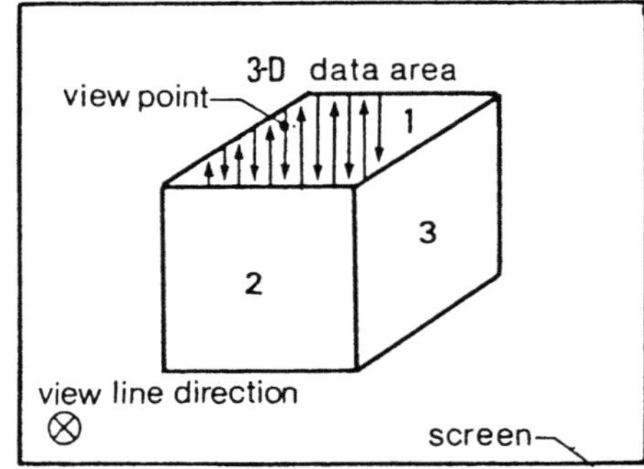

Fig.10   Principle of a technique to reduce
time for searching 3-D image data
in   projecting onto a screen.

DEPTH-SHADING TECHNIQUE FOR DISPLAYING 3-D IMAGES

A few techniques have been reported for determining pixel values
from the obtained values of voxels. These techniques can be broadly
classified into 2 methods:surface display method and transparency display method. In our expeniments, we used the surface display method,
where the surface of the targets are displayed. Since the targets are
three dimensional, it is necessary to add a 3-D effect in displaying
the target surfaces. For the 3-D effect we used a depth-shading technique. In this technique gray levels are assigned to the image data
according to the depth of a point or standard plane in the image space.
Surfaces of the targets become darker as the distance increases from
the point or the standard plane, resulting in a 3-D effect of the displayed images.

As a visual inspection by the naked eye of the different gray levels
especially near the plane where the difference in shading is very slight,
can not determine the depth, calibration is carried out using the curve
shown in Fig.11.

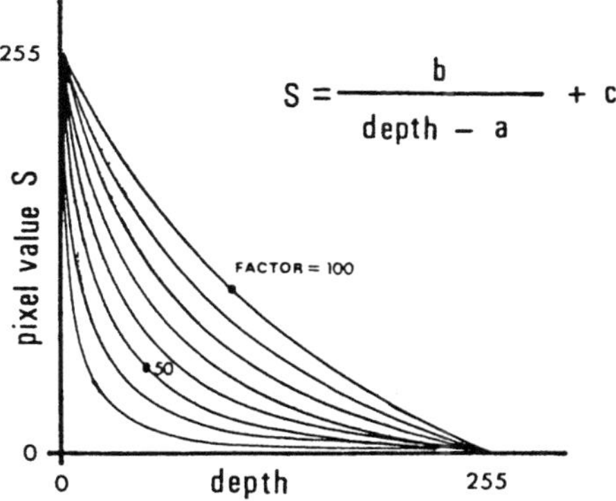

Fig. 11   Compensation curve for
depth-shading method.

CONSTRUCTION OF A SYSTEM FOR PROCESSING 3-D IMAGES

The hardware for recording 3-D image data was constructed using
LSI memory devices, where the memory cells are accessed by X,Y and Z
address signals as shown in Fig.12. These address signals are generated by the address adder which increments the address by the signals
from the depth counter. The output of the depth counter corresponds
to the distance between the standard plane and a voxel along a view line.
After storing 3-D image data in a 3-D RAM, image data displayed on a 2-D
screen are searched by starting the depth counter. The depth counter
stops according to the output of the comparator and the depth of the
image point along the view line is obtained. The memory size of the
3-D RAM is 128 X 128 X 128 X 8 bits = 2 M bytes.

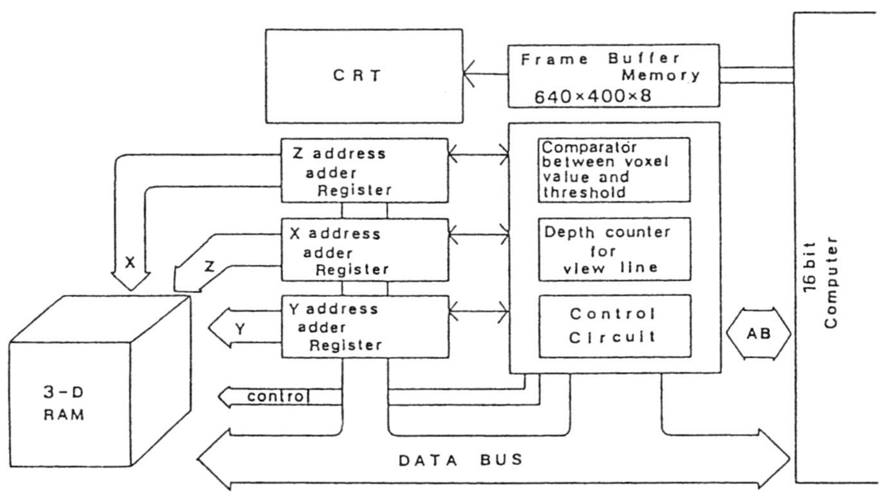

Fig. 12   Block diagram of memory system for storing
and processing 3-D image data.

RESULTS OF DISPLAYING EXPERIMENTS

Experiments on 3-D image display corresponding to the images of
Figs.6,8 were carried out using the techniques mentioned before. Figures
13 and 14 correspond to Figs.6 and 8 respectively. In these figures
the standard planes to measure the depth are 3 rear planes and the
images of metallic balls or schools of fish are displayed enclosed by
the bright edges of the standard planes. In Fig.13, the relatively
large spots are the images of the metallic balls and the small spots
are noises resulting from the display system.
Experimental results in Figs.13, 14 show that our display technique
is useful for diagnosing location and shapes of fish.

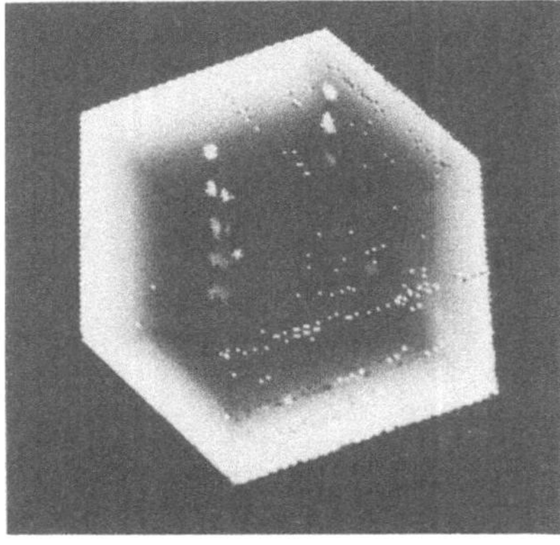

Fig.13 Images of merallic
balls by 3-D image
display technique.

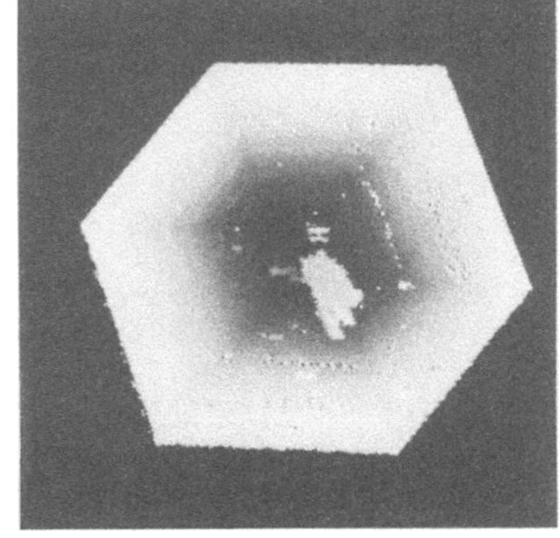

Fig.14 An image of a school of
kokanee by 3-D image
display technique.

CONCLUSION

     In a fish-finder system, both on-line and real-time processing is
desirable for practical use. Thus an array system,like the one we used,
to scan beams electronically is necessary. However, in such a system,
resolution is not high enough due to small number of array elements,
ie number of beams. In such a system image diagnosis is important to
distinguish images of schools of fish from other clutters in the sea.
     Our results show that our 3-D image display thechnique is a promis-
ing technique for use in a fish-finder system using multiple beams.
However, there are problems in the speed of projecting 2-D image data.
Further research is needed to develop an algorithm which can make the
2-D screen display easier to understand. We are continuing to carry out
research in these problem areas in order to develop a pratical system.

ACKNOWLEGEMENT

     This resarch was partially supported by Sapporo Electronics Center
and the authors are grateful for this support.

<u>References</u>

1. K.Iida and T.Suzuki, High resolution fan-beam echo sounder, <u>Bulletin
    of the Faculty of Fisheries, Hokkaido University</u>, Vol.38, No.4, 384
    (1987).
2. K.Iida and T.Suzuki,Four-beam echo sounder system for in situ
    measurement of position and target strength of individual fish,
    <u>Bulletin of the Faculty of Fisheries, Hokkaido University</u>, Vol.38,
    No.4, 375 (1987).
3. Y.Aoki, Y.Takahashi, Y.Sakamoto and M.Ikegami, Display techniques
    of volume images of buried objects in piled snow by acoustical and
    microwave holographic-radar, <u>Acoustical Imaging</u>, Vol.17,p285 (1988).
4. Y.Takahasi, Y.Sakamoto and Y.Aoki, Display method for three-
    dimensional data obtained from snow-search radar,<u>Trans.IEICE(Japan)</u>,
    Vol.J71-D, No.10,2002 (1988).

TUNED ARRAY OF PARABOLOIDAL TRANSDUCERS FOR HIGH-RESOLUTION MARINE PROSPECTING

G.B.Cannelli and E.D'Ottavi

CNR, Istituto di Acustica "O.M.Corbino"
Rome,Italy

INTRODUCTION

High-resolution acoustic techniques to image details of the sub-sea bottom structure, must devote particular care to the duration, the primary to bubble ratio and the repeatability of the acoustic pulse (Parkes et al., 1986). The ideal pulse should be an impulse response representable mathematically by a Dirac delta function, i.e. a very short pulse containing all frequencies. Bubble pulse due to acoustic cavitation (Prosperetti, 1984) is a drawback always present in underwater prospecting and its minimization avoids the need of a complex pulse shape deconvolution.

The electroacoustic paraboloidal source designed at Istituto di Acustica "O.M.Corbino" (Cannelli et al.,1987,1988), presents some important characteristics which allow a proper approach to the high-resolution problem. This transducer can generate a high frequency content pulse by an electric spark gap, set at the focus of the paraboloidal reflector filled with a proper liquid. It has the possibility of frequency spectrum variation of the primary pulse and simultaneously of shifting in time pulse cavitation by electrical capacitance changing. Signal repeatability and its precise time control allow primary pulses to be synchronized and summed constructively to disadvantage of cavitation pulses which are properly shifted in time.

On the grounds of the above electroacoustic effects, some paraboloidal source arrays having different configurations were designed and tested in order to obtain the following improvements, in comparison with a single source: i) primary pulse amplification, ii) cavitation pulse minimization, iii) broadening of frequency band.

After various tests, the optimum solution seemed to be that obtained by means of a nine paraboloidal source array suitably tuned on different capacitance values and assembled along a circular geometry.

Experimental results and a comparison with conventional devices show that the present marine seismic source can give a valid contribution to the high-resolution problem in shallow prospecting.

EXPERIMENTAL TESTS

The first prototype of the paraboloidal source, previously designed for land explorations (Cannelli et al.,1985), is shown in the diagram of Fig.1. The seismic wave is produced by a high energy discharge set at the focus of a paraboloidal aluminum reflector filled with insulating liquid. The acoustic pulse is transmitted to the soil or to the water via a neoprene diaphragm

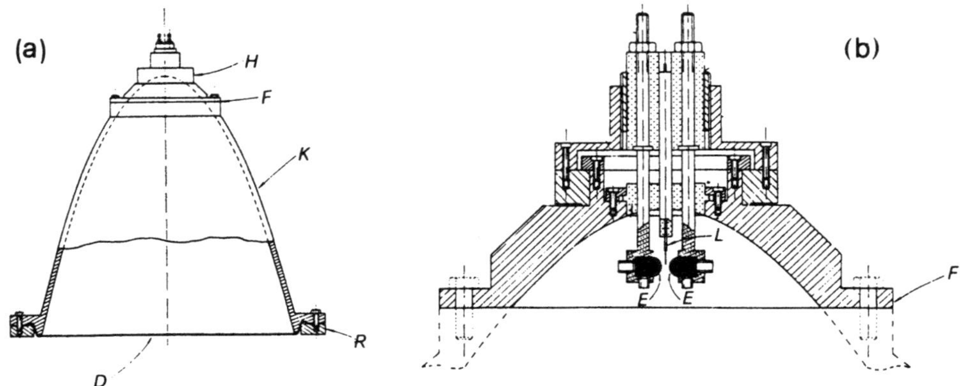

Fig. 1. First prototype of the paraboloidal source. (a) Entire assembly of the source composed of three principal parts. H: head containing the electrodes (F is a flange), K: body of the paraboloid,   R: ring tightening the neoprene diaphragm D.   (b) Section of the head. F: flange, EE: principal electrodes, L: auxiliary spark electrode. The dimensions are: height = 52.1cm, inside base diameter = 50cm, focal length = 3cm.

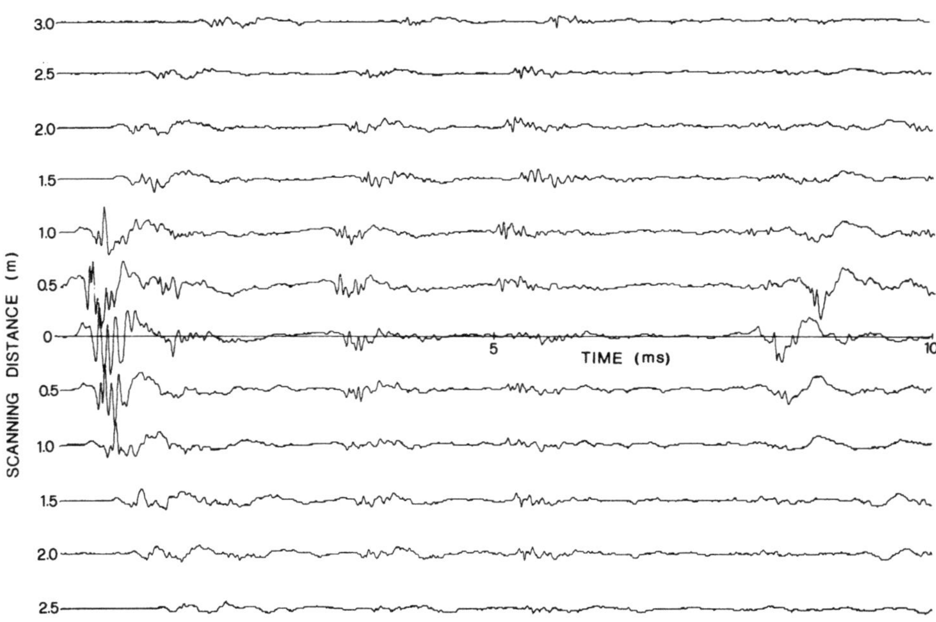

Fig. 2. Acoustic pressure curves of the signals detected in fresh water as a function of time for different scanning distance of the paraboloidal source.

that couples the transducer to the prospecting medium. The discharge is primed between two principal electrodes connected to a bank of capacitors, via a third auxiliary electrode which produces the liquid ionization.

Some preliminary underwater experiments by the above source were made in a towing tank of Rome. The paraboloid was mounted on the motor driving sliding carriage above the towing tank to perform a linear scanning. Acoustic pressure curves as a function of time, for different scanning distance are displayed in Fig.2. In this figure, the first set of pulses is attributable to the direct waves, while the second and third set correspond to the reflected waves on the tank bottom and on the air-water surface, respectively. The last set of pulses is due to the acoustic cavitation. The acoustic pulse signature detected at 1 m beneath the base of the parabolic source, in the starting position 0 of Fig.2, is shown in Fig.3 together with the amplitude spectrum. In the lower diagram, it can be noted the wide range of useful frequencies for high-resolution prospecting.

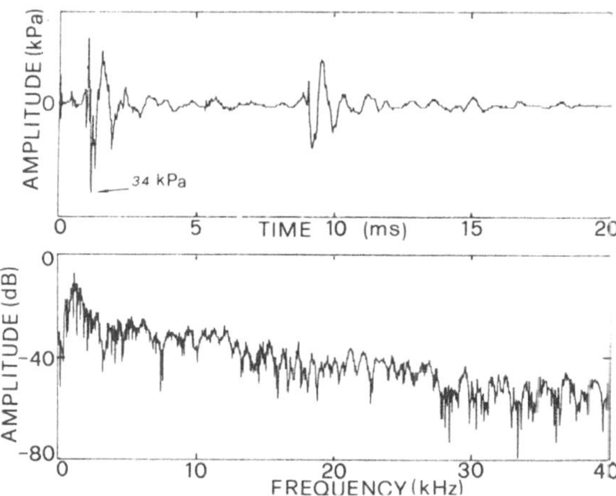

Fig. 3. Pulse signature and amplitude spectrum detected at 1 m beneath the base of the paraboloidal source in fresh-water. Firing energy 100 joule.

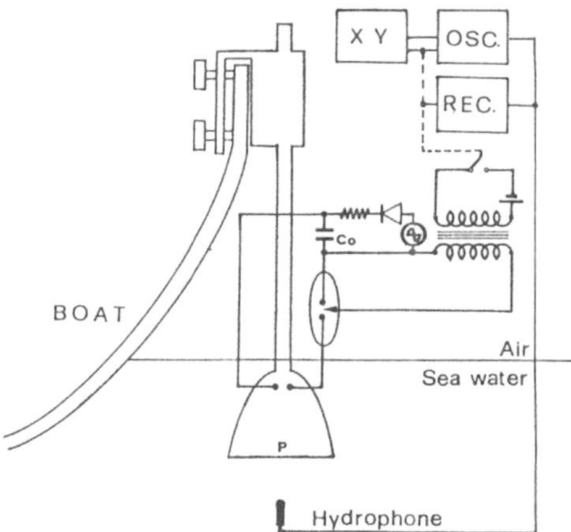

Fig. 4. Schematic diagram of the experiment to pick up the acoustic pulse in sea-water.

A sistematic set of experimental tests were carried out at the sea of Fiumicino (Rome) by mounting the paraboloidal transducer on a boat as shown in Fig.4. The paraboloid was directly dipped in sea-water without the base diaphragm and an auxiliary air spark gap was adopted to trigger the discharge, instead of using insulating liquid together with the third electrode inside the primary spark gap. The discharge energy is supplied by a high-voltage generator connected to a variable set of capacitors Co. By varying of the electrical capacitance, it is possible to change the firing energy.

First experiments aimed at studying the influence of energy on the acoustic cavitation phenomenon. An example is given in Fig.5 which shows the behaviour of the cavitation acoustic pulse at different energy values. In particular, it can be seen that the delay time between the primary pulse and that due to acoustic cavitation increases with energy. This is better illustrated from a quantitative point of view, by plotting the delay time against electrical capacitance is shown in Fig. 6. In this diagram, the energy parameter is replaced by capacitance values since energy is a linear function of capacitance, once the voltage is fixed.

A characteristic of the paraboloidal transducer is the possibility to modify, within certain limits, the frequency content of the acoustic pulse by varying capacitance. In particular, it is possible to change the maximum amplitude frequency component (dominant frequency). An example is given in Fig.7 which shows the dominant frequency as a function of electrical capacitance.

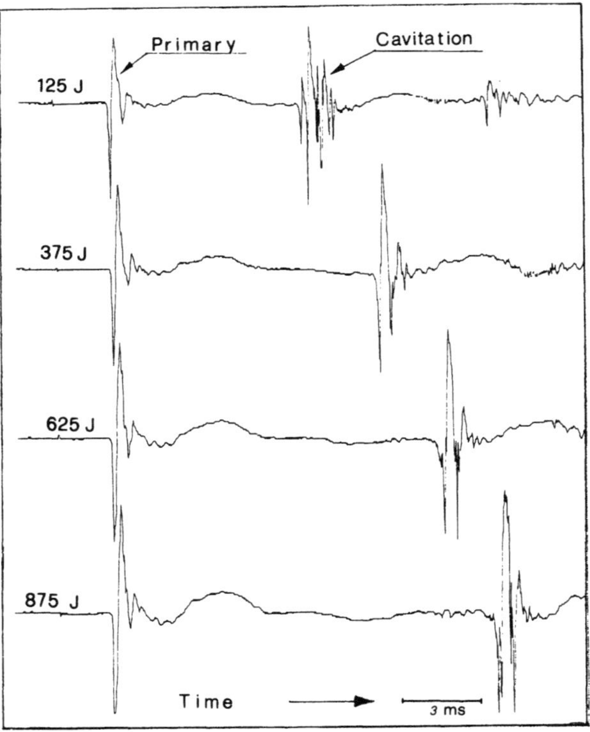

Fig. 5. Behaviour of the cavitation acoustic pulse as the energy changes.

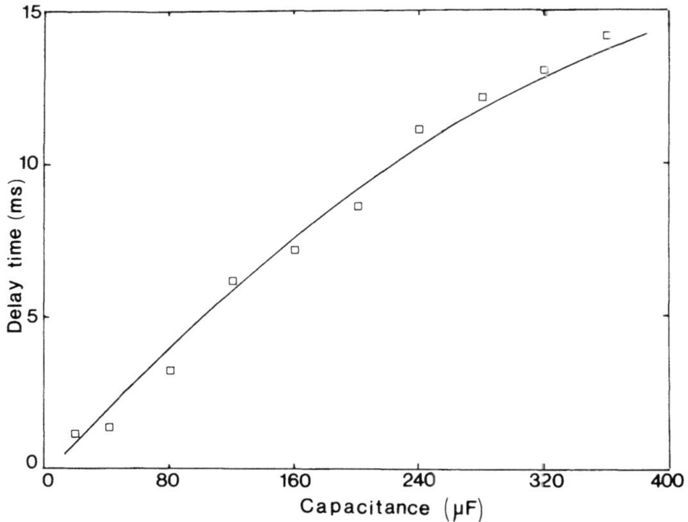

Fig. 6. Delay time between the primary and cavitation acoustic pulse as a function of electrical capacitance.

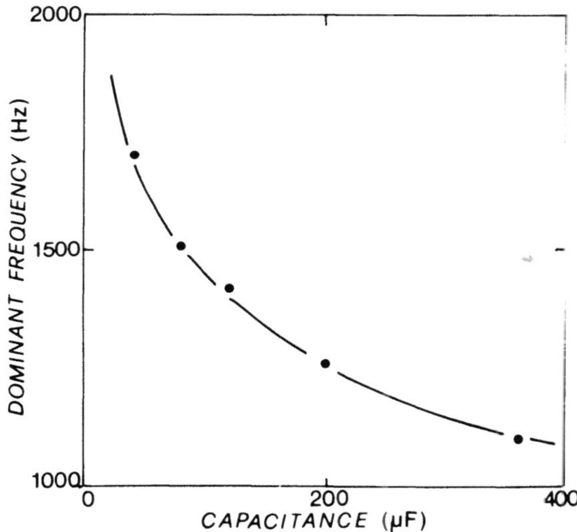

Fig. 7. Dominant frequency as a function of electrical capacitance.

On the grounds of the above observed electroacoustic effects, some paraboloidal source arrays were designed and tested in order to obtain the following improvements, in comparison with a single source: i) primary pulse amplification, ii) cavitation pulse minimization, iii) broadening frequency band.

Each paraboloidal source in the array was designed with dimensions smaller than those of the first prototype, to allow an easier handling of the system in the experimental phase. The dimensions of the single paraboloid are the following: heigth $= 22$ cm, inside base diameter $= 20$ cm, focal length $= 1.2$ cm.

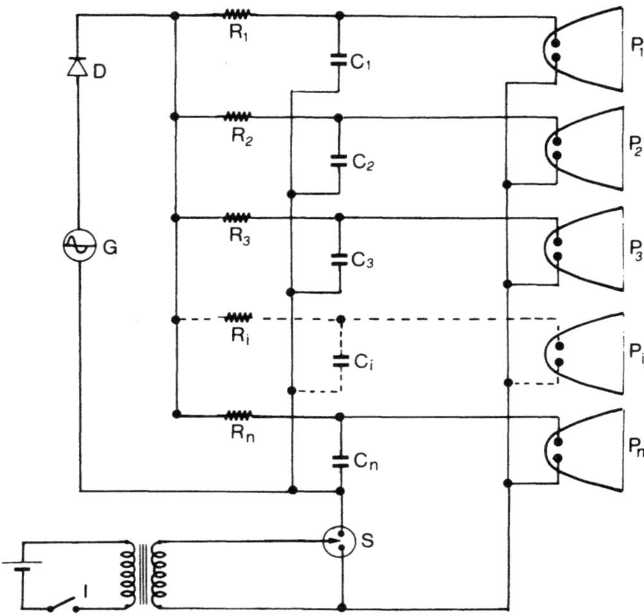

Fig. 8. Block diagram of the electronic system used to power and control the paraboloidal source array. G: high-voltage generator, P_1 to P_n: paraboloidal reflectors, D: diode, R_1 to R_n: resistors, C_1 to C_n : capacitors, I: switch, S: auxiliary air spark-gap.

The block diagram of Fig.8 illustrates the principle of the electronic system to power and control the array of paraboloidal sources. A high-voltage generator G charges the n capacitors $C_1...C_n$ at a peak value of 2.5 kV, via the diode D and resistors from R_1 to R_n. The capacitance values were chosen in such a way as to properly shift in time the cavitation pulses each other, according to the diagram of Fig.6. All the sources of the array are fired in synchronism, each time the switch I is closed. The auxiliary air spark-gap S is used to trigger the primary discharge between the electrodes inside every paraboloid. The firing repetition rate of the array has values around 1 pulse/s, but it can be properly increased by a lower voltage. In this latter case is still valid a diagram similar to that of Fig.6, even if the time shifting between the primary and cavitation pulse is accordingly decreased.

Most trials were carried out in a shallow basin where smooth conditions of the sea water allowed an easier performance of the experiments. Preliminarly, an array of six elements placed at the vertices of a regular hexagon was tested. The acoustic pulse was fired from the sea surface and the corresponding signature was detected at a depth of about 5 m on the axis of the array. The result are shown in Fig.9 where the amplitude spectrum is also given. It can be noted that the primary to cavitation pulse ratio is about 3.3 , moreover the spectrum shows a wide band of high frequencies.

A better configuration seemed to be that obtained by means of nine paraboloidal sources properly tuned on nine different capacitance values and assembled along a circular geometry of middle diameter = 90 cm. A photograph of this array is given in Fig.10.

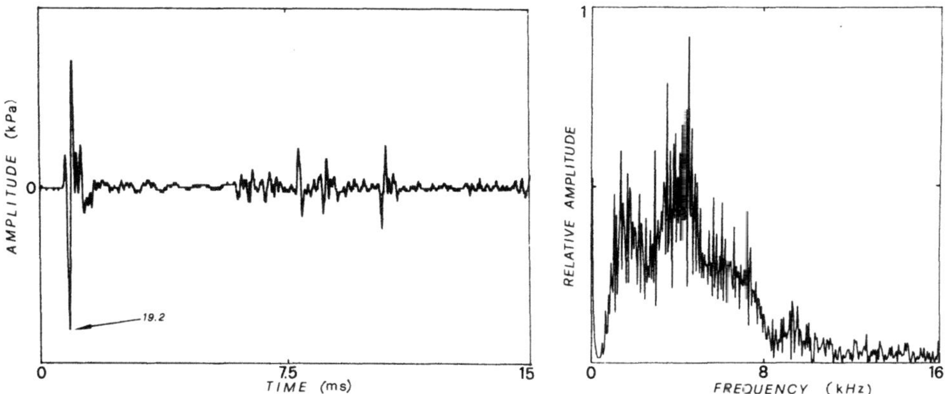

Fig. 9. Signature of a six element paraboloidal source array detected at a depth of 5 m and the corresponding amplitude spectrum.

Fig. 10. Photograph of the nine paraboloidal source array. The middle diameter is 90 cm.

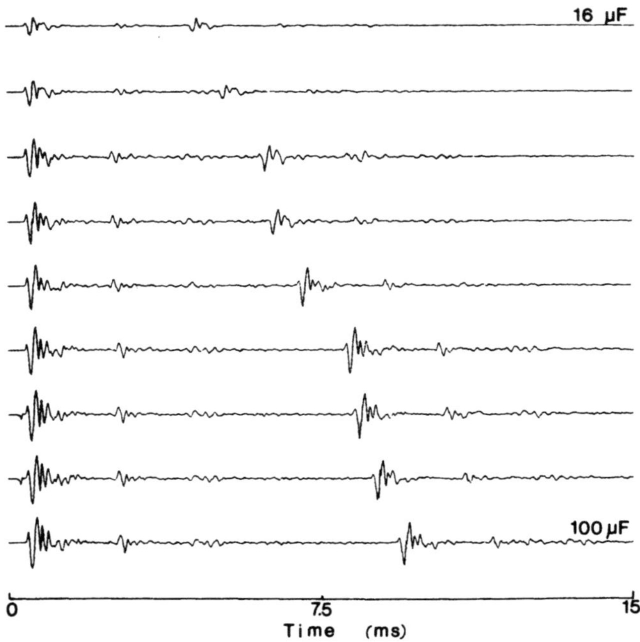

Fig. 11. Signatures of the nine paraboloidal sources of the array, fired in isolation and detected at 1 m from each source. The capacitance values are distributed within the range 16-100 µF. The total electrostatic energy is 1400 Joule.

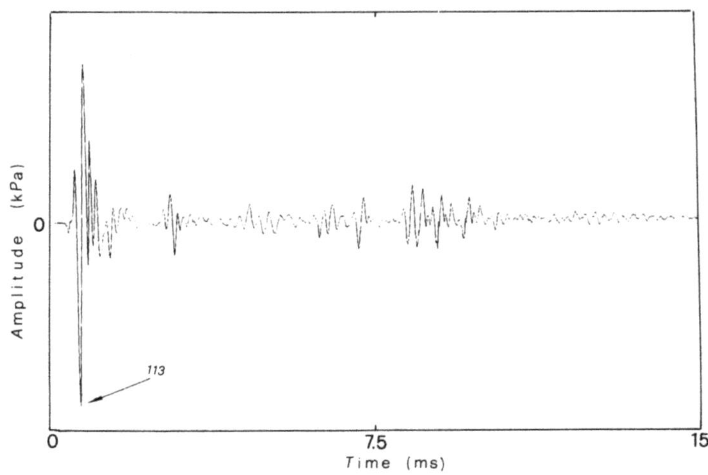

Fig. 12. Signature computed by superposing the nine signatures of Fig. 11. The primary to cavitation ratio is increased from 1:1 to 6:1 in comparison with that of a single source.

508

Each element of the system was tested separately by firing in isolation each source, and detecting the acoustic pulse at 1 m from the paraboloid base on its axis. The capacitance values were changed within the range 16 - 100 µF, for a total electrostatic energy of 1400 Joule. The signatures of the nine sources are displayed in Fig.11. It can be noted the reflected pulses on the bottom and on the air-water surface, both for the primary and the cavitation waves.

An estimate of the goodness of this array is given by the computed signature shown in Fig.12. This was obtained by superposing the nine experimental signatures of Fig.11. It can be noted that the primary to cavitation ratio is increased from 1:1 to 6:1, in comparison with that of a single source. Moreover, the frequency spectrum is accordingly modified and improved in the middle frequency band in comparison, for example, with a single source of 100 Joule (Fig. 13).

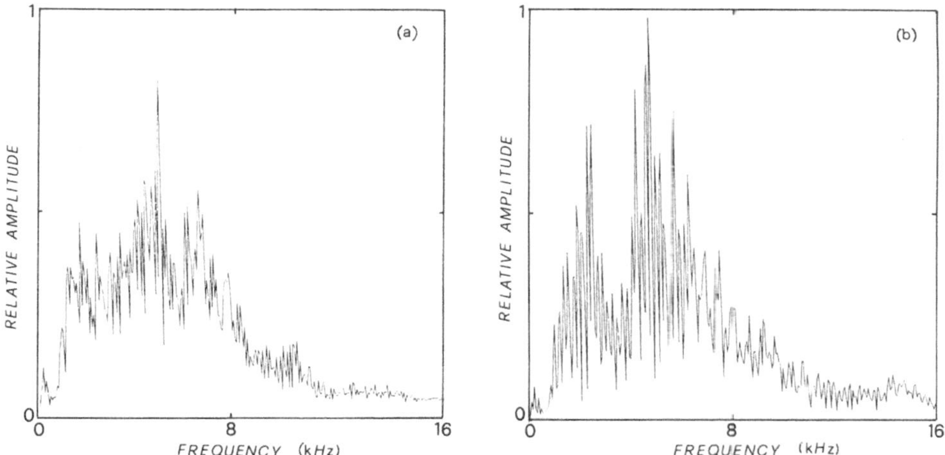

Fig. 13. Comparison between the frequency spectrum of the signature of Fig. 12 (a) and that of a single source of the array at a capacitance value of 40 µF (E=100 Joule) (b).

CONCLUDING REMARKS

The experimental tests carried out on the paraboloidal source array have shown that this device offers the possibility of largely improving the signature and frequency spectrum in comparison with the single source.

The sistem is based on a precise time control and properly tuning of the paraboloidal sources on different capacitance values. So, the primary pulse can be amplified to disadvantage of acoustic cavitation and the frequency spectrum is accordingly improved.

The features of this array are much more suitable for high-resolution explorations than some presently employed sources, as can be seen by comparing, for example, the present results (Fig. 9) with those of a conventional sparkarray in the same experimental conditions (E = 1000 Joule).

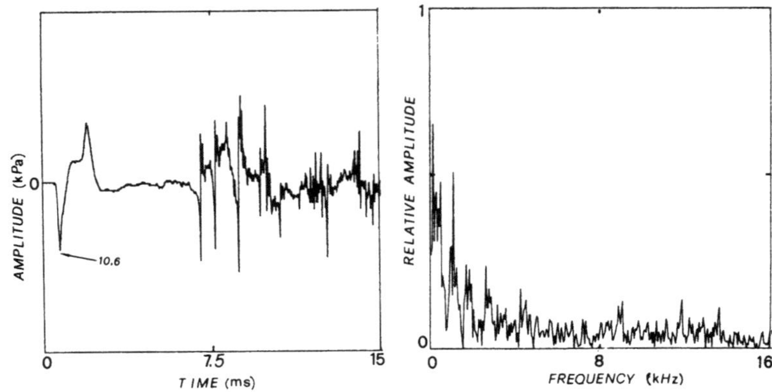

Fig. 14. Signature and amplitude spectrum of a common sparkarray fired in the same experimental conditions as the paraboloidal source array of Fig. 9.

The physical principles used for this system can be conveniently exploited to design other similar arrays having paraboloid sizes and configurations according to particular requirements. For example, larger paraboloids can be used for deep explorations, where lower frequency and higher energy are necessary. While, smaller paraboloids with higher frequency and lower energy, are sufficient for high-resolution shallow prospectings.

ACKNOWLEDGMENTS

The autors wish to thank L. Pitolli and G. Pontuale for their assistance in the laboratory, L. Birarelli and P. Rossi for helpful collaboration in the field.

REFERENCES

Cannelli G. B., D'Ottavi E., and Santoboni S., 1985, Shallow Prospecting on Land by means of a Novel Electroacoustical P-Wave Pulse Generator, Acoustical Imaging Vol. 14, p 585, Plenum, New York.

Cannelli G. B., D'Ottavi E., and Santoboni S., 1987, Elettroacoustic Pulse Source for High-Resolution Seismic Prospecting, Rew. Sci. Instrum., 58, 1254.
U.S.A. Patent n. 4,734,894 issued on March 29, 1988.

Cannelli G. B., D'Ottavi E., and Gasperini M., 1988, Primi risultati di prospezioni marine con sorgenti acustiche paraboloidali, VII Conv. Gruppo Naz. Geofisica della Terra Solida, Roma.

Parkes G., and Hatton L., 1986, The Marine Seismic Source, D. Reidel Publ., Dordrecht, Holland.

Prosperetti A., 1984, Physics of Acoustic Cavitation, Rendiconti della Società Italiana di Fisica, Varenna, Luglio.

MULTI-ELECTRICAL EXCITATION OF A TRANSDUCER FOR ULTRASONIC IMAGING

Jian-yu Lu, Randy Kinnick, James F. Greenleaf, and
Chandra M. Sehgal

Biodynamics Research Unit, Department of Physiology and
Biophysics, Mayo Clinic/Foundation, Rochester, MN 55905

INTRODUCTION

Axial resolution of a transducer is a very important image quality criterion in pulse-echo imaging. It can be improved with many methods, such as using front and back matching material,[1] tapered bar shape design,[2] nonuniform distributions of piezoelectric coefficient,[3] PZT ceramic/polymer composite material,[4] as well as matched active[5,6] or passive filters.[7,8]

In this paper, a simple electrical method, multi-pulse electrical excitation (MPEE) is studied for the improvement of the axial resolution of transducers.

It is well known that a broadband transducer will usually generate a short time duration signal when it is hit by a sharp electrical unipolar pulse. Hence, high axial resolution can be obtained when such a transducer is used in pulse-echo imaging. This paper improves the performance of transducers to make them have a broader effective bandwidth.

In what follows, a theoretical model that applies linear system theory is developed to explain the operation of MPEE. We will then describe the experimental system and provide imaging and nonimaging results to support the theoretical analysis.

THEORETICAL PRELIMINARIES

Let us consider a linear transducer in the pulse-echo mode. If such a system is assumed to be linear, its transfer function is determined by,

$$H(\omega) = T(\omega) \cdot I(\omega) \cdot R(\omega),\tag{1}$$

where $T(\omega)$ and $R(\omega)$ are the transmitting and the receiving transfer function of the transducer, and $I(\omega)$ represents the characteristics of the reflector and the acoustic pathway between the transducer and the reflector.

Double pulse excitation of a transducer can be represented as,

$$S(t) = s_1(t) + ks_1(t-w), \tag{2}$$

where $s_1(t)$ and $s_1(t-w)$ represent the two exciting pulses, separated by the time interval, w, and k is the ratio of the height of the second pulse to that of the first pulse. Taking the Fourier transform of Eq. (2), one obtains

$$\mathscr{F}[S(t)] = \mathscr{F}[s_1(t)] + k\,\mathscr{F}[s_1(t)]\,e^{-j\omega w} \tag{3}$$

or,

$$\mathscr{F}[S(t)] = \mathscr{F}[s_1(t)]\,(1+k\,e^{-j\omega w}), \tag{4}$$

or,

$$|\mathscr{F}[S(t)]| = |\mathscr{F}[s_1(t)]| \cdot \sqrt{(1 + k^2) + 2\,k\,\cos\omega w}. \tag{5}$$

The modulus of the Fourier transform of the system output is given by

$$|\mathscr{F}[o(t)]| = |\mathscr{F}[s_1(t)]| \cdot [\,|H(\omega)| \cdot \sqrt{(1 + k^2) + 2\,k\,\cos\omega w}], \tag{6}$$

where $\mathscr{F}[o(t)]$ is the Fourier transform of the output of the linear system.

If $s_1(t)$ is a sharp electrical pulse, its spectrum is flat within the frequency range of interest, giving

$$|\mathscr{F}[o(t)]| \approx C \cdot [\,|H(\omega)| \cdot \sqrt{(1 + k^2) + 2\,k\,\cos\omega w}], \tag{7}$$

where C is a constant.

According to Eq. (7) the spectrum of the output of the transducer is modified by the weighting function. $\sqrt{(1+k^2) + 2\,k\,\cos\omega w}$. The influence of this function on transfer function is shown in Fig. 1. If w = T/2 (where T is the period corresponding to the center frequency of the system transfer function $H(\omega)$), the peak or the valley of the weighting function will overlap the center frequency portion of $H(\omega)$. The nature of the modification depends on the value of k. When k = 0, i.e., the amplitude of the second pulse is zero, the weighting function is a constant, and the output spectrum is equal to the system transfer function, which corresponds to the case of single pulse excitation. If k = 1.0, it means that the two pulses have the same amplitude, and the output spectrum exhibits a minimum at the center frequency of $H(\omega)$. If k<0, the weighting function has a peak at w = T/2 and the bandwidth of the spectrum of the output will be narrower than that of $H(\omega)$. For the values of k>0, the weighting function resembles, to some degree, the inverse of a "typical" $H(\omega)$ (see Fig. 2). It is, therefore, possible to maximize the bandwidth of the spectrum of the output by adjusting the parameter k. If k >1.0, one can see from Fig. 1 that the modification of the transfer function will be weaker, therefore, the value of k should be within the range 0 to 1, in order that the best performance of the pulse echo system will be achieved. It is noted that if input signal $s_1(t)$ represents double pulses, the theory developed above can be extended to multipulse case in which the Fourier transform of $s_1(t)$ in Eq. (5), $\mathscr{F}[s_1(t)]$, is substituted by a function of the form similar to Eq. (5) itself, and the spectrum of the output of the transducer will be modified by the weighting function of higher power.

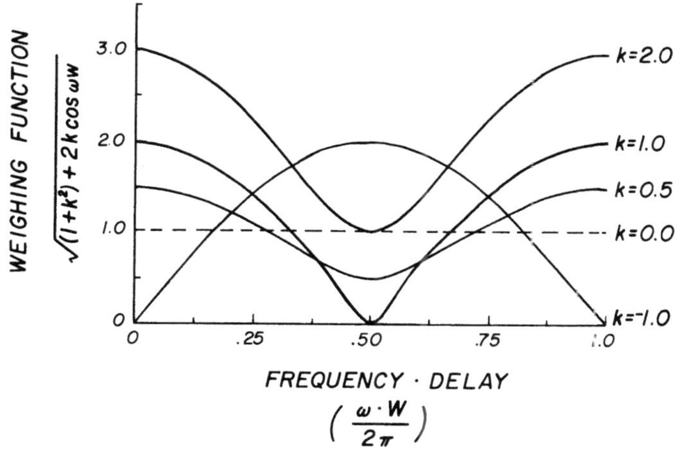

Fig. 1. Plot of the weighing function.

EXPERIMENTAL SYSTEM

Experiments were conducted with air backed and commercial (Panametric) PZT transducers. The latter was provided with matched back and front layers. The diameters for these two transducers were 17.0 mm and 12.7 mm, respectively.

The block diagram of the experimental setup is shown in Figure 3(a). Sharp electrical pulses with desired amplitudes and time delay were generated by the Polynomial Waveform Synthesizer (Data 2020). The pulses then were amplified by an ENI rf power amplifier (Model 350P and Model A300-40P) and used to excite the transducer through an impedance matching circuit.

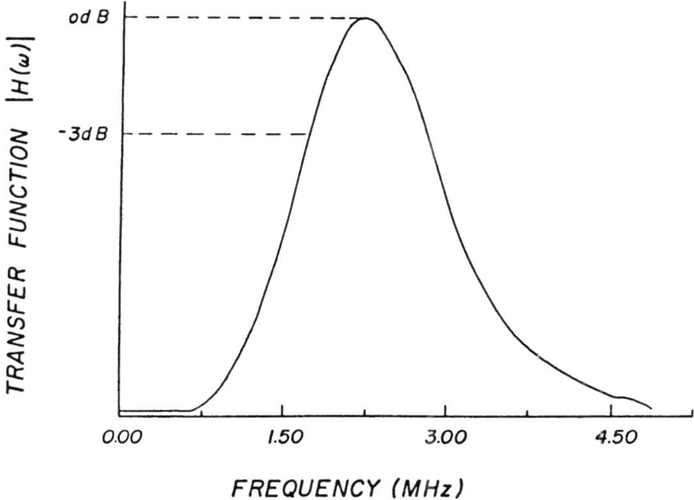

Fig. 2. "Typical" transfer function H(ω) of a commercial PZT transducer.

The reflected signal was received through T/R switch, amplified and digitized at the sampling rate of 50 MHz by a Universal Waveform Analyzer (Data 6000, Model 620). The digital data were transferred via an IEEE-488 interface to a personal computer where it was analyzed by a Scientific System (ASYST 2.0) software. The whole system was synchronized to the 1 KHz pulse sequence generated by the EXACT AM/FM Function generator (Model 7260).

Figure 3(b) shows the water tank and the scanning system. The transducers were scanned along a horizontal axis by a stepping motor. At each position, the transducer was excited and the reflected signal was digitized after a fixed time delay. On completion of the scan, the backscatter data were used to construct B-scan images. To avoid artifacts due to multiple reflection the inside walls of the water tank were lined with absorbing material. In the next section are described results of two experiments: 1) when the object in Fig. 2(b) is a simple reflector and the transducer is at a fixed position, and 2) when the reflector is a wedge phantom.

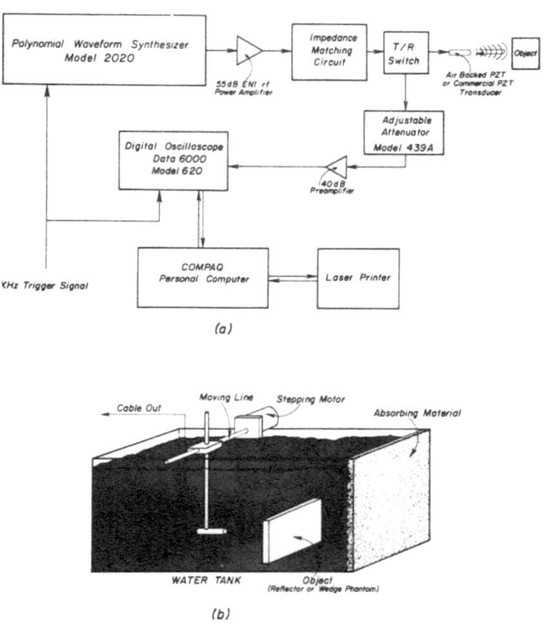

Fig. 3.   Diagram of the experimental system: (a) block diagram, (b) water tank and scanning system.

RESULTS

<u>Bandwidth Measurements</u>

Figures 3 and 4 show the experimental results of the double pulse excitation for an air backed and the commercial PZT transducers, respectively. The center frequencies of the transducers are 2.65 MHz and 2.25 MHz, and the corresponding

514

half periods are 187 ns and 222 ns, respectively. The data were collected by reflecting sound energy from a flat reflector. These figures show the variation of the 3 dB bandwidth of the output signals as a function of w and k.

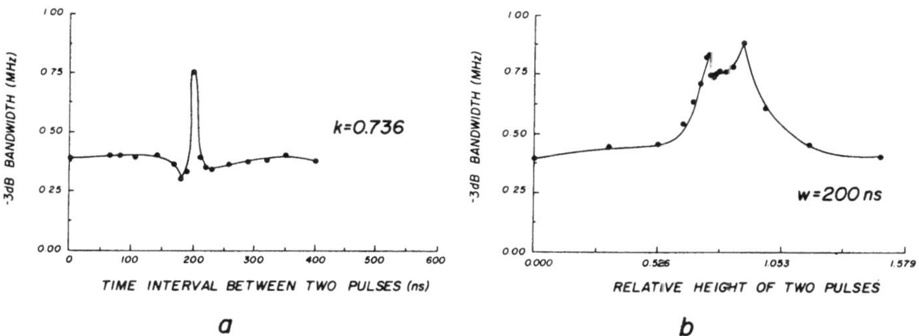

Fig. 4.   Bandwidth (B) changes with w and k for the air backed PZT transducer. (a) B-w curve and (b) B-k curve.

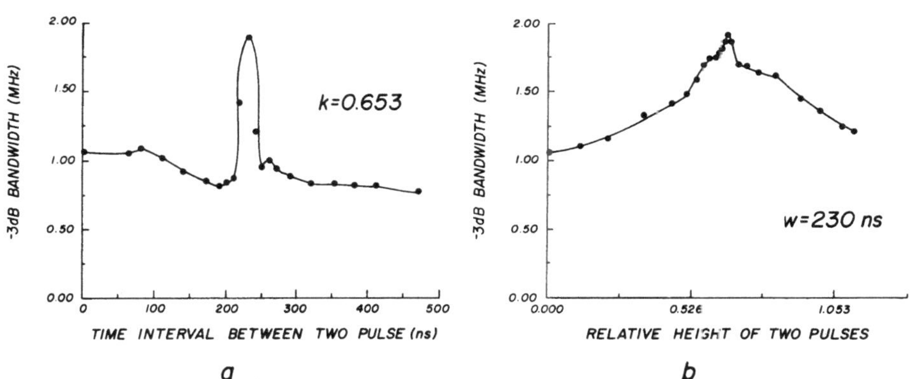

Fig. 5.   Bandwidth (B) changes with w and k for the commercial PZT transducer. (a) B-w curve and (b) B-k curve.

From Figs. 4 and 5, one observes that the broadest bandwidth of the output signal is obtained for the values of w of 200 ns and 230 ns for the air backed and the commercial PZT transducers, respectively. This is consistent with the theoretical prediction that the value of w should be around the half period corresponding to the center frequency of the system transfer function.

Imaging Evaluation of MPEE

The object used for imaging was a 150 mm long wedge phantom with an adjustable gap (see Fig. 6). The phantom consisted of two plexiglas plates and was placed at 150 mm from the transducer. It was scanned in 32 steps, 5 mm apart. All the imaging results below were obtained at a sampling rate of 100 MHz. The image size is 1024 x 32 (1024 points in longitudinal axis (time axis) and 32 points in scanning axis).

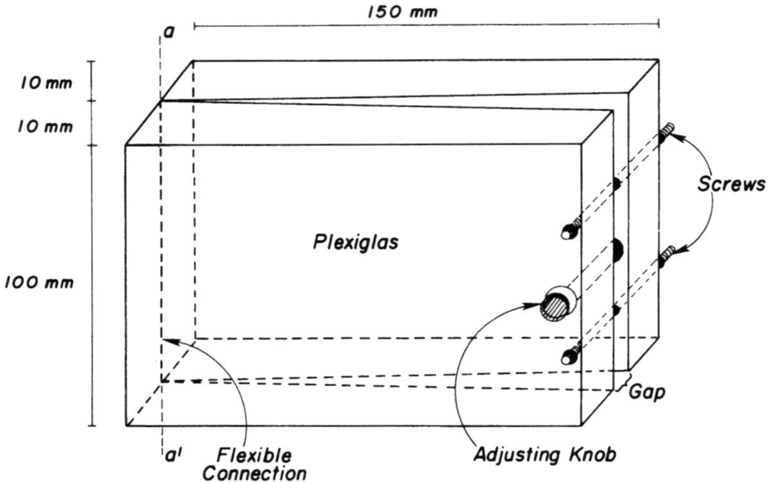

Fig. 6. Wedge phantom used in our imaging experiment.

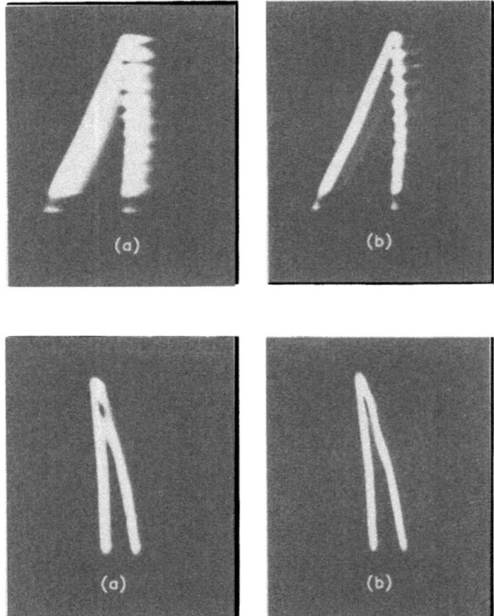

Fig. 7.   Photographic displays of the images of the wedge phantom.   (a) and (b) correspond to the images obtained by the single and double pulse excitation, respectively, and the images in the first row (gap = 3.30 mm) and the second row (gap 1.27 mm) are produced by the air backed and the commercial PZT transducers, respectively.

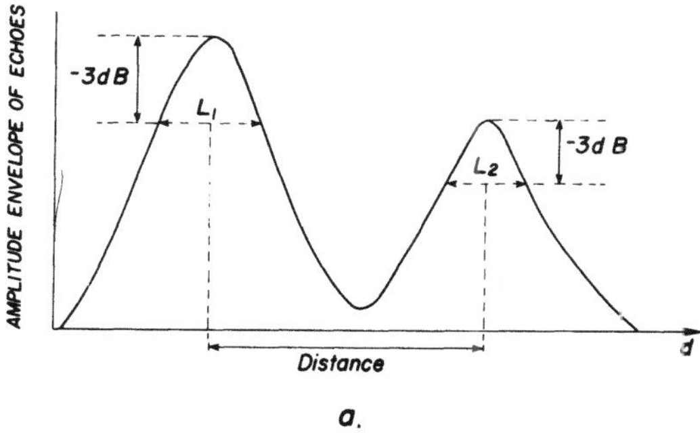

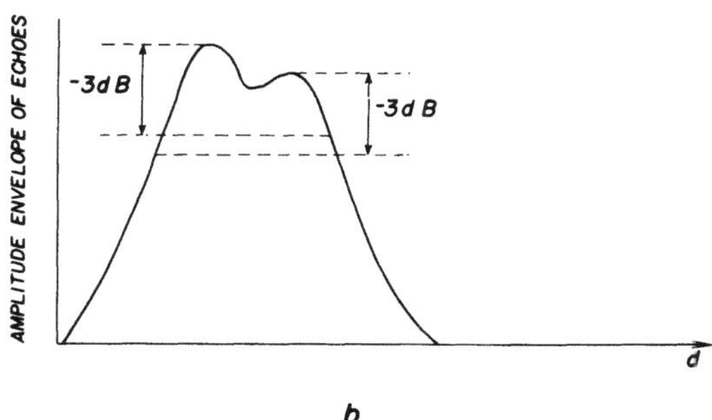

Fig. 8.   Determination of distances from echoes.   (a) distance is measurable, (b) distance is not measurable. The difference between (a) and (b) is the presence or absence of a valley between the -3 dB levels of the two peaks.

Figure 7 is the photographic display of the pulse-echo image of the wedge phantom. The first and second rows of the images in Fig. 7 were constructed by the air backed and the commercial PZT transducers, respectively, and the corresponding gaps of the wedge phantom used were 3.30 mm and 1.27, respectively. It is seen from these images that the MPEE improves the axial resolution significantly.

Figure 8 shows the method for the distances measured from the images. The distances were determined by first converting the output signals of the transducer into analytic envelope signals,[9] then measuring the distances between the centers of the -3 dB widths of two echoes which correspond to the reflections from two inner surfaces of the wedge phantom (see Fig. 8(a)). If the -3 dB width of the echo was not measurable, the distances were simply set to zero (see Fig. 8(b)).

Figure 9 is the comparison of the real distances of the inner walls of the wedge phantom and the distances measured from the images. If the resolution of the images in Fig. 7 is defined as the minimum real distances at which the measured distances are not zero, one can observe from Fig. 9(a) that the axial resolutions of the air backed PZT transducer are 0.99 mm and 0.44 mm with the single and double pulse excitation, respectively. Figure 9(b) shows that the axial resolutions of the commercial PZT transducer are 0.42 mm and 0.17 mm with the single and double pulse excitation, respectively.

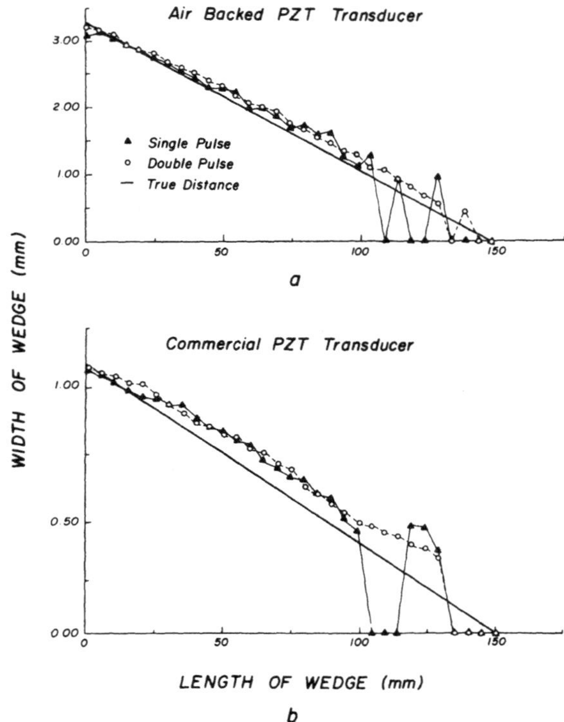

Fig. 9.   Comparison of the real distances and the distances measured from the images. (a) Results of the air backed PZT transducer, (b) results of the commercial PZT transducer. Δ - single pulse and o - double pulses.

CONCLUSION

From the experimental results, Figs. 3 and 4, one can see that the bandwidth of the echo signal is greatly increased by the MPEE with the adjustment of the time interval between the two electrical excitation pulses and the ratio of the heights of the two pulses. This broadens the bandwidth of the effective transfer functions of the transducers (see Eq. (7)) and improves greatly the axial resolution of the transducers in pulse echo imaging. The results of the imaging experiment, Figs. 7 and 9, have confirmed such improvement.

It is worth noting that the MPEE is more important to the commercial PZT transducer. This is because such transducers are usually optimized by using complex backing and front matching techniques and is difficult to be improved further by conventional methods. In addition, applying the MPEE to the commercial PZT transducer does not increase low level noise[10] caused by the long ring-down tail of some transducers (the ring-down tail of the commercial PZT transducer is negligible).

ACKNOWLEDGMENTS

The authors wish to thank Elaine Quarve for secretarial assistance and Christine Welch for graphic assistance. This work was supported in part by grant 43920 from the National Institutes of Health. We also acknowledge the support of Mr. W. Harrison at Honeywell Ceramics Center for transducer design.

REFERENCES

1. G. Kossoff, The effects of backing and matching on the performance of piezoelectric ceramic transducer, IEEE Trans. Sonics Ultrason. SU-13(1):20-31, 1966.
2. P. G. Barthe and P. J. Benkeser, A staircase model of tapered piezoelectric transducers," 1987 IEEE Ultrasonics Symposium Proceedings, pp. 697-700.
3. F. Chapeau-Blondeau and J. F. Greenleaf, A theoretical model of acousto-electric transducer with a nonuniform distribution of piezoelectric coefficient. Application to transducer optimization, J. Acoust. Soc. Am. (In Press).
4. T. R. Gururaja, W. A. Schulze, I. E. Cross, and R. E. Newnham, Piezoelectric composite materials for ultrasonic transducer applications. Part II: Evaluation of ultrasonic medical applications, IEEE Trans. Sonics Ultrason. SU-32:499-513, 1985.
5. B. Mandersson and G. Salomonsson, Weighted least-squares pulse-shaping filters with application to ultrasonic signals, IEEE Trans. Ultrason. Ferroelect. Frequency Control 36(1):109-113, 1989.
6. G. Hayward and J. E. Lewis, A theoretical approach for inverse filter design in ultrasonic applications, IEEE Trans. Ultrason. Ferroelect. Frequency Control 36(3):356-364, 1989.
7. G. A. Hjellen, J. Andersen, and R. A. Sigelmann, Computer-aided design of ultrasonic transducer broadband matching networks, IEEE Trans. Sonics Ultrason. SU-21:302-305, 1974.
8. A. Yamada, On-line deconvolution for the high resolution ultrasonic pulse-echo measurement with narrow-band transducer, 1987 IEEE Ultrasonics Symposium Proceedings, pp. 1027-1030, 1987.
9. A. V. Oppenheim and R. W. Schafer, in: *Digital Signal Processing*. Prentice-Hall, Inc., Englewood Cliffs, New Jersey, Chapter 7, 1975.
10. D. H. Howry, Techniques used in ultrasonic visualization of soft tissue structures of the body, IRE Convention Becoko 3:75-88, June 9, 1955.

ULTRASOUND TRANSDUCER CHARACTERIZATION
USING ANGULAR SPECTRUM BACKPROPAGATION

Mark E. Schafer and Peter A. Lewin[*]

International Sonic Technologies, 101 Gibraltar Rd. Horsham, PA 19044
*Department of Electrical and Computer Engineering, Biomedical
Engineering and Science Institute, Drexel University, Philadelphia, PA 19104

ABSTRACT

This paper presents a measurement method for analyzing the surface velocity patterns of
ultrasonic transmitters, based on the angular spectrum backpropagation method of wavefield
analysis. With this approach, acoustic propagation between parallel planes is modelled by
taking the two-dimensional Fourier transform of the wavefield, multiplying each element in
the spatial frequency domain by the appropriate phase factor, and inverse transforming the
resultant angular spectrum. The basic method was expanded from the monochromatic case to
the wideband pulsed case, which is more typical of diagnostic instrumentation. An
experimental system was designed built to measure the acoustic fields from various
transducers, including single element and multi-element phased arrays. Results are given for
circular planar, circular focussed, and rectangular phase steered transducers. The results
demonstrate the method's ability to reconstruct the surface velocity distributions of complex
ultrasonic radiators.

1. INTRODUCTION

Designers of ultrasound transducers continually seek better methods for directly examining
the effects of transducer construction and assembly on imaging performance. None of the
currently used techniques provide sufficient information on the actual transducer vibration
patterns, which directly affect the transmitted ultrasound beam shape. The purpose of this
work was the development of techniques which could provide this type of vibration pattern
information, based on acoustic measurements of radiated wavefields.

The techniques use the angular spectrum, or Fourier decomposition method[1], one of the
more powerful wavefield propagation techniques. With this method, the monochromatic

pressure distribution measured over a planar surface is decomposed into an equivalent two-dimensional spectrum of plane waves. Wavefield propagation is modeled by applying the appropriate phase factor to each angular spectral component. The method encompasses both forward propagation, from a source plane to a receiver plane, and backpropagation, from a receiver plane to the source. The method has been used to to image scattering objects[2], analyze transducer performance[3], and to examine transducer beam patterns[4]. The basic derivation of the angular spectrum method is described in the next Section for completeness.

This paper presents the experimental work which examined the backpropagated fields from different acoustic sources, including relatively simple single element circular planar transducers and more complicated multi-element phased arrays. Section 3 describes the dedicated experimental system which was developed, along with brief discussions of the acoustic sources and receivers used. The completed experimental system had a sampling rate up to 25MHz, a dynamic range greater than 60dB, and a positional accuracy better than $2\mu m$.

The experimental system was used to examine the reconstructed surface velocity patterns of the following acoustic sources: flat unfocussed single element, spherically focussed single element, modified spherically focussed (with a slot in the face) single element, and phased steered multiple element. Both the continuous wave (CW) and pulsed excitation cases were examined. Section 4 presents selected examples from the measurements, including both single element and phased array transducers.

This work was part of a larger study to examine the effects of propagation through inhomogeneous media using the angular spectrum method[5,6]. The eventual goal is to predict the degradation of transducer performance caused by propagation through tissue. The work described in the present paper was restricted to the measurement of transducers in a homogeneous medium (water).

2. THE ANGULAR SPECTRUM METHOD

The angular spectrum or plane-wave decomposition technique was originally developed in the study of optical diffraction[1]. The brief derivation which follows is similar to that given in a number of references[1,3,7], although the notation has been changed slightly.

Consider the complex pressure wavefield generated by a set of monochromatic sources. Let $u(x,y,z_0)$ represent this wavefield on a plane parallel to the X-Y plane and passing through the point z_0. The two-dimensional Fourier transform of this complex field is denoted $U(\omega_x,\omega_y,z_0)$. The goal is to determine the pressure distribution at a parallel plane passing through z_1. The Fourier transform decomposes the field into a two-dimensional spectrum of plane waves: each (ω_x,ω_y) spatial frequency represents a plane wave travelling with direction cosines $(n_x = \omega_x/k, n_y = \omega_y/k)$. Propagation is then modelled by multiplying each (ω_x,ω_y) component by the appropriate phase propagation factor which account for plane wave travel in the (n_x,n_y) direction. This phase change can be expressed as[3]

522

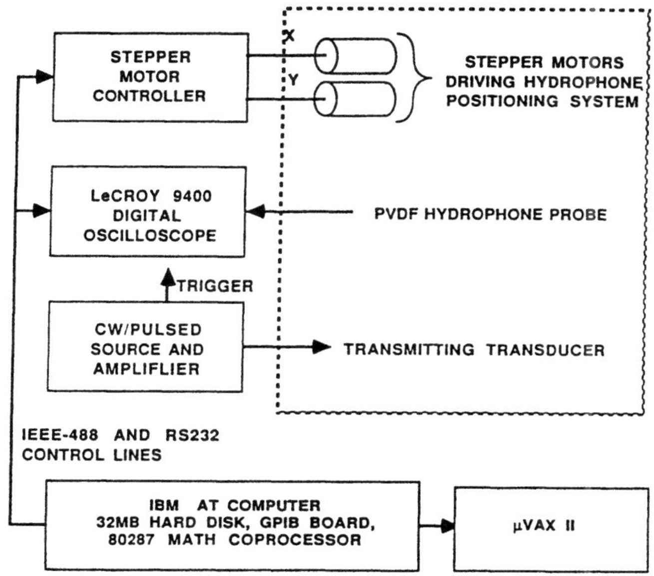

Figure 1. Block diagram of the experimental system. For the phased array experiments, the Varian V3400 unit replaced the source/amplifier as shown.

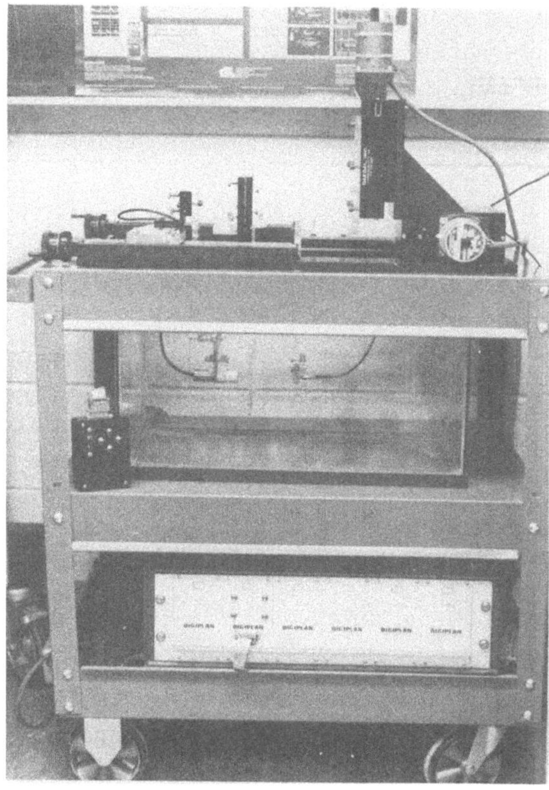

Figure 2. Photograph of the scanning tank. The stepper motor driven stages are to the left, and the manual controls are to the right; the motor controller system is at the bottom. The unit was built on a wheeled cart for increased mobility.

$$G(\omega_x,\omega_y,z_0,z_1) = \exp[j(z_1 - z_0)(k^2 - \omega_x^2 - \omega_y^2)^{\frac{1}{2}}] \equiv \exp[\Phi(\omega_x,\omega_y,z_0,z_1)] \qquad (1)$$

where $G(\omega_x,\omega_y,z_0,z_1)$ is the propagation factor introducing a phase change $\Phi(\omega_x,\omega_y,z_0,z_1)$ to account for the propagation from z_0 to z_1. If each term in the angular spectrum of the field at z_0 is multiplied by the appropriate phase factor Φ, the complex pressure field at the plane z_1 can be found by inverse Fourier transforming the resultant set of two-dimensional components. Since there are no restrictions on the relative position of z_0 and z_1 , the technique can be used for both propagation away from a source plane distribution (forward propagation), or backward from a receiver plane at z_1 towards z_0 (backpropagation). In practice, the restriction to planar distributions does not significantly reduce the technique's utility, because many practical ultrasonic sources can be considered planar.

An important feature of the angular spectrum method is the treatment of propagating and non-propagating (evanescent) waves. Note that if k^2 is greater than $(\omega_x^2+\omega_y^2)$ in Eq. (1), the phase factor Φ is purely imaginary, thus resulting in harmonic wave propagation. If k^2 is less than $(\omega_x^2+\omega_y^2)$, the wave is strongly attenuated in the +Z direction, because Φ becomes a negative real quantity. For backpropagation more than one or two wavelengths, the resulting strong positive exponential term amplifies any noise in the measured signal. As a remedy to this problem, the evanescent wave components are set to zero for backpropagation. Because the (spatial) cutoff frequency for evanescent wave components corresponds to one wavelength, neglecting these spectral components limits the method's resolution to source details larger than a wavelength[8]. Additional features of the angular spectrum technique were thoroughly reviewed by Powers[7] and by Williams and Maynard[8].

3. EXPERIMENTAL METHODS

Figure 1 shows the overall experimental arrangement. The pressure waveforms transmitted by the acoustic sources were sensed by a PDVF needle-type hydrophone, digitized by a LeCroy 9400 oscilloscope, and stored in an IBM-PC/AT compatible microcomputer for subsequent transfer to the μVAX II. The PC/AT also controlled the hydrophone positioning system and the oscilloscope, using an IEEE-488 bus interface, and the phased array scanner, using an RS-232 interface. The design criteria for such precise scanning systems were recently discussed by Schafer and Lewin[9]. Although similar to the system described in that paper, the present scanning tank had to meet additional criteria with regard to a) hydrophone scanning geometry relative to the transmitter; b) positional accuracy and repeatability; and c) angular alignment of the source and receiver planes.

Figure 2 shows the completed scanning tank system, with the hydrophone positioning system on the right, the acoustic source holder and manual positioning adjustments on the left, and the position control electronics (stepper motor control) on the bottom rack. The two motorized positioning stages used to scan the hydrophone had a travel range of 10 cm, a step increment of 1 μm, and a bidirectional repeatability of 1.2 μm. The acoustic sources were

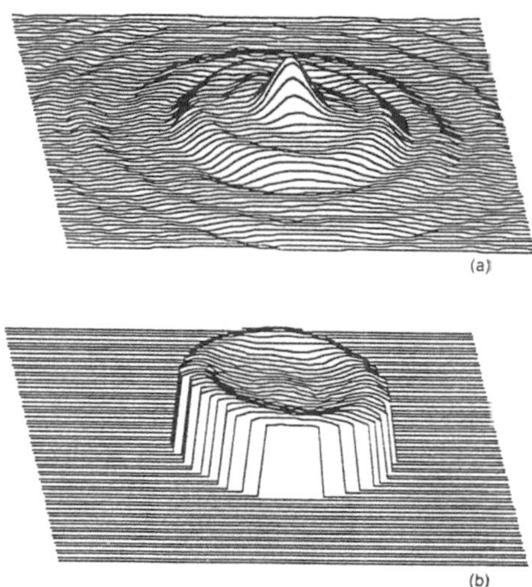

Figure 3. Backpropagation results using a 13mm diameter, 1-MHz focussed transducer, operated in CW mode. (a) Magnitude; (b) Phase.

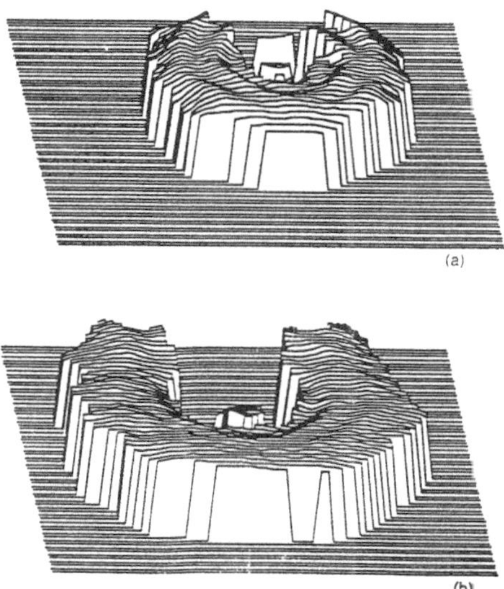

Figure 4. Backpropagation phase results using a 13mm diameter focussed biopsy transducer, at 2MHz (a) and 3MHz (b).

mounted on a manually driven lead screw slider with a readout resolution of 10 μm and a positional repeatability of 20 μm. The source holder incorporated a two-axis precision gymbal mount adjustment system which allowed an angular variation of ±10°; this was used to align the source with the hydrophone scanning plane. Specific alignment routines wre developed to insure the perpendicularity of the source acoustic axis with the scanning plane. The alignmnet accuracy was estimated to be better than 0.1° for focussed transducers and 0.2° for unfocussed transducers. The difference was because planar transducers have broader beam patterns and were therfore more difficult to align. The alignment precision of the phased array transducer was estimated at 1.0°.

Several single element transducers were used in the present study. These included both flat-faced and focussed types manufactured by a commercial firm (KB-Aerotech). In addition, one transducer with a slot in the face from the center to the edge was used to check the resolving power of the angular spectrum implementation. A programmable tone burst signal generator (Hewlett-Packard 8116A), amplified by an RF power amplifier (RF Labs 30-12AC), was used to excite the transducer either in continuous wave (CW) mode at 10V peak-to-peak, or with one or two cycles at 100V peak-to-peak. The other type of acoustic source examined was a phased multi-element array using a commercial scanner (Varian V3400 Ultrasonograph) with two 32-element rectangular transducers (2.25 and 3.5MHz). With this type of system, each element in the transducer array is electrically and acoustically isolated; the elements are driven in specific delay patterns in order to steer and focus the acoustic beam. This permits beam scanning without the need for mechanical movement. A number of specific hardware modifications were developed including an RS-232 port for downloading the desired firing angle, and external connections to internal signal lines for triggering and synchronization.

The measurement system was operated by the PC/AT microcomputer, using specific control and data acquisition programs written in Asyst™. Asyst is a control and analysis software package which has built-in drivers for IEEE-488 and RS-232 control. Specific computer routines were developed for hydrophone scanning, automated oscilloscope gain adjustment, data aquisition, and data storage. The entire data-taking procedure was fully automated, with user input only at the beginning of a measurement run. Unattended operation was critical, since for a 64x64 point array, the largest used in the present work, the pulsed mode measurements took approximately 9 hours to complete; the CW tests took approximately 2 hours. Once the data was taken, it was transferred to a μVAX II computer over a serial port line. A set of FORTRAN programs implemented the angular spectrum technique and the graphics display.

4. EXPERIMENTAL RESULTS AND DISCUSSION

This section presents selected results for single element transducers operating in CW and pulsed mode, and phased array transducers in both the unphased (beam direction normal to the transducer surface) and phase steered modes.

All the backpropagation experiments were conducted with the measurement plane relatively close (<10mm) from the transducer surface. This was done to keep the backpropagation

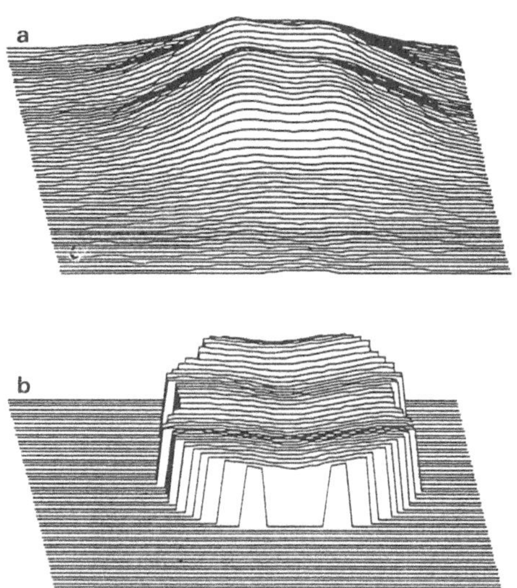

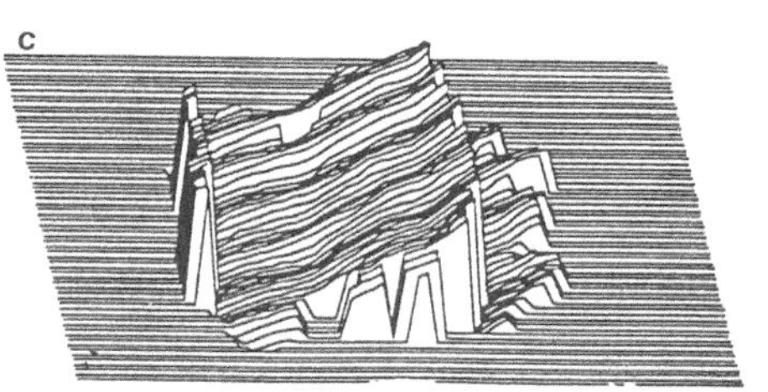

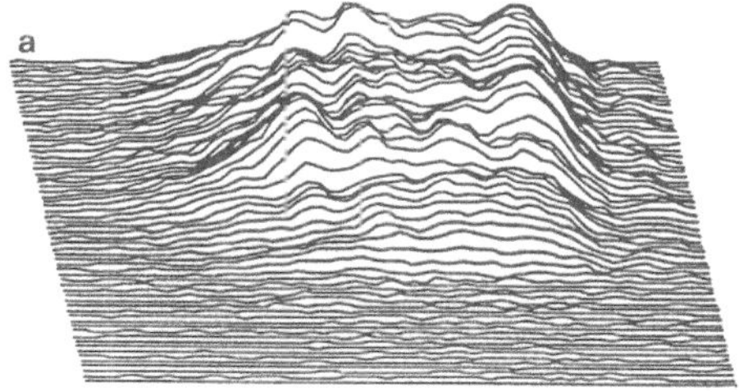

Figure 5. Backpropagation results using 3.5MHz Varian phased array transducer. (a) Magnitude; (b) Phase at 0° phase steering; (c) Phase at 5° phase steering. Shown is the 1.95-MHz frequency bin.

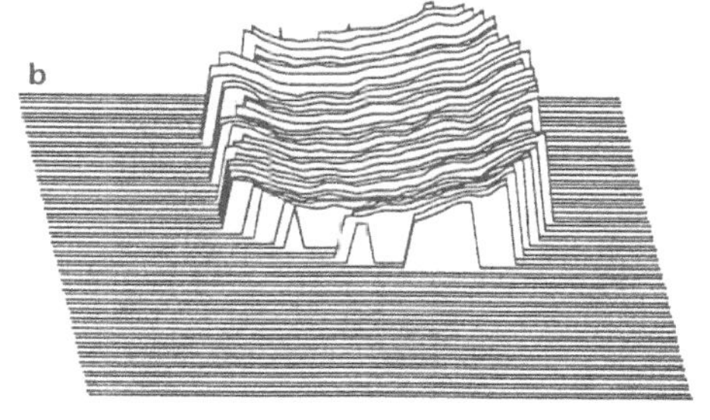

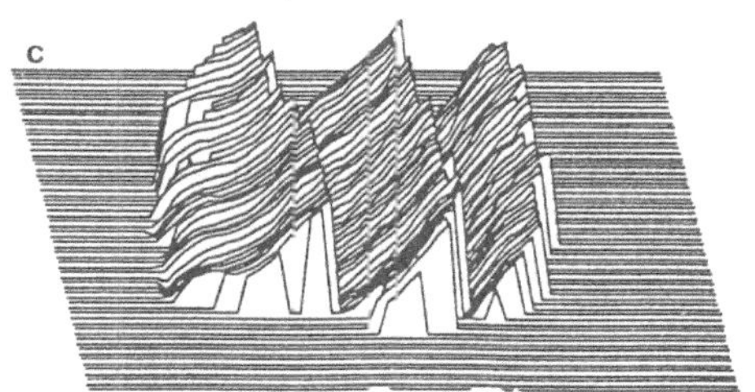

Figure 6. Backpropagation results using 3.5MHz Varian phased array transducer. (a) Magnitude; (b) Phase at 0° phase steering; (c) Phase at 5° phase steering. Shown is the 3.71-MHz frequency bin.

distance short, so as to minimize the influence of replicated source errors[8], and so that the finite measurement aperture would encompass the entire beam.

Figure 3 shows the backpropagation results obtained using a 13-mm-diameter, 1-MHz focused transducer, operated in CW mode. There were 64x64 datapoints taken, at an axial distance of 10-mm from the transducer's surface and at a spacing of 0.372-mm (quarter wavelength). The data was zeropadded to 128x128 points during backpropagation to remove the effects of circular convolution[8]. All phase plots were enhanced using a thresholding algorithm which compared the magnitude of the wavefield at a given location with a preset threshold value (usually set at 25 percent of the peak magnitude). The phase was set to zero if the magnitude value was less than the threshold, thus eliminating the phase noise in regions where there was little true signal. All plots were normalized to the peak value of the data.

The next transducer studied was a biopsy transducer with a slot in the surface. The transducer was first excited at 2MHz (Figure 4a), and scanned at 0.372mm ($\lambda/2$) spacing. In order to determine the resolution of the reconstruction algorithm, the same transducer was tested again using 3MHz excitation, maintaining the half wavelength sampling interval (Figure 4b). The slot is clearly seen on the side of the transducer away from the viewer. The proportions of the slot are slightly distorted: the width of the slot in the phase plots is wider than the actual slot, because the edges of the slot can only be resolved to within a wavelength. Further, the measurements were influenced by the finite size of the measurement aperture, and by spatial averaging effects caused by the 1.0mm hydrophone size. Although increasing the frequency from 2MHz to 3MHz does improve the image, the hydrophone aperture effects limit the degree of resolution improvement possible.

Next, the surface velocity patterns of the phased array transducers were reconstructed to test whether the time (phase) delay pattern across the elements could be detected. The first Varian experiment involved the 3.5MHz transducer. The field was measured at 64x64 points, with a sampling interval of 0.213mm, 5mm from the transducer, to keep the beam completely within the scanning area. The Varian 3400 scanner was set to fire the phased array transducer in "straight ahead" (0° scan angle) direction; the test was then repeated with a 5 degree steering angle. The fields were backpropagated with zeropadding to 128x128 points. The results are shown in Figures 5 and 6, for frequencies 1.95MHz and 3.71MHz. The figures compare only the reconstructed phase patterns for the 0 degree and 5 degree steering conditions, since the magnitude plots were very similar.

The final backpropagation experiment used the 2.25MHz phased array transducer, since its bandwidth and spatial sampling requirements were not as severe as the 3.5MHz transducer. The Varian scanner was set to phase steer the transducer to an angle of 10.5 degrees. The measurement and backpropagation conditions were similar to those used above, except that the sampling interval was 331µm, which is $\lambda/2$ at 2.25MHz, and the axial distance to the transducer was 4mm. Instead of analyzing the data in terms of individual frequency bins, the entire reconstructed image was transformed back into the time domain, and the velocity-time waveforms at selected locations on the transducer surface were reconstructed. The locations selected corresponded to the center of the transducer, and two points perpendicular to the scan direction. The results are shown in Figure 7; the time histories are similar in shape and

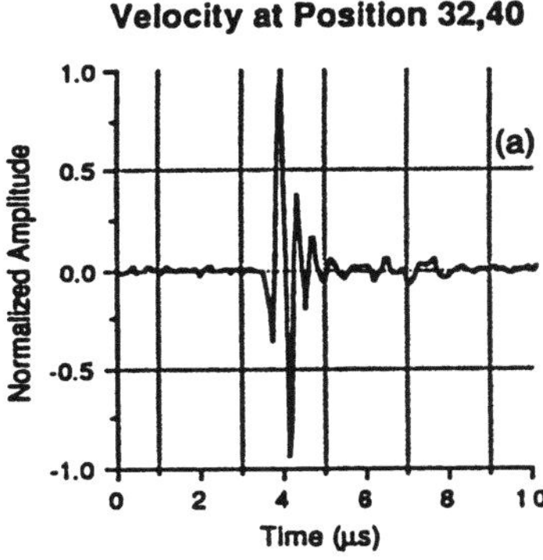

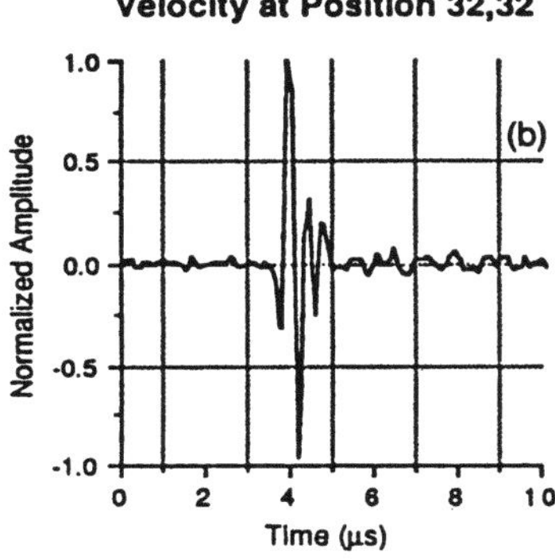

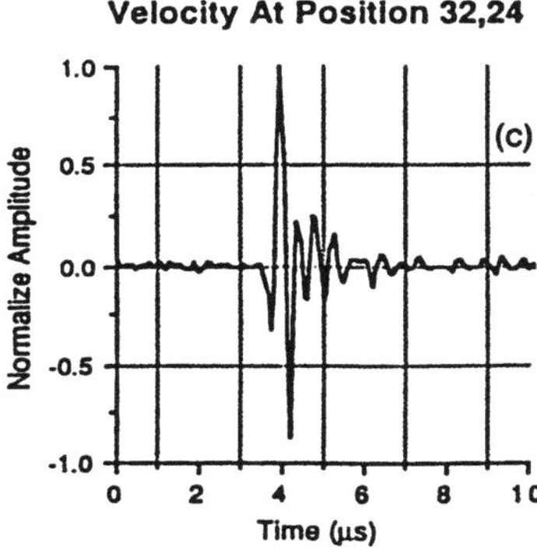

Figure 7.Backpropagation results using a 2.25-MHz Varian phased array transducer. Surface velocity waveforms at three positions along the array, perpendicular to the scan plane. Waveform peaks are time aligned.

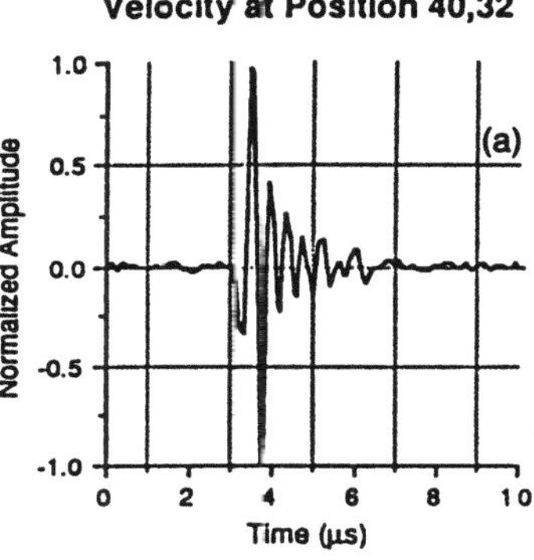

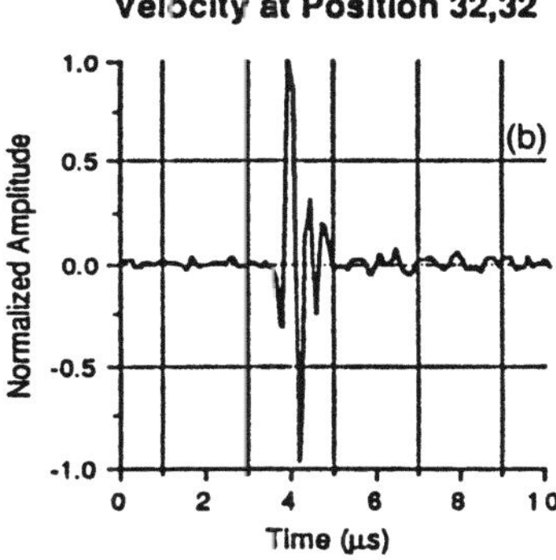

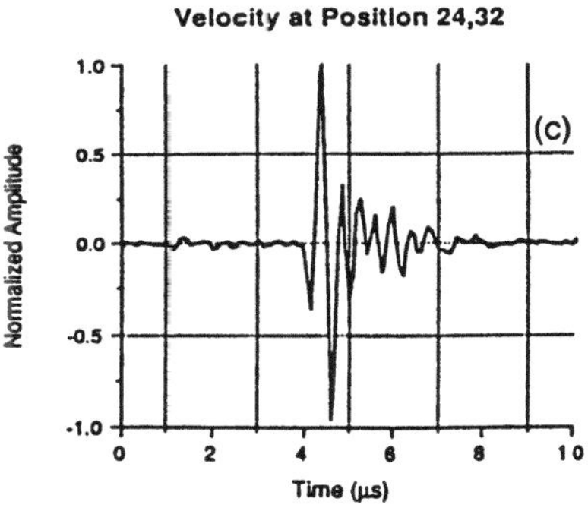

Figure 8.Backpropagation results using a 2.25-MHz Varian phased array transducer. Surface velocity waveforms at three positions along the array, parallel to the scan plane. Waveform peaks are time shifted.

are in time alignment. Two additional locations, in line with the center point but taken along the scan axis, are shown in Figure 8. In this figure, the waveforms show a distinct time shift, indicative of the time delay pattern used to steer the acoustic beam.

The backpropagation reconstruction tests shown in Figures 5 through 7 for the two phased array transducers show an excellent correlation of the phase distribution with the transducer steering angle. The phase plots show the general shape of the rectangular aperture; the phase plots also show clearly the phase weighting used to steer the beam. The jagged appearance of the phase plots is due to phase wrap-around at 2π. As the frequency increases, the phase variation becomes more rapid (Figures 5 and 6). This is to be expected, since the transducer is excited with a uniform time delay from element to element, and this time delay causes an increasing phase delay with frequency. The time delay pattern is evident in Figures 7 and 8.

5. CONCLUSIONS

This work has demonstrated the applicability of the angular spectrum approach to analyzing ultrasonic sources. These measurement techniques, using angular spectrum backpropagation, yield acoustic source velocity information with excellent detail. There were, of course, limits on the reconstruction because the evanescent wave components were neglected, and because of the finite hydrophone size (1.0mm), scanning aperture (64x64 points), and time sampling (128 points). These problems are being addressed through the use of smaller hydrophones, and the use of larger computing facilities.

ACKNOWLEDGEMENTS

This work was supported by Grant ECE-8504602 from the National Science Foundation.

REFERENCES

1. J.W. Goodman, <u>Introduction to Fourier Optics</u> , (McGraw-Hill, New York), 1968.
2. F.P. Higgins, S. J. Norton, and M. Linzer, "Optical Interferometric Visualization and Computerized Reconstruction of Ultrasonic Fields," JASA, **68(4)**, pp 1169-1176, 1980.
3. P.R Stepanishen. and K.C. Benjamin, "Forward and Backward Projection of Acoustic Fields Using FFT Methods," JASA, **71(4)**, pp. 803-812, 1982.
4. M.M. Sondhi, "Reconstruction of Objects from Their Sound-Diffraction Patterns," JASA, **46(5)**, pp. 1158-1164, 1969.
5. M.E. Schafer, P.A. Lewin, and J.M. Reid, "A New Technique for Characterizing Transducers in Inhomogeneous Media," in <u>Acoustical Holography</u>, Vol. 15, H. Jones, ed. (Plenum, New York) pp. 135-146, 1987.
6. M.E. Schafer, "Transducer Characterization in Inhomogeneous Media Using the Angular Spectrum Method," Ph.D. Thesis, Drexel University, 1988.

7. J.P. Powers, "Computer Simulation of Linear Acoustic Diffraction," in <u>Acoustical Holography</u>, Vol. 7, Kesler, L.W., ed. (Plenum, New York) pp. 193-205, 1976.

8. E.G. Williams and J.D. Maynard, "Numerical Evaluation of the Rayleigh Integral for Planar Radiators Using the FFT," JASA, **72(6)**, pp. 2020-2030, 1982.

9. M.E. Schafer and P.A. Lewin, "A Computerized System for Measuring the Acoustic Output from Diagnostic Ultrasound Equipment," IEEE Trans. Ultrasonics, Ferroelectrics, and Freqency Control, **UFFC-35(2)**, pp. 102-109, 1988.

ELECTRONIC FOCUSED ACOUSTIC BEAM SCANNING USING

CHIRPED FRESNEL INTERDIGITAL TRANSDUCER

Tooru Nomura and Tsutomu Yasuda

Department of Electrical Communication
Shibaura Institute of Technology
Minato-ku, Tokyo 108, Japan

ABSTRACT

A new acoustic transducer with electronic focusing and scanning
capabilities is presented. An interdigital transducer radiates acoustic
waves in oblique directions with respect to the substrate surface when
it is placed in water. A linear chirped interdigital transducer for the
acoustic beam scanning was constructed on piezoelectric substrate. The
chirped interdigital transducers consist of interdigital electrodes
whose period and width are linearly changed. Moreover, the electrode is
divided into several sections, each of which is placed to form a one
dimensional Fresnel phase plate. In this structure, the focused acoustic beam scanning was achieved by changing the applied frequency. Experimental verification has been carried out using a chirped Fresnel
phase plate type interdigital transducer which covers the frequency from
20 to 50 MHz. A scan length of 9 mm for a focused acoustic beam about
0.3 mm in width has been obtained.

1. INTRODUCTION

Ultrasonic techniques have been used in medicine and nondestructive
evaluation to form acoustic images. Focused acoustic beams are used in
many acoustic imaging systems. A piezoelectric transducer with an
acoustic lens and a concave transducer are used to produce a focused
acoustic beam. The focused acoustic beam thus generated is scanned over
a sample to make a two-dimensional acoustic image. Most of the present
systems use mechanical scanning techniques for this purpose[1].

However, an acoustic imaging system would benefit from further
refinements. One of them is electronic scanning to increase process
speed. Electronic scanning using arrays of piezoelectric transducers
with suitable electronic circuits have been developed in various laboratories; some have undergone practical evaluation[2].

On the other hand, it is known that an interdigital transducer
(IDT) developed to excite a surface acoustic wave (SAW) radiates an
acoustic wave in oblique directions with respect to the substrate surface when it is placed in water. The wavefronts of the acoustic wave
can be arbitrarily controlled by designing the electrode pattern of the

IDT. Therefore, studies are being conducted on use of the IDT as an
acoustic wave transducer for an acoustic microscope and for nondestructive evaluation. We previously used an IDT for acoustic imaging and for
elastic characterization of materials[3]. IDTs are easily fabricated by
standard photolithographic technology for frequencies in the UHF range.
Many radiation characteristics can be achieved by this technique, and
the devices are particularly attractive for generating complicated waveforms.

In this paper, we have investigated the possibility of electronic
acoustic beam scanning and focusing using the IDT. The beam scanning is
achieved by using a chirped IDT whose electrode is divided to form a one
dimensional Fresnel phase plate (FPP) [3] .   The chirped IDT is well known
for its use in the pulse compressional SAW filter[4]. We call this new
type of acoustic transducer a chirped Fresnel interdigital transducer
(chirped Fresnel IDT).   In the chirped Fresnel IDT, the radiated acoustic beam is brought into focus, and the acoustic beam is shifted laterally by changing the applied frequency. Electronic focusing and
scanning of an acoustic beam is performed simultaneously by using the
chirped Fresnel IDT.   An experiment shows that this technique is very
useful for an acoustic imaging system of high speed scanning.

2. PRINCIPLE OF FOCUSED ACOUSTIC BEAM SCANNING

Figure 1 shows the configuration of the chirped Fresnel IDT for a
focused and scanning acoustic beam.   The coordinates are defined as
shown, and the period and width of the electrodes are gradually changed
along x direction.   The each electrode is divided to make a one dimensional Fresnel phase plate in y direction. In this way it is possible
to control the position along x direction where SAWs are generated
since the IDT is most efficient in generating acoustic waves where the
interdigital spacing is one-half wavelength. Furthermore, focusing in
y direction is achieved with the Fresnel type electrode.

If the IDT is placed in water,   the SAW excited on the piezoelectric substrate becomes a leaky SAW, and converts to an acoustic wave.

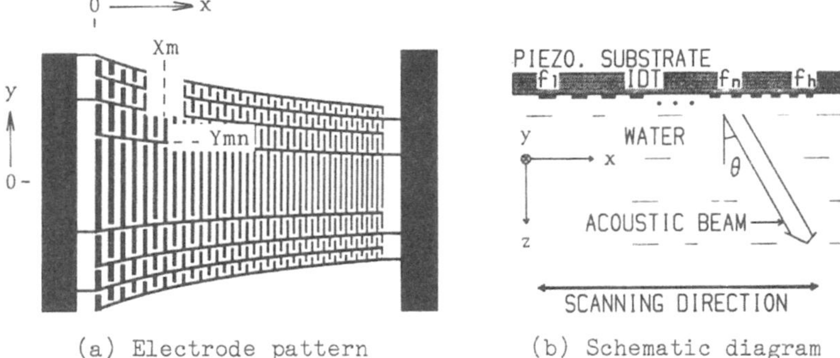

(a) Electrode pattern   (b) Schematic diagram

Fig.1 Configuration of a chirped Fresnel IDT and schematic diagram of
       acoustic beam scan.

In this case, the radiation angle θ is determined from the SAW velocity Vs of the substrate and the acoustic wave velocity Va in water, and is expressed by the equation, $\theta = \sin^{-1}(Va/Vs)$ (Fig.1(b)). The phase and shape of the acoustic beam radiated from the IDT depend upon the alignment and shape of the electrode fingers. Therefore, the chirped Fresnel IDT can radiate acoustic waves whose beam can be scanned by changing the applied frequency and scanned acoustic beam was focused by the effect of diffraction.

We first determine the edge position of the chirped electrode. In order to obtain a linear electronic scanning, consider the design of a chirped IDT with a linear group delay dispersion over a Δf frequency band centered at f_0 MHz. This IDT can also be used as a phase equalizer for pulse compression of chirped pulse with a linear chirp rate $c = df/dt$. The equation for the edge positions of the fingers is[4]

$$Xm = f_1 \, Vs \, / \, (2c)(-1+\sqrt{1+(m-1)(c/f_1^2)} \,) \qquad (1)$$

$$m=1,2,-----M$$

where f_1 is the initial frequency in the IDT design and Vs is the velocity of the surface wave. The number of fingers (M/2) is made large enough to assure that the device has the specified bandwidth.

We next determine the exact edge position of a segmented electrode required to obtain a focused beam. The phase and shape of the acoustic beam radiated from the segmented electrode depend upon the alignment and shape of the electrode fingers. Therefore, the acoustic wave radiated from each finger of the chirped Fresnel IDT is brought into focus in water by diffraction effect. The coordinate Ymn of the segment edge of the n-th zone of m-th electrode is given by the formula based on the Fresnel zone plate theory,

$$Ymn = \sqrt{(2n-1) \, F \, \lambda m \, / \, 2} \qquad (2)$$

$$n=1,2,-----N$$

where F is the focal length, and λm the wavelength of the frequency fm in water.

3. EXPERIMENTAL RESULTS

In order to realize the acoustic beam scanning proposed here, the chirped Fresnel IDT shown in Fig.1 was constructed on a 128° YX-LiNbO$_3$ substrate using a photolithographic technique. Design formula and focusing properties for 128° YX LiNbO$_3$ with water as the liquid are given as well as the construction data for a 35 MHz chirped Fresnel IDT in Table 1.

Table 1. Data for Chirped Fresnel IDT

SAW velocity Vs	3931 m/s
acoustic wave velocity Va	1500 m/s
SAW center frequency f_0	35 MHz
chirp rate c	6 MHz/μsec
number of electrodes M/2	175
frequency band width Δf	30 MHz
number of Fresnel zone N	4
focal length F	16 mm

3-1. Schlieren Observation

The acoustic beam radiated from the chirped Fresnel IDT was observed experimentally using the Schlieren method. The Schlieren images of the radiated acoustic beam from IDT are shown in Fig.2. In the observation the incident laser beam in x direction is required to be vertical to the acoustic beam, so that the piezoelectric substrate is tilted to a horizontal direction by the same angle as radiation angle θ.

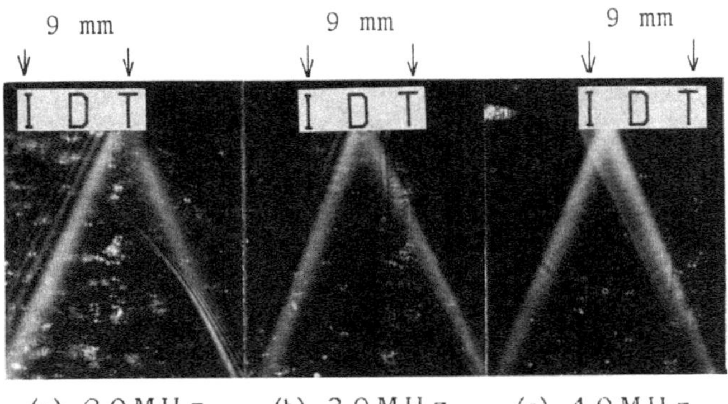

Fig.2 (a) Schlieren images of acoustic beam radiated from the chirped Fresnel IDT in x direction. Frequencies are (a) 20 MHz, (b) 30 MHz and (c) 40 MHz.

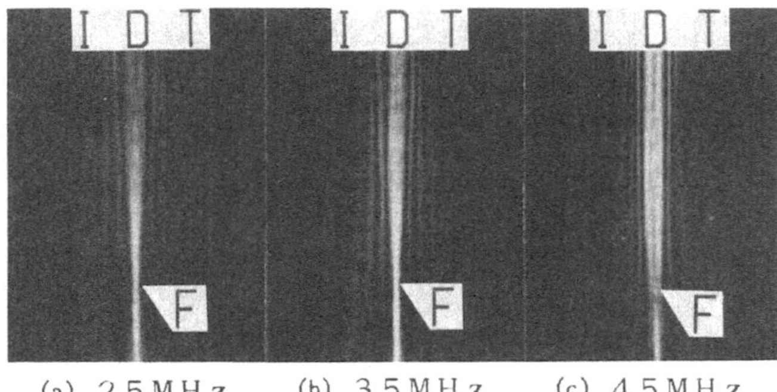

Fig.2 (b) Schlieren images of acoustic beam radiated from the chirped Fresnel IDT in y direction. Frequencies are (a) 25 MHz, (b) 35 MHz and (c) 45 MHz. F denotes the focal point in the predicted in the design.

It is clearly seen that the position of the acoustic waves radiated from the IDT moves with the frequency in x direction (Fig.2(a)). It is also found that the fingers on one end of the IDT resonate at 20 MHz while at the other end they resonate at 40 MHz.

It is found from Figs.2 (b) that the acoustic waves radiated from the IDT were focused around the designed focus points in y direction. The depth of focus is comparatively deep. Figures 2 (b) also show the focus beams when the excitation frequency is changed. It is found that the beam is evidently focused.

3-2. Acoustic Field

The distributions of the acoustic field radiated from the chirped Fresnel IDT were examined. Measurements were carried out by a laser probe method based on the diffraction of light by sound waves[5].

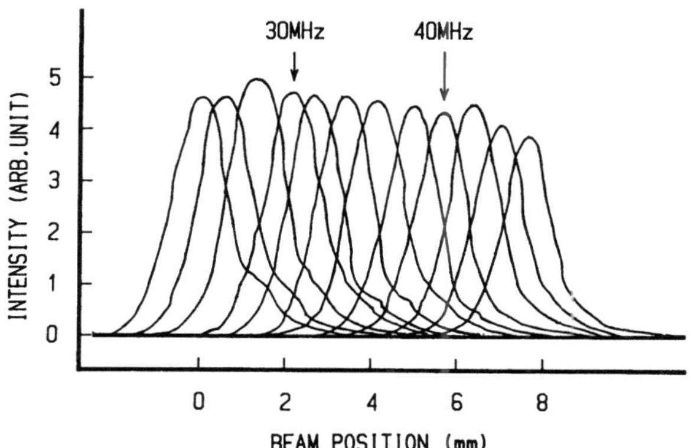

Fig.3   Laser-probed acoustic beam profiles radiated from the chirped Fresnel IDT along x direction.

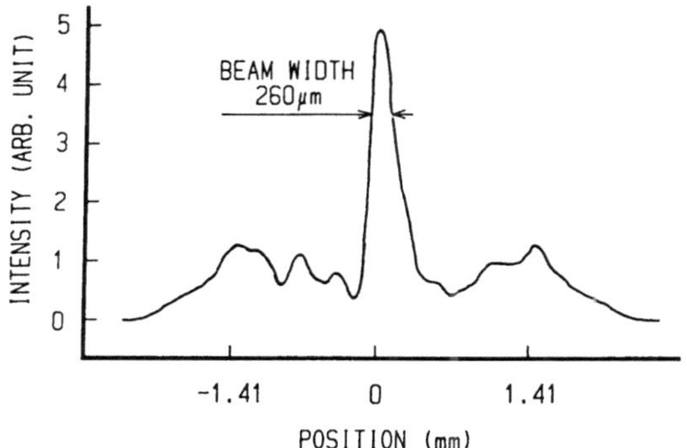

Fig.4   Laser-probed acoustic beam profiles radiated from the chirped Fresnel IDT in y direction.
         Frequency is 40 MHz.

The measured results of the acoustic field distribution at certain frequencies are shown in Fig.3 and Fig.4.   Figure 3 shows the acoustic beam profiles at some frequency.   It is noted that the 3 dB beamwidth is

900 μm in x direction.   The frequency dependence of the acoustic beamwidth and the beam position in x direction are shown in Fig.5. The
beamwidth is constant over the frequency range.   The position of the
acoustic beam corresponds to the applied frequency.   The beam position
moves linearly with the applied frequency.

Figure 4 show the field distribution in y direction.   From the
Fig.4(a), it is noted the 3 dB beamwidth is about 260 μm and the sidelobe is the -9.2 dB in this focused field of 40 MHz.   The field distribution for the other frequency (30 MHz) is shown in Fig.4(b).   It is
evident that the acoustic waves are focused at the frequency.

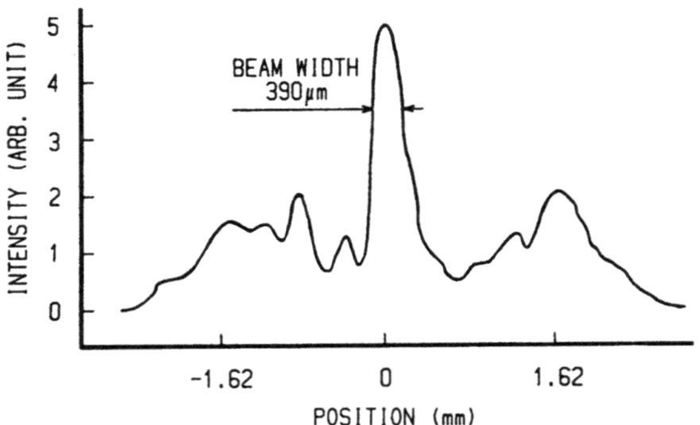

Fig.4   (Continued.) Acoustic beam profiles in y direction.
Frequency is 30 MHz.

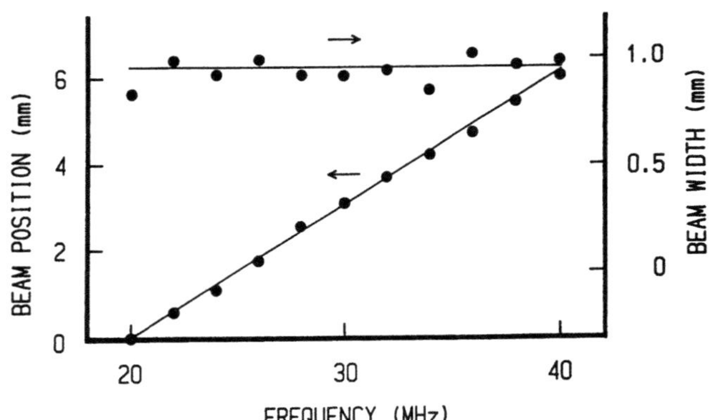

Fig.5 Acoustic beam width and position as a function of frequency.

Frequency response of the chirped Fresnel IDT for an acoustic beam was measured. The schematic diagram of a setup with identical IDTs is shown in Fig.6. The amplitude of the reflected wave was measured while the carrier frequency of the RF tone burst was swept automatically. In order to determine the IDT frequency response and to identify the factors affecting the overall frequency response of the IDT transmission system, the amplitude of RF tone burst reflected from a glass substrate located at the reflection plane was first measured.

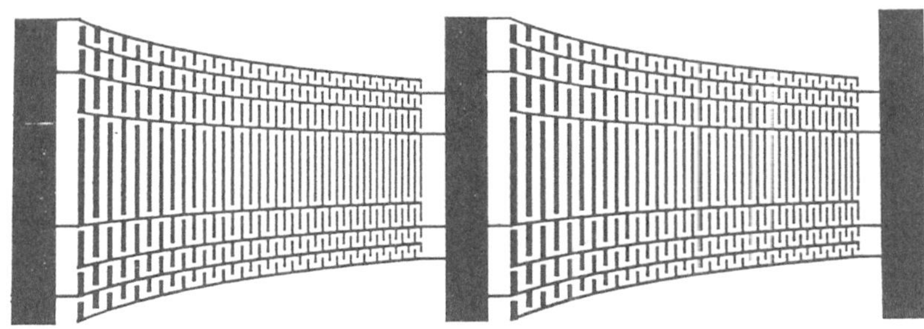

(a) Electrode pattern of a confocal pair of chirped Fresnel IDTs for a reflection imaging system.

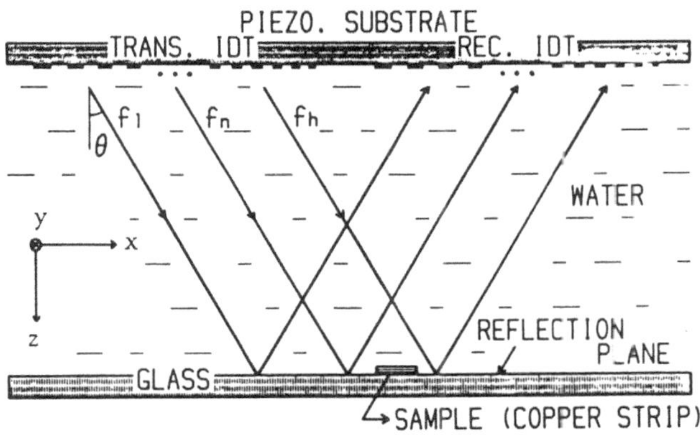

(b) Schematic diagram of electronic acoustic beam scanning using chirped Fresnel IDTs.

Fig.6 Configuration of the setup using chirped Fresnel IDTs.

Figure 7 shows the amplitude response obtained in the water. This
curve clearly indicates that the IDT is a very wide band and covers a
frequency range of 25 to 45 MHz, and the amplitude response is flat
across the specified passband. The maximum difference is about 1 dB and
the flat scan length of 9 mm is obtained. Next, the frequency response
by copper strips (1 mm width) placed on a glass was measured (as
sketched in Fig.6), and is shown in Fig.8. The dip of the frequency
response due to the copper strip is located at 30 MHz. This implies
that the chirped IDT could be used for acoustic imaging system of the
electronic beam scanning.

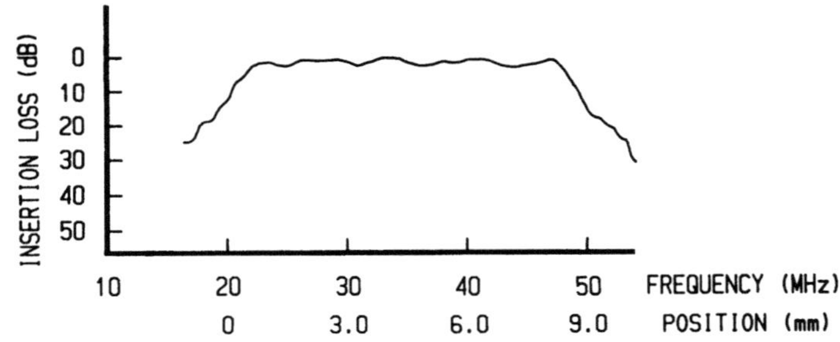

Fig.7   Frequency response of the chirped Fresnel IDT in water.
The beam position corresponds to the applied frequency.

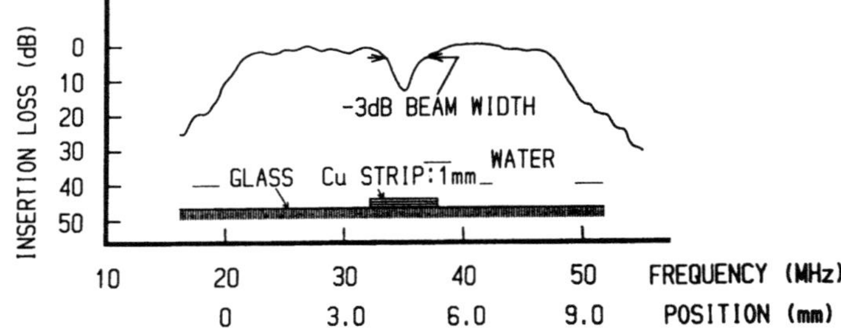

Fig.8   Frequency response of the chirped Fresnel IDT in water
(with copper strip). The dip due to the copper strip
is at 35 MHz.

3-4. Imaging

In order to realize the promising future of the present technique,
acoustic imaging was obtained using the focused acoustic beam from
chirped Fresnel IDTs. Imaging was made in pulse reflection mode using
two chirped Fresnel IDTs on the same substrate. Two chirped IDTs operating as a transmitter and receiver pair thus form an imaging system
that is scanned by the sweeping frequency in one direction and could be
scanned mechanically in the other. Figure 6 shows the configuration of
the device. Acoustic imaging was carried out with an object placed on
the confocal plane, providing a reflection image of the sample.

Figure 9 shows the block diagram for acoustic imaging using the
chirped Fresnel IDT. Electronic line scanning in the axial direction
was perfomed with the chirped Fresnel IDT, with an execution time of 10
millisecond per scan. To produce two-dimensional images, the object was

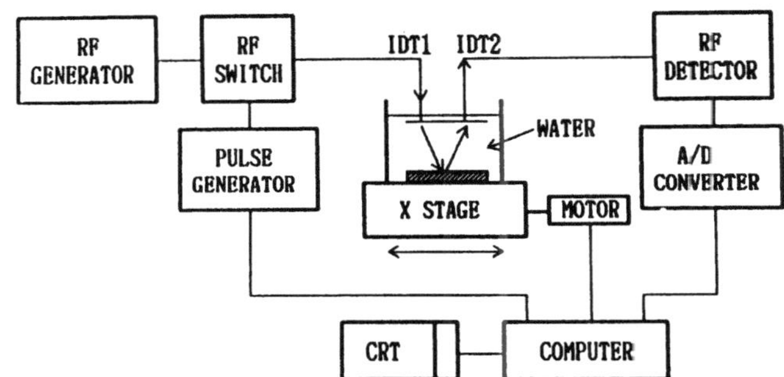

Fig.9 Block diagram of the chirped Fresnel IDT system
for a two-dimensional acoustic imaging.

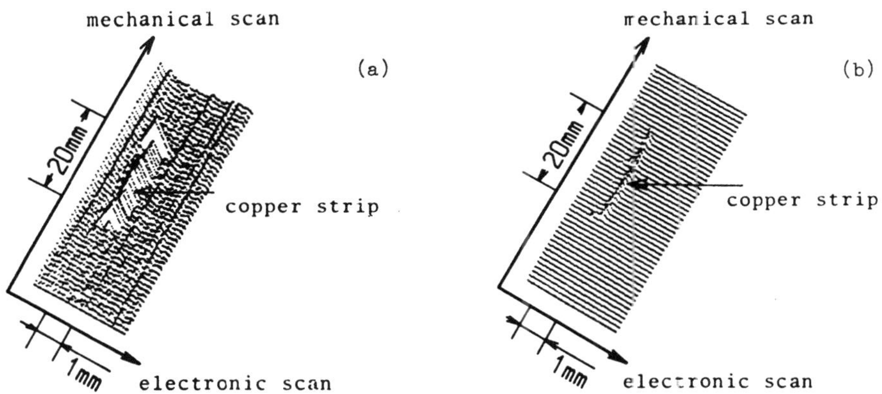

Fig.10 Acoustic images in reflection of a copper strip on glass.
(a) and (b) show the images of the strip with 1.0 mm
and 0.5 mm width, respectively.

translated mechanically in the transverse direction to produce a frame
scan. A complete two-dimensional image was obtained with each mechanical scan. A micro-computer is used for the image processing and the
mechanical operation of the chirped Fresnel IDT imaging system, simultaneously. As illustrated in Fig.9, the computer-controlled RF osillator
is gated to produce a short tone burst and applied to the transmitter
IDT. The reflected signals from the sample amplified and envelope
detected. The resulting signal is recorded by the micro-computer, which
also controls the stepper-motor driven scanner, and the processed signal
is used to modulate a CRT display or a recorder. Figure 10 show two
dimensional images. The samples were copper strips deposited on a glass.
The strip was clearly visible in the acoustic image.

4. CONCLUSION

Focused acoustic beam scanning using IDT has been described, and
electronic focusing and scanning of acoustic beam have been achieved by
a chirped Fresnel IDT. Using the chirped Fresnel IDT, which covers a
frequency from 15 to 45 MHz, a linear scan width of 9 mm for a focused
acoustic beam about 300 μm in width was obtained. Moreover, a tentative
system to obtain the acoustic image of the sample was made using the
chirped Fresnel IDT. As a result it has been found that the electronic
focused beam scanning is suitable for a imaging system with high speed.

ACKNOWLEDGMENT

The authors wish to thank Prof. T.Moriizumi of the Tokyo Institute
of Technology for helpful discussions.

REFERENCES

1. C.E.Quate, A.Atalar and H.K.Wickramasinghe,"Acoustic microscopy with
   mechanical scanning-A review," Proc. IEEE, vol.67, no.2, pp.1092-1114
   1979
2. W.H.Chen, F.C.Fu and W.L.Lu, "Scanning acoustic microscope utilizing
   SAW-BAW conversion," IEEE Trans. Sonics Ultrason., vol.SU-32, no.2,
   pp.181-188, 1985
3. T.Nomura, S.Shiokawa and T.Moriizumi, "Measurement and mapping of
   elastic anisotropy of solids using a leaky SAW excited by an
   interdigital transducer," IEEE Trans. Sonics Ultrason., vol.SU-32,
   no.2, pp.235-240, 1985
4. R.H.Tancrell and M.G.Holland, "Acoustic surface wave filters,"
   Proc.IEEE, vol.59, no.3, pp.393-409, 1971
5. E.G.Lean, J.M.White and C.D.W.Wilkinson, "Thin film acousto-optic
   devices," Proc.IEEE, vol.64, no.5, pp.779, 1976
   (1976) 779.

A COMPUTER-CONTROLLED TRANSDUCER FOR

REAL-TIME THREE-DIMENSIONAL IMAGING

D.G. Bailey†, J.A. Sun‡, A. Meyyappan, G. Wade

Department of Electrical and Computer Engineering
University of California, Santa Barbara, CA 93106, U.S.A.

K.R. Erikson§

InnoVision Medical, Inc.
Irvine, CA 92714, U.S.A.

ABSTRACT

Existing ultrasonic transducers and associated imaging systems are not ideal for high-resolution real-time three-dimensional imaging. Two mechanisms are required in such applications: dynamic focusing, and two-dimensional electronic scanning. These mechanisms are incorporated in a new computer-controlled acoustic transducer. This transducer is divided into a large number of individual acoustic elements, creating a two-dimensional phased array.

Dynamic focusing is accomplished by phasing the separate zones of a circular zone pattern formed on the transducer. Two dimensional scanning may be achieved by controlling the transducer via a two-dimensional shift register.

The advantages of an imaging system using this transducer over existing imagers are that (1) the images are derived from data at sample points in a cube-based matrix as opposed to a stack of sector scans, (2) the resolution is better by an order of magnitude, (3) the transducer patterns are completely programmable enabling the device to be optimized for different depth ranges, and (4) the transducer is fabricated as a single unit as opposed to an array of discrete transducers.

---

† Current Address: Image Analysis Unit, Massey University, Palmerston North, New Zealand.
‡ Current Address: Department of Radio Engineering, Beijing University of Posts and Telecommunications, Beijing, People's Republic of China.
§ Current Address: Diasonics, Inc., 1565 Barber Lane, Milpitas, CA 95035.

Acoustical Imaging, Vol. 18, Edited by H. Lee and G. Wade
Plenum Press, New York, 1991

INTRODUCTION

Acoustic imaging is the technique of using sound waves to obtain an image of the spatial distribution of the acoustic properties of an object. Since acoustic energy yields a view of an object not available with other forms of energy, the exploration of acoustic imaging has attracted many researchers working on a wide variety of applications.

One of the most important applications for an acoustic imaging system is in medical diagnosis. Since biological tissue is semitransparent to sound, it may be imaged with excellent contrast. Cancerous or other diseased tissue can frequently be distinguished from normal tissue in an ultrasonic image[1]. For this reason, there is an increasing enthusiasm among physicians for ultrasonic diagnosis in many fields of medicine involving most parts of the anatomy.

To be practical for medical use, an acoustic imaging system must satisfy a number of requirements. Of particular importance are real-time capability and high sensitivity[2]. In some applications, the ability to obtain three-dimensional images is also important. Real-time capability permits relative motion between the object and the system so that the position of either can be manipulated during imaging to produce the best result. Such capability also allows the study of dynamic biological processes. High sensitivity permits operation at power levels low enough to ensure patient safety.

Existing ultrasonic transducers and associated imaging systems are not well suited for obtaining three-dimensional images in some applications, such as ophthalmic imaging. To obtain three-dimensional images, currently available sensors must be either manually or mechanically scanned[3]. Manual scanning precludes real-time operation, and systems which employ mechanical scanning are too bulky for imaging small objects, such as the eye. Electronic scanning using a phased linear array of transducers is capable of forming good two-dimensional images, but requires mechanical scanning for the third dimension[4]. Such systems also suffer disadvantages stemming from complex system electronics and transducer size. Although transducers which use some form of acoustic lens for focusing are able to achieve high resolution, the depth of focus is usually very poor[5].

THREE-DIMENSIONAL IMAGING

Three-dimensional images may be formed in the following manner. A single pulse of ultrasonic radiation is transmitted and echoes from the various tissue interfaces are received. The position of the reflecting interface in the axial direction may be inferred from the round-trip propagation time of the pulse. This provides one-dimensional depth information. Information about the other two dimensions may be obtained by launching a sequence of pulses in a two-dimensional scanning pattern.

Two mechanisms are required to obtain high resolution three-dimensional images. First, a mechanism for lateral and longitudinal focusing of the pulse is required. Second, a mechanism for scanning the pulse laterally over a two-dimensional pattern is needed.

Dynamic Focusing

To achieve lateral focusing from a planar transducer, an aperture pattern consisting of a set of concentric rings is required. The simplest such arrangement is a Fresnel-zone pattern which has each ring or zone connected to a common signal lead[6] since the

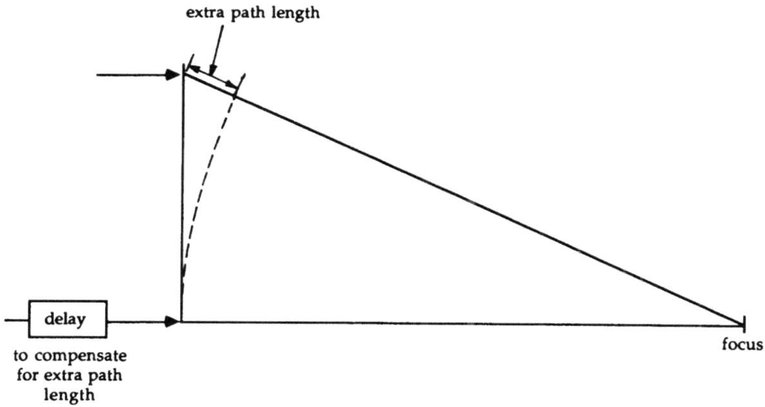

Figure 1.　The use of delays to compensate for different path lengths from each ring to the focus.

difference in path lengths between the focus and the rings is an integral number of wavelengths. However the depth of focus of such strongly focused systems is very small.

An alternate arrangement is to use a pattern consisting of a small number of rings to transmit a weakly focused pulse. The aperture must be small and the pulse weakly focused so that it remains in focus over the complete depth range of interest. Focusing action is accomplished by inserting time delays in the signal feeds to each ring in such a manner that the pulses arrive at the focus point simultaneously (figure 1). A similar focusing action is used while receiving the echoes. As with transmission, delays are inserted into the signal leads from each ring to compensate for the different path lengths from the focus to each ring. By dynamically adjusting these delays as the echoes are being received, it is possible to scan the focus in the axial direction. By scanning the focus in such a manner that the transducer is always focused at the point from which echoes are being received, it is possible to achieve very good lateral resolution, as illustrated in figure 2. By expanding the aperture as the focal length is changed, keeping a constant f number, the resolution remains uniform over the complete scan range[7].

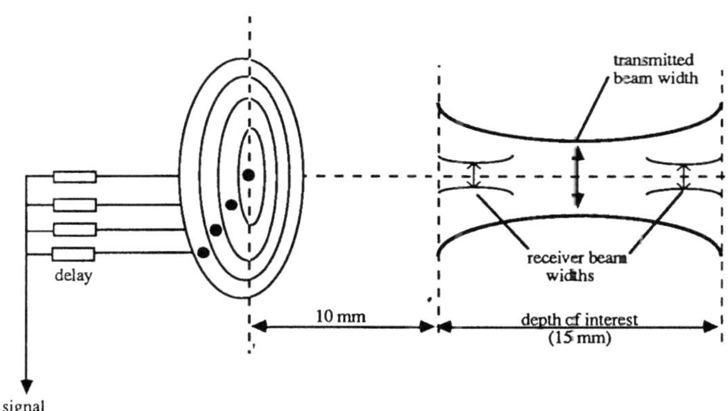

Figure 2.　Schematic diagram illustrating the approach for achieving lateral and long-tudinal focusing.

Figure 3.   A typical aperture pattern showing 10 zones.

Two-Dimensional Scanning

Scanning the pulse may be accomplished by scanning the circular aperture patterns in two dimensions[6]. One way of doing this is to have the transducer fabricated as a two-dimensional array of small elements. Any desired pattern may be synthesized as a union of individual elements. An example of a typical aperture pattern is shown in figure 3. By shifting the pattern across the array, the position of the focus also shifts.

To illustrate this concept in more detail, consider a pattern consisting of only a single ring. A piezoelectric plate would be used for acoustic wave generation and as a substrate. As shown in figure 4, one side of the plate would be covered with a metal ground electrode and a two-dimensional switch array would be fabricated on the other side. One of the two terminals of each switch is connected to an electrode on the plate, while the other terminal is connected to a common lead or signal line. Each switch is controlled separately by a computer-generated signal via a two dimensional shift register. The transducer aperture is created by turning on the switches in the desired pattern through the loading of this pattern into the two-dimensional shift register. By shifting the data within the shift register, the pattern of switches turned on, and consequently the transducer aperture pattern, are shifted. By scanning the data in two-dimensions in the shift registers, it is possible to obtain image information in the two lateral dimensions.

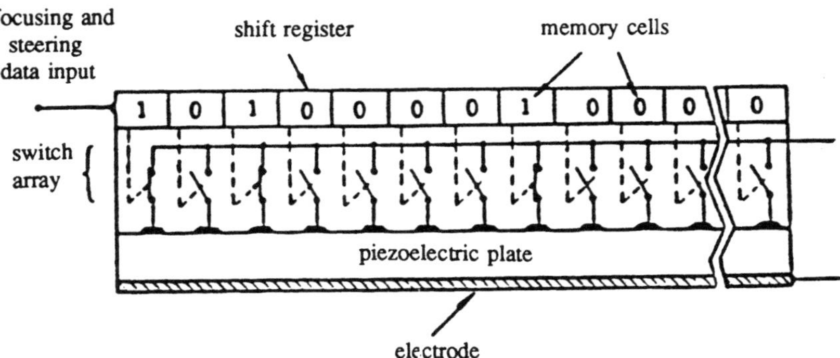

Figure 4.   Schematic diagram showing a switching arrangement for the transducer.

CONTROLLING THE TRANSDUCER

The scheme described above would require a separate set of switches, and a separate two dimensional shift register for each ring. A more practical method of providing multiple signal lines and requiring only three sets of shift registers is as shown in figure 5. A separate signal line is fed to each column within the transducer array. One set of switches, and an associated shift register, connects the column signal line to the transducer elements, as in the case described above. Each column signal line services a separate ring in the transducer pattern, although each ring will have several column lines connecting to it. Since adjacent column lines may be allocated to different signal lines, it is necessary to distribute the signal from the signal lines to adjacent elements. This requires two more sets of switches, and associated shift registers, one set for connecting adjacent elements horizontally, and another for connecting elements vertically. By shifting the three patterns contained in the three shift registers synchronously, it is possible to scan the transducer pattern in two dimensions in the same manner as described earlier.

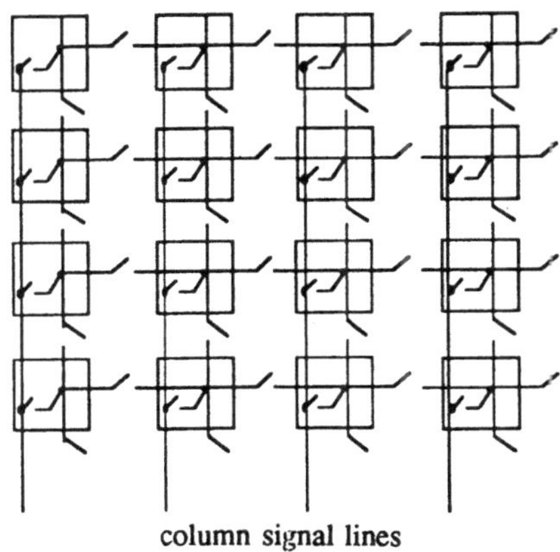

Figure 5. A practical method of providing multiple signal lines.

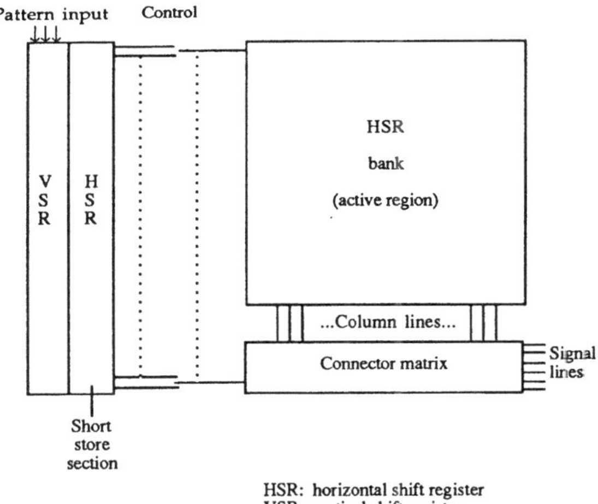

Figure 6. Arrangement of two-dimensional shift registers for loading and scanning the patterns.

The necessary patterns may be loaded and scanned in real time if the two-dimensional shift registers are arranged as shown in figure 6. The two main parts are a bank of horizontal shift registers (HSRs), and a vertical shift register (VSR). The horizontal shift-register bank is used to scan the pattern from left to right in the figure. As the pattern is shifted, a new column of data is loaded in on the left via the vertical shift register. This form of multiplexing simplifies the problem of sending in the large volume of data required to control the patterns. Below the active section is another horizontal shift register bank controlling a crossbar type of connection matrix. The pattern in this bank is used to control switches connecting the column lines to the individual signal lines.

If we assume that the transducer array consists of 100×100 elements, then 299 separate horizontal one-dimensional shift registers are required in the horizontal shift register bank (because there are only 99 rows of switches connecting the elements in the vertical direction). Since each of the three sets of switches for each row of the device are independent, it would be convenient to load the information via three parallel vertical shift registers. The system would work as follows. The first half of the pattern would be loaded into the horizontal shift-register bank which controls the switches permitting a pulse to be launched from the appropriate position and the echoes to be received. The maximum round trip propagation time is about 40 microseconds in ophthalmic inspection applications. During this time the set of vertical shift registers would be used to load two columns of data into the short store section. If there are 15 signal lines, each column of data requires 105 clock cycles (100 cycles for the HSRs and 15/3 for the connector matrix). With a 5 MHz clock, the two columns may be loaded in 42 microseconds. Two columns of data are loaded each time since the lateral resolution is approximately two element widths. The pattern is moved right by two columns, the new data being shifted into the left-hand edge of the horizontal shift-register bank. Then a new pulse is launched. This process is repeated until the pattern center reaches the far side of the array. When the pattern for the next row is shifted into the horizontal shift-register bank, the remaining part of the old pattern is shifted out. The time required to obtain a $50 \times 50 \times 50$ three-dimensional image is 147 ms or a rate of 6.8 images per second. This assumes a 5 MHz clock, 105 cycles to load each column and a pattern radius of 40 elements requiring a total of 140 columns to be loaded.

If a storage section is fabricated on the chip, the 40 columns corresponding to a pattern for a next row may be loaded while the pattern for the current row is being scanned. This reduces the time required to 105 ms or a rate of 9.5 images per second. It is not easy to obtain higher speeds than this, since the round trip propagation time of each pulse is about 40 microseconds.

The operation of the switch registers, and hence the loading and shifting of the patterns on the transducer, and the launching of the acoustic pulses will be controlled by a computer. Thus, a computer-controlled acoustic transducer (ComCAT) will incorporate both the dynamic focusing and the two-dimensional beam scanning described in the previous section. A block diagram of the complete system is shown in figure 7.

SIMULATED RESPONSE

The focusing properties of the transducer were determined by calculating the near-field diffraction patterns of the apertures. In computing the patterns we assume continuous waves rather than pulses. Nevertheless, the calculated patterns provide a reasonably accurate indication of the lateral resolution. Figure 8 shows the amplitude patterns on transmit and receive. Since the effective aperture pattern is the product of the transmit

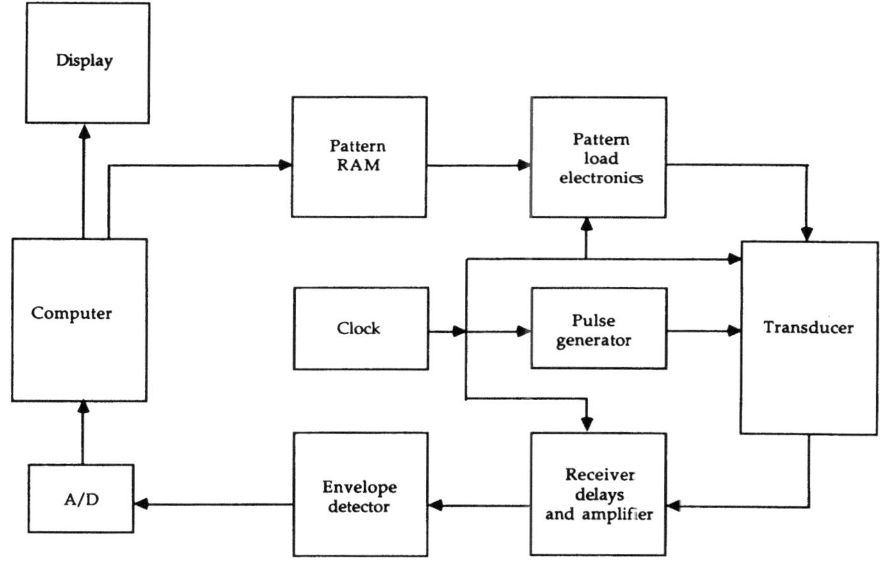

Figure 7.   Block diagram of the complete system.

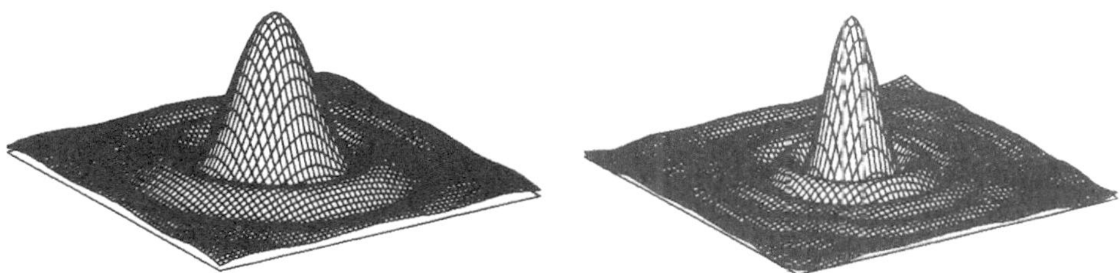

Figure 8.   Simulated aperture patterns at 15 mm range. (a) Transmit. (b) Receive.

and receive patterns, the combined intensity patterns are shown in figure 9, using both a linear and a logarithmic scale.

The lateral resolution is usually defined by the point at which the pressure amplitude falls by 3 dB from its peak value. On transmit, the resolution varies from 0.40 mm to 0.93 mm, depending on the range (from 10 mm to 25 mm). On receive, the effective resolution is approximately constant at 0.33 mm. The combined response gives an effective imaging resolution which varies from 0.25 mm (at 10 mm range) to 0.33 mm (at 25 mm range). The first side lobe is down by −20 dB to −30 dB, compared to the main lobe with the best response at the 15 mm range.

As mentioned, these results only give an approximate indication of the true resolution and side lobe levels since continuous waves rather than pulses were assumed. We are currently analyzing the system to determine the expected performance under pulsed operation.

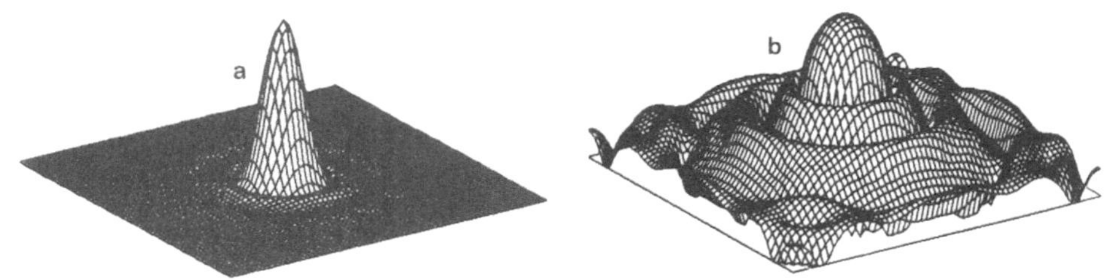

Figure 9.   Simulated transducer response. (a) Amplitude using a linear scale. (b) Intensity using a logarithmic scale.

COMPARISON WITH EXISTING SYSTEMS

In recent years, there has been two tendencies in research to enhance the capabilities of B-scan systems. One of these is to improve the imaging quality and reduce the system complexity of the commercially available linear phased array[8,9]. Systems using this technique are capable of providing electrical beam scanning and focusing in one dimension. However, this technique suffers in at least three respects: (1) Image quality is often disappointing due in part to the existence of high side lobes in the beam pattern[10]. (2) The systems are complicated and are therefore expensive[11]. (3) As mentioned above, this technique is only capable of providing electrical beam scanning in one dimension, rather than in two dimensions. In situations where volumetric images are required, the beam scanning in the direction perpendicular to the scanning plane must be performed either manually or mechanically.

Another research tendency has been to explore the capabilities of systems using annular transducer arrays[12]. This includes the use of transducers constructed with a number of concentric circular rings[13] or a single annular ring aperture consisting of a number of separated segments[14]. The annular array technique is capable of performing two-dimensional beam focusing which provides better sensitivity and lateral resolution than those of linear phased arrays[15,16]. Although such systems are relatively simple, it is very difficult to effectively incorporate electrical beam scanning.

An imaging system that uses ComCAT does not have these difficulties. ComCAT, as described above, employs a pattern of concentric circular rings for two-dimensional focusing and a switch-register bank for electronically scanning the beam in the two lateral dimensions. This enables a system with ComCAT to obtain high-resolution volumetric images in real-time. The inherent flexibility of ComCAT also permits the system to operate in more conventional A- and B-scan modes.

CONCLUSIONS

The advantages of an imaging system that uses ComCAT over existing imagers are that the images are derived from data points on a cube-based matrix instead of a stack of sector scans, the resolution is better by an order of magnitude than that obtainable from other three-dimensional systems, the transducer patterns are completely programmable so that the device can be optimized for different depth ranges, and the transducer is fabricated as a single unit as opposed to an array of discrete transducers.

The techniques available with ComCAT provide a number of powerful features which are frequently needed in medical diagnosis. A ComCAT system can have both two-dimensional beam focusing and two-dimensional electrical scanning and still be inherently simple to construct and operate. The system could be built as a small hand-held acoustic probe making it particularly attractive for ophthalmic inspection.

ACKNOWLEDGEMENTS

This research was funded in part by the University of California and InnoVision Medical, Inc. through California's Microelectronics Innovation and Computer Research Opportunities (MICRO) program.

REFERENCES

1. G. Wade, Ultrasonic Imaging by Reconstructive Tomography, *in* "Acoustical Imaging," 9:379, K. Wang, ed., Plenum, New York (1980).

2. K. R. Erikson, F. J. Fry, and J. P. Jones, Ultrasound in Medicine - A Review, *IEEE Trans. Sonics Ultrason.*, SU-21:144 (1974).

3. J. F. Havlice, and J. C. Taenzer, Ultrasonic Imaging Using Arrays, *Proc. IEEE*, 67:484 (1979).

4. P. N. T. Wells, and M. C. Zilskin, eds., "New Techniques and Instrumentation in Ultrasonography," Churchill Livingston, New York (1980).

5. D. J. Coleman, F. L. Lizzi, and R. L. Jack, "Ultrasonography of the Eye and the Orbit," Lea & Febiger, Philadelphia (1977).

6. J. A. Sun, "Computer Controlled Acoustic Transduction for Real-Time Three-Dimensional Imaging," Ph. D. Dissertation, University of California, Santa Barbara (1989).

7. D. R. Dietz, S. I. Parks, and M. Linzer, Expanding-Aperture Annular Array, *Ultrasonic Imaging*, 1:56 (1979).

8. C. L. Morgon, W. S. Trought, W. M. Clark, O. T. von Ramm, and F. L. Thurstone, Principles and Applications of a Dynamically Focused Phased Array Real Time Ultrasound System, *J. Clinical Ultrasound*, 6:385 (1978).

9. O. T. von Ramm, and S. W. Smith, Beam Scattering with Linear Arrays, *IEEE Trans. Biomed. Engr.*, BME-30:438 (1983).

10. A. Macovski, Ultrasonic Imaging Using Arrays, *Proc. IEEE*, 67:484 (1979).

11. H. E. Karrer, Phased Array Acoustic Imaging Systems, *in* "Physics and Engineering of Medical Imaging," R. Guzzardi, ed., Martinus Nijhoff, Dordrecht (1987).

12. D. Vilkomerson, Acoustic Imaging with Thin Annular Apertures, *in* "Acoustical Holography," 5:283, P. S. Green, ed., Plenum, New York (1974).

13. M. Arditi, W. B. Taylor, F. S. Foster, J. W. Hunt, An Annular Array System for High Resolution Breast Echography, *Ultrasonic Imaging* 4:1 (1982)

14. C. B. Burckhardt, P. -A. Grandchamp, and H. Hoffmann, Focussing Ultrasound Over a Range Depth with an Annular Transducer - An Alternative Method, *IEEE Trans. Sonics Ultrason.*, SU-22:11 (1975).

15. C. B. Burckhardt, P. -A. Grandchamp, and H. Hoffmann, Methods for increasing the lateral resolution of B-Scan, *in* "Acoustical Holography," 5:391, P. S. Green, ed., Plenum, New York (1974).

16. M. S. Patterson, and F. S. Foster, The Improvement and Qualitative Assesment of B-Mode Images Produced by an Annular Array / Cone Hybrid, *Ultrasonic Imaging,* 5:3 (1983)

MIX
Papier aus verantwortungsvollen Quellen
Paper from responsible sources
FSC® C105338

If you have any concerns about our products,
you can contact us on
ProductSafety@springernature.com

In case Publisher is established outside the EU,
the EU authorized representative is:
**Springer Nature Customer Service Center GmbH
Europaplatz 3, 69115 Heidelberg, Germany**

Printed by Libri Plureos GmbH
in Hamburg, Germany